CULTURE OF CELLS FOR TISSUE ENGINEERING

Editors

Gordana Vunjak-Novakovic, PhD
Department of Biomedical Engineering
Columbia University
New York, NY

R. Ian Freshney, PhD
Center for Oncology and Applied Pharmacology
University of Glasgow
Scotland, UK

A JOHN WILEY & SONS, INC., PUBLICATION

Published by John Wiley & Sons, Inc., Hoboken, New Jersey.
Published simultaneously in Canada.

For general information on our other products and services or for technical support, please contact our Customer Care Department within the United States at (800) 762-2974, outside the United States at (317) 572-3993 or fax (317) 572-4002.

Wiley also publishes its books in a variety of electronic formats. Some content that appears in print may not be available in electronic formats. For more information about Wiley products, visit our web site at www.wiley.com.

Library of Congress Cataloging-in-Publication Data is available.

ISBN-13 978-0-471-62935-1
ISBN-10 0-471-62935-9

Printed in the United States of America.

10 9 8 7 6 5 4 3 2 1

CULTURE OF CELLS FOR TISSUE ENGINEERING

Culture of Specialized Cells

Series Editor

R. Ian Freshney

CULTURE OF CELLS FOR TISSUE ENGINEERING
Gordana Vunjak-Novakovic and R. Ian Freshney, Editors

CULTURE OF EPITHELIAL CELLS, SECOND EDITION
R. Ian Freshney and Mary G. Freshney, Editors

CULTURE OF HEMATOPOIETIC CELLS
R. Ian Freshney, Ian B. Pragnell and Mary G. Freshney, Editors

CULTURE OF HUMAN TUMOR CELLS
R. Pfragner and R. Ian Freshney, Editors

CULTURE OF IMMORTALIZED CELLS
R. Ian Freshney and Mary G. Freshney, Editors

DNA TRANSFER TO CULTURED CELLS
Katya Ravid and R. Ian Freshney, Editors

Contents

Preface

Culture of Cells for Tissue Engineering is a new volume in the John Wiley series *Culture of Specialized Cells*, with focus on procedures for obtaining, manipulating, and using cell sources for tissue engineering. The book has been designed to follow the successful tradition of other Wiley books from the same series, by selecting a limited number of diverse, important, and successful tissue engineering systems and providing both the general background and the detailed protocols for each tissue engineering system. It addresses a long-standing need to describe the procedures for cell sourcing and utilization for tissue engineering in one single book that combines key principles with detailed step-to-step procedures in a manner most useful to students, scientists, engineers, and clinicians. Examples are used to the maximum possible extent, and case studies are provided whenever appropriate. We first talked about the possible outline of this book in 2002, at the World Congress of in vitro Biology, encouraged by the keen interest of John Wiley and inspired by discussions with our colleagues.

We made every effort to provide a user-friendly reference for sourcing, characterization, and use of cells for tissue engineering, for researchers with a variety of backgrounds (including basic science, engineering, medical and veterinary sciences). We hope that this volume can also be a convenient textbook or supplementary reading for regular and advanced courses of cell culture and tissue engineering. To limit the volume of the book, we selected a limited number of cells and tissues that are representative of the state of the art in the field and can serve as paradigms for engineering clinically useful tissues. To offer an in-depth approach, each cell type or tissue engineering system is covered by a combination of the key principles, step-by-step protocols for representative established methods, and extensions to other cell types and tissue engineering applications. To make the book easy to use and internally consistent, all chapters are edited to follow the same format, have complementary contents and be written in a single voice.

The book is divided into two parts and contains fifteen chapters, all of which are written by leading experts in the field. ***Part I*** describes procedures currently used for the in vitro cultivation of selected major types of cells used for tissue engineering, and contains five chapters. *Chapter 1* (by Ian Freshney) reviews basic considerations of cell culture relevant to all cell types under consideration in this book. This chapter also provides a link to the Wiley classic *Culture of Animal Cells*, now in its *Fifth Edition. Chapter 2* (by Donald Lennon and Arnold Caplan) covers mesenchymal stem cells and their current use in tissue engineering. *Chapter 3* (by Shulamit Levenberg, Ali Khademhosseini, Mara Macdonald, Jason Fuller, and Robert Langer) covers another important source of cells: embryonic human stem cells. *Chapter 4* (by Brian Johnstone, Jung Yoo, and Matthew Stewart) deals with various cell sources for tissue engineering of cartilage. *Chapter 5* (by Henning Madry) discusses the methods of gene transfer, using chondrocytes and cartilage tissue engineering as a specific example of application.

Part II deals with selected tissue engineering applications by first describing key methods and then focusing on selected case studies. *Chapter 6* (by Gordana Vunjak-Novakovic) reviews basic principles of tissue engineering, and provides a link to tissue engineering literature. *Chapter 7* (by Koichi Masuda and Robert Sah) reviews tissue engineering of articular cartilage, by using cells cultured on biomaterial scaffolds. *Chapter 8* (by Jingsong Chen, Gregory H. Altman, Jodie Moreau, Rebecca Horan, Adam Collette, Diah Bramano, Vladimir Volloch, John Richmond, Gordana Vunjak-Novakovic, and David L. Kaplan) reviews tissue engineering of ligaments, by biophysical regulation of cells cultured on scaffolds in bioreactors. *Chapter 9* (by Jennifer Elisseeff, Melanie Ruffner, Tae-Gyun Kim, and Christopher Williams) reviews microencapsulation of differentiated and stem cells in photopolymerizing hydrogels. *Chapter 10* (by Janet Shansky, Paulette Ferland, Sharon McGuire, Courtney Powell, Michael DelTatto, Martin Nackman, James Hennessey, and Herman Vandenburgh) focuses on tissue engineering of human skeletal muscle, an example of clinically useful tissue obtained by a combination of cell culture and gene transfer methods. *Chapter 11* (by Thomas Eschenhagen and Wolfgang H. Zimmermann) describes tissue engineering of functional heart tissue and its multidimensional characterization, in vitro and in vivo. *Chapter 12* (by Rebecca Y. Klinger and Laura Niklason) describes tissue engineering of functional blood vessels and their characterization in vitro and in vivo. *Chapter 13* (by Sandra Hofmann, David Kaplan, GordanaVunjak-Novakovic, and Lorenz Meinel) describes in vitro cultivation of engineered bone, starting from human mesenchymal stem cells and protein scaffolds. *Chapter 14* (by Peter I. Lelkes, Brian R. Unsworth, Samuel Saporta, Don F. Cameron, and Gianluca Gallo) reviews tissue engineering based on neuroendocrinal and neuronal cells. *Chapter 15* (by Gregory H. Underhill, Jennifer Felix, Jared W. Allen, Valerie Liu Tsang, Salman R. Khetani, and Sangeeta N. Bhatia) reviews tissue engineering of the liver in the overall context of micropatterned cell culture.

We expect that the combination of key concepts, well-established methods described in detail, and case studies, brought together for a limited number of interesting and distinctly different tissue engineering applications, will be of interest for the further growth of the exciting field of tissue engineering. We also hope that the book will be equally useful to a well-established scientist and a novice to a field. We greatly look forward to further advances in the scientific basis and clinical application of tissue engineering.

Gordana Vunjak-Novakovic
R. Ian Freshney

List of Abbreviations

AAF	athymic animal facility
ACL	anterior cruciate ligament
ACLF	human ACL fibroblasts
AIM	adipogenic induction medium
AMP	2-amino-2-methylpropanol
ARC	alginate-recovered-chondrocyte
BAMs	bioartificial muscles
BDM	2,3-butanedione monoxime
bFGF	basic fibroblast growth factor (FGF-1)
BPG	β-glycerophosphate
BDNF	brain-derived growth factor
BSA	bovine serum albumin
BSS	balanced salt solution
CAT	chloramphenicol-acetyl transferase
CBFHH	calcium and bicarbonate-free Hanks' BSS with HEPES
CLSM	confocal light scanning microscopy
CM	cell-associated matrix
CMPM	cardiomyocyte-populated matrices
DA	dopamine
DBH	dopamine-β-hydroxylase
Dex	dexamethasone
DMEM	Dulbecco's modification of Eagle's medium
DMEM-10FB	DMEM with 10% fetal bovine serum
DMEM-HG	DMEM with high glucose, 4.5 g/L
DMEM-LG	DMEM with low glucose, 1 g/L
DMMB	dimethylmethylene blue
DMSO	dimethyl sulfoxide
%dw	percentage by dry weight
E	epinephrine (adrenaline)
EB	embryoid bodies

EC	endothelial cell
ECM	extracellular matrix
EDTA	ethylenediaminetetraacetic acid
EGFP	enhanced green fluorescent protein
EHT	engineered heart tissue
ELISA	enzyme-linked immunosorbent assay
EMA	ethidium monoazide bromide
ES	embryonal stem (cells)
FACS	fluorescence-activated cell sorting
FBS	fetal bovine serum
FM	freezing medium
FRM	further removed matrix
GAG	glycosaminoglycan
GRGDS	glycine-arginine-glycine-aspartate-serine
HARV	high aspect ratio vessel
HBAMs	human bioartificial muscles
HBSS	Hanks' balanced salt solution
HEPES	4-(2-hydroxyethyl)piperazine-1-ethanesulfonic acid
hES	human embryonal stem (cells)
HIV	human immunodeficiency virus
HMEC	human microvascular endothelial cell
hMSC	human mesenchymal stem cell
HPLC	high-performance liquid chromatography
HUVEC	human umbilical vein endothelial cell
IBMX	isobutylmethylxanthine
ID	internal diameter
IM	incubation medium
IP	intraperitoneal
LAD	ligament augmentation devices
L_{max}	length at which EHTs develop maximal active force
MEF	mouse embryo fibroblasts
MRI	magnetic resonance imaging
MSC	mesenchymal stem cell
MSCGM	mesenchymal stem cell growth medium
NASA	National Aeronautics and Space Administration
NE	norepinephrine (noradrenaline)
NGF	nerve growth factor
NT2	NTera-2/clone D1 teratocarcinoma cell line
NT2M	NT2 medium
NT2N	Terminally differentiated NT2
OD	optical density
OD	outer or external diameter
OP-1	osteogenic protein 1 (BMP-7)
PAEC	porcine aortic endothelial cells

PBSA	Dulbecco's phosphate-buffered saline without Ca^{2+} and Mg^{2+}
PECAM	platelet endothelial cell adhesion molecule (CD31)
PEG	polyethylene glycol
PEGDA	polyethylene glycol diacrylate
Pen/strep	penicillin-streptomycin mixture, usually stocked at 10,000 U and 10 mg/ml, respectively
PEO	polyethylene oxide
PET	polyethylene terephthalate
PGA	polyglycolic acid
PITC	phenylisothiocyanate
PLA	poly-L-lactic acid
PLGA	polylactic-co-glycolic acid
PNMT	phenylethanolamine-*N*-methyl-transferase
RCCS	Rotatory Cell Culture Systems™
RGD	arginine-glycine-aspartic acid
RWV	rotating wall vessel bioreactors
SA	sympathoadrenal
SC	Sertoli cells
SDS-PAGE	polyacrylamide gel electrophoresis in the presence of sodium dodecyl (lauryl) sulfate
SMC	smooth muscle cell
SNAC	Sertoli-NT2N-aggregated-cell
SR	sarcoplasmic reticulum
SSEA-3 and 4	stage-specific embryonic antigens 3 and 4
STLV	slow turning lateral vessel (NASA derived)
SZP	superficial zone protein
TBSS	Tyrode's balanced salt solution
TH	tyrosine hydroxylase
TJA	total joint arthroplasty
TRITC	tetramethylrhodamine isothiocyanate
TT	twitch tension
Tween 20	polyoxyethylene-sorbitan mono-laurate
UTS	ultimate tensile strength

Part I

Cell Culture

1

Basic Principles of Cell Culture

R. Ian Freshney

Centre for Oncology and Applied Pharmacology, Cancer Research UK Beatson Laboratories, Garscube Estate, Bearsden, Glasgow G61 1BD, Scotland, UK, i.freshney@ntlworld.com

Culture of Cells for Tissue Engineering, edited by Gordana Vunjak-Novakovic and R. Ian Freshney

1. INTRODUCTION

The bulk of the material presented in this book assumes background knowledge of the principles and basic procedures of cell and tissue culture. However, it is recognized that people enter a specialized field, such as tissue engineering, from many different disciplines and, for this reason, may not have had any formal training in cell culture. The objective of this chapter is to highlight those principles and procedures that have particular relevance to the use of cell culture in tissue engineering. Detailed protocols for most of these basic procedures are already published [Freshney, 2005] and will not be presented here; the emphasis will be more on underlying principles and their application to three-dimensional culture. Protocols specific to individual tissue types will be presented in subsequent chapters.

2. TYPES OF CELL CULTURE

2.1. Primary Explantation Versus Disaggregation

When cells are isolated from donor tissue, they may be maintained in a number of different ways. A simple small fragment of tissue that adheres to the growth surface, either spontaneously or aided by mechanical means, a plasma clot, or an extracellular matrix constituent, such as collagen, will usually give rise to an outgrowth of cells. This type of culture is known as a *primary explant*, and the cells migrating out are known as the *outgrowth* (Figs. 1.1, 1.2, See Color Plate 1). Cells in the outgrowth are selected, in the first instance, by their ability to migrate from the explant and subsequently, if subcultured, by their ability to proliferate. When a tissue sample is disaggregated, either mechanically or enzymatically (See Fig. 1.1), the suspension of cells and small aggregates that is generated will contain a proportion of cells capable of attachment to a solid substrate, forming a *monolayer*. Those cells within the monolayer that are capable of proliferation will then be selected at the first subculture and, as with the outgrowth from a primary explant, may give rise to a *cell line*. Tissue disaggregation is capable of generating larger cultures more rapidly than explant culture, but explant culture may still be preferable where only small fragments of tissue are available or the fragility of the cells precludes survival after disaggregation.,

2.2. Proliferation Versus Differentiation

Generally, the differentiated cells in a tissue have limited ability to proliferate. Therefore, differentiated cells do not contribute to the formation of a primary culture, unless special conditions are used to promote their attachment and preserve their differentiated status. Usually it is the proliferating committed precursor compartment of a tissue (Fig. 1.3), such as fibroblasts of the dermis or the basal epithelial layer of the epidermis, that gives rise to the bulk of the cells in a

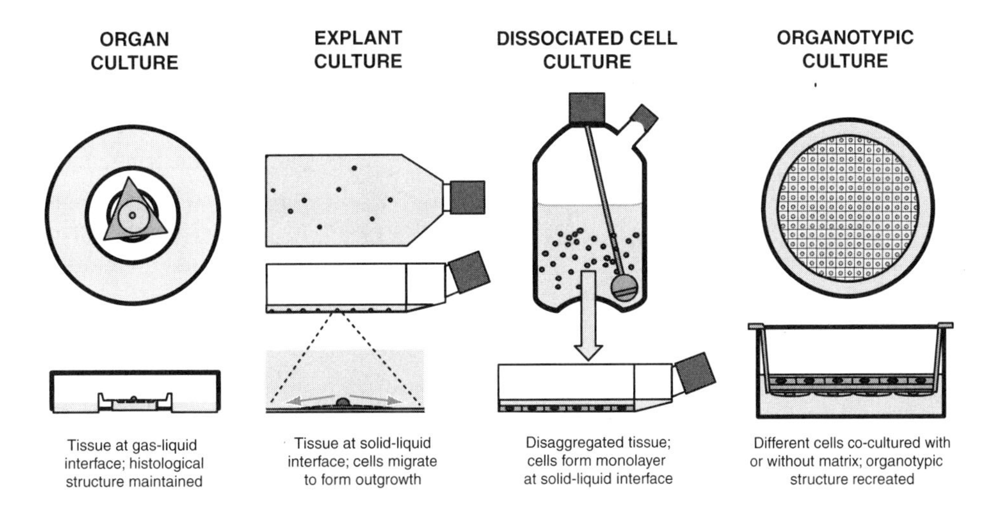

Figure 1.1. Types of culture. Different modes of culture are represented from left to right. First, an organ culture on a filter disk on a triangular stainless steel grid over a well of medium, seen in section in the lower diagram. Second, explant cultures in a flask, with section below and with an enlarged detail in section in the lowest diagram, showing the explant and radial outgrowth under the arrows. Third, a stirred vessel with an enzymatic disaggregation generating a cell suspension seeded as a monolayer in the lower diagram. Fourth, a filter well showing an array of cells, seen in section in the lower diagram, combined with matrix and stromal cells. [From Freshney, 2005.]

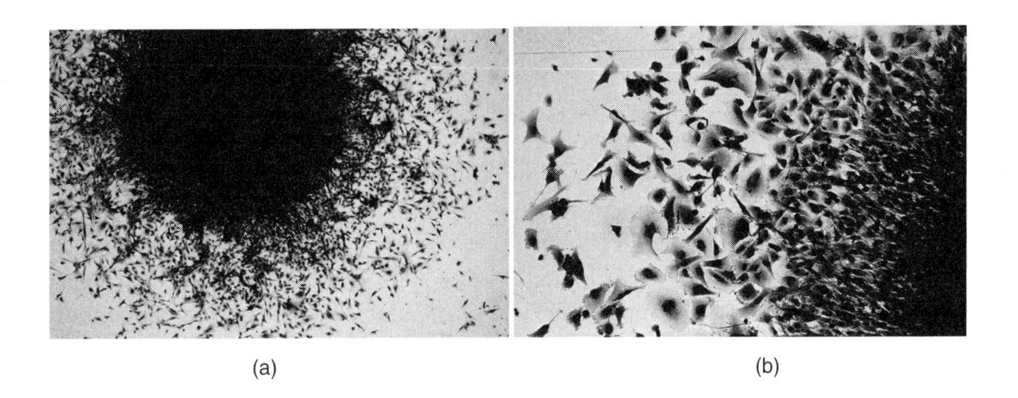

Figure 1.2. Primary explant and outgrowth. Microphotographs of a Giemsa-stained primary explant from human non-small cell lung carcinoma. a) Low-power (4× objective) photograph of explant (top left) and radial outgrowth. b) Higher-power detail (10× objective) showing the center of the explant to the right and the outgrowth to the left. (See Color Plate 1.)

primary culture, as, numerically, these cells represent the largest compartment of proliferating, or potentially proliferating, cells. However, it is now clear that many tissues contain a small population of regenerative cells which, given the correct selective conditions, will also provide a satisfactory primary culture, which may be propagated as stem cells or mature down one of several pathways toward

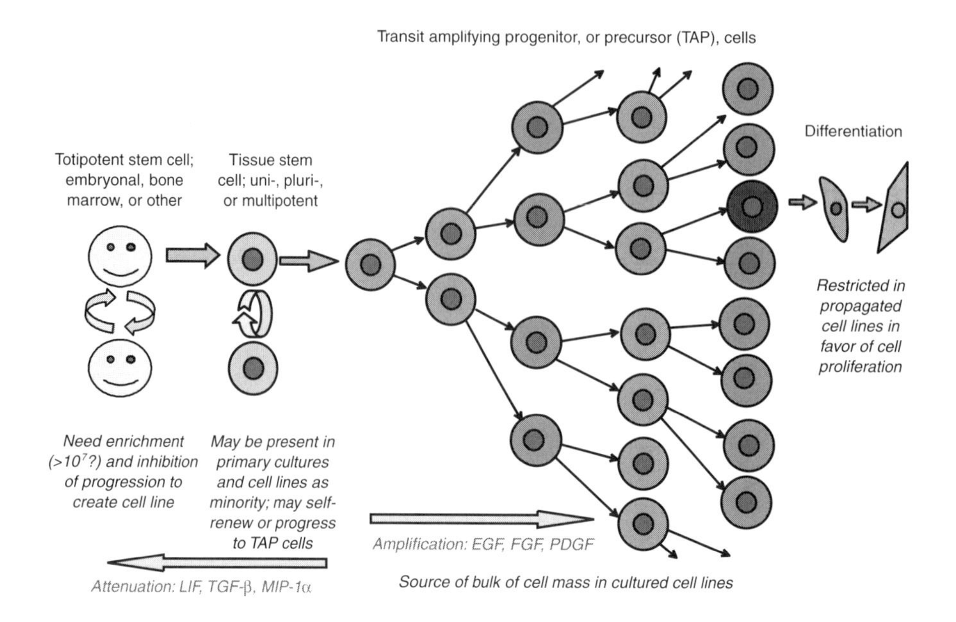

Figure 1.3. Origin of cell lines. Diagrammatic representation of progression from totipotent stem cell, through tissue stem cell (single or multiple lineage committed) to transit amplifying progenitor cell compartment. Exit from this compartment to the differentiated cell pool (far right) is limited by the pressure on the progenitor compartment to proliferate. Italicized text suggests fate of cells in culture and indicates that the bulk of cultured cells probably derive from the progenitor cell compartment, because of their capacity to replicate, but accepts that stem cells may be present but will need a favorable growth factor environment to become a significant proportion of the cells in the culture. [From Freshney, 2005.]

differentiation. This implies that not only must the correct population of cells be isolated, but the correct conditions must be defined to maintain the cells at an appropriate stage in maturation to retain their proliferative capacity if expansion of the population is required. This was achieved fortuitously in early culture of fibroblasts by the inclusion of serum that contained growth factors, such as platelet-derived growth factor (PDGF), that helped to maintain the proliferative precursor phenotype. However, this was not true of epithelial cells in general, where serum growth factors such as transforming growth factor β (TGF-β) inhibited epithelial proliferation and favored differentiation. It was not until serum-free media were developed [Ham and McKeehan, 1978, Mather, 1998, Karmiol, 2000] that this effect could be minimized and factors positive to epithelial proliferation, such as epidermal growth factor and cholera toxin, used to maximum effect.

Although undifferentiated precursors may give the best opportunity for expansion in vitro, transplantation may require that the cells be differentiated or carry the potential to differentiate. Hence, two sets of conditions may need to be used, one for expansion and one for differentiation. The factors required to induce differentiation will be discussed later in this chapter (See Section 7.4) and in later chapters. In general, it can be said that differentiation will probably require a selective medium for the cell type, supplemented with factors that favor differentiation, such as retinoids, hydrocortisone, and planar-polar compounds, such as sodium butyrate (NaBt). In addition, the correct matrix interaction, homotypic and heterotypic cell interaction, and, for epithelial cells, the correct cellular polarity will need to be established, usually by using an organotypic culture. This assumes, of course, that tissue replacement will require the graft to be completely or almost completely differentiated, as is likely to be the case where extensive tissue repair is carried out. However, there is also the option that cell culture will only be required to expand a precursor cell type and the process of implantation itself will then induce differentiation, as appears to be the case with stem cell transplantation [Greco and Lecht, 2003].

2.3. Organotypic Culture

Dispersed cell cultures clearly lose their histologic characteristics after disaggregation and, although cells within a primary explant may retain some of the histology of the tissue, this will soon be lost because of flattening of the explant with cell migration and some degree of central necrosis due to poor oxygenation. Retention of histologic structure, and its associated differentiated properties, may be enhanced at the air/medium interface, where gas exchange is optimized and cell migration minimized, as distinct from the substrate/medium interface, where dispersed cell cultures and primary outgrowths are maintained. This so-called *organ culture* (See Fig. 1.1) will survive for up to 3 weeks, normally, but cannot be propagated. An alternative approach, with particular relevance to tissue engineering, is the amplification of the cell stock by generation of cell lines from specific cell types and their subsequent recombination in *organotypic culture*. This allows the synthesis of a tissue equivalent or construct on demand for basic studies on cell-cell and cell-matrix interaction and for in vivo implantation. The fidelity of the construct in terms of its real tissue equivalence naturally depends on identification of all the participating cell types in the tissue in vivo and the ability to culture and recombine them in the correct proportions with the correct matrix and juxtaposition. So far this has worked best for skin [Michel et al., 1999, Schaller et al., 2002], but even then, melanocytes have only recently been added to the construct, and islet of Langerhans cells are still absent, as are sweat glands and hair follicles, although some progress has been made in this area [Regnier et al., 1997; Laning et al., 1999].

There are a great many ways in which cells have been recombined to try to simulate tissue, ranging from simply allowing the cells to multilayer by perfusing a monolayer [Kruse et al., 1970] to highly complex perfused membrane (Membroferm [Klement et al., 1987]) or capillary beds [Knazek et al., 1972]. These are termed *histotypic cultures* and aim to attain the density of cells found in the tissue from which the cells were derived (Fig. 1.4). It is possible, using selective media, cloning, or physical separation methods (See Section 3.4), to isolate purified cell strains from disaggregated tissue or primary culture or at first subculture. These purified cell populations can then be combined in organotypic culture to recreate both the tissue cell density and, hopefully, the cell interactions and

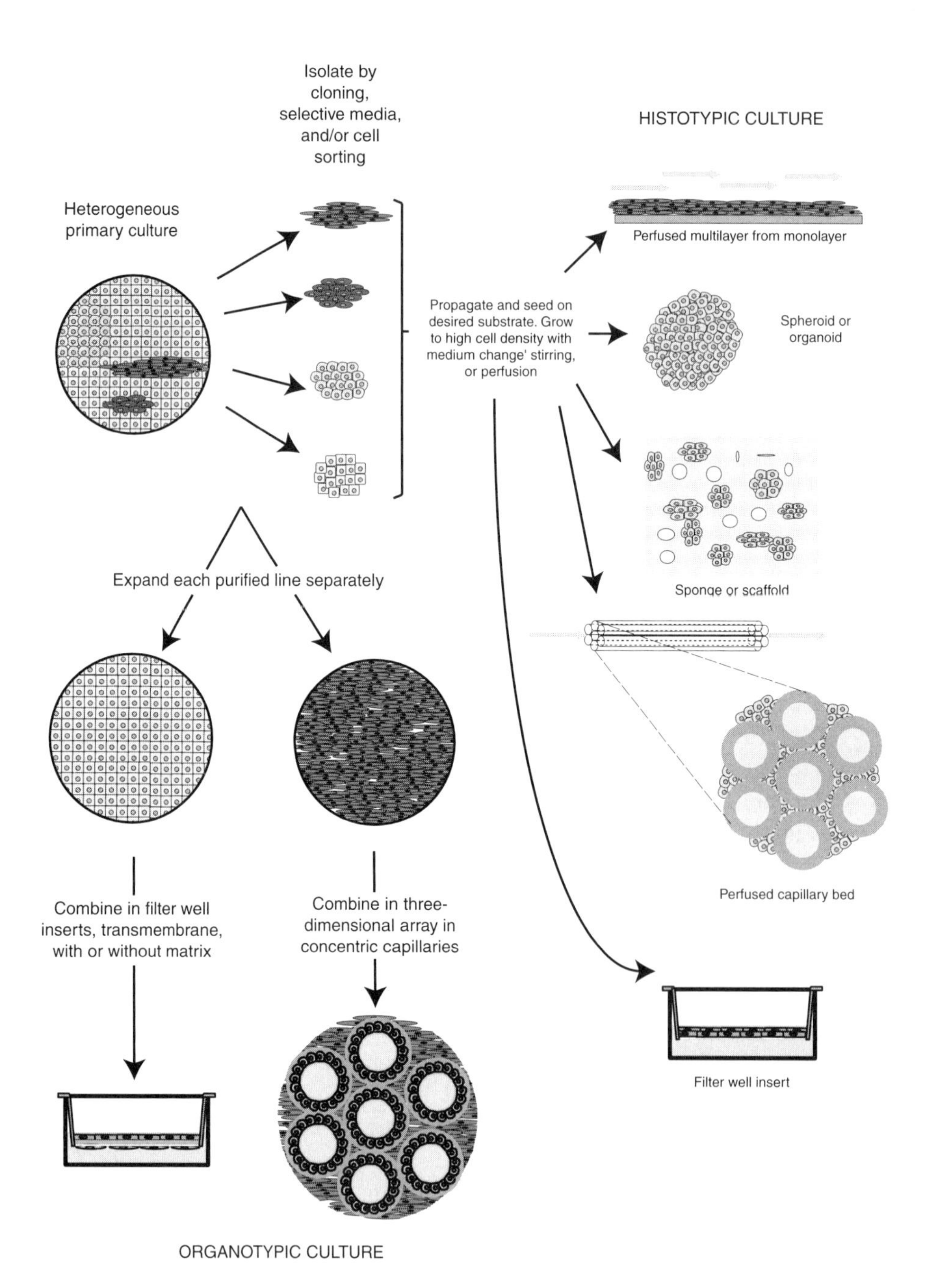

Figure 1.4. Histotypic and organotypic culture. Indicates the heterogeneity of a primary culture (top left), how this might be purified to give defined cell populations, which, if expanded and seeded into appropriate conditions can give high-density cultures of one cell type in perfused multilayers (top right), spheroids or organoids in stirred suspension (second top right), three-dimensional multilayers in perfused capillaries (third top right), or monolayers or multilayers in filter well inserts (bottom right). Expansion of purified populations and recombination can generate organotypic cultures, in filter well inserts (bottom left) or on concentric microcapillaries (bottom center). This last seems to be suggested by the architecture of the device (CellGro Triac), but the author has no knowledge of its use in this capacity.

matrix generation found in the tissue (See Fig. 1.4). Filter well inserts provide the simplest model system to test such recombinants, but there are many other possibilities including porous matrices, perfused membranes, and concentric double microcapillaries (Triac hollow fiber modules, [www.spectrapor.com/1/1/9.html]).

2.4. Substrates and Matrices

Initially, cultures were prepared on glass for ease of observation, but cells may be made to grow on many different charged surfaces including metals and many polymers. Traditionally, a net negative charge was preferred, such as found on acid-washed glass or polystyrene treated by electric ion discharge, but some plastics are also available with a net positive charge (e.g., Falcon Primaria), which is claimed to add some cell selectivity. In either case, it is unlikely that the cell attaches directly to synthetic substrates and more likely that the cell secretes matrix products that adhere to the substrate and provide ligands for the interaction of matrix receptors such as integrins. Hence it is a logical step to treat the substrate with a matrix product, such as collagen type IV, fibronectin, or laminin, to promote the adhesion of cells that would otherwise not attach.

The subject of scaffolds will be dealt with in detail in later chapters (See Part II). Suffice it to say at this stage that scaffolds have the same requirements as conventional substrates in terms of low toxicity and ability to promote cell adhesion, often with the additional requirement of a three-dimensional geometry. If the polymer or other material does not have these properties, derivatization and/or matrix coating will be required.

Most studies suggest that cell cultivation on a three-dimensional scaffold is essential for promoting orderly regeneration of engineered tissues in vivo and in vitro. Scaffolds investigated to date vary with respect to material chemistry (e.g., collagen, synthetic polymers), geometry (e.g., gels, fibrous meshes, porous sponges, tubes), structure (e.g., porosity, distribution, orientation, and connectivity of the pores), physical properties (e.g., compressive stiffness, elasticity, conductivity, hydraulic permeability), and degradation (rate, pattern, products).

In general, scaffolds should be made of biocompatible materials, preferentially those already approved for clinical use. Scaffold structure determines the transport of nutrients, metabolites, and regulatory molecules to and from the cells, whereas the scaffold chemistry may have an important role in cell attachment and differentiation. The scaffold should biodegrade at the same rate as the rate of tissue assembly and without toxic or inhibitory products. Mechanical properties of the scaffold should ideally match those of the native tissue being replaced, and the mechanical integrity should be maintained as long as necessary for the new tissue to mature and integrate.

3. ISOLATION OF CELLS FOR CULTURE

3.1. Tissue Collection and Transportation

The first, and most important, element in the collection of tissue is the cooperation and collaboration of the clinical staff. This is best achieved if a member of the surgical team is also a member of the culture project, but even in the absence of this, time and care must be spent to ensure the sympathy and understanding of those who will provide the clinical material. It is worth preparing a short handout explaining the objectives of the project and spending some time with the person likely to be most closely involved with obtaining samples. This may be the chief surgeon (who will need to be informed anyway), or it may be a more junior member of the team willing to set up a collaboration, one of the nursing staff, or the pathologist, who may also require part of the tissue. Whoever fulfils this role should be identified and provided with labeled containers of culture medium containing antibiotics, bearing a contact name and phone number for the cell culture laboratory. A refrigerator should be identified where the containers can be stored, and the label should also state clearly **DO NOT FREEZE!** The next step is best carried out by someone from the laboratory collecting the sample personally, but it is also possible to leave instructions for transportation by taxi or courier. If a third party is involved, it is important to ensure that the container is well protected [See, for example, www.ehs.ucsf.edu/Safety%20Updates/Bsu/Bsu5.pdf], preferably double wrapped in a sealed polythene bag and an outer padded envelope provided with the name, address, and phone number of the recipient at the laboratory. Refrigeration during transport is not usually necessary, as long as the sample is not allowed to get too warm, but if delivery will take more than an hour or two, then one or two refrigeration packs, such as used in picnic chillers, should be included but kept out of direct contact.

If the tissue sample is quite small, a further tissue sample (any tissue) or a blood sample should be obtained for freezing. This will be used ultimately to corroborate the origin of any cell line that is derived from the sample by DNA profiling. A cell line is the culture that is produced from subculture of the primary, and every additional subculture after this increases the possibility of cross-contamination, so verification of origin is important (See Section 6). In addition, the possibility of misidentification arises during routine subculture and after recovery from cryopreservation (See Section 5).

3.2. Biosafety and Ethics

All procedures involved in the collection of human material for culture must be passed by the relevant hospital ethics committee. A form will be required for the patient to sign authorizing research use of the tissue, and preferably disclaiming any ownership of any materials derived from the tissue [Freshney, 2002, 2005]. The form should have a brief layman's description of the objectives of the work and the name of the lead scientist on the project. The donor should be provided with a copy.

All human material should be regarded as potentially infected and treated with caution. Samples should be transported securely in double-wrapped waterproof containers; they and derived cultures should be handled in a Class II biosafety cabinet and all discarded materials autoclaved, incinerated, or chemically disinfected. Each laboratory will its own biosafety regulations that should be adhered to, and anyone in any doubt about handling procedures should contact the local safety committee (and if there is not one, create it!). Rules and regulations vary among institutions and countries, so it is difficult to generalize, but a good review can be obtained in Caputo [1996].

3.3. Record Keeping

When the sample arrives at the laboratory, it should be entered into a record system and assigned a number. This record should contain the details of the donor, identified by hospital number rather than by name, tissue site, and all information regarding collection medium, time in transit, treatment on arrival, primary disaggregation, and culture details, etc. [Freshney, 2002, 2005]. This information will be important in the comparison of the success of individual cultures, and if a long-term cell line is derived from the culture, this will be the first element in the cell line's provenance, which will be supplemented with each successive manipulation or experimental procedure. Such records are best maintained in a computer database where each record can be derived from duplication of the previous record with appropriate modifications. There may be issues of data protection and patient confidentiality to be dealt with when obtaining ethical consent.

3.4. Disaggregation and Primary Culture

Detailed information on disaggregation as a method for obtaining cells is provided in the appropriate chapters. Briefly, the tissue will go through stages of rinsing, dissection, and either mechanical disaggregation or enzymatic digestion in trypsin and/or collagenase. It is often desirable not to have a complete single-cell suspension, and many primary cells survive better in small clusters. Disaggregated tissue will contain a variety of different cell types, and it may be necessary to go through a separation technique [See Chapter 15, Freshney 2005], such as density gradient separation [Pretlow and Pretlow, 1989] or immunosorting by magnetizable beads (MACS), using a positive sort to select cells of interest [Carr et al., 1999] or a negative sort to eliminate those that are not required [Saalbach et al., 1997], or by using fluorescence-activated cell sorting (FACS) [See, e.g. Swope et al., 1997]. The cell population can then be further enriched by selection of the correct medium (e.g., keratinocyte growth medium (KGM) or MCDB 153 for keratinocytes [Peehl and Ham, 1980]), many of which are now available commercially (See Sources of Materials), and supplementing this with growth factors. Survival and enrichment may be improved in some cases by coating the substrate with gelatin, collagen, laminin, or fibronectin [Freshney, 2005].

4. SUBCULTURE

Frequently, the number of cells obtained at primary culture may be insufficient to create constructs suitable for grafting. Subculture gives the opportunity to expand the cell population, apply further selective pressure with a selective medium, and achieve a higher growth fraction and allows the generation of replicate cultures for characterization, preservation by freezing, and experimentation. Briefly, subculture involves the dissociation of the cells from each other and the substrate to generate a single-cell suspension that can be quantified. Reseeding this cell suspension at a reduced concentration into a flask or dish generates a secondary culture, which can be grown up and subcultured again to give a tertiary culture, and so on. In most cases, cultures dedifferentiate during serial passaging but can be induced to redifferentiate by cultivation on a 3D scaffold in the presence of tissue-specific differentiation factors (e.g., growth factors, physical stimuli). However, the cell's ability to redifferentiate decreases with passaging. It is thus essential to determine, for each cell type, source, and application, a suitable number of passages during subculture. Protocols for subculture of specific cell types are given in later chapters, and a more general protocol is available in Chapter 13, Freshney [2005].

4.1. Life Span

Most normal cell lines will undergo a limited number of subcultures, or passages, and are referred to as *finite cell lines*. The limit is determined by the number of doublings that the cell population can go through before it stops growing because of senescence. Senescence is determined by a number of intrinsic factors regulating cell cycle, such as Rb and p53 [Munger and Howley, 2002], and is accompanied by shortening of the telomeres on the chromosomes [Wright and Shay, 2002]. Once the telomeres reach a critical minimum length, the cell can no longer divide. Telomere length is maintained by telomerase, which is downregulated in most normal cells except germ cells. It can also be higher in stem cells, allowing them to go through a much greater number of doublings and avoid senescence. Transfection of the telomerase gene hTRT into normal cells with a finite life span allows a small proportion of the cells to become immortal [Bodnar et al., 1998; Protocol 18.2, Freshney, 2005], although this probably involves deletion or inactivation of other genes such as p53 and *myc* [Cerni, 2000].

4.2. Growth Cycle

Each time that a cell line is subcultured it will grow back to the cell density that existed before subculture (within the limits of its finite life span). This process can be described by plotting a growth curve from samples taken at intervals throughout the growth cycle (Fig. 1.5), which shows that the cells enter a latent period of no growth, called the *lag period*, immediately after reseeding. This period lasts from a few hours up to 48 h, but is usually around 12–24 h, and allows the cells to recover from trypsinization, reconstruct their cytoskeleton, secrete matrix to aid attachment, and spread out on the substrate, enabling them to reenter cell cycle. They then enter exponential growth in what is known as the *log phase*, during which the cell population doubles over a definable period, known as the *doubling time* and characteristic for each cell line. As the cell population becomes crowded when all of the substrate is occupied, the cells become packed, spread less on the substrate, and eventually withdraw from the cell cycle. They then enter the *plateau* or *stationary phase*, where the growth fraction drops to close to zero. Some cells may

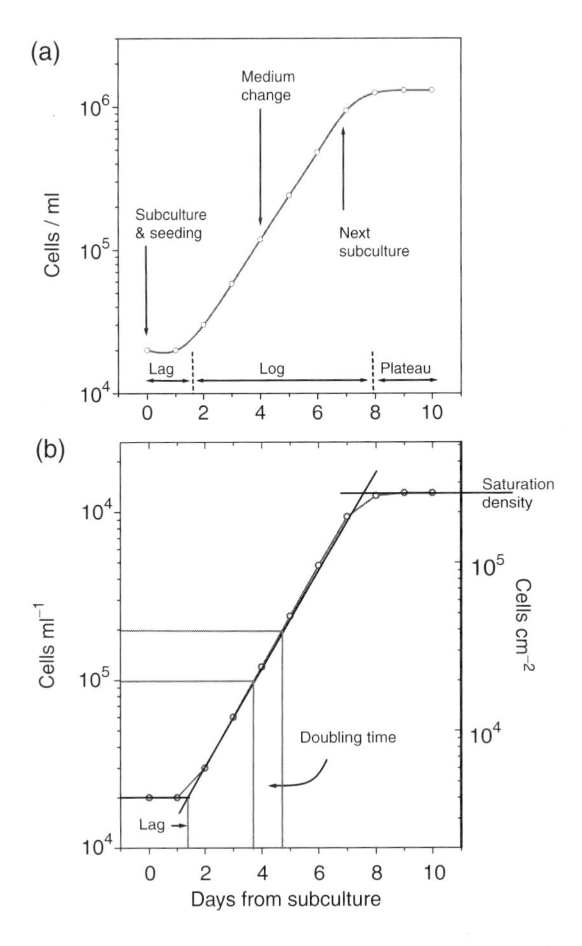

Figure 1.5. Growth curve. Increase in cell number on a log scale plotted against days from subculture. a) Defines the lag, log (exponential), and plateau phases, and when culture should be fed and subcultured after the indicated seeding time. b) Shows the kinetic parameters that can be derived from the growth curve: *lag* from the intercept between a line drawn through the points on the exponential phase and the horizontal from the seeding concentration; *doubling time* from the time taken, in the middle of the exponential phase, for the cell population to double; *saturation density* from the maximum (stable) cell density achieved by the culture, under the prevailing culture conditions. This is determined in cells/cm^2 (cell density rather than cell concentration) and is not absolute, as it will vary with culture conditions. It is best determined (as characteristic of the cell type) in conditions that are nonlimiting for medium, e.g., a small area of high-density cells in a large reservoir of medium (such as a coverslip, or a filter well insert, in a non-tissue culture-grade dish) or under continuous perfusion of medium. [Adapted from Freshney, 2005.]

differentiate in this phase; others simply exit the cell cycle into G_0 but retain viability. Cells may be subcultured from plateau, but it is preferable to subculture before plateau is reached, as the growth fraction will be higher and the recovery time (lag period) will be shorter if the cells are harvested from the top end of the log phase.

Reduced proliferation in the stationary phase is due partly to reduced spreading at high *cell density* and partly to exhaustion of growth factors in the medium at high *cell concentration*. These two terms are not interchangeable. Density implies that the cells are attached, and may relate to monolayer density (two-dimensional) or multilayer density (three-dimensional). In each case there are major changes in cell shape, cell surface, and extracellular matrix, all of which will have significant effects on cell proliferation and differentiation. A high density will also limit nutrient perfusion and create local exhaustion of peptide growth factors [Stoker, 1973; Westermark and Wasteson, 1975]. In normal cell populations this leads to a withdrawal from the cycle, whereas in transformed cells, cell cycle arrest is much less effective and the cells tend to enter apoptosis.

Cell concentration, as opposed to cell density, will exert its main effect through nutrient and growth factor depletion, but in stirred suspensions cell contact-mediated effects are minimal, except where cells are grown as aggregates. Cell concentration per se, without cell interaction, will not influence proliferation, other than by the effect of nutrient and growth factor depletion. High cell concentrations can also lead to apoptosis in transformed cells in suspension, notably in myelomas and hybridomas, but in the absence of cell contact signaling this is presumably a reflection of nutrient deprivation.

4.3. Serial Subculture

Each time the culture is subcultured the growth cycle is repeated. The number of doublings should be recorded (Fig. 1.6) with each subculture, simplified by reducing the cell concentration at subculture by a power or two, the so-called *split ratio*. A split ratio of two allows one doubling per passage, four, two doublings, eight, three doublings, and so on (See Fig. 1.6). The number of elapsed doublings should be recorded so that the time to senescence (See Section 4.1) can be predicted and new stock prepared from the freezer before the senescence of the existing culture occurs.

5. CRYOPRESERVATION

If a cell line can be expanded sufficiently, preservation of cells by freezing will allow secure stocks to be maintained without aging and protect them from problems of contamination, incubator failure, or medium and serum crises. Ideally, $1 \times 10^6 - 1 \times 10^7$ cells should be frozen in 10 ampoules, but smaller stocks can be used if a surplus is not available. The normal procedure is to freeze a token stock of one to three ampoules as soon as surplus cells are available, then to expand remaining cultures to confirm the identity of the cells and absence of contamination, and freeze down a seed stock of 10–20 ampoules. One ampoule, thawed from this stock, can then be used to generate a using stock. In many cases, there may not be sufficient doublings available to expand the stock as much as this, but it is worth saving some as frozen stock, no matter how little, although survival will tend to decrease below 1×10^6 cells/ml and may not be possible below 1×10^5 cells/ml.

Factors favoring good survival after freezing and thawing are:

(i) High cell density at freezing ($1 \times 10^6 - 1 \times 10^7$ cells/ml).

(ii) Presence of a preservative, such as glycerol or dimethyl sulfoxide (DMSO) at 5–10%.

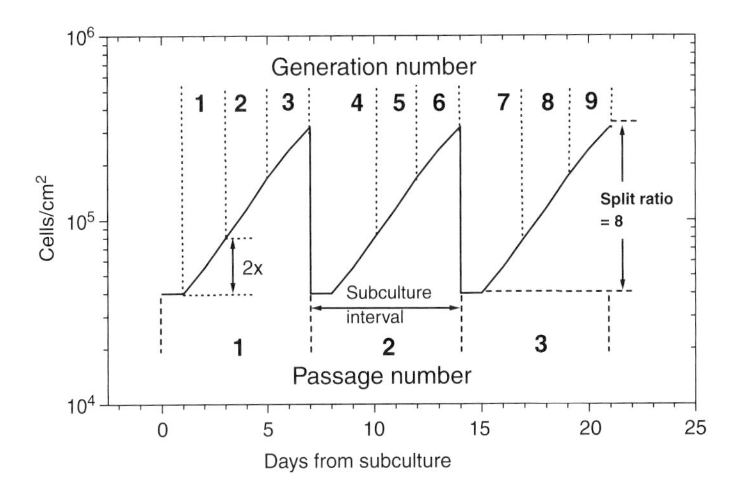

Figure 1.6. Serial subculture. Recurring growth curves during serial subculture, not necessarily recorded by daily cell counts, but predicted from one or two detailed growth curve analyses of these cells. Each cycle should be a replicate of the previous one, such that the same terminal cell density is achieved after subculture at the same seeding concentration. The lower number represents the passage number, i.e., the number of times the culture has been subcultured. The upper numbers represent the generation number, i.e., the number of times the population has doubled. In this example, the cell population doubles three times between each subculture, suggesting that the culture should be split 1:8 to regain the same seeding concentration each time. [From Freshney, 2005.]

(iii) Slow cooling, 1 °C/min, down to −70 °C and then rapid transfer to a liquid nitrogen freezer.

(iv) Rapid thawing.

(v) Slow dilution, ~20-fold, in medium to dilute out the preservative.

(vi) Reseeding at 2- to 5-fold the normal seeding concentration. For example, if cells are frozen at 5×10^6 cells in 1 ml of freezing medium with 10% DMSO and then thawed and diluted 1:20, the cell concentration will still be 2.5×10^5 cells/ml at seeding, higher than the normal seeding concentration for most cell lines, and the DMSO concentration will be reduced to 0.5%, which most cells will tolerate for 24 h.

(vii) Changing medium the following day (or as soon as all the cells have attached) to remove preservative. Where cells are more sensitive to the preservative, they may be centrifuged after slow dilution and resuspended in fresh medium, but this step should be avoided if possible as centrifugation itself may be damaging to freshly thawed cells.

There are differences of opinion regarding some of the conditions for freezing and thawing, for example, whether cells should be chilled when DMSO is added or diluted rapidly on thawing, both to avoid potential DMSO toxicity. In the author's experience, chilling diminishes the effect of the preservative, particularly with glycerol, and rapid dilution reduces survival, probably due to osmotic shock. Culturing in diluted DMSO after thawing can be a problem for some cell lines if they respond to the differentiating effects of DMSO, for example, myeloid leukemia cells, neuroblastoma cells, and embryonal stem cells; in these cases it is preferable to centrifuge after slow dilution at thawing or use glycerol as a preservative.

6. CHARACTERIZATION AND VALIDATION

6.1. Cross-Contamination

There has been much publicity about the very real risks of cross-contamination when handling cell lines [Marcovic and Marcovic, 1998; Macleod et al., 1999; Masters et al., 2001; van Bokhoven et al., 2001; Masters, 2002], particularly continuous cell lines. This is less of a problem with short-term cultures, but the risk remains that if there are other cell lines in use in the laboratory, they can cross-contaminate even a primary culture, or misidentification can arise during subculture or recovery from the freezer. If a laboratory focuses on one particular human cell type, superficial observation of lineage characteristics will be inadequate to ensure the identity of each line cultured. Precautions must be taken to avoid cross-contamination:

(i) Do not handle more than one cell line at a time, or, if this is impractical, do not have culture vessels and medium bottles for more than one cell line open at one time, and never be tempted to use the same pipette or other device for different cell lines.

(ii) Do not share media or other reagents among different cell lines.

(iii) Do not share media or reagents with other people.

(iv) Ensure that any spillage is mopped up immediately and the area swabbed with 70% alcohol.

(v) Retain a tissue or blood sample from each donor and confirm the identity of each cell line by DNA profiling: (a) when seed stocks are frozen, (b) before the cell line is used for experimental work or transplantation.

(vi) Keep a panel of photographs of each cell line, at low and high densities, above the microscope, and consult this regularly when examining cells during maintenance. This is particularly important if cells are handled over an extended period, and by more than one operator.

(vii) If continuous cell lines are in use in the laboratory, handle them after handing other, slower-growing, finite cell lines.

6.2. Microbial Contamination

Antibiotics are often used during collection, transportation, and dissection of biopsy samples because of the intrinsic contamination risk of these operations. However, once the primary culture is established, it is desirable to eliminate antibiotics as soon as possible. If the culture grows well, then antibiotics can be removed from the bulk of the stocks at first subculture, retaining one culture in antibiotics as a precaution if necessary. Antibiotics can lead to lax aseptic technique, can inhibit some eukaryotic cellular processes, and can hide the presence of a microbial contamination. If a culture is contaminated this must become apparent as soon as possible, either to indicate that the culture should be discarded before it can spread the contamination to other cultures or to indicate that decontamination should be attempted. The latter should only be used as a last resort; decontamination is not always successful and can lead to the development of antibiotic-resistant organisms.

Most bacterial, fungal, and yeast infections are readily detected by regular careful examination with the naked eye (e.g., by a change in the color of culture medium) and on the microscope. However, one of the most serious contaminations is mycoplasma, which is not visible by routine microscopy. Any cell culture laboratory should have a mycoplasma screening program in operation, but those collecting tissue for primary culture are particularly at risk. The precautions that should be observed are as follows:

(i) Treat any new material entering the laboratory from donors or from other laboratories as potentially infected and keep it in quarantine. Ideally, a separate room should be set aside for receiving samples and imported cultures. If this is not practicable, handle separately from other cultures, preferably last in the day and in a designated hood, and swab the hood down after use with 2% phenolic disinfectant in 70% alcohol. Use a separate incubator, and adhere strictly to the rules given above regarding medium sharing.

(ii) Screen new cultures as they arrive, and existing stocks at regular intervals, e.g., once a month. There are a number of tests available, but the most reliable and sensitive are fluorescence microscopy after staining with Hoechst 33258 [Chen et al., 1977; Protocol 19.2, Freshney, 2005] or PCR with a pan-mycoplasma primer [Uphoff and Drexler, 2002a; Protocol 19.3, Freshney, 2005]. The latter is more sensitive but depends on the availability of PCR technology, whereas the former is easier, will detect any DNA-containing contamination, but requires a fluorescence microscope. Both techniques are best performed with so-called *indicator cultures*. The test culture is refreshed with antibiotic-free medium and, after 3–5 days, the medium is transferred to an antibiotic-free, 10% confluent indicator culture of a cell line such as 3T6 or A549 cells, which are well spread and known to support mycoplasma growth. After a further 3 days (the indicator cells must not be allowed to reach confluence) the indicator culture is fixed in acetic methanol and stained with Hoechst 33258, or harvested by scraping for PCR (trypsinization may remove the mycoplasma from the cell surface).

(iii) Discard all contaminated cultures. If the culture is irreplaceable, decontamination may be attempted (under strict quarantine conditions) with agents such as Mycoplasma Removal Agent (MRA), ciprofloxacin, or BM-cycline [Uphoff and Drexler, 2002b]. Briefly, the culture is rinsed thoroughly, trypsinized (wash by centrifuging three times after trypsinization), and subcultured into antibiotic-containing medium. This procedure should be repeated for three subcultures and then the culture should be grown up antibiotic-free and tested after one, two, and four further antibiotic-free subcultures, whereupon the culture reenters the routine mycoplasma screening program.

6.3. Characterization

Most laboratories will have, as an integral part of the research program, procedures in place for the characterization of new cultures. Species identification (in case the cells are misidentified or cross-contaminated) will be unnecessary if DNA profiling is being used to confirm cell line identity; otherwise, chromosome analysis or isoenzyme electrophoresis [Hay et al., 2000] can be used. The lineage or tissue or origin can be determined by using antibodies to intermediate filament proteins, for example, cytokeratins for epithelial cells, vimentin for mesodermal cells such as fibroblasts, endothelium, and myoblasts, desmin for myocytes, neurofilament protein for neuronal and some neuroendocrine cells, and glial fibrillary acidic protein for astrocytes. Some cell surface markers are also lineage specific, for example, EMA in epithelial cells, A2B5 in glial cells, PECAM-1 in endothelial cells, and N-CAM in neural cells, and have the additional advantage that they can be used in cell sorting by magnetic sorting or flow cytometry. Morphology can also be used, but can be ambivalent as similarities can exist between cells of very different origins.

Spontaneous transformation is unlikely in normal cells of human origin, but indicators are a more refractile appearance under phase contrast with a lower cytoplasmic/nuclear ratio, piling up of the cells and loss of contact inhibition and density limitation of growth, increased clonogenicity in agar, and the ability to form tumors in immune-deprived hosts, such as the Nude or SCID mouse. Where transformation is detected, it is more likely to be due to cross-contamination, although it is possible that the tissue sample may have contained some preneoplastic cells that have then progressed in culture.

6.4. Differentiation

As stated above, a prerequisite for sustained growth in culture is the ability for the cells to proliferate, and this may preclude differentiation. If differentiation is required, then it is generally necessary for the cells to withdraw from the cell cycle. This can be achieved by removing, or changing, the growth factor supplementation; for example, the O2A common precursor of astrocytes and oligodendrocytes remains as a proliferating precursor cell in PDGF and bFGF, whereas combining bFGF with ciliary neurotropic factor (CNTF) results in differentiation into a type 2 astrocyte [Raff, 1990], and embryonal stem cells, which remain as proliferating primitive cells in the presence of leukemia inhibitory factor (LIF), will differentiate in the absence of LIF and in the presence of a positively acting factor such as phorbol myristate acetate (PMA, also known as TPA) [Rizzino, 2002].

There are four main parameters governing the entry of cells into differentiation:

(i) Soluble factors such as growth factors (e.g., EGF, KGF, TGF-β and HGF, NGF), cytokines (IL-6, oncostatin-M, GM-CSF, interferons), vitamins (e.g., retinoids, vitamin D_3, and vitamin K) and calcium [Table 16.1, Freshney, 2000], and planar polar compounds (e.g., DMSO and NaBt) [Tables 17.1, 17.2, Freshney, 2005].

(ii) Interaction with matrix constituents such as collagen IV, laminin, and proteoglycans. Heparan sulfate proteoglycans (HSPGs), in particular, have a significant role not only in binding to cell surface receptors but also in binding and translocating growth factors and cytokines to high-affinity cell surface receptors [Lopez-Casillas et al., 1993; Filla, 1998].

(iii) Enhanced cell-cell interaction will also promote differentiation. Homotypic contact interactions can act via gap junctions, which tend to coordinate the response among many like cells in a population by allowing free intercellular flow of second messenger molecules such as cyclic adenosine monophosphate (cAMP) and via cell adhesion molecules such as E-cadherin or N-CAM, which signal via anchorage to the cytoskeleton [Juliano, 2002]. Heterotypic interactions, in solid tissues at least, will tend to act across a basal lamina and are less likely to involve direct cell-cell contact. Signaling is achieved by, on the one hand, modification of the matrix by the mutual contribution of both cell types, and, on the other, by reciprocal transmission of cytokines and growth factors across the basal lamina, such as the transfer of KGF and GM-CSF from dermal fibroblasts to the basal layer of the epidermis in response to IL-1α and -β diffusing from the epidermis to the dermal fibroblasts [Maas-Szabowski et al., 2002].

(iv) The position, shape, and polarity of the cells may induce, or at least make the cells permissive for the induction of, differentiation. Epidermal keratinocytes [Maas-Szabowski et al., 2002] and bronchial epithelial cells [Petra et al., 1993] require to be close to the air/liquid interface, presumably to enhance oxygen availability, and secretory cells, such as thyroid epithelium, need the equivalent of the acinar space, that is, no direct access to nutrient or hormones, above them in a thin fluid space [Chambard et al., 1983, 1987]. When cells are grown on collagen at this location, and the collagen gel is allowed to retract, a shape change can occur, for example, from a flat squamous or cuboidal cell into a more columnar morphology, and this, combined with matrix interaction, allows the establishment of polarity in the cells, such that secretory products are released apically and signaling receptors and nutrient transporters locate basally.

Combining these effects in vitro may require strict attention to culture geometry, for example, by growing cells in a filter well insert on a matrix incorporating stromal fibroblasts, and providing differentiation inducers basally in a defined, nonmitogenic medium. Similar conditions may be created in a perfused capillary bed or scaffold.

SOURCES OF MATERIALS

Suppliers of specific materials are listed at the end of each chapter, but the following list provides information on general tissue culture suppliers [See Freshney, 2000 for more detailed list].

Item	*Supplier*
Antibodies	R & D Systems; Dako; Upstate Biotechnology
BM-cycline	Roche
CellFlo	Spectrum Laboratories
Ciprofloxacin	Bayer
CO_2 automatic switch-over device	Air Products and Chemicals; Gow-Mac; Lab-Impex; Shandon
CO_2 incubators	Hereaus; IEC (Hotpak); Lab-Line; Napco; NuAire; Precision Scientific; Thermo-Forma
Collagenase	Sigma; Worthington
Dispase	Roche
DNA profiling or fingerprinting	ATCC; Cellmark Diagnostics; ECACC; Laboratory of the Government Chemist
Electronic cell counter	Beckman-Coulter; Schärfe Systems
FACS	BD Biosciences; Beckman-Coulter
Filter well inserts	BD Biosciences; Corning; Millipore
Growth factors	BD Biosciences; Cambridge Biosciences; GIBCO; ICN; Roche; Sigma; Upstate Biotechnology
Heparan sulfate proteoglycans (HSPGs)	Sigma
Hydrocortisone	Sigma
Inverted microscope	Leica; Olympus; Nikon; Zeiss
Laminar flow hoods (Class II biosafety cabinets)	Atlas Clean Air; Baker; Heto-Holten; Medical Air Technology; NuAire
Liquid nitrogen freezer	Aire Liquide; Cryo-Med; Integra Biosciences; MVE; Taylor Wharton; Thermolyne
MACS	Miltenyi Biotec
Matrix products: heparan sulfate, fibronectin, laminin, collagen, Matrigel	BD Biosciences; Biofluids; Pierce; R & D Systems; Sigma; TCS
Membroferm	Polymun
Mycoplasma Removal Agent (MRA)	ICN
Retinoids	Sigma
Sample containers for transportation (See also www.uos.harvard.edu/ehs/bio_bio_shi.shtml and www.ehs.ucsf.edu/Safety%20Updates/Bsu/Bsu5.pdf)	Saf-T-Pak: 1-800-814-7484; Nalge: 1-716-586-8800; Air Sea Atlanta: 1-404-351-8600; Cellutech: 1-800-575-5945; Cin-Made Corp: 1-513-681-3600
Serum-free media	Cambrex; PromoCell; Cascade Biologicals; Sigma; GIBCO
Sodium butyrate (NaBt)	Sigma
Trypsin	Cambrex; GIBCO; ICN; Sigma; Worthington

REFERENCES

Bodnar, A.G., Ouellette, M., Frolkis, M., Holt, S.E., Chiu, C.-P., Morin, G.B., Harley, C.B., Shay, J.W., Lichsteiner, S., Wright, W.E. (1998) Extension of life-span by introduction of telomerase into normal human cells. *Science* 279: 349–352.

Caputo, J.L. (1996) Safety Procedures. In Freshney, R.I., Freshney, M.G., eds., *Culture of Immortalized Cells*. New York, Wiley-Liss, pp. 25–51.

Carr, T., Evans, P., Campbell, S., Bass, P., Albano, J. (1999) Culture of human renal tubular cells: positive selection of kallikrein-containing cells. *Immunopharmacology* 44: 161–167.

Cerni, C. (2000) Telomeres, telomerase, and myc. An update. *Mutat Res*. 462: 31–47.

Chambard, M., Vemer, B., Gabrion, J., Mauchamp, J., Bugeia, J.C., Pelassy, C., Mercier, B. (1983) Polarization of thyroid cells in culture; evidence for the basolateral localization of the iodide "pump" and of the thyroid-stimulating hormone receptor-adenyl cyclase complex. *J. Cell Biol.* 96: 1172–1177.

Chambard, M., Mauchamp, J., Chaband, O. (1987) Synthesis and apical and basolateral secretion of thyroglobulin by thyroid cell monolayers on permeable substrate: Modulation by thyrotropin. *J. Cell Physiol.* 133: 37–45.

Chen, T.R. (1977) In situ detection of mycoplasma contamination in cell cultures by fluorescent Hoechst 33258 stain. *Exp. Cell Res.* 104: 255.

Filla, M.S., Dam, P., Rapraeger, A.C. (1998) The cell surface proteoglycan syndecan-1 mediates fibroblast growth factor-2 binding and activity. *J. Cell Physiol.* 174: 310–321.

Freshney, R.I. (2002) Cell line provenance. *Cytotechnology* 39: 3–15.

Freshney, R.I. (2005) *Culture of Animal Cells, a Manual of Basic Technique*, 5th Ed. Hoboken NJ, John Wiley & Sons.

Greco, B., Recht, L. (2003) Somatic plasticity of neural stem cells: Fact or fancy? *J. Cell. Biochem.* 88: 51–56.

Ham, R.G., McKeehan, W.L. (1978) Development of improved media and culture conditions for clonal growth of normal diploid cells. *In Vitro* 14: 11–22.

Hay, R.J., Miranda-Cleland, M., Durkin, S., Reid, Y.A. (2000) Cell line preservation and authentication. In Masters, J.R.W, ed., *Animal Cell Culture*, 3rd Ed. Oxford, Oxford University Press, pp. 69–103.

Juliano, R.L. (2002) Signal transduction by cell adhesion receptors and the cytoskeleton: functions of integrins, cadherins, selectins, and immunoglobulin-superfamily members. *Annu. Rev. Pharmacol. Toxicol.* 42: 283–323.

Karmiol, S. (2000) Development of serum-free media. In Masters, J.R.W, ed., *Animal Cell Culture*, 3rd Ed. Oxford, Oxford University Press, pp. 105–121.

Klement, G., Scheirer, W., Katinger, H.W. (1987) Construction of a large scale membrane reactor system with different compartments for cells, medium and product. *Dev Biol Stand.*, 66: 221–226.

Knazek, R.A., Gullino, P., Kohler, P.O., Dedrick, R. (1972) Cell culture on artificial capillaries. An approach to tissue growth in vitro. *Science* 178: 65–67.

Kruse, P.F., Jr., Keen, L.N., Whittle, W.L. (1970) Some distinctive characteristics of high density perfusion cultures of diverse cell types. *In Vitro* 6: 75–78.

Laning, J.C., DeLuca, J.E., Hardin-Young, J. (1999) Effects of immunoregulatory cytokines on the immunogenic potential of the cellular components of a bilayered living skin equivalent. *Tissue Eng*. 5: 171–181.

Lopez-Casillas, F., Wrana, J.L., Massague, J. (1993) Betaglycan presents ligand to the TGF-β signalling receptor. *Cell* 73: 1435–1444.

Maas-Szabowski, N., Stark, H.-J., Fusenig, N.E. (2002) Cell interaction and differentiation. In Freshney. R.I. and Freshney. M.G., eds, *Culture of Epithelial Cells*, 2nd Ed. New York, John Wiley & Sons, pp. 31–63.

MacLeod, R.A., Dirks, W.G., Matsuo, Y., Kaufmann, M., Milch, H., Drexler, H.G. (1999) Widespread intraspecies cross-contamination of human tumor cell lines arising at source. *Int. J. Cancer* 83: 555–63.

Marcovic, O., Marcovic, N. (1998) Cell cross-contamination in cell cultures: the silent and neglected danger. *In Vitro Cell Dev Biol.* 34: 108.

Masters, J.R. (2002) HeLa cells 50 years on: the good, the bad and the ugly. *Nat. Rev. Cancer* 2: 315–9.
Masters, J.R.W., Thomson, J.A., Daly-Burns, B., Reid, Y.A., Dirks, W.G., Packer, P., Toji, L.H., Ohno, T., Tanabe, H., Arlett, C.F., Kelland, L.R., Harrison, M., Virmani, A., Ward, T.H., Ayres, K.L., Debenham, P.G. (2001) STR profiling provides an international reference standard for human cell lines. *Proc. Natl. Acad. Sci. USA* 98: 8012–8017.
Mather, J.P. (1998) Making informed choices: medium, serum, and serum-free medium. How to choose the appropriate medium and culture system for the model you wish to create. *Methods Cell Biol.* 57: 19–30.
Michel, M., L'Heureux, N., Pouliot, R., Xu, W., Auger, F.A., Lucie, G. (1999) Characterization of a new tissue-engineered human skin equivalent with hair. *In Vitro Cell. Dev. Biol. Anim.* 35: 318–326.
Munger, K., Howley, P.M. (2002) Human papillomavirus immortalization and transformation functions. *Virus Res.* 89(2): 213–228.
Peehl, D.M., Ham, R.G. (1980) Clonal growth of human keratinocytes with small amounts of dialysed serum. *In Vitro* 16: 526–540.
Petra, M., de Jong, P.M., van Sterkenburg, A.J.A., Kempenaar, J.A., Dijkman, J.H., Ponec, M. (1993) Serial culturing of human bronchial epithelial cells derived from biopsies. *In Vitro Cell Dev. Biol.* 29A: 379–387.
Pretlow, T.G., and Pretlow, T.P. (1989) Cell separation by gradient centrifugation methods. *Methods Enzymol.* 171: 462–482.
Raff, M.C. (1990) Glial cell diversification in the rat optic nerve. *Science* 243: 1450–1455.
Regnier, M., Staquet, M.J., Schmitt, D., Schmidt, R. (1997) Integration of Langerhans cells into a pigmented reconstructed human epidermis. *J. Invest. Dermatol.* 109: 510–512.
Rizzino, A. (2002) Embryonic stem cells provide a powerful and versatile model system. *Vitam Horm.* 64: 1–42.
Saalbach, A., Aust, G., Haustein, U.F., Herrmann, K., Anderegg, U. (1997) The fibroblast-specific MAb AS02: A novel tool for detection and elimination of human fibroblasts. *Cell Tissue Res.* 290: 593–599.
Schaller,. M., Mailhammer, R., Grassl, G., Sander, C.A., Hube, B., Korting, H.C. (2002) Infection of human oral epithelia with *Candida* species induces cytokine expression correlated to the degree of virulence. *J. Invest. Dermatol.* 118: 652–657.
Stoker, M.G.P. (1973) Role of diffusion boundary layer in contact inhibition of growth. *Nature* 246: 200–203.
Swope, V.B., Supp, A.P., Cornelius, J.R., Babcock, G.F., Boyce, S.T. (1997) Regulation of pigmentation in cultured skin substitutes by cytometric sorting of melanocytes and keratinocytes. *J, Invest. Dermatol.* 109: 289–295.
Uphoff, C.C., Drexler, H.G. (2002a) Comparative PCR analysis for detection of mycoplasma infection in continuous cell lines. *In Vitro Cell. Dev. Biol. Anim.* 38: 79–85.
Uphoff, C.C., and Drexler, H.G. (2002b) Comparative antibiotic eradication of mycoplasma infections from continuous cell lines. *In Vitro Cell. Dev. Biol. Anim.* 38: 86–89.
van Bokhoven, A., Varella-Garcia, M., Korch, C., Hessels, D., Miller, G.J. (2001). Widely used prostate carcinoma cell lines share common origins. *Prostate* 47: 36–51.
Westermark, B., Wasteson, A. (1975) The response of cultured human normal glial cells to growth factors. *Adv. Metab. Disord.* 8: 85–100.
Wright, W.E., Shay, J.W. (2002) Historical claims and current interpretations of replicative aging. *Nat. Biotechnol.* 20: 682–688.

2

Mesenchymal Stem Cells for Tissue Engineering

Donald P. Lennon and Arnold I. Caplan

Skeletal Research Center, Department of Biology, Case Western Reserve University, Cleveland, Ohio 44106

Corresponding author: dpl@po.cwru.edu

Culture of Cells for Tissue Engineering, edited by Gordana Vunjak-Novakovic and R. Ian Freshney

1. BACKGROUND AND LOGIC

1.1. Introduction

Tissue engineering, a multidiscipline approach to reconstructing biological tissues, is generally considered to include three main components: cells that fabricate the lost or damaged tissue, materials intended to serve as delivery vehicles and scaffolds for the cells, and cytokines and other bioactive factors consistent with appropriate cell proliferation and differentiation [Ringe et al., 2002]. Although differentiated cells have been used successfully in both experimental and clinical protocols [Bell et al., 1983; Aigner et al., 1998; Brittberg et al., 2001], numerous advantages attend the use of autologous stem cells as the source of donor cells in tissue engineering: These include low donor site morbidity, diminished or absent immune response, high proliferative potential, and relative ease of access to the cell repository [Bruder et al., 1994; Ringe et al., 2002]. Thus, in the past decade, stem cells, generally defined as cells that have the capacity both to self-renew and to generate differentiated progeny [Morrison et al., 1997; Jackson et al., 2002], have generated substantial interest and have been the object of intensive research efforts.

A wide array of studies supports the contention that various tissues may be repositories of mesenchymal stem or progenitor cells. These tissues include bone marrow [Caplan, 1991; Haynesworth et al., 1992a,b; Prockop, 1997; Bianco et al., 1999], periosteum [Nakahara et al., 1991], adipose tissue [Halvorsen et al., 2000; Erickson et al., 2002; Safford et al., 2002], dermis [Toma et al., 2001], muscle [Lee et al., 2000; Jankowski et al., 2002], and vasculature (pericytes) [Brighton et al., 1992].

1.2. History

The notion that stem cells are present in bone marrow was suggested by Friedenstein [1976], based on characterization of clonal populations of marrow stromal cells isolated from mouse, guinea pig, and rabbit bone marrow and on implantation of the cells in diffusion chambers in syngeneic host animals [Friedenstein et al., 1970; Friedenstein, 1976; See Phinney, 2002 for review]. Owen [1985, 1988], and Owen and Friedenstein [1988] proposed a model of differentiation for progenitor cells in the stromal system analogous to that of the hematopoietic system. According to this hypothesis, marrow stromal stem cells give rise to committed

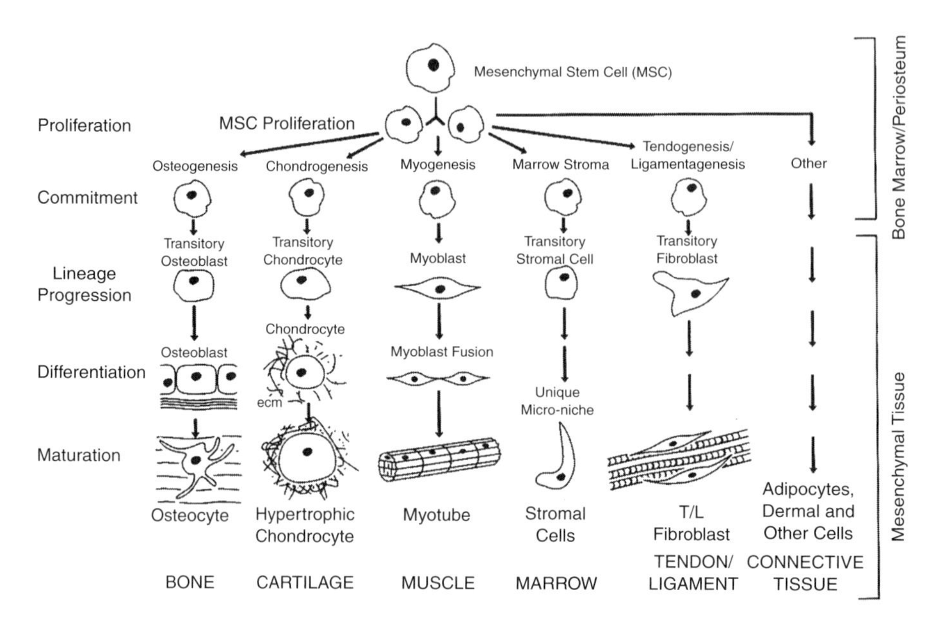

Figure 2.1. The mesengenic process. A schematic representation of some of the differentiation capabilities of mesenchymal stem cells and a simplified illustration of the cellular transitions from stem cell to well-differentiated cells. Reprinted from Clinics in Plastic Surgery, volume 21, number 3, Arnold I. Caplan, "The Mesengenic Process", pp. 429–435, copyright 1994, with permission from Elsevier.

progenitors of fibroblastic, reticular, adipogenic, osteogenic, and possibly other cells. Our vision of the lineage pathways available to mesenchymal stem cells was originally depicted in the mesengenic process diagram seen in Fig. 2.1. In this scheme, the pathways are arranged so that the best-understood are on the left, and the least established are on the right [Caplan, 1994].

Our laboratory's interest in marrow-derived mesenchymal stem cells (MSCs) can be viewed as a logical consequence of our earlier in vitro studies involving mesenchymal cells of the embryonic chick limb bud system [Caplan, 1977, 1984]. Although there are obvious differences between the two systems, there are also certain parallels between the limb bud and marrow stromal systems. Most importantly, both bone and cartilage development in the embryonic limb bud and bone turnover and repair in the adult limb result from the division of small numbers of progenitor cells (MSCs) and the ultimate commitment and differentiation of the progeny of these cells [Caplan, 1991; Bruder et al., 1994]. Thus Ohgushi et al. [1989, 1990] demonstrated that intact or disaggregated rat marrow forms bone and cartilage when loaded into porous ceramic (hydroxyapatite-tricalcium phosphate) cubes and implanted into subcutaneous sites in syngeneic hosts. Goshima et al. [1991a] achieved similar results when they loaded the same type of ceramics with cultured rat marrow cells, referred to as mesenchymal stem cells.

An important development occurred when Haynesworth et al. [1992b] demonstrated that cultured human marrow MSCs also generate bone, but rarely cartilage, when introduced into the same assay system. Interestingly, companion human cells loaded into diffusion chambers did not form bone; accordingly, we consider the ceramic cube assay to be the standard for testing MSCs. In these experiments, Haynesworth and co-workers (1992b) used marrow cells isolated by two different techniques. In the first, cancellous bone marrow from femoral heads was mechanically disrupted and a single-cell suspension was isolated and seeded into tissue culture dishes in serum-containing medium. In the second method, bone marrow aspirated from the iliac crest was rinsed, resuspended in serum-containing medium, loaded onto a Percoll density gradient, and centrifuged. In preliminary experiments, it was determined that the majority of the adherent cells were restricted to the top 25% of the gradient. Cells recovered from this fraction were rinsed in serum-containing medium, counted, and seeded into tissue culture dishes. Ultimately, the second method was found to be superior to the isolation of cells by direct plating and has been used extensively by our laboratory and many others; this method is described in detail in Protocol 2.3.

Although the isolated cell fraction is seeded at a rather high density (1×10^7 nucleated cells per 10-cm dish), the total cell inoculum includes, along with a small fraction of erythrocytes, many nucleated cells, probably from the hematopoietic lineage, that are not capable of attaching to the culture substrate. The unattached cells are eventually removed during the course of routine changes of medium, and a smaller subset of cells becomes anchored to the substrate. These primarily fibroblast-like cells begin to proliferate and form colonies, which can first be seen around days 4–6 of culture. The ability of these cells to attach to the culture substrate represents the first and most important step in the selection of MSCs from the total nucleated cell population. Because of variability among individual marrow donors, the number of colonies per dish may vary; in our experience a typical 10-cm dish may contain 100 to 200 colonies. Assuming, as suggested by Friedenstein [Friedenstein et al., 1970; Friedenstein, 1976], that a colony of these cells arises from a single attached cell, we would conclude that these cells are rare indeed (1–2 per 10^5 nucleated cells from the marrow of young donors).

1.3. Assays for Phenotypic Potencies

Having isolated cells from an aspirate of human bone marrow, how does an investigator verify the "stem" nature of the cells? That is to ask, what assays and cell markers are relevant to identifying mesenchymal stem cells? As indicated above, we believe that the in vivo ceramic cube assay, described in Section 6, is the standard for identifying MSCs. Human MSC-ceramic composites implanted subcutaneously in immunocompromised host animals almost always produce bone, whereas bone and cartilage are never present when dermal fibroblasts are used in the same manner [Lennon et al., 2000]. Moreover, empty cubes used as a negative control do not contain bone in the central portion of the cubes, although long-term implants rarely include small portions of bone in the periphery of the cube. It should be noted that marrow-derived MSCs from different species produce different proportions of bone and cartilage. Cartilage, almost never seen in human MSC-ceramic composites, is frequently observed with rat MSCs, but much less so than bone [Dennis and Caplan, 1993; Lennon et al., 2001]. On the other hand, cartilage is more commonly observed than bone for rabbit MSCs.

The in vivo ceramic cube assay system is complemented by the use of a number of in vitro assays, each of which is specific for different differentiated cells. Elevated alkaline phosphatase activity, although not unique to osteogenic cells, is a useful early marker for osteogenic differentiation. Accordingly, biochemical and cytochemical assays for alkaline phosphatase are used early in the culture of MSCs in osteoinductive conditions. On the other hand, calcium biochemistry and von Kossa staining to detect mineralization are used later in the culture period. These assays and the conditions that promote osteogenic differentiation are described in Section 7.

Aggregate or pellet cultures can be established in a defined medium to promote chondrogenic differentiation of MSCs [Johnstone et al., 1998]. Such cultures will not be covered in this section as they are treated thoroughly in Chapter 4 of this book.

Differentiation of MSCs to adipocytes can be induced through the use of a unique medium in cultures seeded at a slightly higher density than that used for osteogenic differentiation. Flow cytometric analysis after Nile Red staining is used to assess expression of the adipocytic phenotype after induction. This methodology is described in Section 8.

The ability of MSCs to differentiate along these various phenotypic lines is strongly suggestive of their stem cell nature. MSCs, although sometimes referred to with different terminology and derived from a number of species, have also been shown to differentiate into other mesenchymal lineages, including skeletal muscle [Wakitani et al., 1995], cardiac muscle [Tomita et al., 1999], and hematopoietic supportive tissue [Koç et al., 2000], in addition to nonmesenchymal tissues including neurons [Woodbury et al., 2000; Black and Woodbury, 2001] and retinal cells [Tomita et al., 2002]. Importantly, differentiation of clonal populations of human [Pittenger et al., 1999] and murine [Dennis et al., 1999] MSCs along multiple lineages has been demonstrated.

Although no cell surface marker unique to MSCs has been identified to date, an extensive expression profile of cytokines and their receptors and adhesion and extracellular matrix molecules shared by these cells has been described [Mosca et al., 1997; Majumdar et al., 1998; Minguell et al., 2001; Shur et al., 2002]. Moreover, a number of monoclonal antibodies reactive with MSCs have been developed; antibodies SH-2, SH-3, and SH-4 are reactive with MSCs, but not with hematopoietic or differentiated bone cells [Haynesworth et al., 1992a].

1.4. Key Technical Details

Selection of the proper lot of fetal bovine serum (FBS) for use in culturing MSCs is perhaps the most important parameter in MSC technology [Lennon et al., 1996] and is described briefly in Section 9. It has been our experience that it is not good practice to purchase serum "off the shelf" (that is, without testing). We have also found that a batch of serum optimal for MSCs from one species will probably not support proliferation of MSCs from another. Although the selection of the proper batch of FBS is extremely important, all of the technical details involved in the isolation, expansion, and analysis of MSCs are also important. We believe that these details influence the outcome of our experimentation. With this in mind, we have organized the remainder of this chapter to provide all of these technical details.

2. PREPARATION OF MEDIA AND REAGENTS

2.1. Density Gradients

Protocol 2.1. Preparation of Percoll Density Gradient

Reagents and Materials

Sterile

- ❑ Percoll
- ❑ Sodium chloride, 1.5 M
- ❑ Tyrode's or Hanks' balanced salt solution (TBSS or HBSS)
- ❑ Polycarbonate high-speed centrifuge tubes, 50 ml

Nonsterile

- ❑ High-speed centrifuge (20,000 *g*)
- ❑ Fixed-angle rotor for 50-ml tubes

Protocol

(a) Combine the following sterile solutions (or any convenient multiples thereof) in a sterile container:
   i) 22.05 ml Percoll
   ii) 2.45 ml 1.5 M sodium chloride
   iii) 10.5 ml TBSS or HBSS

(b) Mix the solutions thoroughly, and then add 35 ml per sterile 50-ml polycarbonate centrifuge tube. Place caps on tubes.

(c) Centrifuge at 20,000 *g* for 15 min at room temperature in an appropriate centrifuge (preferably in a fixed-angle rotor).

(d) Remove tubes and store at 4 °C until they are needed (See Protocol 2.3).

2.2. Solutions for In Vitro Osteogenic Induction

2.2.1. Dexamethasone (dex)

1. Prepare a stock solution of 1×10^{-3} M dex in 100% ethanol (3.92 mg dexamethasone in 10 ml 100% ethanol).
2. Prepare a solution of 1×10^{-5} M dex by making a 1:100 dilution of the 10^{-3} M dex in serum-free medium (the medium appropriate for the cells being cultured).
3. Filter the solution through a 0.22-μm filter and store at −20 °C

2.2.2. β-Glycerophosphate (BGP)

(i) Prepare a 200 mM solution by dissolving 0.216 *g* BGP in 5 ml TBSS or other balanced salt solution (BSS).

(ii) Sterilize by passing solution through a 0.22-μm filter and store at 4 °C.

2.2.3. Ascorbic Acid 2-Phosphate

(i) Prepare a 5 mM solution of ascorbic acid 2-phosphate by dissolving 0.0347 g ascorbic acid 2-phosphate in 10 ml TBSS or other BSS.

(ii) Sterilize with a 0.22-μm filter and store at 4 °C.

2.3. Solutions for Quantitative Biochemical Alkaline Phosphatase Assay

2.3.1. Substrate Buffer

Glycine, 50 mM, $MgCl_2$, 1 mM, pH 10.5.: Dissolve 1.88 g of glycine and 0.1017 g of $MgCl_2 \cdot 6H_2O$ in 500 ml of water. Adjust to pH 10.5 with 1 N NaOH.

2.3.2. Substrate

Dissolve 1 tablet (5 mg) of phosphatase substrate (*p*-nitrophenyl phosphate) per 5 ml of substrate buffer.

2.3.3. *p*-Nitrophenol for Standard Curve

Prepare a 50 nmol/ml solution of *p*-nitrophenol by combining 50 μl of 10 μmol/ml *p*-nitrophenol standard solution with 9.95 ml of 0.02N NaOH. Prepare further dilutions as illustrated in the following table:

p-Nitrophenol Concentration (nmol/ml)	Volume (ml) of Diluted *p*-Nitrophenol Solution (50 nmol/ml)	Volume (ml) of 0.02 M NaOH
4.5	0.1	1.0
9.0	0.2	0.9
18	0.4	0.7
27	0.6	0.5
36	0.8	0.3
45	1.0	0.1

2.4. Solutions for Qualitative Cytochemical Alkaline Phosphatase Assay

2.4.1. Fast Violet Stain

Dissolve 1 fast violet capsule in 48 ml of water. This solution can be stored as 12-ml aliquots at 4 °C.

2.4.2. Citrate Working Solution

Add 2 ml of citrate concentrated solution to 98 ml of water.

2.4.3. Citrate Buffered Acetone

Combine 60% citrate working solution and 40% acetone.

2.5. Solutions for Adipogenic Induction and Flow Cytometry

2.5.1. Adipogenic Induction Medium

3-Isobutyl-1-methylxanthine (IBMX) and indomethacin:

(i) Combine 0.1789 g indomethacin and 0.555 g IBMX in a 15- or 50-ml centrifuge tube.

(ii) Add dimethyl sulfoxide (DMSO) to a volume of 5 ml. This gives a concentration of 0.1 M indomethacin and 0.5 M IBMX.

(iii) Add 0.5 ml to a 500-ml bottle of serum-supplemented DMEM-HG to give final concentrations of 100 μM indomethacin and 500 μM IBMX.

(iv) Store unused stock solution at −20 °C.

2.5.2. Insulin

(i) Prepare a 10 mg/ml solution of bovine insulin in 0.01 N HCl.

(ii) Add 0.6 ml to 11.4 ml DMEM-HG (0.5 mg/ml).

(iii) Add 10 ml per 500 ml DMEM-HG for a final concentration of 10 μg/ml.

(iv) Sterilize by passing solution through a 0.22-μm filter and store at 4 °C.

2.5.3. Dexamethasone

(i) Prepare a stock solution of 10^{-3} M dex in 100% ethanol (0.00392 g dexamethasone in 10 ml 100% ethanol).

(ii) Add 0.5 ml per 500 ml of Dulbecco's modification of Eagle's medium with high glucose, 4.5.g/l (DMEM) to give a final concentration of 10^{-6} M dex.

(iii) Sterilize by passing solution through a 0.22-μm filter and store at 4 °C.

2.5.4. Nile Red Working Solution

(i) Prepare a 1 mg/ml solution of Nile Red in DMSO.

(ii) Add 250 μl of this solution per final volume of 10 ml in phosphate-buffered saline without Ca^{2+} and Mg^{2+} (PBSA)

3. ISOLATION OF HUMAN MARROW-DERIVED MESENCHYMAL STEM CELLS (HMSCS)

3.1. Aspiration of Human Bone Marrow

Aspiration of bone marrow from the posterior superior iliac crest is carried out by physicians of the Department of Hematology-Oncology at University Hospitals of

Cleveland, which is affiliated with Case Western Reserve University. Normally, we receive the bone marrow sample in a 20-ml syringe and proceed to the steps listed in Protocol 2.3. The details of Protocol 2.2 were provided by Dr. Omer Koç of the Department of Hematology-Oncology; although this protocol is outside the scope of cell culture, it is provided here for the reader's information. The Institutional Review Board of the hospital must approve research protocols involving human marrow, and donors must give informed consent.

Protocol 2.2. Collection of Human Bone Marrow

Reagents and Materials

Sterile

❑ Betadine
❑ Lidocaine (1%)
❑ Scalpel with # 15 blade
❑ Jamshidi needle, 11-gauge
❑ Syringe, 20 ml, containing 2 ml heparin (preservative-free, 400 units per ml)

Protocol

(a) Marrow donors who have given informed consent lie in a lateral decubitus position.
(b) The donor's skin is wiped with Betadine.
(c) The posterior superior iliac crest is located.
(d) Lidocaine (1%) is used to anesthetize the skin and subcutaneous tissue superficial to the iliac crest.
(e) A small cut (the width of a scalpel blade) is made though the skin and subcutaneous tissue with a scalpel blade.
(f) An 11-gauge Jamshidi needle is inserted though the cut and is anchored into the posterior superior iliac crest.
(g) Once the needle is anchored, the hub is removed.
(h) A 20-ml syringe containing 2 ml heparin (preservative-free, 400 units per ml) is attached to the needle.
(i) The marrow is aspirated by pulling the syringe plunger back briskly.
(j) The needle is rotated 90 degrees clockwise several times, and marrow is aspirated at each new position.
(k) After the needle is removed, pressure is applied to the skin until bleeding stops.

3.2. Enrichment of Mesenchymal Stem Cells from Human Marrow

A cell fraction enriched for mesenchymal stem cells is isolated by density gradient centrifugation.

Protocol 2.3. Isolation and Seeding of Human Mesenchymal Stem Cells (hMSCs)

Reagents and Materials

Sterile

- ❑ DMEM-LG-10FB: Dulbecco's modified Eagle's medium with 1 g/l glucose (DMEM-LG) supplemented with 10% FBS. The FBS is preelected [Lennon et al., 1996], as described in Section 9, to support proliferation and differentiation (given the appropriate conditions) of hMSCs.
- ❑ Polypropylene centrifuge tubes, 50 ml
- ❑ Centrifuge tube, 15 ml
- ❑ Tissue culture flasks, 75 cm^2, or Petri dishes, 10 cm

△ *Safety note.* Personnel wearing the proper personal protective equipment, including a lab coat, goggles or a face shield, gloves, and a surgical mask, process the marrow sample in a Class II biological safety cabinet.

The marrow sample and all cells derived from it are treated with standard biohazard precautions. That is, it is assumed that the sample is contaminated with hepatitis B or HIV. Use appropriate waste containers for all sharp and nonsharp disposable supplies that come into contact with human tissue or cells, or with medium that has contacted these cells. All liquid waste generated must be treated with bleach before it is disposed of in a sink; bleach is added to produce a 20% concentration, and the solution may be disposed of after 30 min. These safety precautions apply to all procedures involving unfixed human cells described in this chapter.

Protocol

(a) The marrow sample is usually delivered in a 20-ml syringe. Eject the contents of the syringe into a 50-ml polypropylene centrifuge tube.

(b) Add 20 to 30 ml DMEM-LG-10FB to the tube.

(c) Pipette up and down thoroughly to mix the medium and the marrow sample, and then transfer a small aliquot (about 200 μl) of the suspension to a 15-ml centrifuge tube.

(d) Centrifuge the suspension in the 50-ml tube at 450 *g* for 5 min in a bench-top centrifuge.

(e) While the sample is being centrifuged, conduct a preliminary count of the cells with the suspension in the small tube.

  i) Transfer 50 μl of this suspension to another suitable small tube.

  ii) Add 50 μl DMEM-LG-10FB.

  iii) Add 100 μl 4% acetic acid (to lyse the red blood cells).

  iv) Count the nucleated cells with a hemocytometer, and determine the total number of such cells in the 50-ml tube.

(f) After the sample has been centrifuged, remove the supernate.
(g) Determine the number of tubes of preformed Percoll (density 1.03–1.12 g/ml) (See Protocol 2.1) that will be required to fractionate the nucleated cells. The number of tubes of Percoll to be used depends on the number of nucleated cells determined in step (e) and on the volume of the cell pellet. The maximum number of nucleated cells per tube of Percoll is 2×10^8. However, if the cell number is lower than this figure, but the pellet volume is greater than 5 ml, more than one tube of Percoll will also be required.
(h) Adjust the volume of the pellet with DMEM-LG-10FB to allow 5 ml cell suspension per tube of Percoll.
(i) Carefully load 5 ml of the cell suspension per tube of Percoll with a pipette. Transfer the suspension slowly, so that it remains at the top of the gradient.
(j) Carefully transfer the tubes to a centrifuge, and spin at 480 g for 15 min with the brake off, preferably in a fixed-angle rotor.
(k) Return the sample to the biological safety cabinet and transfer the top 10 to 14 ml of each Percoll tube to a sterile 50-ml polypropylene centrifuge tube.
(l) Increase the volume in the tube to 50 ml with DMEM-LG-10FB and mix completely by pipetting up and down.
(m) Centrifuge the tube at 450 g in a bench-top centrifuge.
(n) Remove the supernatant and resuspend the cell pellet in 10 ml DMEM-LG-10FB.
(o) Determine the final number of nucleated cells in the same manner as indicated in step (e)
(p) Adjust the volume as necessary with DMEM-LG-10FB and seed the cells at a density of 1.8×10^5 per cm^2 in tissue culture dishes or flasks.
(q) Place dishes in a humidified tissue culture incubator at 37 °C/5% CO_2.
(r) Change the medium after 3 days, and every 3 to 4 days thereafter.

3.3. Primary Culture

Primary cultures of MSCs are seeded at 1×10^7 cells per 10-cm culture dish. This cell inoculum contains a mixture of cells, including red blood cells, unidentified nucleated cells of the hematopoietic lineage, monocytes, macrophages, and fibroblast-like cells. Erythrocytes and leukocytes are not capable of attaching to the culture substrate and are eventually removed during the course of routine changes of culture medium. Nonadherent cells are not rigorously removed by rinsing. Instead, medium is simply pipetted or aspirated from the dish without vigorous swirling or rinsing, as it is assumed that the nonadherent cells provide paracrine factors needed for the optimal growth of the attached cells.

Attachment of cells to the negatively charged culture dish is, in fact, considered a method of selection for fibroblastoid cells [Phinney, 2002]. Only a relatively small number of cells attach to the dish. The fibroblast-like cells begin to proliferate and form loose colonies of spindle-shaped cells that can usually be identified between days 4 and 6 of culture (Fig. 2.2). The colonies greatly increase in size

Figure 2.2. Phase-contrast photomicrographs of human mesenchymal stem cells (MSCs) on day 6 of primary culture. A) A low-magnification view of two adjacent colonies of MSCs. B) MSCs from another colony of the same preparation of cells at higher magnification.

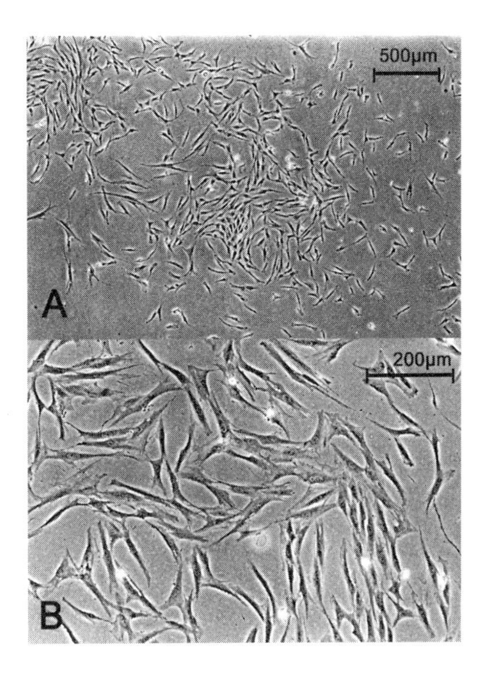

over the next 7 days and should be subcultured before the cells become dense and multilayered (proliferation of the cells is not contact inhibited).

4. PROPAGATION OF MESENCHYMAL STEM CELLS

The density of the colonies is the primary consideration in determining when the cells should be passaged. Thus, if the cells in the colonies are becoming densely associated, the cultures should be passaged, even though the cells throughout the dish may not be confluent.

Protocol 2.4. Subculture of Human Mesenchymal Stem Cells (hMSCs)

Reagents and Materials

Sterile

- ❑ BSS: balanced salt solution, e.g., Tyrode's or Hanks' BSS
- ❑ Trypsin-EDTA: 0.25% trypsin, 1 mM EDTA in Hanks' BSS lacking Ca^{2+} and Mg^{2+}
- ❑ BCS: bovine calf serum
- ❑ DMEM-LG-10FB

Protocol

(a) Primary isolates of hMSCs should be subcultured before individual cell colonies become overly dense. The cultures are usually trypsinized around day 14 of culture (± 3 days).

(b) Remove the culture medium.
(c) Rinse the cell layer with 5 ml BSS (for 10-cm tissue culture dishes).
(d) Add 4 ml trypsin-EDTA and return the vessel to the incubator for 5–10 min. Keep the time of exposure as brief as possible. (See (e) and (f) below.)
(e) When the majority of cells have become well rounded or have detached from the culture dish, stop the reaction by adding a volume of BCS equal to 1/2 the volume of the trypsin.
(f) Draw up the cell suspension with a pipette and, with the same pipette, use the suspension to gently wash the remaining cells from the dish. It is not necessary to remove all of the cells from the dish, as most of the nonfibroblastoid cells (which are not likely to be MSCs) in these cultures are more trypsin resistant than the spindle-shaped cells. Thus trypsinization represents, along with separation of nucleated cells by density centrifugation and attachment of fibroblastic cells to plastic, an important component of the process of the selection of MSCs from the total marrow cell population.
(g) Transfer the cell suspension from all of the cultures to an appropriate-size centrifuge tube or tubes. In our laboratory, we prefer to keep groups of culture dishes from a single preparation segregated from one another throughout the subculture process as a form of insurance against contamination of the entire group, because we have found that low levels of contamination are not always easily detectable. For example, a preparation consisting of 12 culture dishes might be divided into 4 groups of 3 dishes; we would attempt to ensure that there was no cross-contamination among the groups by using separate pipettes and tubes for each group.
(h) Centrifuge the tubes at 400 *g*.
(i) Remove the supernatant with a pipette or other suitable device.
(j) Resuspend the cell pellet in a suitable volume of DMEM-LG-10FB (usually 5 or 10 ml).
(k) Remove a sample of the suspension with a Pasteur pipette or micropipettor and count the cells with a hemocytometer.
(l) Adjust the volume of medium as necessary, and seed the cells at 3500 to 4000 cells per cm^2.
(m) Change the culture medium every 3 to 4 days.
(n) Further subculture of hMSCs is conducted in essentially the same manner except for the following considerations:
    i) Subcultured hMSCs are evenly distributed on the tissue culture vessel surface and are not in colonies as for primary cultures. Therefore, the key criterion for determining when the cells should be trypsinized is the degree of confluence; basically, hMSCs should be trypsinized before they become confluent.
    ii) Passaged hMSCs are more easily trypsinized than primary cultures, so exposure of these cells to trypsin is usually limited to 5 min.

Mesenchymal stem cells have a high proliferative capacity and may be subcultured repeatedly. During the process of cell expansion, MSCs maintained in DMEM-LG supplemented only with FBS remain in an undifferentiated state, as indicated by negative results in in vitro assays for osteogenesis (See Section 7). Bruder et al. [1997] maintained hMSCs through as many as 15 passages. At each passage some cells were continued in serum-supplemented medium, while others were placed into an osteoinductive medium also containing dexamethasone (See Section 7). The conservation of osteogenic potential at each passage was demonstrated by increased alkaline phosphatase activity and the formation of mineralized matrix in cells of the latter group. Careful documentation of cell numbers at each passage revealed that hMSCs could undergo an average of 38 population doublings before reaching senescence (as indicated by arrested cell growth and a broad, flattened cell morphology).

Subcultured MSCs are similar to primary isolates of the cells in that they remain spindle-shaped fibroblastoid cells. As indicated above, however, passaged MSCs are distributed evenly over the culture dish rather than being organized into colonies. They are also slightly wider than for primary cultures, a feature that becomes more pronounced with additional subcultivation.

5. CRYOPRESERVATION AND THAWING OF HMSCS

Another interesting aspect of the report by Bruder et al. [1997] is that a portion of the cells trypsinized at the end of primary culture was cryopreserved in liquid nitrogen and then thawed. These cells were then taken through the same process of extensive subcultivation and exposure to osteoinductive treatment as for the unfrozen cells. Cryopreservation was not found to have an adverse effect on further cell expansion or on osteogenic differentiation at each passage. The procedures involved in freezing and thawing hMSCs are described in Protocol 2.5 and Protocol 2.6, respectively.

Protocol 2.5. Cryopreservation of Human Mesenchymal Stem Cells (hMSCs)

Reagents and Materials

Sterile

- ❑ Trypsin-EDTA: 0.25% trypsin, 1 mM EDTA in Hanks' BSS lacking Ca^{2+} and Mg^{2+}
- ❑ DMSO
- ❑ FBS (from the same lot of serum selected for growth of hMSCs)

Nonsterile

- ❑ Nalgene freezing container, with a 1 °C/min cooling rate when filled to the appropriate level with isopropanol

Protocol

(a) Prepare freezing medium consisting of 10% DMSO and 90% FBS. Use only freshly prepared freezing medium.

(b) The cells must be in log phase (rapidly dividing) for the cryopreservation to be successful.

(c) Trypsinize the cells in the manner described in Protocol 2.4.

(d) Count the resuspended cells with a hemocytometer or electronic cell counter to determine the total cell yield.

(e) Centrifuge the cell suspension at 400 *g* for 5 min at room temperature.

(f) Gently resuspend the cells in freezing medium at a concentration of 1×10^6 cells per ml.

(g) Label cryovials with the appropriate information, including complete cell identification (cell type, donor number, passage number, and cell density), date, and initials of the operator.

(h) Transfer 1 ml of the cell suspension to each labeled cryovial, and then seal the vial with the cap.

(i) Place the cryovial(s) into the Nalgene freezing container.

(j) Place the freezing container in a $-70\,^{\circ}$C freezer. The Nalgene containers are designed to reduce the temperature of the cell suspension by $1\,^{\circ}$C/min under these conditions.

(k) After 24 h transfer the vials to a liquid nitrogen freezer. (See Safety note.)

△ ***Safety note.*** Wear safety goggles or, preferably, a mask, in addition to gloves and protective clothing when handling liquid nitrogen, and ensure the room is properly ventilated.

Protocol 2.6. Thawing Cryopreserved Human Mesenchymal Stem Cells (hMSCs)

Reagents and Materials

Sterile

- ❑ DMEM-LG-10FB
- ❑ Plastic tissue culture dishes, 10 cm

Nonsterile

- ❑ Trypan Blue viability stain
- ❑ Water bath at $37\,^{\circ}$C

Protocol

(a) Warm DMEM-LG-10FB to room temperature.

(b) Remove cryovials to be thawed from liquid nitrogen (See Safety note above).

(c) Partially thaw in a $37\,^{\circ}$C water bath (i.e., some ice should still be present).

△ *Safety note.* Wear safety goggles and place a lid on the water bath while thawing the cells. If liquid nitrogen has entered the vial, it may explode on thawing. Ideally, store vials in the vapor phase of liquid nitrogen.

(d) Transfer gently into a tube containing 35 ml DMEM-LG-10FB (the sample will have completed thawing by this time).
(e) Centrifuge at 250 *g* for 5 min.
(f) Remove supernate from cells.
(g) Add 2 to 5 ml DMEM-LG-10FB and resuspend cells gently.
(h) Combine 100 μl of the cell suspension with an equal volume of Trypan Blue and count the unstained cells with a hemocytometer.
(i) Seed cells into a 10-cm tissue culture dish or dishes in 7 ml medium per dish at a density of 2.5×10^5.
(j) Change medium every 3–4 days. Plates should be confluent in 7–10 days.
(k) Cells may be subcultured as described in Protocol 2.4.

Thawed hMSCs are similar in morphology to those that have never been frozen and then thawed. They may have elongated, slender processes or may be somewhat more compact 1 day after being seeded but, after readapting to culture, become identical to unfrozen cells. A variable number of cells fail to attach to the culture dish and remain as floating cells; these are removed in the course of routine medium changes. Some presumably donor-dependent variability in recovery from cryopreservation has been observed.

6. IN VIVO ASSAY FOR OSTEOGENESIS

We consider the in vivo assay, in which ceramic cubes are loaded with isolated and culture-expanded putative stem cells and then implanted subcutaneously into immunocompromised host animals, to be the definitive test of the osteogenic and chondrogenic potential of stem cells in general. However, as indicated in Section 1, cartilage formation is almost never observed when human MSCs are used in this assay. Although bone formation has been observed in diffusion chambers loaded with cultured nonhuman marrow-derived stromal cells, the frequency of bone-positive results and the rate of bone formation are higher when the same cells are implanted in ceramic vehicles [Dennis and Caplan, 1993].

Our laboratory routinely uses biphasic ceramics consisting of 60% tricalcium phosphate and 40% hydroxyapatite, but we have also had success with coral-based ceramics. Pretreatment of the ceramics with fibronectin or laminin facilitates attachment of the cells to the ceramic surface and, in turn, results in the development of bone at earlier time points than for untreated cubes [Dennis and Caplan, 1993].

Subcutaneously implanted cell-loaded ceramic cubes provide a critical control for cells used in experimental tissue engineering designs. For example, Kadiyala et al. [1997] used ceramic cubes loaded with rat marrow-derived MSCs as a positive control in experiments testing the efficacy of these cells in aiding the repair of a critical-size defect in rat femora. We have also used this assay to test the effect of in vitro factors, such as reduced oxygen tension, on differentiation of MSCs [Lennon et al., 2001] and to attempt to detect the presence of MSCs or osteoprogenitor cells in tissues other than bone marrow, including dental pulp [Mann et al., 1996] and peripheral blood [Lazarus et al., 1997].

Protocol 2.7. Preparation and Cell Loading of Ceramic Cubes

Reagents and Materials

Sterile

- ❑ Fibronectin: Prepare fibronectin as instructed by the manufacturer. Store as 100-μl aliquots at a concentration of 1 mg/ml in 12 × 75-mm tubes at −20 °C
- ❑ Tyrode's or Hanks' BSS
- ❑ Trypsin-EDTA: 0.25% trypsin, 1 mM EDTA in Hanks' BSS lacking Ca^{2+} and Mg^{2+}
- ❑ BCS
- ❑ DMEM-LG, serum-free
- ❑ Additional caps for 12 × 75-mm tube
- ❑ Conical centrifuge tubes, 15 or 50 ml
- ❑ Syringes, 20 ml and 30 ml, with 20-gauge needles
- ❑ Tissue culture Petri dish, 10 cm

Nonsterile

- ❑ Ceramic rods
- ❑ Glass beaker, 250 ml, with ultrapure water (UPW)
- ❑ Glass Petri dish
- ❑ Paper towel

Protocol

1. Cube Preparation and Coating

(a) Cut ceramics into 3-mm cubes. Trim to ideal size with a razor after cutting to approximate size with a hacksaw.

(b) Place cubes in the 250-ml glass beaker containing UPW, and swirl to remove small ceramic particles from the cubes.

(c) Let cubes settle to the bottom of the beaker, and then pour off the water containing the small ceramic particles.

(d) Repeat steps (b) and (c) 3 or 4 times, or until most of the debris has stopped coming off the cubes.

(e) Blot cubes on a dry paper towel, and then transfer them to a glass Petri dish and dry under a bright light.

(f) Autoclave for 30 min.

(g) Dilute fibronectin to a concentration of 100 μg/ml with sterile Tyrode's or Hanks' BSS in a 12 × 75-mm sterile tube.

(h) Add ceramics, but only enough so they are totally immersed (usually about 16 cubes per ml).
(i) Attach a 20-gauge needle to a 30-ml syringe. Insert the needle through the cap of the tube, and create a partial vacuum in the tube by pulling the plunger back completely. This helps the fibronectin enter the cube pores. Leave under negative pressure for about 60 s. Flick the tubes with your finger to help free air bubbles trapped in the cubes.
(j) Replace the punctured tube cap with a new one.
(k) Leave cubes in fibronectin at room temperature for 2 h.
(l) Aspirate liquid and transfer the cubes to a 10-cm tissue culture dish. Dry overnight in a laminar flow hood.

2. Cell Preparation and Loading
(a) Rinse cultures with 5 ml of TBSS (volumes indicated here are for 10-cm dishes) and remove liquid.
(b) Add 4 ml trypsin-EDTA. Incubate at 37 °C for 5–6 min.
(c) Add 2 ml BCS. Pipette resulting cell suspension up and down and gently rinse remaining cells off plate. Do not try to get all the cells off, just the easily detachable ones.
(d) Transfer suspension to 15- or 50-ml conical tube and centrifuge at 350 g for 5 min.
(e) Remove supernate and resuspend cell pellet in 10 ml serum-free DMEM-LG.
(f) Centrifuge as in (d) above.
(g) Remove supernate and resuspend pellet in 5 ml serum-free medium.
(h) Determine the number of cells with a hemocytometer or electronic cell counter.
(i) Centrifuge and resuspend pellet in serum-free medium at 5×10^6 cell/ml.
(j) Transfer cell suspension to a sterile 12 × 75-mm polystyrene tube.
(k) Add an appropriate number of fibronectin-coated cubes to the cell suspension.
(l) Draw off air from the tube as above (step (1)(i)), but with a 20-ml syringe. Flick tube to eliminate bubbles.
(m) Replace cap with a new one.
(n) Place tube in a 37 °C incubator with cap loose (to allow for CO_2 equilibration) for at least 2 h.

Protocol 2.8. Implantation of Ceramic Cubes into SCID Mice

Reagents and Materials

Sterile

- ❑ Ceramic cubes, prepared and loaded with cells as indicated in Protocol 2.7
- ❑ Anesthetic: "rodent cocktail," 1.5 parts ketamine (100 mg/ml), 1.5 parts xylazine (20 mg/ml), and 0.5 parts acepromazine (10 mg/ml). Diluted 1:4 in sterile water
- ❑ Local anesthetic: 0.5% Marcaine

- ❑ Syringes, 1 ml
- ❑ Syringe needles, 25 and 27 gauge
- ❑ Surgical instruments: fine scissors, blunt forceps and sharp forceps
- ❑ Betadine and alcohol swabs
- ❑ Staple sutures and forceps for applying staples
- ❑ Ear-notching forceps or other marking device

Nonsterile

- ❑ Athymic mice
- ❑ Styrofoam block
- ❑ Paper towels

Protocol

(a) Prepare a diagram of the scheme that you will use for implanting the ceramic cubes. The rest of the procedure is done at a facility designed for housing and working with athymic rodents. Adhere to the rules for the facility, including proper attire, use of laminar flow hoods, use of disinfectant solutions, etc. Many of the steps listed here are those used at the Athymic Animal Facility (AAF) at Case Western Reserve University and are included for reference.

(b) Before entering the interior of the AAF put on shoe covers, mask, cap, and sterile gown and gloves.

(c) Once inside the clean room, turn the laminar flow hood fan to the high setting.

(d) Spray entire inside working area of the hood with Clidox disinfectant.

(e) Leave the spray in hood and get the mouse cages from the room in which they are housed. The general rule is to use the oldest available mice first. Use all of the mice in one cage before getting a second cage (if necessary).

(f) Wipe down hood, and place the cage inside the hood.

(g) Spray hands with Clidox, and take a mouse out of the cage.

(h) Inject mouse with 0.1 to 0.15 ml anesthetic per 25 g body weight. Injections are made intraperitoneally (IP) with a new 25-gauge needle for each mouse.

(i) Return mouse to top of cage. Inject another mouse if needed. If more than 2 mice will be used, inject 1 additional mouse just before you begin the implants on 1 of the previously anesthetized mice.

(j) Fill out cage card with the cube location and identification as indicated on your implant sheet.

(k) Place a Styrofoam block in the center of the hood and cover it with a sterile towel. This will serve as the operating field. Place a second sterile towel adjacent to the block and place the instruments on it.

(l) Spray hands with Clidox.

(m) Remove a mouse from the cage and place on the towel over the Styrofoam block. Wipe the mouse's back with Betadine and alcohol swabs.

(n) Inject 0.1 ml 0.5% Marcaine (a local anesthetic) subcutaneously with a 27-g needle over the area to be incised with scissors. If adequate general anesthesia has not been obtained, you may inject an additional 0.05 ml rodent cocktail.

(o) Open skin with scissors and create subcutaneous pockets with blunt forceps.
(p) Insert cubes, up to 12 per mouse.
(q) Close skin with staples.
(r) Wipe ear with alcohol and make ear notch for identification.
(s) Return mouse to cage, and repeat the process with as many mice as needed.
(t) When finished, return cage to its original location. Wipe hood surface with water. Wipe any carts you used with Clidox, then with water.
(u) Turn off hood.

Host animals are normally harvested at 3 and 6 weeks after implantation, although little bone is usually present at the early harvest time. The cubes are fixed overnight with 10% phosphate-buffered formalin, decalcified in RDO (a low-pH rapid bone decalcifying agent), embedded in paraffin, and cut into sections, which are placed on a glass slide, stained with Toluidine Blue or Mallory–Heidenhain, and cover-slipped. When viewed by brightfield microscopy the decalcified ceramic material appears as a gray to white granular, amorphous material. Cells can be seen occupying the pore areas of the cube. In some pores a mineralized material consistent in morphology with that classically ascribed to bone can be seen. The bone present within the cubes takes on a medium blue color with Toluidine Blue staining and a blue or red color when stained with Mallory–Heidenhain. The amount of bone can be quantified by histomorphometric techniques or by simpler visual estimates of the number or percentage of bone-containing pores [Dennis et al., 1998]. Formation of bone begins with the development of a sheet of cuboidal cells on the inner surfaces of the pores (Fig. 2.3). The cells, shown to be osteoblasts, begin to proliferate and fabricate extracellular matrix. Some of the osteoblasts become enmeshed in the matrix so that they elaborate and become osteocytes. These osteogenic cells have been shown to be of donor origin in appropriate marking experiments [Goshima et al., 1991b; Allay et al., 1997]. As this process continues, the developing bone progresses toward the center of the individual pores. In the case of human MSCs, most of the bone-containing pores are situated toward the periphery of the cube, but some are more centrally located. Vasculature from the host animal is always present in these cubes and is always associated with developing bone therein.

7. IN VITRO OSTEOGENIC INDUCTION AND ASSAYS FOR OSTEOGENESIS

7.1. In Vitro Osteogenic Induction

Although the in vivo model has been considered the optimal assay for osteogenesis, the in vitro assay is another valuable tool for evaluating hMSCs. The in vivo ceramic cube assay is expensive, includes technology not needed in the in vitro assay (histologic processing), and requires facilities for housing and working with athymic rodents. In vitro assays are free of these requirements and, furthermore, avoid possible undesirable host-contributed factors.

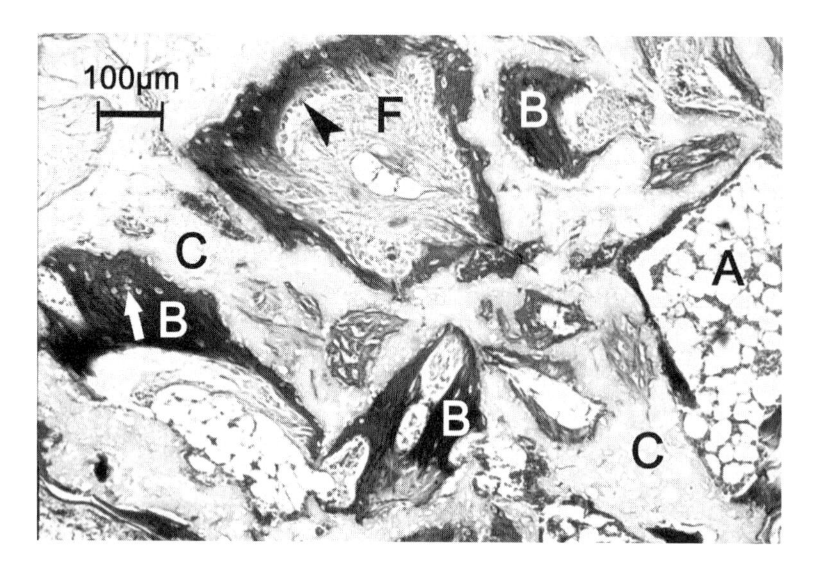

Figure 2.3. Brightfield photomicrograph of a histologic section of an MSC-containing ceramic cube harvested after 6 weeks in vivo and stained with Mallory–Heidenhain. Demineralized ceramic (C) appears as a gray amorphous material. Bone (B), deposited by osteoblasts (arrowhead), can be seen as a dark stained tissue adjacent to the inner surface of a number of pores. A group of osteocytes (arrow) is enmeshed within the osseous matrix. Adipose (A) or fibrous (F) tissues occupy pore space adjacent to bone.

As indicated in the Introduction, hMSCs remain in an undifferentiated state through repeated subculture. They can, however, be induced to differentiate along osteogenic lines by exposure to dexamethasone and ascorbic acid, and to mineralize by the later addition of β-glycerophosphate.

Protocol 2.9. Osteogenic Differentiation in Human Mesenchymal Stem Cells (hMSCs)

Reagents and Materials

Sterile or Aseptically Prepared

- ❑ DMEM-LG-10FB
- ❑ Trypsin-EDTA: 0.25% trypsin, 1 mM EDTA in Hanks' BSS lacking Ca^{2+} and Mg^{2+}
- ❑ Dexamethasone (dex) (See Section 2.2.1)
- ❑ β-Glycerophosphate (BGP) (See Section 2.2.2)
- ❑ Ascorbic acid 2-phosphate, (See Section 2.2.3)
- ❑ OS medium: supplement DMEM-LG-FB with 1% (v/v) each of 10 µM dexamethasone and 5 mM ascorbic acid 2-phosphate. Thus the final concentrations are 0.1 µM dexamethasone and 50 µM ascorbic acid 2-phosphate.
- ❑ Control medium: supplement DMEM-LG-FB with 1% (v/v) 5 mM ascorbic acid 2-phosphate
- ❑ Culture vessels: 3.5-cm tissue culture dishes or multiwell plates, 6-, 12-, or 24 well

Protocol

(a) Calculate the number of cultures and the volume of medium needed for the experiment. Osteogenic induction involves treating cells with or without dexamethasone. The former are referred to as OS (osteogenic supplemented) cultures, and the latter are referred to as control cultures. Equal numbers of culture dishes are established for each condition.
(b) Prepare OS and control media.
(c) Subculture hMSCs as described in Protocol 2.4. Second-passage cells are typically used, but later-passage hMSCs can be used.
(d) Count the cells and seed them at 3×10^3 per cm^2 in serum-containing medium, usually into 35-mm dishes or 12- or 24-well plates.
(e) On day 1 of culture (approximately 24 h after cells are seeded), the medium in which the cells were seeded is removed and immediately replaced with an appropriate volume of either control or OS medium.
(f) Media are changed every 3–4 days.
(g) On day 10 and thereafter, the medium for both groups of cultures should be further augmented with 1% 200 mM β-glycerophosphate (for a final concentration of 2 mM).
(h) Cells are maintained in culture up to 21 days.

Human MSCs respond to exposure to dexamethasone by becoming more compact; typically, these cells change from a spindle-shaped morphology to one that is more cuboidal or polygonal, although some fibroblast-like cells remain. After the addition of β-glycerophosphate, granular mineral-like deposits can be seen on the surface of the OS cultures. Unlike rat MSCs, which form discrete multilayered nodules in the presence of dexamethasone, presumptive osteoblasts differentiated from human MSCs are distributed rather evenly throughout the culture dish, although irregularly shaped multilayered areas are seen.

Through at least day 9 of culture, cell proliferation, as determined by quantification of DNA, is lower in OS than in control cultures, although by day 28 the values for the two groups are equivalent [Lennon et al., 2000]. Techniques used to examine osteoinduced MSCs are described in Sections 7.2 through 7.4.

7.2. Determination of Alkaline Phosphatase Activity

Alkaline phosphatase activity, on a per cell basis, reaches its peak between days 9 and 12 of culture for passaged cells. Accordingly, this time period is optimal for assays of the enzyme, although earlier and later determinations of alkaline phosphatase activity may be useful. The biochemical and cytochemical assays are described below. Both assays may be conducted on individual culture dishes as long as the sequence below is followed.

In the biochemical assay, alkaline phosphatase produced by the cells cleaves the phosphate ion from the substrate, *p*-nitrophenyl phosphate. The resulting

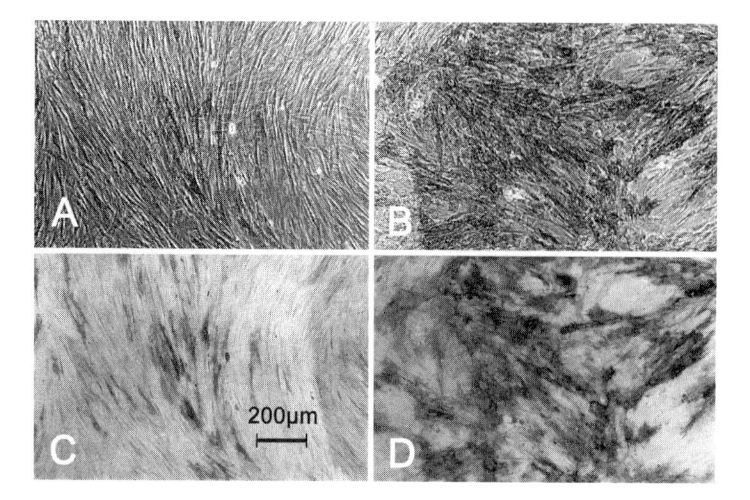

Figure 2.4. Human mesenchymal stem cells stained with Fast Violet B for cytochemical detection of alkaline phosphatase activity on day 10 of second passage. Cells in control medium are shown via phase-contrast microscopy (A) and brightfield optics (C). More intense staining is apparent in cells cultured in the presence of dexamethasone, as seen in phase-contrast (B) and brightfield (D) micrographs. Scale (bar in panel C) is the same for all panels.

p-nitrophenol can be measured colorimetrically by the addition of an alkaline solution. The quantity of *p*-nitrophenol liberated from the substrate can be determined by comparison to a curve generated from known concentrations of *p*-nitrophenol standards. A 5- to 10-fold increase in the specific activity of the enzyme is typical for OS cultures (compared with control cultures).

The substrate used in the cytochemical assay is naphthol AS-MX phosphate. Again, the phosphate group is removed, and the resulting naphthol AS-MX combines with fast violet or fast blue salts to produce violet or blue color at cellular sites of alkaline phosphatase activity. Dark to medium red stain develops in OS cultures stained with fast violet. Most of the stain can be seen in the multilayered areas. Control cultures stain much less intensely (Fig. 2.4).

7.2.1. Alkaline Phosphatase Assays (Biochemical and Cytochemical)

Protocol 2.10A. Biochemical Assay of Alkaline Phosphatase Activity

Reagents and Materials

Nonsterile

- ❑ TBSS
- ❑ Substrate buffer (See Section 2.3.1)
- ❑ Substrate (See Section 2.3.2): Calculate the volume of substrate required, and add substrate tablets to the buffer solution to yield this volume
- ❑ 1 M NaOH
- ❑ *p*-Nitrophenol for standard curve (See Section 2.3.3)

Protocol

(a) Turn on spectrophotometric plate reader in time to have it properly warmed up when samples are ready to read.
(b) Label a 12 × 75-mm tube for each tissue culture dish or well of multiwell plates to be analyzed. Add a volume of 1 M NaOH equal to the volume of substrate used (See step (d)) to each tube.
(c) Rinse cells twice with TBSS.
(d) Add an appropriate volume of substrate solution to tissue culture dishes or wells (1 ml per 3.5-cm dish, 0.5 ml per well of 12-well plate, etc.) Time the addition of substrate so that cells in each well or dish are in contact with the solution for the same length of time.
(e) Incubate for 5 to 15 min, depending on the rate of the reaction (as seen by the density of yellow developing in the substrate; the ultimate OD when measured must fit within the linear range of the standard curve).
(f) Transfer the substrate solution to the NaOH-containing tubes set up earlier. Do this in the same sequence in which substrate was added to the dishes or wells, and remove substrate so that each dish or well is exposed to the substrate for the same length of time.
(g) If cultures are to be used for additional assays (alkaline phosphatase cytochemistry, DNA, etc.) add TBSS to the cells immediately, and do additional rinses and fixation as soon as time permits.
(h) Set up and label a 96-well plate.
    i) To the first row add 200 μl of a mixture of equal parts of 1 M NaOH and substrate buffer to serve as blanks.
    ii) Transfer 200 μl of appropriately diluted (or undiluted) experimental samples to corresponding labeled wells (usually 4 replicate wells per sample). As a rough guide for determining whether samples require dilution, compare the intensity of color of the samples to that of the highest concentration of *p*-nitrophenol. Dilutions are made with the same solution used for blanks.
    iii) Set up a standard curve by transferring 200 μl diluted *p*-nitrophenol standards to wells (See Section 2.3).
(i) Read the absorbance at 405 nm on a microplate reader. Do further dilutions for samples that give readings beyond the linear range of the standard curve and read samples again. Note any dilutions made.

Protocol 2.10B. Cytochemical Assay for Alkaline Phosphatase

Reagents and Materials

Nonsterile

- ❑ Fast violet stain (See Section 2.4.1)
- ❑ Citrate-buffered acetone (See Section 2.4.3)
- ❑ Naphthol AS-MX Alkaline Solution (Sigma 85-5)
- ❑ TBSS

Protocol

(a) Rinse cultures twice with TBSS.
(b) Fix for 30 s with citrate-buffered acetone.
(c) Rinse cultures twice with distilled water. Leave water from the second rinse on the cultures until you are ready to proceed with the next step.
(d) Add 0.5 ml Naphthol AS-MX Alkaline Solution per 12 ml of fast violet solution. Cover with aluminum foil to protect solution from the light.
(e) Remove water from cultures and add an appropriate volume of fast violet-naphthol solution to each dish or well.
(f) Incubate at room temperature in the dark for 45 min.
(g) Remove solution and rinse twice with distilled water. Keep the cultures covered with water and store in the dark. The cultures can be further stained according to other protocols (von Kossa or other stain).

7.3. Von Kossa Staining

Staining with the von Kossa method is a qualitative assay for mineralization. In the case of human MSCs, mineralization is usually not readily detectable until day 21. A positive reaction is manifested by brown or black staining, which, in the case of hMSCs, is diffuse and is usually distributed fairly evenly throughout the culture (Fig. 2.5). This contrasts with mineralization in cultures of rat MSCs, in which intense staining can be identified in discrete bone nodules.

Protocol 2.11. Staining for Mineralization in Cultured Mesenchymal Stem Cells (MSCs)

Reagents and Materials

Nonsterile

- ❑ Silver nitrate, 2% in distilled water. Make up only what you will need because the shelf life of the solution is approximately 1 week
- ❑ Tyrode's or Hanks' BSS
- ❑ Phosphate-buffered formalin, 10%
- ❑ Distilled water
- ❑ Ethanol, 100%

Protocol

(a) Rinse cultures twice with cold TBSS (or HBSS).
(b) Fix cultures with 10% phosphate-buffered formalin for 30 min. All steps involving formalin must be carried out in a fume hood.
(c) Rinse cultures twice with distilled water.
(d) Add 2% silver nitrate to cover the cells; place the dishes or flasks in a dark environment for 10 min.

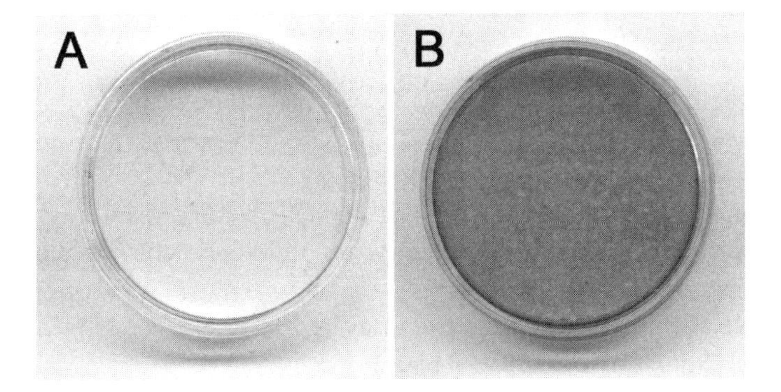

Figure 2.5. Human mesenchymal stem cells stained by the von Kossa method on day 21 of second passage. Second-passage human mesenchymal stem cells were cultivated in 35-mm dishes in control medium (A) or medium supplemented with dexamethasone (B). Media for cultures in both conditions were further augmented with 2 mM β-glycerophosphate, beginning on day 10 of culture. Cultures were fixed on day 21 and stained by the von Kossa method to reveal mineralization. Positive staining (gray to black) is apparent only in the dexamethasone-treated cultures.

(e) Rinse three times with distilled water. Leave the water from the final rinse on the culture dish for step (f).
(f) With water still covering the cells, expose the cultures to bright light for 15 min. Use a white background beneath the dishes to help reflect the light.
(g) Remove the water covering the cells, then rinse two more times with distilled water.
(h) Dehydrate with 100% ethanol; remove the ethanol after 1 min, and then allow the cultures to air dry.

7.4. Calcium Assay

Protocol 2.12. Calcium Assay in Cultured Mesenchymal Stem Cells (MSCs)

Reagents and Materials

Nonsterile

- ❑ TBSS
- ❑ Formalin, 10%, phosphate buffered
- ❑ Distilled water
- ❑ HCl, 0.6 N
- ❑ $CaCl_2$ standards: 6.25, 12.5, 25, 50, 100, and 200 μg/ml
- ❑ Calcium binding reagent and calcium buffer: Sigma calcium diagnostic kit (#587-A)
- ❑ Polystyrene tubes, 12 × 75 mm
- ❑ Rotary shaker
- ❑ Microtitration plate reader

Protocol

(a) Rinse cultures twice with TBSS.

(b) Fix with 10% phosphate-buffered formalin for 30 min (use fume hood for all steps involving the use of formalin).

(c) Remove formalin by aspiration and save for proper disposal.

(d) Rinse three times with distilled water, and then air dry. The culture dishes may be stored at 4°C so that the assay of all plates from an experiment may be completed at the same time.

(e) After all cultures have been collected, begin the assay by adding 1 ml 0.6 N HCl per 3.5-cm dish (0.5 ml per well of a 24-well plate).

(f) Place dishes on a rotary shaker overnight at about 50 rpm to extract calcium.

(g) Collect the HCl and transfer it to labeled 12 × 75-mm polystyrene tubes.

(h) Set up and label a 96-well plate as follows:

   i) To the first row on the left add 20 μl 0.6 N HCl per well to serve as blanks.

   ii) Add 20 μl per well of dilutions of $CaCl_2$ in 0.6 N HCl to establish a standard curve. The standard curve should include the following concentrations of $CaCl_2$: 6.25, 12.5, 25, 50, 100, and 200 μg/ml. Use 4 replicates for each concentration.

   iii) Add 20 μl of the experimental samples to each of 4 wells. If necessary, first dilute the samples with 0.6 N HCl in separate tubes.

(i) Prepare the required volume of reagents from the Sigma calcium diagnostic kit by combining equal volumes of the calcium binding reagent and calcium buffer. The required volume equals the number of wells × 180 μl, plus about 10%.

(j) Turn on the plate reader to have it properly warmed up.

(k) Add 180 μl of the calcium binding reagent-calcium buffer to each well.

(l) Read the samples on a microtitration plate reader at 575-nm absorbance. Record the results, and calculate a standard curve and the concentration of calcium in the experimental samples by linear regression analysis. Do further dilutions for samples that give readings beyond the linear range of the standard curve, and then read the samples again.

---

8. IN VITRO ADIPOGENIC INDUCTION AND ASSAYS FOR ADIPOGENESIS

Adipogenic induction involves culturing cells (starting 24 h after they are seeded) with either control or adipogenic induction medium (AIM). The method described below is a variation of that described by Smyth and Wharton [1992, 1993] and by Pittenger et al. [1999]. Accumulation of lipid droplets can be seen in the cytoplasm of induced cells. Approximately 30% to 40% of the cells may include lipid droplets by day 12. After that time the amount of lipid may increase, but few additional cells will differentiate into adipocytes.

8.1. Adipogenic Induction

Protocol 2.13. Induction of Adipogenesis in Mesenchymal Stem Cells (MSCs)

Reagents and Materials

Sterile

- ❑ hMSC cultures, first or second passage
- ❑ DMEM-LG-10FB
- ❑ Adipogenic induction medium (AIM), consisting of DMEM-HG-10FB, 1 μM dexamethasone, 100 μM indomethacin, 500 μM 3-isobutyl-1-methylxanthine (IBMX), and 10 μg/ml insulin (See Section 2.5)
- ❑ Adipogenic maintenance medium (AMM): DMEM-HG-10FB with 10 μg/ml insulin
- ❑ Trypsin-EDTA (See Protocol 2.4)
- ❑ Tissue culture dishes, 3.5 cm, or 6-well plates

Protocol

(a) Calculate the number of cultures needed for the experiment.

(b) Trypsinize first- or second-passage hMSCs as described in Protocol 2.4.

(c) Seed cells into 3.5-cm dishes (or 6-well plates) at a density of 2×10^5 cells in DMEM-LG-10FB.

(d) Incubate at 37 °C overnight.

(e) On day 1, switch medium for cultures to be induced to AIM, and replace medium in control cultures with fresh DMEM-LG-10FB. Set up the desired number of cultures for each condition.

(f) Media are changed twice a week.

(g) On day 10, replace AIM with AMM.

Cells are usually fixed for Nile Red staining on days 0, 2, 12, and 21.

Protocol 2.14. Fixing Mesenchymal Stem Cells (MSCs) for Flow Cytometric Assay of Adipogenesis

Reagents and Materials

Sterile

- ❑ PBSA: phosphate-buffered saline without Ca^{2+} and Mg^{2+}
- ❑ EDTA, 0.1 mM
- ❑ Trypsin-EDTA: 0.25% trypsin, 1 mM EDTA
- ❑ FBS
- ❑ Trypsin inhibitor, Type II-O: chicken egg white, 100 mg/ml in DMEM
- ❑ Pasteur pipettes
- ❑ Polypropylene tubes, 12×75 mm

Nonsterile
❑ Paraformaldehyde, 4%

Protocol

(a) Remove medium and rinse 3 times with phosphate-buffered saline (PBSA). Leave the third addition of PBSA on the cultures for 3–5 min.
(b) Remove PBSA and add 1 ml 0.1 mM EDTA per well or dish. Incubate at 37 °C for 10–20 min, checking cells at 5-min intervals to see whether they are ready for the addition of trypsin-EDTA, as indicated by retraction of the cells (cells should not be completely round). If the cells are being released on days 1 or 2 of culture, this step can be skipped.
(c) Remove EDTA and add 475 μl trypsin-EDTA per well and incubate at 37 °C for 5 to 7 min.
(d) Return plate to room temperature. Agitate by hand if any cells remain adherent.
(e) Add 25 μl FBS and 25 μl trypsin inhibitor. Swirl or rock plate to mix.
(f) With a Pasteur pipette, combine suspensions from 3 wells (of a 6-well plate) into 1 well. Gently flush empty wells once or twice with the cell suspension.
(g) Transfer cell suspension to a 12 × 75-mm polypropylene tube.
(h) Rinse empty wells with a total volume of 500 ml PBSA and add the rinse to the polypropylene tube.
(i) Cells may be counted at this point.
(j) Bring suspension to a concentration of 0.5% paraformaldehyde with the addition of a suitable volume of 4% paraformaldehyde.
(k) Invert the tube several times to mix contents, and store at 4 °C until the day of staining.

Protocol 2.15. Staining Mesenchymal Stem Cells (MSCs) with Nile Red for Flow Cytometric Assay of Adipogenesis

Reagents and Materials

Nonsterile
❑ Nile Red working solution. (See Section 2.5)
❑ Pasteur pipette
❑ Nitex nylon filter, 100 μm
❑ Polypropylene tubes, 12 × 75 mm
❑ Aluminum foil
❑ Ice tray
❑ Flow cytometer

Protocol

(a) Resuspend each sample with a Pasteur pipette.
(b) Collect 300 μl of each sample and filter through a 100-μm Nitex nylon filter into a fresh polypropylene tube.

(c) Add an equal volume of Nile Red working solution.
(d) Cover tube with foil and store on ice until time of assay.
(e) Cells are analyzed with a flow cytometer. Gold fluorescence is emitted when Nile Red is dissolved in neutral lipids, and can be collected between 560 and 590 nm with a band-pass filter. Ten thousand cells are analyzed from each sample [Smyth and Wharton, 1992, 1993].

9. SELECTION OF FETAL BOVINE SERUM

The process of selecting the best lot of FBS from the lots available has been described previously [Lennon et al., 1996] and will not be reviewed in detail here. It should, however, be emphasized that we regard the process of serum selection as an extremely important aspect of the culture of MSCs. Our experience with testing FBS dates to the time when we were working primarily with stage 24 embryonic chick limb bud mesenchymal cells. We were fortunate that the lot selected for these cells also supported the proliferation of marrow-derived MSCs, although it may be argued that the ability of a single lot of serum to support both cell types is related to the concept that both sets of cells are MSCs (See Section 1). Be that as it may, we do not use stage 24 limb bud cells as part of our current assay for FBS.

We consider the ability of a lot of FBS to promote the initial adherence of MSCs to the culture dish, to support cell proliferation, and to maintain the multipotentiality of the cells, as indicated by their ability to promote osteogenesis in the ceramic cube assay, to be the most important criteria for selection of serum. Theoretically, an ideal lot of serum would meet these standards, and would also promote in vitro chondrogenesis, osteogenesis, and adipogenesis in serum-supplemented media further augmented with the appropriate inductive agents (or in the defined medium appropriate for chondrogenic differentiation). In practice, we have found that it is not always possible to select such an ideal serum. For example, one lot of FBS that supported proliferation and gave good results in the in vivo assay was far from ideal for in vitro osteogenesis. Thus, although we have considered the in vivo assay to be our "gold standard" for serum selection, it may be prudent for investigators to adopt the assay or inductive method that will be most widely utilized with MSCs in their laboratory as their most stringent criterion in the selection process.

Moreover, it should be noted that although in vitro chondrogenic induction for MSCs takes place in a medium devoid of serum (See Chapter 4), exposure to the proper lot of FBS in monolayer culture before the cells are introduced into aggregate or pellet culture is very important.

A brief description of the process of selecting FBS for use in culturing and inducing differentiation in MSCs is outlined below. We usually screen 8 to 10 lots of serum (including the control lot) in a given serum screen.

Protocol 2.16. Selecting FBS for Mesenchymal Stem Cells (MSCs)

Reagents and Materials

Sterile

- ❑ Materials and reagents for Protocols 2.3 and 2.4
- ❑ MSCs as isolated in Protocol 2.3
- ❑ Tissue culture dishes, 10 cm

Protocol

(a) Isolate MSCs as described in Protocol 2.3, but using a concentration of 5% FBS.

(b) Determine the final cell concentration.

(c) Adjust the cell suspension (still in 5% of the FBS currently in use, or the control serum) to a volume sufficient to seed 1 or 2 ml per culture dish.

(d) Before the cells are seeded, pipette 5 or 6 ml of medium supplemented with 10% of the sera being tested into 10-cm dishes

(e) Add the cell suspensions and mix to distribute the cells evenly. In this way, the cells are exposed to the test sera from the beginning of their time in culture.

(f) Maintain the cells in culture as described in Protocol 2.3 and in Section 4. Examine cultures frequently by phase microscopy to evaluate cell morphology, to gain a qualitative assessment of cell proliferation, and to examine cell attachment, as indicated by a comparison of the number of colonies.

(g) Subculture cells as described in Protocol 2.4. Cell yields are determined individually for cultures in the various test sera, with at least two cell counts being taken for each sample.

(h) Seed the cells into 10-cm dishes (all at the same density and again in control or test serum).

(i) Repeat the process of trypsinizing and counting the cells when the culture is just preconfluent. Detailed records of cell yields are kept to assess cell proliferation.

Cells harvested at the end of the first passage (or later passages if an insufficient number of cells is available at the end of the first passage) are used for the assays that are deemed appropriate. In our laboratory, highest priority is given to loading ceramic cubes for implantation. We do this for two reasons. First, as indicated above, we regard this as the definitive assay for stem cells. The second reason is a practical matter relating to the length of the in vivo assay (6 weeks in vivo plus time for histologic processing and evaluation) and the time that companies that provide serum are willing to hold the product on reserve. Because we are usually pushing the limit of the reserve time, we try to get the cubes implanted as quickly as possible.

Additional cell preparations may be needed to provide enough cells to complete all of the required assays, and because duplicate or triplicate data sets are desirable. Among the other assays that we routinely include in the testing of FBS for human MSCs are the in vitro assays for osteogenesis and chondrogenesis. For the former, we use von Kossa staining to assess mineralization, but do not routinely examine alkaline phosphatase activity. For in vitro chondrogenesis, sera are tested for their ability to support pellet formation as indicated by toluidine blue staining of fixed and embedded pellets.

Data reflecting the ability of the various lots of serum to support proliferation and to promote differentiation in the induction assays are compiled and evaluated. The results are compared with the lot of serum currently in use. If no serum is currently in use for MSCs, the lot providing the best results is selected (if acceptable).

ACKNOWLEDGMENTS

The authors wish to thank Dr Omer Koç for his valuable input regarding aspiration of human bone marrow and M-Danielle Mackay and Dr. Steven Haynesworth for their important contributions relating to adipogenic induction and flow cytometric analysis of MSCs.

SOURCES OF MATERIALS

Materials	Supplier	Catalog Number
Acepromazine	Henry Schein	356–7290
Alkaline phosphatase assay materials	Sigma	85L-3R or individual components
Alkaline phosphatase substrate	Sigma	104–105
Ascorbic acid 2-PO_4	Wako	013–12061
Biphasic ceramics consisting of 60% tricalcium phosphate and 40% hydroxyapatite	Zimmer	97-1109-531-00
Buffered formalin	Fisher	SF 100–4
Calcium assay kit	Sigma	587-A
Calf serum	Hyclone	SH30073-03
Citrate concentrated solution	Sigma	85-4C
Clidox disinfectant	Pharmacal Research Laboratories	95120 (Activator) 96120 (Base)
Coral-based ceramics	Interpore	
Cryofreezer	Nalgene	5100-0001
Cryovials	Nalgene	5000-0020
Dexamethasone	Sigma	D 4902
Dimethyl sulfoxide	Sigma	D 2560
Dulbecco's modified Eagle's medium	Sigma	D 5523
	GIBCO	31600-083
Electronic cell counter	Beckman Coulter	
Fast Violet B Salt	Sigma	85-1
Fetal bovine serum	Best available	
Fibronectin	Becton Dickinson	354008
β-Glycerophosphate	Sigma	G 9891

Materials	Supplier	Catalog Number
Halothane	Henry Schein	982-0753
IBMX	Sigma	I 5879
Indomethacin	Sigma	I 7378
Insulin	Sigma	I 1882
Ketamine	Henry Schein	995–5770
Marcaine	NLS Animal Health	108435
Microtitration plate reader	Bio-Rad	2550
Naphthol AS-MX Alkaline Solution	Sigma	85-5
Nile Red	Sigma	N 3013
Nitex nylon filter, 100 μm	TETKO, Inc.	3-100/47
p-Nitrophenol standard solution	Sigma	104-1
p-Nitrophenyl phosphate:	Sigma	104–105
Percoll	Sigma	P 1644
Polycarbonate tubes	Nalgene/Oakridge	3118-0050
RDO decalcifying agent	Apex Engineering Products Corp.	RDO-04
Rotary shaker	New Brunswick	
SCID mice	Charles River	CB17
Silver nitrate	Sigma	S 0139
Toluidine Blue	Sigma	T 3260
Trypan blue	Gibco	15250-061
Trypsin inhibitor	Sigma	T 9253
Trypsin-EDTA: 0.25% trypsin, 1 mM EDTA	Gibco	25200-072
Tyrode's salt solution (TBSS)	Sigma	T 2145
Xylazine	NLS Animal Health	105650

REFERENCES

Aigner, J., Tegeler, J., Hutzler, P., Campoccia, D., Pavesio, A., Hammer, C., Kastenbauer, E., Naumann, A. (1998) Cartilage tissue engineering with novel nonwoven structured biomaterial based on hyaluronic acid benzyl ester. *J. Biomed. Mater. Res.* 42 (2): 172–181.

Allay, J.A., Dennis, J.E., Haynesworth, S.E., Majumdar, M.K., Clapp, D.W., Shultz, L.D., Caplan, A.I., Gerson, S.L. (1997) LacZ and interleukin-3 expression in vivo after retroviral transduction of marrow-derived human osteogenic mesenchymal progenitors. *Hum. Gene Therapy* 8: 1417–1427.

Bell, E., Sher, S., Hull, B., Merrill, C., Rosen, S., Chamson, A., Asselineau, D., Dubertret, L., Coulomb, B., Lapiere, C., Nusgens, B., Neveux, Y. (1983) The reconstitution of living skin. *J. Invest. Dermatol.* 81(1 Suppl): 2s–10s.

Bianco, P., Riminucci, M., Kuznetsov, S., Robey, P.G. (1999): Multipotential cells in the bone marrow stroma: regulation in the context of organ physiology. *Crit. Rev. Eukaryot. Gene Expr.* 9(2): 159–173

Black, I.B., Woodbury, D. (2001): Adult rat and human bone marrow stromal stem cells differentiate into neurons. *Blood Cells Mol. Dis.* 27: 632–636.

Brighton, C.T., Lorich, D.G., Kupcha, R., Reilly, T.M., Jones, A.R., Woodbury, R.A. II (1992): The pericyte as a possible osteoblast progenitor cell. *Clin. Orthop.* 275: 287–299.

Brittberg, M., Tallheden, T., Sjögren-Jansson, B., Lindahl, A., Peterson, L. (2001): Autologous chondrocytes used for articular cartilage repair: an update. *Clin. Orthop.* 391 Suppl: S337–348

Bruder. S.P., Fink, D.J., Caplan, A.I. (1994): Mesenchymal stem cells in bone development, bone repair, and skeletal regeneration therapy. *J. Cell. Biochem.* 56: 283–294.

Bruder, S.P., Jaiswal, N., Haynesworth, S.E. (1997): Growth kinetics, self-renewal, and the osteogenic potential of purified human mesenchymal stem cells during extensive subcultivation and following cryopreservation. *J. Cell. Biochem.* 64: 278–294.

Caplan, A.I. (1977): Muscle, cartilage and bone development and differentiation from chick limb mesenchymal cells. In Ede, D.A., Hinchliffe, J.R., Balls, M., eds., *Vertebrate Limb and Somite Morphogenesis*, Cambridge, Cambridge University Press, pp. 199–213.

Caplan, A.I. (1984): Cartilage. *Sci. Am.* 251: 84–94.

Caplan, A.I. (1991): Mesenchymal stem cells. *J. Orthop. Res.* 9: 641–650.

Caplan, A.I. (1994): The mesengenic process. *Clin. Plastic Surg.* 21: 429–435.

Dennis, J.E., Caplan, A.I. (1993): Porous ceramic vehicles for rat-marrow-derived (*Rattus norvegicus*) osteogenic cell delivery: effects of pre-treatment with fibronectin or laminin. *J. Oral Implantol.* 19 (2): 106–115.

Dennis, J.E., Konstantakos, E.K., Arm, D.M., Caplan, A.I. (1998): In vivo osteogenesis assay: a rapid method for quantitative analysis. *Biomaterials* 19: 1323–1328.

Dennis, J.E., Merriam, A., Awadallah, A., Yoo, J.U., Johnstone, B., Caplan, A.I. (1999): A quadripotential mesenchymal progenitor cell isolated from the marrow of an adult mouse. *J. Bone Miner. Res.* 14: 700–709.

Erickson, G.R., Gimble, J.M., Franklin, D.M., Rice, H.E., Awad, H., Guilak, F. (2002): Chondrogenic potential of adipose tissue-derived stromal cells in vitro and in vivo. *Biochem. Biophys. Res. Commun.* 290: 763–769.

Friedenstein, A.J., Chailakhjan, R.K., Lalykina, K.S. (1970): The development of fibroblast colonies in monolayer cultures of guinea-pig bone marrow and spleen cells. *Cell Tissue Kinet.* 3: 393–403.

Friedenstein, A.J. (1976): Precursor cells of mechanocytes. *Int. Rev. Cytol.* 47: 327–359.

Goshima, J., Goldberg, V.M., Caplan, A.I. (1991a): The osteogenic potential of culture-expanded rat marrow derived mesenchymal cells assayed in vivo in calcium phosphate ceramic blocks. *Clin. Orthop.* 262: 298–311.

Goshima, J., Goldberg, V.M., Caplan, A.I. (1991b): The origin of bone formed in composite grafts of porous calcium phosphate ceramic loaded with marrow cells. *Clin. Orthop.* 269: 274–283.

Halvorsen, Y.C., Wilkison, W.O., Gimble, J.M. (2000): Adipose-derived stromal cells-their utility and potential in bone formation. *Int. J. Obes. Related Metab. Disord.* 24, Suppl 4: S41–S44.

Haynesworth, S.E., Baber, M.A., Caplan, A.I. (1992a): Cell surface antigens on human marrow-derived mesenchymal cells are detected by monoclonal antibodies. *Bone* 13: 69–80.

Haynesworth, S.E., Goshima, J., Goldberg, V.M., Caplan, A.I. (1992b): Characterization of cells with osteogenic potential from human marrow. *Bone* 13: 81–88.

Jackson, K.A., Majka, S.M., Wulf, G.G., Goodell, M.A. (2002): Stem cells: a minireview. *J. Cell Biochem. Suppl.* 38: 1–6.

Jankowski, R.J., Deasy, B.M., Huard, J. (2002): Muscle-derived stem cells. *Gene Ther.* 9: 642–647.

Johnstone, B., Hering, T.M., Caplan, A.I., Goldberg, V.M., Yoo, J.U. (1998): In vitro chondrogenesis of bone marrow-derived mesenchymal progenitor cells. *Exp. Cell Res.* 238: 265–272.

Kadiyala, S., Jaiswal, N., Bruder, S.P. (1997): Culture-expanded, bone marrow-derived mesenchymal stem cells can regenerate a critical-sized segmental bone defect. *Tissue Eng.* 3(2): 173–185.

Koç, O.N., Gerson, S.L., Cooper, B.W., Dyhouse, S.M., Haynesworth, S.E., Caplan, A.I., Tainer, N., Lazarus, H.M. (2000): Rapid hematopoietic recovery after co-infusion of autologous blood stem cells and culture-expanded marrow mesenchymal stem cells in advanced breast cancer patients receiving high dose chemotherapy. *J. Clin. Oncol.* 18: 307–316.

Lazarus, H.M., Haynesworth, S.E., Gerson, S.L., Caplan, A.I. (1997): Human bone marrow-derived mesenchymal (stromal) progenitor cells (MPCs) cannot be recovered from peripheral blood progenitor cell collections. *J. Hematol.* 6: 447–455.
Lee, J.Y., Qu-Petersen, Z., Cao, B., Kimura, S., Jankowski, R., Cummins, J., Usas, A., Gates, C., Robbins, P., Wernig, A., Huard, J. (2000): Clonal isolation of muscle-derived cells capable of enhancing muscle regeneration and bone healing. *J. Cell Biol.* 150: 1085–1099.
Lennon, D.P., Haynesworth, S.E., Bruder, S.P., Jaiswal, N., Caplan, A.I. (1996): Human and animal mesenchymal progenitors from bone marrow: identification of serum for optimal selection and proliferation. *In vitro Cell Dev. Biol. Anim.* 32: 602–611.
Lennon, D.P., Haynesworth, S.E., Arm, D.M., Baber, M.A., Caplan, A.I. (2000): Dilution of human mesenchymal stem cells with dermal fibroblasts and the effects on in vitro and in vivo osteochondrogenesis. *Dev. Dyn.* 219: 50–62.
Lennon, D.P., Edmison, J.M., Caplan, A.I. (2001): Cultivation of rat marrow-derived mesenchymal stem cells in reduced oxygen tension: effects on in vitro and in vivo osteochondrogenesis. *J. Cell Physiol.* 187: 345–355.
Majumdar, M.K., Thiede, M.A., Mosca, J.D., Moorman, M., Gerson, S.L. (1998): Phenotypic and functional comparison of cultures of marrow-derived mesenchymal stem cells (MSCs) and stromal cells. *J. Cell Physiol.* 176: 57–66.
Mann, L.M., Lennon, D.P., Caplan, A.I. (1996): Cultured rat pulp cells have the potential to form bone, cartilage, and dentin in vivo. In Davidovitch, Z., Norton, L.A., eds., *Biological Mechanisms of Tooth Movement and Craniofacial Adaptation*, Harvard Society for the Advancement of Orthodontics, Boston, pp 7–16.
Minguell, J.J., Erices, A., Conget, P. (2001): Mesenchymal stem cells. *Exp. Biol. Med.* 226: 507–520.
Morrison, S.J., Shah, N.M., Anderson, D.J. (1997): Regulatory mechanisms in stem cell biology. *Cell* 88: 287–296.
Mosca, J.D., Majumdar, M.K., Hardy, W.B., Pittenger, M.F., Thiede, M.A. (1997): Initial characterization of the phenotype of the human mesenchymal stem cells and their interaction with cells of the hematopoietic lineage. *Blood* 88: 186a.
Nakahara, H., Goldberg, V.M., Caplan, A.I. (1991): Culture-expanded human periosteal-derived cells exhibit osteochondral potential in vivo. *J. Orthop. Res.* 9: 465–476.
Ohgushi, H., Goldberg, V.M., Caplan, A.I. (1989): Heterotopic osteogenesis in porous ceramics induced by marrow cells. *J. Orthop. Res.* 7: 568–578.
Ohgushi, H., Okumura, M., Tamai, S., Shors, E.C., Caplan, A.I. (1990): Marrow cell induced osteogenesis in porous hydroxyapatite and tricalcium phosphate: a comparative histomorphometric study of ectopic bone formation. *J, Biomed. Mater. Res.* 24: 1563–1570.
Owen, M. (1985): Lineage of osteogenic cells and their relationship to the stromal system. In Peck, W.A. (ed): *Bone and Mineral Research* 3RD ed., Amsterdam, pp 1–25.
Owen, M. (1988): Marrow stromal stem cells. *J. Cell. Sci. Suppl.* 10: 63–76.
Owen, M., Friedenstein, A.J. (1988): Stromal stem cells: marrow-derived osteogenic precursors. *Ciba Found. Symp.* 136: 42–60.
Phinney, D.G. (2002): Building a consensus regarding the nature and origin of mesenchymal stem cells. *J. Cell. Biochem. Suppl.* 38: 7–12.
Pittenger, M.F., Mackay, A.M., Beck, S.C., Jaiswal, R.K., Douglas, R., Mosca, J.D., Moorman, M.A., Simonetti, D.W., Craig, S., Marshak, D.R. (1999): Multilineage potential of adult human mesenchymal stem cells. *Science* 284: 143–147.
Prockop, D.J. (1997): Marrow stromal cells as stem cells for nonhematopoietic tissues. *Science* 276: 71–74.
Ringe, J., Kaps, C., Burmester, G.-R., Sittinger, M. (2002): Stem cells for regenerative medicine: advances in the engineering of tissues and organs. *Naturwissenschaften* 89: 338–351.
Safford, K.M., Hicok, K.C., Safford, S.D., Halvorsen, Y.-D.C., Wilkison, W.O., Gimble, J.M., Rice, H.E. (2002): Neurogenic differentiation of murine and human adipose-derived stromal cells. *Biochem. Biophys. Res. Commun.* 294: 371–379.
Shur, I., Marom, R., Lokiec, F., Socher, R., Benayahu, D. (2002): Identification of cultured progenitor cells from human marrow. *J. Cell. Biochem.* 87: 51–57.
Smyth, M.J., Wharton, W. (1992): Differentiation of A31T6 proadipoctyes to adipocytes: a flow cytometric analysis. *Exp. Cell Res.* 199: 29–38.

Smyth, M.J., Wharton, W. (1993): Multiparameter flow cytometric analysis of the effects of indomethacin on adipocyte differentiation in A31T6 cells. *Cell Prolif.* 26: 103–114.
Toma, J.G., Akhavan, M., Fernandes, K.J., Barnabe-Heider, F., Sadikot, A., Kaplan, D.R., Miller, F.D. (2001): Isolation of multipotent adult stem cells from the dermis of mammalian skin. *Nat. Cell Biol.* 3: 778–784.
Tomita, M., Adachi, Y., Yamada, H., Takahashi, K., Kiuchi, K., Oyaizu, H., Ikebukuro, K., Kaneda, H., Matsumura, M., Ikehara, S. (2002): Bone marrow-derived stem cells can differentiate into retinal cells in injured rat retina. *Stem Cells* 20: 279–283.
Tomita, S., Li, R.-K., Weisel, R.D., Mickle, D.A.G., Kim,, E.-J., Sakai, T., Jia, Z.-Q. (1999): Autologous transplantation of bone marrow cells improves damaged heart function. *Circulation* 100[Suppl II]: II-247–II-256.
Wakitani, S., Saito, T., Caplan, A.I. (1995): Myogenic cells derived from rat bone marrow mesenchymal stem cells exposed to 5-azacytidine. *Muscle Nerve* 18: 1417–1426.
Woodbury, D., Schwarz, E.J., Prockop, D.J., Black, I.B. (2000): Adult rat and human bone marrow stromal cells differentiate into neurons. *J. Neurosci. Res.* 61: 364–370.

3

Human Embryonic Stem Cell Culture for Tissue Engineering

Shulamit Levenberg[1], Ali Khademhosseini[2], Mara Macdonald[1], Jason Fuller[2], and Robert Langer[1,2]

Department of Chemical Engineering[1], and Division of Biological Engineering[2], Massachusetts Institute of Technology, 77 Massachusetts Avenue, E25-342, Cambridge, Massachusetts 02139

Corresponding author: rlanger@mit.edu

Culture of Cells for Tissue Engineering, edited by Gordana Vunjak-Novakovic and R. Ian Freshney

1. EMBRYONIC STEM CELLS

Embryonic stem (ES) cells are typically derived from the inner cell mass of a blastocyst [Czyz and Wobus, 2001; Evans and Kaufman, 1981]. From the information gathered from murine and other systems, human embryonic stem (hES) cells were successfully derived and characterized in 1998 [Shamblott et al., 1998; Thomson et al., 1998].

ES cells have three unique characteristics. The first is that they can maintain an undifferentiated phenotype [Czyz and Wobus, 2001; Evans and Kaufman, 1981]. The second is that these cells are able to renew themselves continuously through many passages, leading to the claim that they are immortal [Czyz and Wobus, 2001;

Evans and Kaufman, 1981]. The third characteristic is that these cells are pluripotent, meaning that they are able to create all three germ layers (the endoderm, ectoderm, and mesoderm) of the developing embryo and thus can be manipulated to differentiate to form every cell type of an adult organism [Czyz and Wobus, 2001; Evans and Kaufman, 1981]. This third characteristic is what makes ES cells such a powerful tool in regenerative medicine. ES cells may potentially provide an unlimited supply of any of the hundreds of highly specialized cells that can be afflicted with disease in the human body, creating raw material for cell therapy and tissue engineering applications.

2. MAINTENANCE AND EXPANSION OF HUMAN EMBRYONIC STEM (HES) CELLS

To generate a therapeutically valuable tissue mass, it is crucial to maintain and expand hES cells in an undifferentiated state. Human ES cells require two to three environmental factors to prevent spontaneous differentiation. The first of these requirements are the factors derived from embryonic fibroblast feeder cells. Embryonic fibroblasts can be derived from murine [Shamblott et al., 1998; Thomson et al., 1998] or human [Richards et al., 2002] sources to sustain the undifferentiated phenotype. Thus far, most analysis on the use of cocultures to prevent hES cell differentiation has been performed on murine embryonic fibroblasts (MEFs). Before use, MEFs are mitotically inactivated, through either irradiation or application of mitomycin C, to prevent overgrowth of the feeder cells in the cocultures. The factors that the MEF cells add to the medium are unknown, but they are crucial for maintaining hES cells in an undifferentiated state [Itskovitz-Eldor et al., 2000; Reubinoff et al., 2000; Thomson et al., 1998].

There are two methods for using MEFs to prevent the differentiation of hES. In the first method, hES cells are cultured directly on a monolayer of mitotically inactivated MEFs [Itskovitz-Eldor et al., 2000; Thomson et al., 1998], allowing for free exchange of MEF-secreted factors with the ES cells. In the second method, hES cells are grown on a layer of diluted Matrigel, which simulates the extracellular matrix that the MEFs provide [Xu et al., 2001]. The ES medium that is fed to these cells is derived from medium conditioned by mitotically inactivated MEFs for approximately 24 hours. Thus the unknown factors that MEFs provide are available to the hES cells without their direct contact [Xu et al., 2001]. This second scenario is useful for applications when MEF contamination in subsequent steps is undesirable.

Two other factors that may play a role in the self-renewal of hES cells are basic fibroblast growth factor (bFGF; FGF-2) and leukemia inhibitory factor (LIF). LIF is of particular interest because its presence has been shown to be sufficient for self-renewal of mouse ES cells [Boeuf et al., 1997; Pease et al., 1990; Williams et al., 1988]. However, although the mouse and human models are similar in many ways, LIF alone does not prevent the differentiation of hES cells and its effect on the human system is not yet clear [Schuldiner et al., 2000].

To ensure that hES cells are pluripotent, their ability to differentiate into all germ layers must be confirmed in vivo. Currently, the most common technique to test for the pluripotency of hES cells is through injection of ES cells in a SCID mouse [Xu et al., 2001]. Because of the lack of an immune system in these mice, the injected ES cells are not rejected and grow into tumors that can then be examined to ensure that the cells can differentiate into all germ layers. Alternatively, morphological and immunochemical assays can be performed in vitro to confirm ES cell potency. The first step is to check the culture to ensure that the cells are morphologically similar to undifferentiated cells (i.e., tight colonies with high ratio of nucleus to cytoplasm). Human ES cells grow in compact colonies that, when undifferentiated, have a bright, even border. The colonies are rounded, with no jagged points or invaginations. The cells inside the colony should be homologous; no structures or variation should be noticeable. Colonies should be checked under a microscope before every passage. The second step is to perform immunohistochemical or immunofluorescent assays to test the expression of ES cell-specific markers [Ling and Neben, 1997; Richards et al., 2002; Shamblott et al., 1998; Xu et al., 2001; Zandstra et al., 2000] such as stage-specific embryonic antigens 3 and 4 (SSEA-3 and -4), Tra-1-60, Oct-4. and alkaline phosphatase (Fig. 3.1, See Color Plate 2).

2.1. Preparation of Media and Reagents

2.1.1. Stock Solutions

1. Gelatin (1% w/v) in sterile water. The solution should be autoclaved before use to ensure sterility.
2. bFGF: 10 μg of bFGF in 1 ml of 0.1% bovine serum albumin (BSA) in Dulbecco's phosphate-buffered saline lacking Ca^{2+} and Mg^{2+} (PBSA). 1 ml of the BSA solution should be used to resuspend the lyophilized bFGF. Immediately after resuspension, 250-μl aliquots should be stored at −20 °C.
3. LIF: 10^6 units/ml. (Optional; See above discussion.)

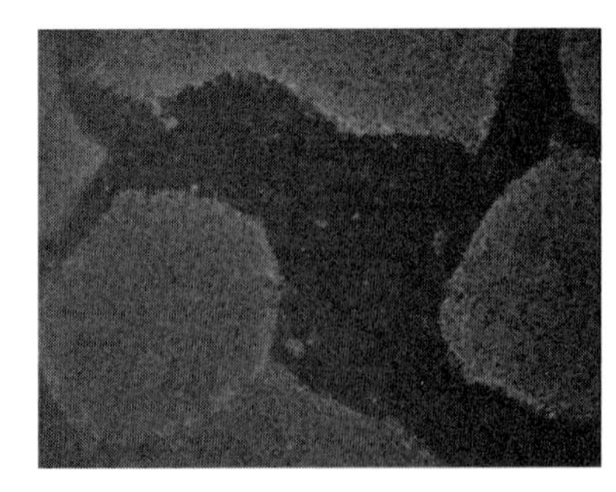
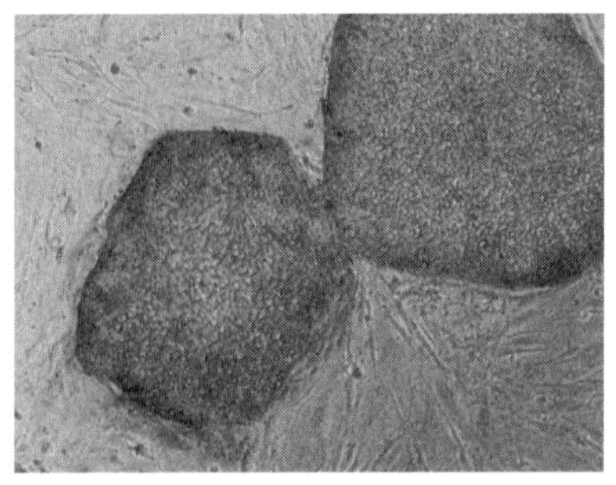

Figure 3.1. hES cell colonies grown on inactivated mouse embryonic fibroblasts. Colonies are stained with undifferentiated cell markers (See Color Plate 2): SSEA-4 (in red, left) and alkaline phosphatase (in blue, right).

4. Mitomycin C: 8 μg/ml in DMEM. Care should be taken when handling the powdered form of mitomycin. Use a syringe to add 5 ml of DMEM to the closed bottle of mitomycin by puncturing the top. Once mitomycin is resuspended in the bottle, use the syringe to draw the DMEM and mitomycin out of the bottle, and then dilute it to 8 μg/ml in DMEM and sterilize with a 0.22-μm low-protein-binding filter. Aliquot and freeze at −20 °C until needed.
5. Collagenase type IV: 200 units of collagenase per ml of DMEM, filtered through a 0.22-μm filter. Solution can be used for up to 2 weeks.
6. Trypsin, 0.5% in PBSA.

2.1.2. Murine Embryonic Fibroblast (MEF) Medium

(i) Fetal bovine serum (FBS) 10%

(ii) Dulbecco's modified Eagle medium (DMEM) 90%

(iii) Filter the solution, using a 0.22-μm filter bottle.

2.1.3. ES Cell Medium

(i) Knockout (KO) serum, 20%

(ii) KO DMEM, 78.3%

(iii) Nonessential amino acid solution, 1%

(iv) 2-Mercaptoethanol, 0.2% (55 mM in PBSA)

(v) L-Glutamine, 0.5% (200 mM in 0.85% NaCl)

(vi) LIF (10^3 units/ml)

(vii) bFGF from a stock of 5 ng/ml

(viii) Filter the solution, using a 0.22-μm filter bottle.

2.1.4. Embryoid Body (EB) Cell Media

(i) Knockout (KO) serum, 20%

(ii) KO DMEM medium, 78.3%

(iii) Nonessential amino acid solution, 1%

(iv) 2-Mercaptoethanol, 0.2% of a 55 mM stock in PBSA

(v) L-Glutamine, 0.5% of a 200 mM stock in 0.85% NaCl

(vi) Filter the solution, using a 0.22-μm filter bottle.

Serum and stock solutions should be stored at −20 °C.

2.2. Preparation of MEF Cells

Protocol 3.1. Seeding Cryopreserved Murine Embryonic Fibroblast (MEF) Cells

Reagents and Materials

Sterile

- ❑ MEF cells, frozen stock, 1×10^6 cells/ampoule
- ❑ MEF medium
- ❑ Conical centrifuge tubes, 15 ml
- ❑ Culture flasks, 75 cm^2 (or similar)

Protocol

(a) Take the cells from storage in a liquid nitrogen tank.

△ *Safety notes.* Gloves must be worn to protect from frostbite. Great care must be taken with ampoules that have been immersed in liquid nitrogen as they may explode on warming if they have leaked and inspired liquid nitrogen. Wear a face mask, and thaw in a covered bath.

(b) Leave the cryogenic vial in a 37 °C water bath until the contents begin to melt.
(c) Transfer to a sterile hood.
(d) Put the cells in a 15-ml centrifuge tube.
(e) In a drop-by-drop manner add 5 ml MEF medium to the tube.
(f) Centrifuge the tube at 1000 rpm (125 *g*) for 5 min.
(g) Add 15 ml MEF medium to a 75-cm^2 tissue culture flask.
(h) Remove the medium from the tube, and resuspend the cells in 2 ml medium.
(i) Seed the cells in the flask, and place them in the CO_2 incubator.
(j) Replace the medium with fresh MEF medium the next day.
(k) The MEFs can be maintained in culture for a few passages (4–5 passages) before losing their proliferative potential.

Protocol 3.2. Seeding Mitotically Inactivated Murine Embryonic Fibroblasts (MEF) for hES Culture

Reagents and Materials

Sterile

- ❑ Mitomycin C, 8 μg/ml (See Section 2.1.1. (iv))
- ❑ Gelatin, 1% (See Section 2.1.1. (i))
- ❑ PBSA
- ❑ Trypsin, 0.25% in PBSA (GIBCO-Invitrogen)
- ❑ Petri dishes, tissue culture grade, 10 cm
- ❑ Centrifuge tubes, 15 ml

Protocol

(a) Remove the medium that is in the flask of confluent MEF.
(b) Add 7 ml mitomycin C solution.
(c) Leave the flask in the CO_2 incubator for 2 h.
(d) While the MEF are incubating with the mitomycin, put 3–4 ml 1% gelatin in the bottom of four 10-cm Petri dishes. Spread the gelatin to cover the entire surface and incubate at 37 °C until needed.
(e) After 2 h, aspirate the mitomycin.
(f) Wash 4 times with PBSA.
(g) Remove the last PBSA wash.
(h) Add 2 ml trypsin solution.
(i) Put the flask in the CO_2 incubator until the cells are free-floating.
(j) Take the flask from the CO_2 incubator and add 5–6 ml MEF medium to stop the trypsin.
(k) Pipette up and down ~10 times to break cell clumps, and then move the contents to a 15-ml centrifuge tube.
(l) Spin down at 1400 rpm (250 *g*) for 5 min.
(m) While cells are centrifuging, take the gelatin-coated Petri dishes from the CO_2 incubator, remove the gelatin, and add 10 ml MEF medium to each dish.
(n) Resuspend the cells and divide equally among the pretreated dishes: usually 1 flask into 4 10-cm dishes.

2.3. ES Cell Expansion and Passaging

Collagenase is the preferred enzyme for passaging hES cells because it selectively removes the ES cell aggregates from the cocultures, without disturbing the MEF monolayer. Thus it is possible to enrich for hES cells during the passaging process.

Protocol 3.3. Passaging Human Embryonic Stem Cells (hES) with Collagenase

Reagents and Materials

Sterile

❑ Collagenase, 200 U/ml (See Section 2.1.1 (v))
❑ ES medium (See Section 2.1.3)

Protocol

(a) Aspirate the medium from the dishes.
(b) Add 4 ml 200 U/ml collagenase solution to each dish.
(c) Leave the dish in the CO_2 incubator for 30–45 min.
(d) Add 5 ml ES medium.
(e) Wash the plate gently to remove the ES colonies without removing MEF from the bottom of the dishes.
(f) Move the ES colonies to a 15-ml centrifuge tube.

(g) Wash the plate a second time with 3 ml ES medium to collect any ES colonies that were not taken the first time. Add to the 9 ml of collagenase and medium already in the 15-ml tube.
(h) Spin down at 800 rpm (80 g) for about 3 min.
(i) During centrifugation, take plates prepared with mitomycin-inactivated MEF from the CO_2 incubator, remove the MEF medium, and add 10 ml ES medium to each plate.
(j) Resuspend the ES cell pellet, and pipette strongly to break the colonies into smaller pieces.
(k) Spin down at 800 rpm (80 g) for 3 min.
(l) Aspirate the medium, add new medium, and pipette up and down to resuspend the colonies.
(m) Add the resuspended colonies to the mitomycin-treated MEF plates. Split in the range from 1:4 to 1:10.
(n) Put the dishes in the CO_2 incubator.
(o) Change the medium daily.

Protocol 3.4. Passaging Human Embryonic Stem (hES) Cells with Trypsin

Sterile

- ❑ PBSA
- ❑ Trypsin, 0.1%
- ❑ TNS: Trypsin neutralization solution
- ❑ ES medium
- ❑ MEF cells in gelatin-coated dishes (usually split 1 to 4; See Protocol 3.2, Step (n))
- ❑ Conical centrifuge tubes, 15 ml

Protocol

(a) Aspirate the medium from the dishes.
(b) Wash once with PBSA.
(c) Add 3 ml trypsin.
(d) Leave the flask in the CO_2 incubator 5 min.
(e) Add 6 ml TNS.
(f) Pipette the contents several times to remove all the cells.
(g) Put them in a 15-ml conical centrifuge tube and pipette vigorously.
(h) Spin down at 700 rpm (60 g) for about 3 min.
(i) During centrifugation, take new dishes from the CO_2 incubator, remove the gelatin, and add 10 ml ES medium.
(j) Resuspend the cells and add to the MEF cells in gelatin-coated dishes.
(k) Return the plates to the CO_2 incubator.

---

3. INDUCTION OF DIFFERENTIATION IN ES CELLS

Human ES cells can be induced to differentiate in culture by a number of different techniques. These techniques involve the removal of the chemical signals and molecular cues that induce stem cell self-renewal (See Section 2), while at the same time providing molecular signals that induce differentiation [Assady et al., 2001; Itskovitz-Eldor et al., 2000; Kaufman et al., 2001; Kehat et al., 2001; Levenberg et al., 2002; Mummery et al., 2003; Reubinoff et al., 2001; Schuldiner et al., 2000, 2001; Xu et al., 2002; Zhang et al., 2001]. Typically, stem cells are induced to differentiate in two-dimensional cultures or within a suspension culture of cell aggregates or spheroids that can be derived clonally or from aggregation of many ES cells. These cell aggregates are called embryoid bodies (EBs) because they mimic and recapitulate many aspects of normal embryonic development (Fig. 3.2). Another method that we have developed is the induction of the differentiation and organization of the cells on three-dimensional polymer scaffolds [Levenberg et al., 2003].

3.1. Embryoid Body Formation

EBs can be formed by a number of methods including suspending cells in gels that restrict the migration of the cells, placing cells within nonadhesive dishes, and seeding cells within hanging drops that induce aggregate formation of the cells.

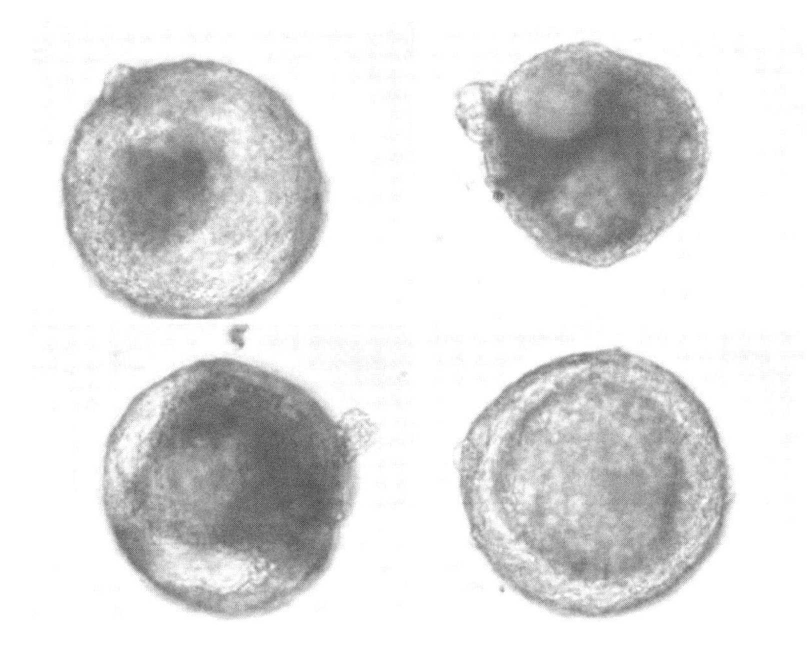

Figure 3.2. Human EBs (hEBs). hEBs grown in suspension in differentiation medium form spheres.

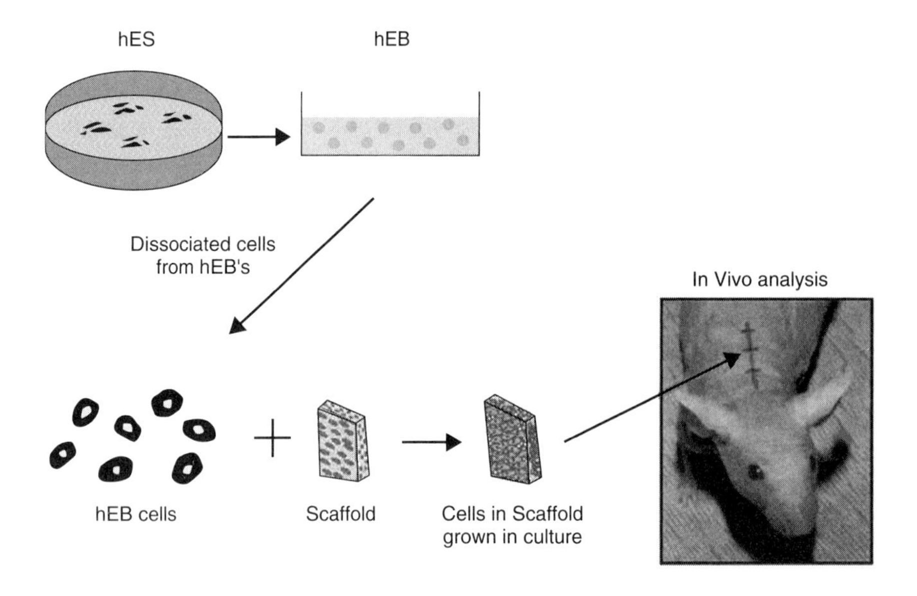

Figure 3.3. Differentiation of hES cells on 3D scaffolds. Cells are partially differentiated in EB. EB cells are dissociated, seeded into polymer scaffolds, and cultured in vitro. After culture in vitro and formation of tissue structure, the constructs are then implanted in vivo.

Protocol 3.5. Formation of Embryoid Bodies in Methylcellulose

Reagents and Materials

Sterile

- ❑ Methylcellulose in EB medium [Wiles and Keller, 1991] (can be bought from Stem Cell Technologies, Inc.)
- ❑ Trypsin, 0.1% in PBSA

Protocol

(a) Trypsinize hES cells as previously described (See Protocol 3.4).
(b) Mechanically disperse the cells into a single-cell suspension.
(c) Suspend the ES cells in methylcellulose in EB medium at 1×10^5 cells/ml and immediately mix the contents vigorously by vortexing.
(d) The methylcellulose solution is viscous and vortexing will form bubbles within the gel, so allow 5–10 min for the bubbles to rise to the top.
(e) Dispense the medium and ES cells into 6-cm Petri dishes, using a 3-ml disposable syringe attached to a 16-gauge blunt-ended needle.
(f) Put the plates in the CO_2 incubator.

Protocol 3.6. Formation of Embryoid Bodies in Nonadhesive Dishes

Reagents and Materials

Sterility

- ❑ EB medium (See Section 2.1.4)

- ❑ Collagenase (See Section 2.1.1(v))
- ❑ Petri dishes, non-tissue culture grade, 10 cm

Protocol

(a) Detach the ES cells with collagenase as previously described (See Protocol 3.3).

(b) Resuspend the cells in EB medium and add to non-tissue culture-grade polystyrene dishes.

(c) Usually the cells are seeded so that one 10-cm ES dish is split into three 10-cm EB dishes (or similar).

(d) Put the dishes in the CO_2 incubator.

After one or two days, the cells typically form clusters that range in size from 50 to 1000 μm.

Protocol 3.7. Formation of Embryoid Bodies in Hanging Drop Cultures

Reagents and Materials

Sterile

- ❑ Trypsin, 0.1% in PBSA

Protocol

(a) Trypsinize ES cells as previously described (See Protocol 3.4).

(b) To initiate the cultures, place a drop of medium containing the cells on the inside of the lid of a Petri dish while the lid is turned upside down. Place more than one drop on the lid of the Petri dish if there is sufficient space that the drops will not touch each other. The lid is then quickly inverted right-side up while ensuring that the drops do not merge.

(c) Slowly place the lid back on the Petri dish so that the drop is suspended in the middle of the dish. It is suggested to examine the culture under phase-contract microscope to ensure that each drop contains cells.

(d) Carefully return the dishes to the CO_2 incubator.

3.2. Two-Dimensional Confluent Cultures

ES cells can also be induced to differentiate within 2D cultures on removal of the factors that induce their self-renewal. Thus for mouse ES cells, LIF would be removed from the ES cells. However, in the case of hES cells care must be taken to remove the feeder cells from the cultures. This is typically done by dissociating the ES cells from the dishes by a collagenase protocol (similar to Protocol 3.3). The suspended ES cells can then be seeded directly onto tissue culture dishes that have been coated with gelatin.

3.3. Three-Dimensional Cultures on Polymer Scaffolds

Recently we demonstrated that hES cells can also be induced to differentiate within biodegradable polymer scaffolds [Levenberg et al., 2003]. Polymer scaffolds [Langer and Vacanti, 1999; Lavik et al., 2002; Niklason and Langer, 2001] represent a promising system for allowing formation of complex 3D tissues during differentiation. They provide physical cues for cell orientation and spreading, and pores provide space for remodeling of tissue structures [Vacanti and Langer, 1999]. In addition, directed degradation of scaffolds can be used as a tool for localized and controlled growth factor supplementation [Richardson et al., 2001]. Ultimately, in vitro-differentiated constructs can potentially be used for transplantation.

4. ISOLATION OF SPECIFIC CELL TYPES FROM CULTURES ORIGINATING FROM ES CELLS

So far no ES cell differentiation protocol has resulted in a pure population of cells. The heterogeneity in the ES cell-derived cultures necessitates the isolation of the desired cell types from a heterogeneous population of cells. There are a number of ways in which cells can be isolated for therapeutic or research applications. These methods range from purely genetic approaches to approaches based on morphological and physical properties of the cells.

4.1. Immunostaining Followed by Cell Fluorescence-Activated Cell Sorting (FACS)

Individual or combinations of various membrane-bound proteins can be used to distinguish different cell types from each other. Thus labeling cells with antibodies that are specific for particular surface proteins and then sorting the desired cells from the population is an approach that may be used for selecting desired cell types. FACS is readily used to isolate such distinct populations of cells at a rapid and reproducible rate [Eiges et al., 2001]. In this technique, cell surface markers are labeled with fluorescent antibodies. With a FACS flow sorter, the positive population can then be purified from a heterogeneous mixture of cells. The advantage of using FACS is the ability to use a combination of markers, each with a distinctive fluorescent label, for a multiparametric sort. Thus cells that coexpress three or four distinguishing proteins can be labeled and isolated, providing a robust method of isolating desired cells. This approach has been used clinically for characterizing and isolating bone marrow cells [Jurecic et al., 1993; Katayama et al., 1993]. In theory, if cell surface markers that define any cell population are known, that population can be isolated from developing EBs, making FACS a potentially powerful technique.

Despite this power there are several practical limitations of using FACS for cell isolation. For example, a distinctive set of cell surface markers may not be known, or even exist, for a desired cell type. In addition, internal markers such as proteins that reside within the cell cannot be used. Currently, cell permeabilization is required to mark internal cell proteins, but this kills the cells. Furthermore, the fraction of cells in the desired population may be small (sometimes less than half a percent of the total number of cells), making subsequent expansion of the culture difficult. Finally, completely pure populations of cells are difficult to achieve. Thus if target cells that take a long time to go through population doublings are contaminated with even a few cells of a type that double quickly, within a few passages the culture will be overwhelmed with "weeds," or the undesired, quickly repopulating contaminant cells.

Nevertheless, as the body of knowledge of cell surface markers and techniques for sorting cells improves, FACS will only become more attractive as a method for isolating rare cell populations, both for study and for clinical applications.

Protocol 3.8. Separating Endothelial Cells from EBs by Immunostaining and Flow Sorting

Note: The antibody that is utilized in this protocol has already been conjugated to a fluorescent marker. However, it is possible that under different circumstances, the cell surface marker would be bound by an antibody that would then be attached to a secondary antibody containing the fluorescent signal. Protocol variations for this have been noted below.

Reagents and Materials

Sterile

- ❑ EBs, 13- to 15-day culture(with medium changed every 2nd–3rd day)
- ❑ Trypsin, 0.1% in PBSA
- ❑ TNS: trypsin neutralizing solution
- ❑ FBS, 5% in PBSA
- ❑ Conical centrifuge tubes, 15 ml

(a) Take EBs, which have been cultured on nonadherent dishes for 13–15 days, from the CO_2 incubator and place in the hood.

(b) Remove the suspension of cells in medium and place in 15-ml conical centrifuge tubes.

(c) Allow the EBs to settle out of the medium (5–10 min).

(d) Aspirate the medium. Try to maximize the amount removed without disturbing the pellet.

(e) Add 7 ml trypsin to 3 to 4 15-ml tubes. Repeat as necessary for additional tubes. Cap tubes very tightly and put on an xyz shaker in an CO_2 incubator for 5 min.

(f) Remove tubes from the incubator and pipette up and down strongly to dissociate EBs. If necessary, place back in the incubator for an additional 2 min.

(g) Add 7 ml TNS to each tube.

(h) If necessary, pour the cells through a cell filter to remove any clumps.

(i) Centrifuge the cells for 3 min at 800 rpm (80 *g*).
(j) Resuspend the cells in a small (1–2 ml) volume of 5% FBS in PBSA.
(k) Count the cells.
(l) Reserve approximately 0.5–1 $\times$ 10^6 cells for a negative control in a sterile Eppendorf tube. (If your antibody is not already conjugated to the fluorescent signal and you will have to apply a secondary antibody containing the fluorescent marker, also reserve a fraction for secondary antibody only). Place this tube on ice.
(m) Spin the rest of the cells down and aspirate the medium.
(n) Add an appropriate amount of the fluorescently labeled antibody to 100–200 μl 5% FBS in PBSA.
(o) Resuspend the pellet in this minimal volume and place on ice.
(p) Every 10 min, flick the tubes to make sure that mixing occurs.
(q) After 30-min incubation, dilute the 100 μl with 10 ml 5% FBS-PBSA.
(r) Spin down the cells.
(s) Resuspend in 5 ml 5% FBS-PBS and spin down (wash).
(t) Resuspend the cells in the volume of 5% FBS-PBS recommended by your sorting facility (1 ml or so) and place in a polypropylene tube. Repeat for the control cells.
(u) Take to the cell sorting machine, along with collection tubes $\frac{3}{4}$-filled with medium.

Magnetic sorting is an alternative approach that can be used to isolate the desired cells by positive or negative selection [Luers et al., 1998; Wright et al., 1997; Zborowski et al., 1999]. In this approach, instead of using a fluorescent label, small magnetic beads that attach to the primary antibody can be used to label particular cells. The beads are typically attached to the primary antibody by a biotin-streptavidin linkage. In this approach, all cells are then passed through a electromagnetic column. The cells that express the marker of interest are held within the column because of the magnetic attraction of the beads to the column. Thus cells that do not express the desired antigen are washed through the column and collected. Subsequently, the cells that are retained within the column can be collected by switching off the electromagnet.

4.2. Genetically Engineered Selectable Markers

A technique that is currently under development to enrich particular cell types is to engineer a cell's gene expression so that the desired progeny is enriched [Friedrich and Soriano, 1991; Moritoh et al., 2003; Soria et al., 2000]. This process of enrichment can be induced either through the activation of suicide genes on the expression of particular genes or the expression of genes that maintain the cells. For example, neomycin resistance (*neo*, aminoglycoside phosphotransferase) can be engineered into ES cells in a construct with a lineage-specific promoter. The expression of neomycin resistance can then be regulated by the promoters that are activated for the desired cells. The use of this technique and similar approaches promises to be a powerful tool for directed differentiation of ES cells and is an area of active research.

4.3. Preferential Detachment and Attachment

Different cell types express various levels of a number of cell adhesion molecules such as integrins. Thus the cell's adherence properties can be used to isolate a specific cell type. This has been used extensively in the isolation of mesenchymal stem cells from bone marrow populations [Pittenger et al., 1999]. However, its utility in ES cell culture has not been tested vigorously.

4.4. Hand Enrichment (Mechanical Isolation of Defined Structures)

Hand enrichment of desired cell types in the form of colonies or mechanical isolation of defined structures is another method of isolating the desired cells. The use of such a technique requires visibly distinct morphological properties of the desired cells. For example, beating cardiomyocytes can easily be dissected from a culture of heterogeneous cells. Thus it is possible to isolate the desired cells from a heterogeneous culture based on distinctive morphological properties. However, it is anticipated that such methods will not be efficient for the scale-up that is required for therapeutic applications.

5. CHARACTERIZATION OF ISOLATED ENDOTHELIAL PRECURSOR CELLS

As described above (See Section 3), several techniques are available for inducing differentiation of ES cells in the absence of self-renewing agents, and the resulting mixture of cells can be enriched for a specific combination of surface receptor expression by one of the above isolation techniques. The cells derived by these means must be characterized to validate gene expression, phenotype, and in vivo functionality.

Recently, we established the successful isolation of endothelial cells from human ES cells [Levenberg et al., 2002]. The isolation procedure was as follows. The hES cells were grown on gelatin-coated dishes over mitomycin-treated MEF. The growth medium consisted of 80% KO DMEM and 20% KO serum-free formulation, with glutamine, and supplements of β-mercaptoethanol, bFGF, LIF, and nonessential amino acids [Schuldiner et al., 2000]. To form EB aggregates, the cells were dissociated with 1 mg/ml collagenase type IV and grown in Petri dishes. EBs at 13–15 days were dissociated with trypsin and incubated with fluorescently labeled CD31 antibody for 30 min before cell sorting with a FACStar flow sorter. The $CD31^{+}$ cells were replated and grown in vitro in endothelial cell growth medium.

5.1. Expression of Endothelial Markers

Through studies in animal models, and more recently in humans, a number of related markers, transcriptional factors, adhesion molecules, and growth factor receptors for endothelial cells have been identified including endothelial cell adhesion molecules such as PECAM1/CD31, vascular endothelial cadherin and CD34; growth factor receptors such as vascular endothelial growth factor receptor-2 and Tie-2; and transcription factors GATA-2 and GATA-3. These molecules have been used to characterize endothelial cells by RNA/gene expression assays (RT-PCR, Northern blot, in situ hybridization) or by immunostaining for protein expression and localization in cell structures [Levenberg et al., 2002].

5.2. LDL Incorporation

Functional characterization of endothelial cells involves measuring the uptake of acetylated low-density lipoprotein (ac-LDL) using the fluorescent probe 1,1′-dioctadecyl-3,3,3′,3′-tetramethylindocarbocyanine perchlorate (Dil-Ac-LDL). This assay apparently has no effect on endothelial cell growth rate at incubation conditions of 10 μg/ml Dil-Ac-LDL for 4 hours at 37 °C {Voyta, 1984}. We have shown that human embryonic-derived $CD31^+$ cells stained brightly for Dil-ac-LDL [Levenberg et al., 2002].

5.3. Analysis of In Vitro Tube Formation

Three-dimensional matrices such as collagen or Matrigel are often used to analyze endothelial cell differentiation, vascularization potential, and organization into tubelike structures in vitro. In this method, cells are seeded either on or in the gel (either by mixing the cells with the gel or seeding in between two layers of the gel) [Balconi et al., 2000; Hatzopoulos et al., 1998; Levenberg et al., 2002; Yamashita et al., 2000]. Capillary tube formation can be evaluated by phase-contrast microscopy after seeding the cells for several hours or up to a few days. The effect of growth factors on these processes can also be studied by the addition of growth factors to the culture medium. The structure of the capillary network, stability of the cords over time, and lumen formation (by electron microscopy of the tube cross sections) can be used to characterize the tube structure and lumen size [Grant et al., 1991; Vernon et al., 1995].

5.4. Analysis of In Vivo Vessel Formation

In vivo testing is useful for studying the therapeutic potential of ES cell-derived endothelial cells. Various methods have been used to analyze involvement of implanted endothelial cells in the host vasculogenesis and angiogenesis processes. One method involves injecting endothelial cells into chicken embryos to analyze the vasculogenesis potential of the cells and incorporation into vascular structure in the developing embryo [Hatzopoulos et al., 1998]. Endothelial precursors have also been injected into infarcted myocardium and ischemic hindlimb to analyze the effects of the cells on neovascularization and angiogenesis processes [Kocher et al., 2001]. Another method involves seeding endothelial cells into polymer scaffolds and then implanting the cell-scaffold construct in vivo to analyze vessel formation within the implant [Nor et al., 2001]. This technique has been used to characterize the endothelial cells derived from hES cells. The cells were seeded on highly porous biodegradable polymer scaffolds fabricated from poly-L-lactic acid (PLLA) and poly(lactic acid-co-glycolic acid) (PLGA) that are commonly used as scaffolds for tissue engineering. Sponges seeded with embryonic-derived $CD31^+$ cells were implanted in the subcutaneous tissue of SCID mice and analyzed by immunostaining with human-specific endothelial markers after 1 week and 2 weeks of implantation. We have shown that the implanted cells formed blood vessels in vivo that appeared to anastomose with the mouse vasculature [Levenberg et al., 2002].

Briefly, $PECAM1^+$ cells (1×10^6) were resuspended in 50 μl of a 1:1 mix of culture medium and Matrigel and allowed to absorb into the PLLA/PLGA polymer sponges. After a 30-min incubation at 37 °C to allow for gelation of the Matrigel, the cells plus scaffolds were implanted subcutaneously in the dorsal region of 4-week-old SCID mice. After transplantation (7 or 14 days), the implants were retrieved, fixed overnight in 10% (v/v) buffered formalin at 4 °C, embedded in paraffin, and sectioned for histological examination.

6. SCALE-UP OF ES CELLS IN TISSUE ENGINEERING

The widespread clinical use of ES cells as tissue engineering precursors will require optimization and standardization of large-scale production of these cells. Fortunately, cells can be expanded nearly indefinitely in the undifferentiated state, but some question remains as to whether it is best to expand these undifferentiated cells to large numbers, or if it is more beneficial to induce differentiation of the cells and then to expand them once differentiated. Regardless of the order, a bioreactor capable of overcoming the nutritional and metabolic limitations characteristic of large cell numbers will be required.

6.1. Expansion of Cells in Undifferentiated State

One major distinction between the handling of undifferentiated and differentiated cells is the requirement of undifferentiated cells for feeder cells. This requirement complicates the use of a steady-state chemostat reactor or other such stirred bioreactors. However, some promising results on the growth of hematopoietic [Zandstra et al., 1994], neural [Kallos and Behie, 1999], and ES and progenitor cell numbers in stirred suspension bioreactors cultures have been obtained [Zandstra and Nagy, 2001].

Steady-state stirred suspension reactors are easily scalable and relatively simple. Their relatively homogeneous nature makes them uniquely suited for investigations of different culture parameters (e.g., O_2 tension, cytokine concentration, serum components, medium exchange rates) that may influence the viability and turnover of specific stages and types of stem cells.

Stem cell properties are the result of the expression of a specific subset of genes, changes in the expression of which determine exit from the stem cell compartment into functional cell lineages. Although there is still much to learn about the genes involved in such changes (as well as how they are regulated), it is clear that stem cells interact with many molecules in their extracellular milieu via transmembrane receptors (or receptor complexes) to maintain their viability, and to effect change in their cell cycle progression and differentiated state. A key feature of any stem cell culture system is the combination of cytokines it delivers to the microenvironment of the cells, and how the concentrations of these cytokines and their associated receptors are maintained over time.

Significant efforts have been made to define cytokine and growth factor supplementation strategies to control stem cell responses. The cytokine composition of the medium is particularly challenging to optimize in stem cell cultures because multiple cell types compete for several cytokines that each influence stem cell fate directly or indirectly.

7. PROTOCOLS FOR USING ES CELLS IN TISSUE ENGINEERING

One of the major goals of isolating hES cells is their future use as precursor cells for tissue engineering. One option is direction of the differentiation of these cells followed by isolation of the desired cell type. These differentiated cells are theoretically identical to their somatic cell counterparts and therefore can be seeded into scaffolds and implanted identically to any other somatic cells. However, it has been shown that coculture with adult cells directs the differentiation and integration of ES cells with their surrounding cells. This discovery leads to the interesting concept of seeding and implanting undifferentiated ES cells, allowing them to differentiate in vivo.

7.1. Seeding Differentiated Cells onto Scaffolds

Seeding differentiated cells into scaffolds will be identical to seeding any cell into the corresponding scaffold (See Section 5.4).

7.2. Seeding Undifferentiated ES Cells for In Vivo Differentiation

Adult cells are known to express and excrete some of the proteins and factors that induce the differentiation of ES cells. In addition, ES cells have been shown to fuse with somatic cells and to repair or replace the adult cells. Logically, if undifferentiated ES cells are seeded into scaffolds and transplanted into the site of tissue damage they may differentiate to regenerate the damaged tissue. This process has not yet been attempted in humans but has been successful in treating mice with spinal cord injuries [Langer and Vacanti, 1999; Lavik et al., 2002; Niklason and Langer, 2001].

8. CONCLUSION AND FUTURE PERSPECTIVES

ES cells have generated a great deal of interest as a source of cells for tissue engineering. Protocols for growing hES cells are established but will need to be modified for use of the cells in the clinic. These modifications include establishing growth conditions without feeder cells, to generate a pure population of ES cells, and scale-up of the cell culture. In addition, protocols for the induction of specific differentiated phenotypes are required, including methods for isolating the desired cell types and their characterization. Other challenges in the use of hES in tissue engineering include ensuring the safety of the cells in vivo, that is, ensuring that the cells are immunologically compatible with the patient and will not form tumors, and enhancing efficacy by improving current tissue engineering methods. We are getting close to the day when ES cells can be manipulated in culture to produce fully differentiated cells that can be used to create and repair specific tissues and organs.

ACKNOWLEDGMENT

NIH Grant HL-60435

SOURCES OF MATERIALS

Item	Catalog No.	Supplier
bFGF	13256-029	Invitrogen (GIBCO)
Cell filter		BD Biosciences (Falcon)
Centrifuge tubes, 15 ml		BD Biosciences (Falcon)
Collagenase IV		
DMEM	11965-118	Invitrogen (GIBCO)
FBS	16000-044	Invitrogen (GIBCO)
Glutamine		
Knockout DMEM	10829-018	Invitrogen (GIBCO)
Knockout serum	10828	Invitrogen (GIBCO)
LIF	ESG1106	Chemicon
2-Mercaptoethanol		
Nonessential amino acid solution	11140-035	Invitrogen (GIBCO)
Trypsin Neutralizing Solution (TNS)		Cambrex (Clonetics)

REFERENCES

Assady, S., Maor, G., Amit, M., Itskovitz-Eldor, J., Skorecki, K.L., Tzukerman, M. (2001) Insulin production by human embryonic stem cells. *Diabetes* 50: 1691–1697.

Balconi, G., Spagnuolo, R., Dejana, E. (2000) Development of endothelial cell lines from embryonic stem cells: A tool for studying genetically manipulated endothelial cells in vitro. *Arterioscler. Thromb. Vasc. Biol.* 20: 1443–1451.

Boeuf, H., Hauss, C., Graeve, F.D., Baran, N., Kedinger, C. (1997) Leukemia inhibitory factor-dependent transcriptional activation in embryonic stem cells. *J. Cell Biol.* 138: 1207–1217.

Czyz, J., and Wobus, A. (2001) Embryonic stem cell differentiation: the role of extracellular factors. *Differentiation* 68: 167–174.

Eiges, R., Schuldiner, M., Drukker, M., Yanuka, O., Itskovitz-Eldor, J., Benvenisty, N. (2001) Establishment of human embryonic stem cell-transfected clones carrying a marker for undifferentiated cells. *Curr. Biol.* 11: 514–518.

Evans, M.J., and Kaufman, M.H. (1981) Establishment in culture of pluripotential cells from mouse embryos. *Nature* 292: 154–156.
Friedrich, G., and Soriano, P. (1991) Promoter traps in embryonic stem cells: a genetic screen to identify and mutate developmental genes in mice. *Genes Dev.* 5: 1513–1523.
Grant, D.S., Lelkes, P.I., Fukuda, K., Kleinman, H.K. (1991) Intracellular mechanisms involved in basement membrane induced blood vessel differentiation in vitro. *In Vitro Cell. Dev. Biol.* 27A: 327–336.
Hatzopoulos, A.K., Folkman, J., Vasile, E., Eiselen, G.K., Rosenberg, R.D. (1998) Isolation and characterization of endothelial progenitor cells from mouse embryos. *Development* 125: 1457–1468.
Itskovitz-Eldor, J., Schuldiner, M., Karsenti, D., Eden, A., Yanuka, O., Amit, M., Soreq, H., Benvenisty, N. (2000) Differentiation of human embryonic stem cells into embryoid bodies compromising the three embryonic germ layers. *Mol. Med.* 6: 88–95.
Jurecic, R., Van, N.T., Belmont, J.W. (1993) Enrichment and functional characterization of Sca-1 + WGA+, Lin-WGA+, Lin-Sca-1+, and Lin-Sca-1 + WGA+ bone marrow cells from mice with an Ly-6a haplotype. *Blood* 82: 2673–2683.
Kallos, M.S., and Behie, L.A. (1999) Inoculation and growth conditions for high-cell-density expansion of mammalian neural stem cells in suspension bioreactors. *Biotechnol. Bioeng.* 63: 473–483.
Katayama, N., Shih, J.P., Nishikawa, S., Kina, T., Clark, S.C., Ogawa, M. (1993) Stage-specific expression of c-kit protein by murine hematopoietic progenitors. *Blood* 82: 2353–2360.
Kaufman, D.S., Hanson, E.T., Lewis, R.L., Auerbach, R., Thomson, J.A. (2001) Hematopoietic colony-forming cells derived from human embryonic stem cells. *Proc. Natl. Acad. Sci. USA* 98: 10,716–10,721.
Kehat, I., Kenyagin-Karsenti, D., Snir, M., Segev, H., Amit, M., Gepstein, A., Livne, E., Binah, O., Itskovitz-Eldor, J., Gepstein, L. (2001) Human embryonic stem cells can differentiate into myocytes with structural and functional properties of cardiomyocytes. *J. Clin. Invest.* 108: 407–414.
Kocher, A.A., Schuster, M.D., Szabolcs, M.J., Takuma, S., Burkhoff, D., Wang, J., Homma, S., Edwards, N.M., Itescu, S. (2001) Neovascularization of ischemic myocardium by human bone-marrow-derived angioblasts prevents cardiomyocyte apoptosis, reduces remodeling and improves cardiac function. *Nat. Med.* 7: 430–436.
Langer, R.S., and Vacanti, J.P. (1999). Tissue engineering: the challenges ahead. *Sci. Am.* 280: 86–89.
Lavik, E., Teng, Y.D., Snyder, E., Langer, R. (2002) Seeding neural stem cells on scaffolds of PGA, PLA, and their copolymers. *Methods Mol. Biol.* 198: 89–97.
Levenberg, S., Golub, J.S., Amit, M., Itskovitz-Eldor, J., Langer, R. (2002) Endothelial cells derived from human embryonic stem cells. *Proc. Natl. Acad. Sci. USA* 99: 4391–4396.
Levenberg, S., Huang, N.F., Lavik, E., Rogers, A.B., Itskovitz-Eldor, J., Langer, R. (2003). Differentiation of human embryonic stem cells on three-dimensional polymer scaffolds. *Proc. Natl. Acad. Sci. USA* 100: 12,741–12,746.
Ling, V., and Neben, S. (1997) In vitro differentiation of embryonic stem cells: immunophenotypic analysis of cultured embryoid bodies. *J. Cell. Physiol.* 171: 104–115.
Luers, G.H., Hartig, R., Mohr, H., Hausmann, M., Fahimi, H.D., Cremer, C., Volkl, A. (1998) Immuno-isolation of highly purified peroxisomes using magnetic beads and continuous immunomagnetic sorting. *Electrophoresis* 19: 1205–1210.
Moritoh, Y., Yamato, E., Yasui, Y., Miyazaki, S., Miyazaki, J. (2003) Analysis of insulin-producing cells during in vitro differentiation from feeder-free embryonic stem cells. *Diabetes* 52: 1163–1168.
Mummery, C., Ward-van Oostwaard, D., Doevendans, P., Spijker, R., van den Brink, S., Hassink, R., van der Heyden, M., Opthof, T., Pera, M., de la Riviere, A.B., et al. (2003) Differentiation of human embryonic stem cells to cardiomyocytes: role of coculture with visceral endoderm-like cells. *Circulation* 107: 2733–2740.
Niklason, L.E., and Langer, R. (2001) Prospects for organ and tissue replacement. *JAMA* 285: 573–576.

Nor, J.E., Peters, M.C., Christensen, J.B., Sutorik, M.M., Linn, S., Khan, M.K., Addison, C.L., Mooney, D.J., Polverini, P.J. (2001) Engineering and characterization of functional human microvessels in immunodeficient mice. *Lab. Invest.* 81: 453–463.

Pease, S., Braghetta, P., Gearing, D., Grail, D., Williams, R.L. (1990) Isolation of embryonic stem (ES) cells in media supplemented with recombinant leukemia inhibitory factor (LIF). *Dev. Biol.* 141: 344–352.

Pittenger, M.F., Mackay, A.M., Beck, S.C., Jaiswal, R.K., Douglas, R., Mosca, J.D., Moorman, M.A., Simonetti, D.W., Craig, S., Marshak, D.R. (1999) Multilineage potential of adult human mesenchymal stem cells. *Science* 284: 143–147.

Reubinoff, B.E., Itsykson, P., Turetsky, T., Pera, M.F., Reinhartz, E., Itzik, A., Ben-Hur, T. (2001) Neural progenitors from human embryonic stem cells. *Nat. Biotechnol.* 19: 1134–1140.

Reubinoff, B.E., Pera, M.F., Fong, C.Y., Trounson, A., Bongso, A. (2000) Embryonic stem cell lines from human blastocysts: somatic differentiation in vitro. *Nat. Biotechnol.* 18: 399–404.

Richards, M., Fong, C.Y., Chan, W.K., Wong, P.C., Bongso, A. (2002) Human feeders support prolonged undifferentiated growth of human inner cell masses and embryonic stem cells. *Nat. Biotechnol.* 20: 933–936.

Richardson, T.P., Peters, M.C., Ennett, A.B., Mooney, D.J. (2001) Polymeric system for dual growth factor delivery. *Nat. Biotechnol.* 19: 1029–1034.

Schuldiner, M., Eiges, R., Eden, A., Yanuka, O., Itskovitz-Eldor, J., Goldstein, R.S., Benvenisty, N. (2001) Induced neuronal differentiation of human embryonic stem cells. *Brain Res.* 913: 201–205.

Schuldiner, M., Yanuka, O., Itskovitz-Eldor, J., Melton, D.A., Benvenisty, N. (2000) Effects of eight growth factors on the differentiation of cells derived from human embryonic stem cells. *Proc. Natl. Acad. Sci. USA* 97: 11,307–11,312.

Shamblott, M.J., Axelman, J., Wang, S., Bugg, E.M., Littlefield, J.W., Donovan, P.J., Blumenthal, P.D., Huggins, G.R., Gearhart, J.D. (1998) Derivation of pluripotent stem cells from cultured human primordial germ cells. *Proc. Natl. Acad. Sci. USA* 95: 13,726–13,731.

Soria, B., Roche, E., Berna, G., Leon-Quinto, T., Reig, J.A., Martin, F. (2000) Insulin-secreting cells derived from embryonic stem cells normalize glycemia in streptozotocin-induced diabetic mice. *Diabetes* 49: 157–162.

Thomson, J.A., Itskovitz-Eldor, J., Shapiro, S.S., Waknitz, M.A., Swiergiel, J.J., Marshall, V.S., Jones, J.M. (1998) Embryonic stem cell lines derived from human blastocysts. *Science* 282: 1145–1147.

Vacanti, J.P., Langer, R. (1999) Tissue engineering: the design and fabrication of living replacement devices for surgical reconstruction and transplantation. *Lancet* 354 *Suppl* 1: SI32–SI34.

Vernon, R.B., Lara, S.L., Drake, C.J., Iruela-Arispe, M.L., Angello, J.C., Little, C.D., Wight, T.N., Sage, E.H. (1995) Organized type I collagen influences endothelial patterns during "spontaneous angiogenesis in vitro": planar cultures as models of vascular development. *In Vitro Cell. Dev. Biol. Anim.* 31: 120–131.

Wiles, M.V., and Keller, G. (1991) Multiple hematopoietic lineages develop from embryonic stem (ES) cells in culture. *Development* 111: 259–267.

Williams, R.L., Hilton, D.J., Pease, S., Willson, T.A., Stewart, C.L., Gearing, D.P., Wagner, E.F., Metcalf, D., Nicola, N.A., Gough, N.M. (1988) Myeloid leukaemia inhibitory factor maintains the developmental potential of embryonic stem cells. *Nature* 336: 684–687.

Wright, A.P., Fitzgerald, J.J., Colello, R.J. (1997) Rapid purification of glial cells using immunomagnetic separation. *J. Neurosci. Methods* 74: 37–44.

Xu, C., Inokuma, M.S., Denham, J., Golds, K., Kundu, P., Gold, J.D., Carpenter, M.K. (2001) Feeder-free growth of undifferentiated human embryonic stem cells. *Nat. Biotechnol.* 19: 971–974.

Xu, R.H., Chen, X., Li, D.S., Li, R., Addicks, G.C., Glennon, C., Zwaka, T.P., Thomson, J.A. (2002) BMP4 initiates human embryonic stem cell differentiation to trophoblast. *Nat. Biotechnol.* 20: 1261–1264.

Yamashita, J., Itoh, H., Hirashima, M., Ogawa, M., Nishikawa, S., Yurugi, T., Naito, M., Nakao, K. (2000) Flk1-positive cells derived from embryonic stem cells serve as vascular progenitors. *Nature* 408: 92–96.

Zandstra, P.W., Eaves, C.J., Piret, J.M. (1994) Expansion of hematopoietic progenitor cell populations in stirred suspension bioreactors of normal human bone marrow cells. *Biotechnology (NY)* 12: 909–914.
Zandstra, P.W., Le, H.V., Daley, G.Q., Griffith, L.G., Lauffenburger, D.A. (2000) Leukemia inhibitory factor (LIF) concentration modulates embryonic stem cell self-renewal and differentiation independently of proliferation. *Biotechnol. Bioeng.* 69: 607–617.
Zandstra, P.W., and Nagy, A. (2001) *Stem Cell Bioeng.* 3: 275–305.
Zborowski, M., Sun, L., Moore, L.R., Chalmers, J.J. (1999) Rapid cell isolation by magnetic flow sorting for applications in tissue engineering. *Asaio J.* 45: 127–130.
Zhang, S.C., Wernig, M., Duncan, I.D., Brustle, O., Thomson, J.A. (2001) In vitro differentiation of transplantable neural precursors from human embryonic stem cells. *Nat. Biotechnol.* 19: 1129–1133.

4

Cell Sources for Cartilage Tissue Engineering

Brian Johnstone et al [1], Jung Yoo[1], and Matthew Stewart[2]

Department of Orthopaedics and Rehabilitation[1], Oregon Health and Science University, Portland, Oregon 97239; Department of Veterinary Clinical Medicine[2], Veterinary Teaching Hospital, University of Illinois, Urbana, Illinois 61802

Corresponding author: johnstob@ohsu.edu

Culture of Cells for Tissue Engineering, edited by Gordana Vunjak-Novakovic and R. Ian Freshney

1. INTRODUCTION

There are many sites of permanent cartilage within the body. However, because cartilage has specific and distinct functions depending on its location, no two cartilages are the same. Each cartilage has a specific extracellular matrix that is produced by cells termed chondrocytes, which are defined by their production of type II collagen, the major collagen of most cartilage. However, there are differences between chondrocytes, both within and among different cartilaginous tissues. All cartilage extracellular matrices have common constituent molecules, but they are present in different proportions, with some molecules unique to certain types of cartilage. Furthermore, each cartilage has a distinct matrix organization. For these reasons, it is argued that any project for cartilage tissue engineering should include a consideration of the specific type of cartilage one is seeking to repair or regenerate. As a caveat to that statement, it is noted that it is presently unclear how specific any tissue-engineered cartilage needs to be before its implantation, because there are few data to date indicating what form of initial cartilage implant is acceptable, given that the implant may remodel in vivo into the desired cartilage type.

Classically, cartilage is divided into three types: hyaline, elastic, and fibrocartilage. However, for any given type, cartilage differs with site, age, and species. The interest in cartilage tissue engineering is due to the fact that many cartilage types have poor intrinsic regenerative capabilities after injury. Furthermore, degenerative diseases, such as osteoarthritis, cause loss of cartilage from many sites in the body, for which there are presently few biological repair or regenerative therapies. Tissue engineering offers the possibility to replace or repair the lost cartilage. A basic requirement for this goal is consideration of the possible cell types that could be used for cartilage tissue engineering. Chondrocytes and cells with chondrogenic differentiation potential from embryonic and postnatal sources are currently being used for cartilaginous tissue repair and regeneration studies. This chapter discusses possible cell sources, the isolation of cells from different tissues, the options for their culture in vitro, and the methods and assays of their chondrogenic capacity.

2. PREPARATION OF MEDIA AND REAGENTS

2.1. Dissection Medium

Hanks' balanced salt solution (HBSS) supplemented with concentrates, penicillin (10,000 units/ml)-streptomycin (10 mg/ml) and Fungizone™ 250 μg/ml

For 500 ml of dissection medium:

1. Add 10 ml penicillin/streptomycin and 5 ml Fungizone to 500 ml HBSS to give final concentrations of 200 U/ml penicillin, 200 μg/ml streptomycin and 5 μg/ml Fungizone.
2. Cool on ice before use.

2.2. Predigestion Medium for Human Articular Cartilage

HBSS with penicillin/streptomycin, Fungizone and Pronase

For 100 ml of digestion medium, sufficient for 10 g of cartilage:

(i) Add 750 mg pronase to 96 ml HBSS.

(ii) Warm the solution in a 37 °C water bath with occasional agitation to dissolve the pronase.

(iii) Filter the solution through a 0.22 μm filter.

(iv) Add 2 ml penicillin/streptomycin and 1 ml Fungizone.

2.3. Digestion Medium for Human Articular Cartilage

Collagenase A in Opti-MEM 1 reduced-serum medium supplemented with fetal bovine serum, penicillin/streptomycin, and Fungizone. Opti-MEM 1 is a modification of Eagle's minimum essential medium and is supplemented with insulin, transferrin, and selenous acid (ITS). The exact constitution is proprietary information. Alternative ITS supplements are available for addition to media.

For 100 ml of digestion medium, sufficient for 10 g of cartilage:

(i) Add 100 mg of collagenase A to 94 ml Opti-MEM.

(ii) Warm the solution in a 37 °C water bath with occasional agitation to dissolve the collagenase.

(iii) Sterilize by vacuum filtration through a 0.22-μm filter.

(iv) Add 2 ml penicillin/streptomycin, 1 ml Fungizone, and 2 ml sterile FBS.

2.4. Digestion Medium for Bovine Articular Cartilage

Collagenase type II in Opti-MEM 1 reduced-serum medium supplemented with fetal bovine serum, penicillin/streptomycin, and Fungizone

For 100 ml of digestion medium, sufficient for 10 g of cartilage:

(i) Add 100 mg of collagenase type II to 94 ml Opti-MEM.

(ii) Warm the solution in a 37 °C water bath with occasional agitation to dissolve the collagenase.

(iii) Add 2 ml FBS.

(iv) Sterilize by vacuum filtration through a 0.22 μm filter.

(v) Add 2 ml penicillin/streptomycin and 2 ml Fungizone.

2.5. Ascorbic Acid Stock (5 mg/ml)

L-Ascorbic acid phosphate in ultrapure water (UPW)

For 20 ml of ascorbic acid stock:

(i) Add 100 mg L-ascorbic acid phosphate powder to 20 ml distilled, deionized water.

(ii) Agitate intermittently until the powder is dissolved.

(iii) Sterilize by vacuum filtration through a 0.22 μm filter.

2.6. Differentiation Medium

Opti-MEM 1 reduced-serum medium supplemented with ascorbic acid stock (See Section 2.5), penicillin/streptomycin, and Fungizone

For 515.45 ml of primary differentiation medium, to 500 ml of Opti-MEM 1:

(i) Add 5.15 ml penicillin/streptomycin and 5.15 ml Fungizone concentrates as in Section 2.1.

(ii) Add 5.15 ml ascorbic acid stock.

2.7. Growth Medium

High-glucose DMEM, containing pyruvate and glutamate, 1:1 with Ham's F-12 nutrient medium, and supplemented with fetal bovine serum, ascorbic acid, penicillin-streptomycin, and Fungizone

For 500 ml of growth medium:

(i) Combine 217.5 ml DMEM and 217.5 ml Ham's F-12 medium.

(ii) Add 50 ml fetal bovine serum.

(iii) Add 5 ml penicillin/streptomycin and 5 ml Fungizone concentrates as in Section 2.1.

(iv) Add 5 ml ascorbic acid stock.

(v) Sterilize by vacuum filtration through a 0.22 μm filter.

2.8. Alginate Solution (1.2% Alginate in 150 mM NaCl)

Low-viscosity alginate and sodium chloride in UPW
For 100 ml of alginate solution:

(i) Dissolve 0.87 g sodium chloride in 90 ml water.

(ii) Sterilize by vacuum filtration through a 0.22 μm filter.

(iii) Add 1.2 g sterile, low-viscosity alginate and dissolve.

(iv) Bring to a final volume of 100 ml with sterile water.

2.9. $CaCl_2$, 102 mM

Calcium chloride in UPW
For 500 ml of $CaCl_2$ solution:

(i) Dissolve 5.66 g $CaCl_2$ in 450 ml UPW.

(ii) Bring final volume to 500 ml.

(iii) Sterilize by vacuum filtration through a 0.22 μm filter.

2.10. Sterile Saline, 0.9%

Sodium chloride in UPW
For 500 ml of sterile saline solution:

(i) Dissolve 4.50 g NaCl in 450 ml UPW.

(ii) Bring final volume to 500 ml.

(iii) Sterilize by vacuum filtration through a 0.22 μm filter.

2.11. Alginate Depolymerization Solution

Sodium citrate, 55 mM, sodium chloride, 0.15 M, pH 6.8 in UPW
For 500 ml of alginate depolymerization solution:

(i) Dissolve 8.09 g sodium citrate and 4.35 g sodium chloride in 450 ml UPW.

(ii) Adjust pH to 6.8 with hydrochloric acid.

(iii) Bring volume to 500 ml.

(iv) Sterilize by vacuum filtration through a 0.22 μm filter.

2.12. L-Ascorbate-2-Phosphate Stock

Ascorbic acid 2-phosphate, Mg salt in Tyrode's salt solution
For 10 ml of stock:

(i) Weigh out 40.0 mg ascorbic acid 2-phosphate Mg salt.

(ii) Add 10 ml Tyrode's salt solution.

(iii) Use a 0.22 μm filter to sterilize.

(iv) Store 1- to 2-ml aliquots in sterile cryotubes at −20 °C. This solution is stable for up to 30 days at 4 °C and for 1 year at −20 °C.

2.13. Dexamethasone (10^{-3} M) Stock

Dexamethasone in ethanol
For 10 ml of stock:

(i) To 3.92 mg dexamethasone add 10 ml 100% ethanol

(ii) Use a 0.22 μm filter to sterilize.

(iii) Store at −20 °C for up to one year.

(iv) Make a 1×10^{-5} M working solution by 1/100 dilution of 1×10^{-3} M stock in DMEM.

(v) Aliquot (1–2 ml in sterile cryotubes) and store at −20 °C. This solution is stable for up to 30 days at 4 °C and for 1 year at −20 °C.

2.14. Transforming Growth Factor-β1 (TGF-β1) Stock

TGF-β1 with bovine serum albumin (BSA) in dilute HCl
For a 10 ng/ml solution:

(i) Prepare buffered solvent by combining 12.5 mg BSA, 12 ml UPW, and 0.5 ml 0.1 N HCl.

(ii) Filter to sterilize.

(iii) Reconstitute 10 μg TGF-β1 with 10 ml prepared buffer.

(iv) Make aliquots of appropriate volumes and store at −70 °C.

2.15. Basal Medium

Dulbecco's modified Eagle's medium (DMEM) with 4.5 g/l glucose, 4 mM glutamine, and supplemented with penicillin/streptomycin, ITS + Premix, and 100 mM sodium pyruvate.

For 500 ml of basal medium, to DMEM-high glucose (475 ml) add:

(i) L-Glutamine (200 mM or 29.2 g/l), 10 ml.

(ii) Penicillin 1×10^4 U/ml and streptomycin 10 mg/ml, 5 ml.

(iii) ITS + Premix, 5 ml.

(iv) Sodium pyruvate 100 mM, 5 ml.

3. ARTICULAR CHONDROCYTES FOR CARTILAGE TISSUE ENGINEERING

3.1. Human Articular Cartilage

The predominant focus of cartilage tissue engineering has been on regeneration or repair of articular cartilage defects. Therefore, this section is limited to protocols addressing articular chondrocyte isolation and culture. It should be noted that several other cartilaginous structures have been considered as sources of chondrocytes for tissue engineering purposes, including auricular cartilage [Mandl et al., 2002] and nasal septum [Homicz et al., 2003; Kafienah et al., 2002; Rotter et al., 2002]. These chondrocyte sources offer possibilities for use in reconstructive procedures [Chang et al., 2001].

In light of issues related to potential disease transmission and immunological incompatibilities, the current clinical approaches to human articular cartilage tissue engineering involve collection and ex vivo expansion of autogenous chondrocytes. In this approach, approximately 300–450 mg of healthy cartilage is arthroscopically removed from a non-weight-bearing area of the joint, usually from the abaxial surface of the femoral trochlear ridges [Brittberg et al., 1994]. For research purposes, human articular cartilage can also be obtained from tissues discarded during total joint arthroplasty (TJA) and limb amputation procedures and from organ donor sources [Reginato et al., 1994; Stokes et al., 2002]. TJA specimens can generally be obtained within hours of removal from the patient but are overtly pathological material and are usually acquired from older individuals. In contrast, organ donor sources provide healthy specimens from a broader age range. It should be noted that allogeneic chondrocytes have been used in a number of in vivo studies in animal models of articular cartilage repair. Although the outcomes of these studies vary considerably, immunological reactivity does not appear to be a significant factor in the host response to chondrocyte transplantation. In the future, allogeneic sources of articular chondrocytes may prove to be clinically acceptable for human procedures.

Protocol 4.1 details the steps involved in isolating articular chondrocytes from a TJA tissue specimen. The protocol is equally relevant to collection from other sources, such as postmortem, amputation, and organ donor specimens.

Protocol 4.1. Isolation of Human Articular Cartilage Chondrocytes

Reagents and Materials

Sterile

- ❑ Dissection medium (See Section 2.1)
- ❑ Predigestion medium for human articular cartilage (See Section 2.2)
- ❑ Digestion medium for human articular cartilage (See Section 2.3)
- ❑ Scalpel with #10 or #15 blade
- ❑ Plastic centrifuge tube (e.g., Falcon), 50 ml, or small Erlenmeyer flask
- ❑ Steriflip filters
- ❑ Bottle filters
- ❑ Filters, 40 μm

Nonsterile

- ❑ Trypan Blue
- ❑ Ice bath

Protocol

(a) Weigh the container (such as a 50-ml Falcon tube or small Erlenmeyer flask) that will be used to hold the cartilage.

(b) Excise the cartilage from the joint surface in 2- to 3-mm-thick blocks, using a #10 or #15 scalpel blade. Areas of unacceptable pathology should be excluded from the collection.

(c) Place the explants into dissection medium and hold on ice until completion of the cartilage dissection.

(d) Wash the explants several times in dissection medium. Swirl the explants in 10–15 volumes of dissection medium, then aspirate the liquid as the explants settle to the bottom of the container. Repeat this until the supernatant remains clear after agitation.

(e) After completing the wash steps, remove as much liquid as possible and reweigh the container. The difference between the first and second weights gives a reasonable estimate of the weight of cartilage collected. In practice, between 3.0 and 6.0 g of cartilage can be collected from a single TJA specimen, depending primarily on the degree of pathology (4.86 ± 1.97 g, $n = 20$).

(f) Incubate the explants in predigestion medium for human articular cartilage (1 g/10 ml) in a shaking incubator at 37 °C for 1 h.

(g) Rinse the explants in dissection medium.

(h) Add the digestion medium for human articular cartilage (1 g/10 ml).

(i) Replace the flask in a shaking incubator at 37 °C overnight.

(j) After overnight digestion, filter the medium through a 40 μm filter.

(k) Pellet the isolated cells by centrifugation at 250 *g* for 10 min.

(l) Aspirate the supernatant and resuspend the cells in 10 volumes of Ham's F-12 medium.

(m) After 2–3 rounds of centrifugation and resuspension, determine the cell number with a hemocytometer and test for viability by Trypan Blue exclusion.

Approximately $2-5 \times 10^6$ viable cells can be isolated per gram of cartilage ($3.48 \pm 1.50 \times 10^6$ cells/g, $n = 20$), varying with donor age and the degree of pathology.

Collection of articular cartilage from human tissues should be carried out with all appropriate personal protection required for handling biohazardous material. Similarly, all plasticware, blades, and liquids used in the procedure should be handled appropriately as biohazardous waste.

For a shorter (3–6 h) digestion, the collagenase concentration can be increased up to 0.40%. The FBS can be omitted with shorter digestion times. Note that with higher collagenase concentrations, filtration can be difficult.

3.2. Articular Chondrocytes from Other Species

Several animal species are routinely used for articular cartilage and chondrocyte research: rabbits, pigs, goats, dogs, horses, and cattle. Sufficient quantities of articular cartilage can be collected from the joints of these species for in vitro experiments. It is possible to isolate up to 1×10^8 articular chondrocytes from extensive collections from equine or bovine limbs. These sources are not compromised by preexisting pathology, delays in acquisition, or potential biohazardous risks. Of particular relevance to tissue engineering research, articular cartilage defect models for in vivo analyses of engineered constructs are feasible in these species.

Protocol 4.2 details the steps involved in isolating articular chondrocytes from young adult (15–18 month old) bovine metacarpophalangeal joints. This source is commonly obtained from abattoirs, because the lower limbs are removed from bovine carcasses early in the butchering process. The protocol is equally applicable to other experimental species, such as dog, horse, goat, and rabbit.

Protocol 4.2. Isolation of Articular Chondrocytes from Other Species

Reagents and Materials

Sterile

- ❑ Dissection medium (See Section 2.1)
- ❑ Trypsin-EDTA: Trypsin, 0.25%, EDTA, 1 mM in phosphate-buffered saline lacking Ca^{2+} and Mg^{2+} (PBSA), supplemented with penicillin 200 U/ml and streptomycin 200 μg/ml
- ❑ Digestion medium for bovine articular cartilage (See Section 2.4)
- ❑ Erlenmeyer flasks, 500 ml and 50 ml
- ❑ Scalpel with #10 or #15 blade
- ❑ Filters, 40 μm
- ❑ Petri dish, 10 cm

Nonsterile

- ❑ Trypan Blue
- ❑ Ice bath

Protocol

(a) Weigh the container (500 ml Erlenmeyer flask) to be used to hold the sample.

(b) Excise the cartilage from the joint surface with a #10 or #15 scalpel blade and place the explants into dissection medium on ice. Approximately 7 g of cartilage can be excised from the articular metacarpophalangeal joint surfaces of a yearling (15–18 month old) steer.

(c) After excision of the articular cartilage, wash the explants several times in dissection medium. Repeat the washes until the supernatant remains clear after agitation.

(d) At the completion of the wash steps, remove as much liquid as possible and reweigh the container. The difference between the first and second weights gives a reasonable estimate of the weight of cartilage collected.

(e) Transfer the cartilage into a sterile 10 cm Petri dish and immerse in dissection medium.

(f) Dice the samples into approximately 2 mm-thick slices.

(g) Transfer the diced cartilage into an Erlenmeyer flask.

(h) Predigest the explants in trypsin-EDTA (10 ml/g) containing penicillin-streptomycin in a shaking incubator at 37 °C for 1 h.

(i) Rinse the explants in dissection medium, then add the digestion medium for bovine articular cartilage. Replace the flask in a shaking incubator at 37 °C overnight.

(j) After overnight digestion, filter the digestion medium through 40 μm filters.

(k) Pellet the isolated cells by centrifugation at 250 g for 10 min.

(l) Aspirate the supernatant and resuspend the cells in 10 volumes of Ham's F-12 medium.

(m) After 2–3 rounds of centrifugation and resuspension, cell number can be determined with a hemocytometer and assessed for viability by Trypan Blue exclusion. Approximately 1×10^6 chondrocytes ($9.69 \times 10^5 \pm 1.85 \times 10^5$, $n = 8$) can be isolated from each gram of articular cartilage obtained from the metacarpophalangeal joints of yearling steers.

3.3. Expansion Culture

With the current autologous chondrocyte implantation (ACI) approach, insufficient chondrocytes are obtained from primary isolation for direct implantation. Therefore, the primary chondrocyte population requires ex vivo expansion before use for tissue repair or engineering. Based on information derived from Brittberg et al. [1994] and from literature provided by Genzyme Tissue Repair, the current ACI protocol involves in vitro expansion of 180,000–455,000 primary cells over 2–5 weeks to provide sufficient cells for implantation ($2.6–5.0 \times 10^6$ chondrocytes

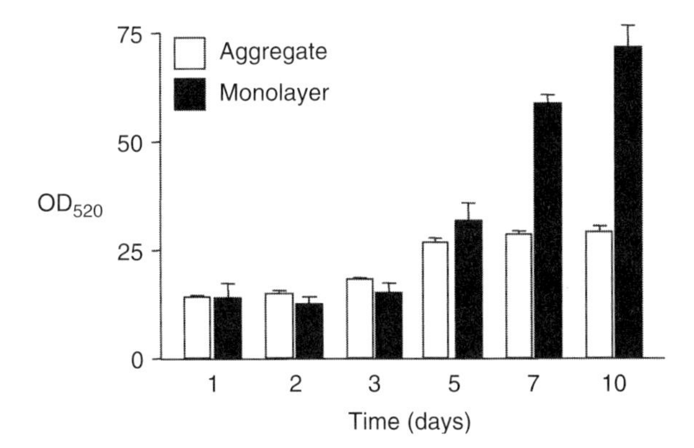

Figure 4.1. In vitro proliferation of articular chondrocytes. Bovine articular chondrocytes were cultured as aggregates (white bars) or monolayers (black bars) in expansion medium for up to 10 days. DNA content was measured by Pico Green fluorescence (Molecular Probes, Eugene, OR). Chondrocytes cultured as aggregates undergo a single population doubling in the first 5 days in culture and then cease proliferative activity. In contrast, monolayer cultures maintain proliferative activity throughout the culture period, with approximately sixfold increase in cell number by day 10 ($n = 4$).

in Brittberg et al., 1.2×10^7 chondrocytes in Genzyme's "Carticel" protocol). This requires 25–30 cell doublings.

Conditions that maintain the differentiated articular chondrocytic phenotype in vitro (See below) do not support sustained proliferation (Fig. 4.1). The standard technique for articular chondrocyte expansion involves seeding isolated chondrocytes as subconfluent monolayers in the presence of serum and/or specific mitogens. Articular chondrocyte monolayers become confluent at a density of $2.5–3.0 \times 10^5$ cells/cm^2. Therefore, for expansion, primary cells are usually seeded at an initial density of $1.0–2.5 \times 10^4$ cells/cm^2, allowing for 10- to 25-fold increases in cell number before confluence. Monolayer expansion results in loss of expression of genes that characterize the articular chondrocytic phenotype [Stewart et al., 2000]. The rate at which the differentiated phenotype is lost is dependent on seeding density.

Protocol 4.3 details establishment of primary chondrocyte monolayer cultures and subculture of chondrocyte monolayers. The particular culture vessels used (cell culture dishes, capped flasks, etc) will depend on the starting cell number, required expansion indices, biological safety requirements, and laboratory preference. The protocol assumes use of tissue culture flasks.

Protocol 4.3. Monolayer Culture of Articular Chondrocytes

Reagents and Materials

Sterile

- ❑ Ascorbic acid stock (5 mg/ml) (See Section 2.5)
- ❑ Differentiation medium (See Section 2.6)

- ❑ Growth medium (See Section 2.7)
- ❑ PBSA
- ❑ Trypsin-EDTA: trypsin, 0.25%, EDTA, 1 mM in PBSA
- ❑ Plastic centrifuge tube, 50 ml (e.g., Falcon)

Protocol

(a) Resuspend the chondrocytes in growth medium by gentle trituration. The volume of medium will depend on the starting cell number and the required cell concentration for the experiment. For example, if chondrocytes are to be seeded in 75 cm^2 flasks (20 ml/flask) at an initial density of 2.5×10^4 cells/cm^2, each flask will require 1.875×10^6 cells, suspended in 20 ml of culture medium. Periodic resuspension of the chondrocytes is recommended to keep the cell density consistent.

(b) Maintain cultures at 37 °C in a 95% air-5% CO_2 humidified incubator.

(c) Refeed the cells with fresh growth medium every 2–3 days, as required, until confluence. The time required to reach confluence varies considerably, depending on donor age. A minimum of 4 days would be expected for a fourfold increase in cell density.

(d) At confluence, aspirate the medium and rinse the cell layer with PBSA.

(e) Add approximately 1.0 ml trypsin/EDTA per 20 cm^2. Tip the flask to distribute the trypsin-EDTA across the entire surface of the cell layer and incubate at 37 °C for 5–10 min, or until the cells begin to detach.

(f) Dislodge the chondrocytes from the surface by lightly tapping the bottom of the flask against a hard surface.

(g) Add 10 volumes of growth medium and complete the cell dislodgement and resuspension by vigorous pipetting.

(h) Transfer the suspension to a Falcon tube. Rinse the flask with 3 ml medium and transfer the rinse to the Falcon tube.

(i) Pellet the cells by centrifugation at 250 *g* for 5 min.

(j) Resuspend the chondrocytes in fresh culture medium and count the cells with a hemocytometer.

(k) In some instances, chondrocyte monolayers detach as a single sheet and do not dissociate into single cells. Secondary digestion of the chondrocyte layer in 0.05% collagenase A for 1–2 h, similar to the protocol provided for primary chondrocyte isolation (Protocol 4.1), provides single-cell suspensions.

(l) Reseed the passaged chondrocytes into flasks at the appropriate density.

In the expansion protocol detailed above, serum is used as the source of mitogenic stimuli for chondrocyte proliferation. A considerable amount of research has been carried out to identify specific growth and differentiation factors [Bradham and Horton, 1998; Pei et al., 2002] and culture conditions [Domm et al., 2002; Kuriwaka et al., 2003; Murphy and Sambanis, 2001] that can be used with, or in place of, serum to stimulate chondrocyte proliferation while mitigating effects on the differentiated phenotype and/or facilitating restoration of the differentiated phenotype after expansion. In particular, fibroblast growth factor seems to have a specific phenotype-sparing activity during chondrocyte expansion [Jakob et al., 2001; Mandl et al., 2002; Martin et al., 1999]. It should also be noted that recent studies have demonstrated that the capacity of monolayer-expanded chondrocytes to regain the differentiated chondrocytic phenotype is lost as the number of passages increases [Dell'Accio et al., 2001; Schulze-Tanzil et al., 2002].

3.4. Differentiation/Redifferentiation Culture Models

Three-dimensional (3D) culture conditions are required for the maintenance or restoration of a differentiated articular chondrocytic phenotype. Low-serum or serum-free conditions are also preferable, but the effects of serum supplementation in 3D culture are considerably less than in monolayer culture [Stewart et al., 2000]. Reexpression of the differentiated phenotype generally requires 3D culture for 2–3 weeks. Several approaches have been developed for the 3D maintenance of chondrocytes (Fig. 4.2). Micromass [Bradham and Horton, 1998; Schulze-Tanzil et al., 2002], aggregate [Reginato et al., 1994; Stewart et al., 2000], and pellet [Ballock and Reddi, 1994] models are simple procedures that do not require addition of a substrate to the cells. The micromass model involves seeding small numbers of chondrocytes (approximately 2.0×10^5) in small volumes at superconfluent densities, inducing multicellular nodule formation. In practice, this technique generates a mixed population of 3D and monolayer cells and is not ideal for redifferentiation applications. The aggregate model involves culturing chondrocytes in nonadherent wells or dishes. These can be purchased (Corning-Costar ultra-low-attachment plates) or made by layering the bottom of regular dishes with methacrylate [Reginato et al., 1994; Stokes et al., 2002] or agarose [Archer et al., 1990; Kolettas et al., 1995]. The aggregate model is ideal for maintaining the differentiated phenotype of primary chondrocytes. It is less effective for restoring the differentiated phenotype of monolayer-expanded chondrocytes because the characteristic aggregation that occurs with primary cells is attenuated after monolayer expansion. For this application, the pellet model (Protocol 4.4) is more effective.

Alginate [Bonaventure et al., 1994; Gagne et al., 2000; Häuselmann et al., 1994; Homicz et al., 2003; Mandl et al., 2002; Masuda et al., 2003; Murphy and Sambanis, 2001], agarose [Benya and Shaffer, 1982; Lee et al., 1998; Weisser et al., 2001], fibrin [Fortier et al., 2002; Hendrickson et al., 1994], and collagen [Kuriwaka et al., 2003; Lee et al., 2003] models involve suspension of the chondrocytes within a 3D substrate or gel. The substrate-based models are particularly suitable for tissue engineering applications because phenotypic redifferentiation and generation of an engineered construct for implantation can occur simultaneously [Chang et al., 2001; Marijnissen et al., 2002; Masuda et al., 2003; Miralles et al., 2001; Weisser et al., 2001]. The alginate model is particularly versatile in that the chondrocytes can be reisolated from the construct by calcium chelation [Masuda et al., 2003] (Protocol 4.5).

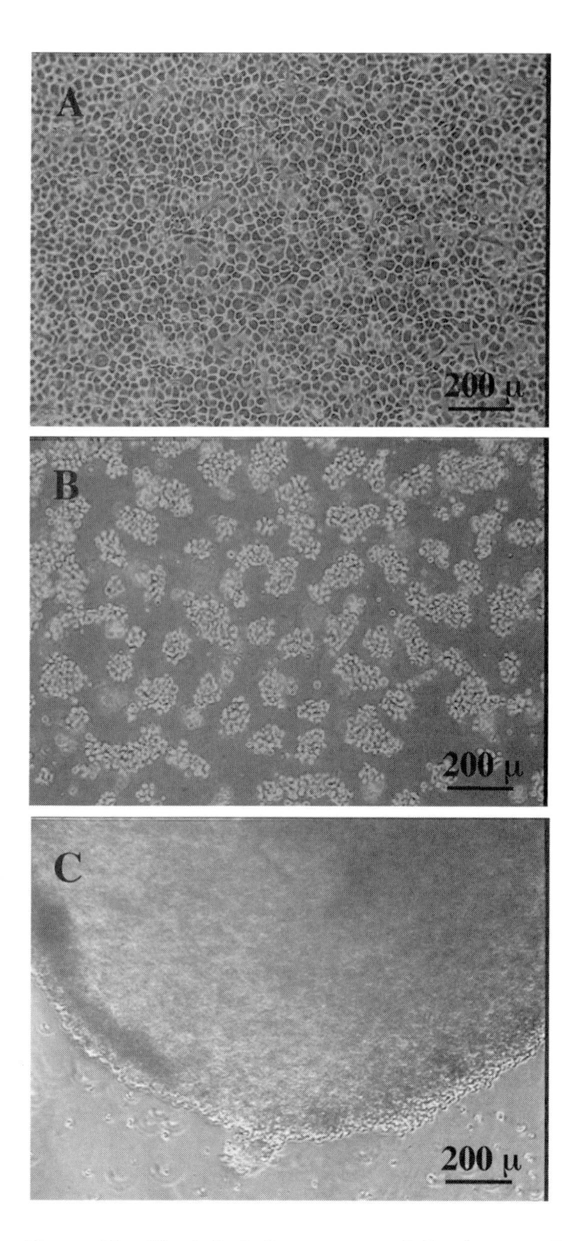

Figure 4.2. Morphological appearance of chondrocyte cultures. Images were obtained 48 hours after bovine articular chondrocyte cultures were established as high-density monolayers (A), aggregates (B), or pellets (C).

Protocol 4.4. Pellet Culture for Chondrocytes

Reagents and Materials

Sterile

- ❑ 15 ml conical polypropylene centrifuge tubes

❑ Ultra-low-attachment plates
❑ Differentiation medium (See Section 2.6)

Protocol

(a) Resuspend the chondrocytes in differentiation medium at 160,000–250,000 cells/ml.
(b) Dispense 1-ml aliquots of the resuspended cells into 15-ml conical polypropylene centrifuge tubes.
(c) Centrifuge the tubes at 250 *g* for 5 min at 4 °C.
(d) Loosen the tube caps to allow gas exchange and maintain at 37 °C in a 95% air-5% CO_2 humidified incubator.
(e) Refeed the pellets every 2–3 days by gently aspirating 80% of the medium. Particular care should be taken to avoid disturbing the pellet during refeeding.
(f) After 3–5 days, the cell pellets will have consolidated sufficiently to permit direct handling.

Pellet cultures can also be established by adding 2.5×10^5 cells in 500 μl of medium to 1.5 ml microfuge tubes and then centrifuging the tubes in a Beckman 12 horizontal centrifuge at 250 *g* for 5 min. After centrifugation, the microfuge tube caps are pierced with an 18 g needle to allow gas exchange and the tubes are maintained in racks at 37 °C in a 95% air-5% CO_2 humidified incubator. The pellets can be maintained in the microfuge tubes for the duration of the experiment, or transferred to nonadherent plates after 3 days.

Protocol 4.5. Alginate Culture of Chondrocytes

Reagents and Materials

Sterile

❑ Alginate solution (See Section 2.8)
❑ Syringe, 5 or 10 ml
❑ Hypodermic needle, 22 g
❑ $CaCl_2$ solution, 102 mM (See Section 2.9)
❑ Growth medium (See Section 2.7) or differentiation medium (See Section 2.6)
❑ Saline, 0.9% (See Section 2.10)
❑ Alginate depolymerization solution (See Section 2.11)

Protocol

(a) Resuspend the cells in alginate solution. Cell concentrations ranging from 1×10^3 to 1×10^7 cells/ml have been reported in the scientific literature [Gagne et al., 2000], with 4×10^6 cells/ml being standard [Binette et al., 1998; Häuselmann et al., 1994; Masuda et al., 2003].
(b) Aspirate the suspension into a syringe.

(c) Dispense the cell suspension as drops through a 22 g hypodermic needle into 102 mM $CaCl_2$ solution.
(d) Allow the alginate beads to polymerize for 10 min.
(e) Rinse the beads twice with 0.9% sterile saline and then with culture.
(f) Immerse the beads in culture medium and maintain at 37 °C in a 95% air-5% CO_2 humidified incubator.
(g) Change culture medium every 2–3 days for the duration of the experiment.
(h) The chondrocytes and associated pericellular matrix can be released by dissolving the alginate in alginate depolymerization solution for 20 min.
(i) Transfer the solution to a centrifuge tube and spin at 200 g for 10 min at 4 °C to pellet the cells with associated pericellular matrices.

The assays routinely used to assess in vitro phenotypic modulation of chondrocytes and chondrocyte progenitors primarily assess cell morphology and the expression of genes, proteins, and proteoglycans representative of the cartilaginous matrix. The use of these assays for tissue engineering purposes assumes that the qualitative (re)appearance of specific marker expression indicates comprehensive restoration of the differentiated chondrocytic phenotype and that in vitro behavior reflects in vivo reparative or regenerative capacity. Recent studies suggest these assumptions may not be correct. First, "redifferentiation" does not restore expression profiles of primary chondrocytes [Benz et al., 2002; Domm et al., 2002]. Further, in vitro phenotypic modulation is not directly indicative of in vivo behavior [Dell'Accio et al., 2001]. Clearly, caution must be exercised in extrapolating selective in vitro data to general assumptions about "phenotype" and anticipated in vivo performance.

4. CELLS WITH CHONDROGENIC DIFFERENTIATION POTENTIAL

Many postnatal tissues contain cells that have chondrogenic differentiation potential, but they are not necessarily easily accessible for harvesting chondroprogenitors if autograft tissue engineering is the goal. Tissues such as bone marrow, fat, skin, and possibly synovium may be of practical use for the isolation of chondroprogenitors for cartilage tissue engineering. Cells isolated from each of these tissues have been shown to possess chondrogenic differentiation potential [Erickson et al., 2002; Johnstone et al., 1998; Nishimura et al., 1999; Yates et al., 2001]. The majority of the progenitor cell isolation procedures for tissue engineering used to date have been relatively straightforward because they have not involved steps to specifically purify the chondroprogenitor cells from other cell types. One reason for this has been the lack of specific markers for these cells. As with any new scientific field, the procedures are becoming more sophisticated, with enhanced purification schemes for progenitor cells based on various parameters becoming available. Use of cell sorting, differential plating and/or altered medium conditions have produced enrichment of cells with differing differentiation potentials [Gronthos et al., 2003; Jiang et al., 2002; Jones et al., 2002; Reyes et al., 2001; Sekiya et al., 2002; Vacanti et al., 2001]. Although these procedures are now available, there are not yet any published studies demonstrating the benefit of the purification schemes to tissue-engineered implant production.

Most of the purification schemes have been worked out with bone marrow as the source of cells. The ease of harvest and relative ease of expansion of marrow-derived cells make them good candidates for tissue engineering of musculoskeletal tissues. Bone marrow can be accessed in most patients as an autologous cell source. The isolation of mesenchymal progenitor cells from bone marrow is described extensively in Chapter 2. It is known that the number of marrow-derived progenitor cells has been shown to decrease with age [Bergman et al., 1996; D'Ippolito et al., 1999; Nishida et al., 1999]. Although this has been reported to be a potential limitation of marrow as a source of these cells, it may also be true of all other tissues, too; it has been also noted in periosteum but not determined for other sources to date [O'Driscoll et al., 2001].

Periosteum contains osteochondral progenitor cells that can differentiate into osteoblasts and chondrocytes during normal bone growth and fracture healing. Detailed experimental knowledge of periosteal chondrogenesis has been gained from in vitro culture and chondrogenic induction of periosteal explants, in a system developed by O'Driscoll and others [Fukumoto et al., 2002; Miura et al., 1994, 2002; Mizuta et al., 2002; Mukherjee et al., 2001; O'Driscoll et al., 1994, 1997, 1998, 2001; Sanyal et al., 2002; Saris et al., 1999]. This system of culture has been used as an assay for the chondrogenic capacity of other tissues [Nishimura et al., 1999]. Isolated periosteal cells can be induced to undergo chondrogenesis [Izumi et al., 1992; Nakahara et al., 1991], and both periosteum and isolated periosteal cells with chondrogenic capacity have been used for articular cartilage and meniscus repair experiments [Kobayashi et al., 2002; Mason et al., 1999; O'Driscoll, 1999; O'Driscoll and Fitzsimmons, 2001; Perka et al., 2000; Rubak et al., 1982; Walsh et al., 1999]. Periosteal tissue is also used as a cover for articular cartilage defects into which chondrocytes are placed [Brittberg et al., 1994, 1996; Minas and Nehrer, 1997]. Periosteal cells are not an obvious choice as a source for cells for in vitro tissue engineering strategies unless exposure of bone is done in preliminary surgery.

Although termed synoviocytes, the population of cells isolated and expanded from synovial tissue contains several cell types. The cells of the synovial intima are thought to be a mixture of bone marrow-derived macrophages (type A cells) and specialized fibroblasts (type B cells) [Edwards, 1994]. In the subintimal layers there is connective tissue with blood vessels, nerves, and lymphatics. Although some researchers use approaches that attempt to isolate only the intima, it is difficult to do this precisely. Within the mixed population of cells termed synoviocytes are chondroprogenitor cells. Synovium-derived cells undergo chondrogenic differentiation when cultured on bone morphogenetic protein-coated plates [Iwata et al., 1993] and in in vitro chondrogenesis assays of the type described in Protocol 4.1 [Nishimura et al., 1999]. Recently, these cells have been shown to also have osteogenic, adipogenic, and myogenic potential [De Bari et al., 2001, 2003]. The presence of progenitor cells in synovium fits with the clinically noted formation of multiple cartilaginous nodules in the synovium (synovial chondromatosis), a benign reactive metaplasia of synovial cells with unknown etiology [Crone and Watt, 1988]. Furthermore, rheumatoid pannus has been shown to contain chondrocyte-like cells, which might also arise from synovial progenitor cells [Allard et al., 1988; Xue et al., 1997]. Biopsy procedures may allow the use of autograft synovium for isolation and expansion of synovial progenitor cells for use in tissue engineering.

Both fat [Erickson et al., 2002; Tholpady et al., 2003; Winter et al., 2003; Young et al., 1993; Zuk et al., 2001, 2002] and dermis [Mizuno and Glowacki, 1996a,b; Yates et al., 2001; Young et al., 2001] have been demonstrated to contain chondroprogenitor cells. Both of these tissues could be practically harvested for isolation of autologous progenitor cells for tissue engineering. The chondrogenic extracellular matrix formed by cells from these sources has not been shown to be extensive. It is noted that dilution of bone marrow-derived mesenchymal progenitor cells with dermal fibroblasts decreases chondrogenesis [Lennon et al., 2000], so it may that the relatively simple isolation procedures used to date may include too many contaminating nonchondrogenic cells. Characterization of the progenitor cells present in these tissues [Gronthos et al., 2001] may allow better purification of the subpopulation with chondrogenic differentiation potential, which may then provide for extensive cartilage production.

5. CHONDROGENESIS OF PROGENITOR CELLS IN VITRO

Until the late 1990s, the chondrogenic assays for mammalian progenitor cells isolated from postnatal tissues consisted of various methods for implantation of the cells in vivo, generally in scaffolds or diffusion chambers [Brown et al., 1985; Cassiede et al., 1996; Goshima et al., 1991; Harada et al., 1988; Jaroma and Ritsila, 1988; Nakahara et al., 1990]. The implants were then harvested and submitted for histologic evaluation of the tissues formed. This implantation method has been used for "proof of concept" studies of chondrocytes in various scaffold types [Cao et al., 1997; Isogai et al., 1999; Paige et al., 1996; Sims et al., 1996]. However, although it can be used for assessing the differentiation potential of cells, it is limited for tissue engineering with chondroprogenitor cells. Thus the development of in vitro systems for inducing chondrogenesis was important.

A method based on the pellet culture system that was first used for the culture of differentiated chondrocytes was developed [Johnstone et al., 1998; Yoo et al., 1998]. Pellet culture was first described by Holtzer et al. [1960] for culturing chondrocytes in conditions that maintained their differentiated phenotype. It was later adapted for use with growth plate chondrocytes and allowed the study of the differentiation of proliferative chondrocytes into the hypertrophic state [Ballock et al., 1993; Chen et al., 1995; Inoue et al., 1990; Iwamoto et al., 1991; Kato et al., 1988].

The appeal of the system for use with progenitor cells was its use of two parameters long known to be important for chondrogenesis: high cell density and lack of cell-substratum interactions, such as would occur in monolayer conditions [Solursh, 1991]. Ballock et al. published a method for pellet culture of growth plate chondrocytes with a serum-free defined medium [Ballock et al., 1993; Ballock and Reddi, 1994]. This formed the basal medium in which successful chondrogenesis of progenitor cells was first achieved [Johnstone et al., 1998]. The addition to the medium of dexamethasone [Johnstone et al., 1998] and TGF-β1 [Johnstone et al., 1998; Yoo et al., 1998] facilitated the differentiation of chondroprogenitor cells within the adherent cell population isolated from bone marrow (Fig. 4.3).

The sequence of chondrogenesis induced in these conditions begins with the aggregation of the cells, forming a free-floating ball, generally within 12 hours of initiation (Fig. 4.4). There is a condensation of this cell mass within the first 2–3 days and then a gradual expansion in size as the cells undergo chondrogenic differentiation and produce extracellular matrix (Fig. 4.5, See Color Plate 3). At day 1, the aggregated cells vary greatly in morphology. Some of the cells are arranged in a syncytium, whereas others will be individually delineated. At the periphery of the aggregates a few of the cells are already elongated (fusiform). By day 5, there is an increase in fusiform cells throughout the aggregate, but especially at the periphery, where they form a multilayered zone. Matrix production increases especially in the middle of the aggregate. By day 7, the center of the aggregate consists of more rounded cells. By day 14, the cells of the center of the pellet stand out as individual entities surrounded by extracellular matrix, which exhibits metachromasia with Toluidine Blue staining.

Peripheral layers of flattened cells are noted in other types of culture that allow chondrogenic differentiation, such as the micromass cultures of chick and mouse cells used for studies of the mechanism of chondrogenesis [Ahrens et al., 1977; Cottrill et al., 1987; Gluhak et al., 1996; Merker et al., 1984]. Although these layers resemble a perichondrium in appearance, there is no evidence for any activity that would be associated with a perichondrial layer. It is noted that the formation of the

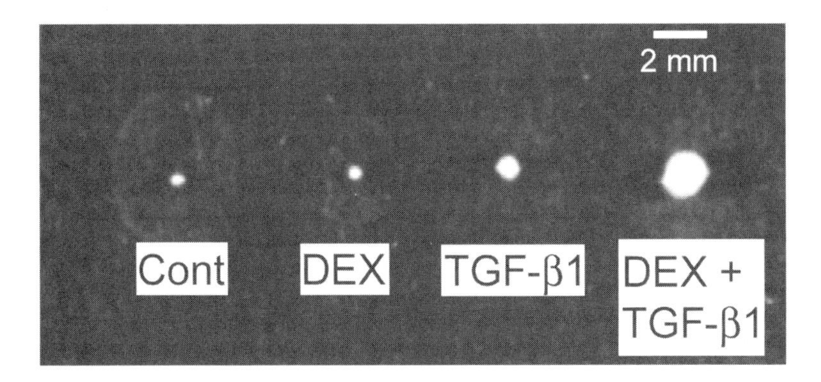

Figure 4.3. Aggregates of bone marrow-derived cells. Culture after 14 days in defined medium (Cont) without dexamethasone or with additions as labeled (Dex: dexamethasone).

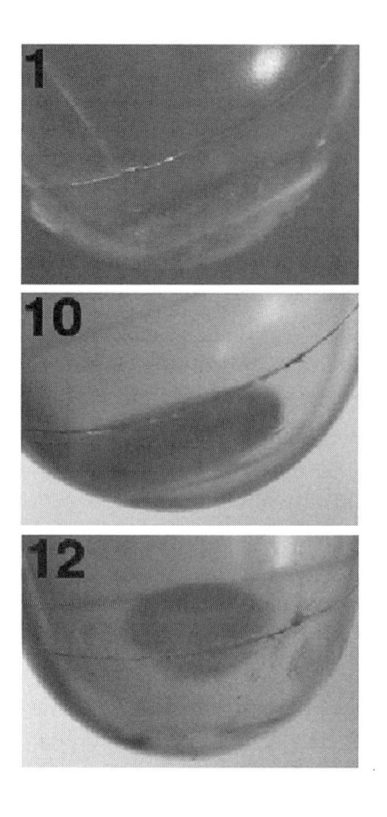

Figure 4.4. Sequence of aggregation. The sequence of aggregation from 1 hour to 12 hours after centrifugation of human marrow-derived cells in chondrogenic medium.

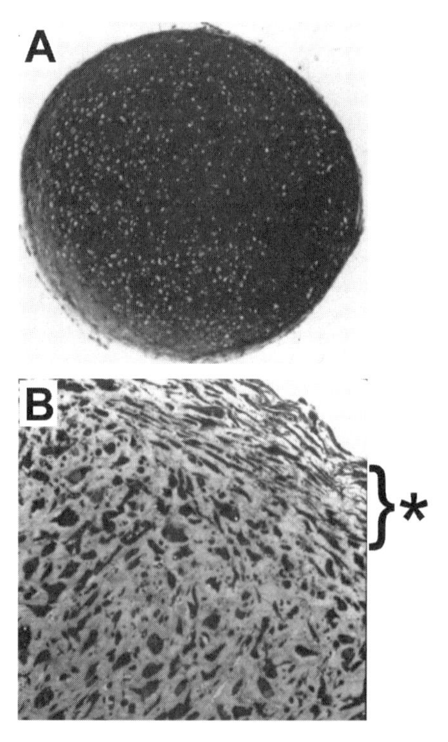

Figure 4.5. The Toluidine Blue metachromatic matrix of cartilaginous aggregates after 14 days in chondrogenic medium. A) Section of paraffin-embedded whole aggregate of human marrow-derived cells. B) Higher-magnification image of edge of a methyl methacrylate embedded section with the region of flattened cells indicated (*). The preservation in the methyl methacrylate embedding process allows a better appreciation of the different regions and the extent of cartilage extracellular matrix production. (See Color Plate 3.)

flattened peripheral layers is also seen when clonal chondroprogenitor cell lines are cultured in the same system [Dennis et al., 1999], arguing against the idea that cell sorting is the basis for the development of these layers of cells that do not undergo differentiation.

Protocol 4.6. In Vitro Chondrogenesis

Reagents and Materials

Sterile

- ❑ Trypsin-EDTA: 0.25% trypsin, 1 mM EDTA in PBSA
- ❑ L-Ascorbate-2-phosphate stock (See Section 2.12)
- ❑ Dexamethasone (10^{-3} M) stock (See Section 2.13)
- ❑ Transforming growth factor β-1 (TGF-β1) stock (See Section 2.14)
- ❑ Basal medium (See Section 2.15)
- ❑ 15-ml conical polypropylene centrifuge tubes

Protocol

(a) Trypsinize cells from monolayer and count.

(b) Calculate the number of pellets to be made (typically 2.0×10^5 cells per pellet, 4.0×10^5 cells/ml) and make up the appropriate amount of differentiation medium by adding L-ascorbate-2-phosphate stock, dexamethasone stock, and TGF-β1 stock in a 1:100 ratio to basal medium.

(c) Aspirate trypsin solution and resuspend cells at 4.0×10^5 cells/ml.

(d) Dispense 0.5-ml aliquots of the resuspended cells into 15 ml conical polypropylene centrifuge tubes.

(e) Centrifuge the tubes at 500 *g* for 5 min at 4 °C.

(f) Loosen the tube caps to allow gas exchange and maintain at 37 °C in a 95% air-5% CO_2 humidified incubator.

(g) Refeed the pellets every 2–3 days by gently aspirating 80% of the medium. Particular care should be taken to avoid disturbing the pellet during feeding.

(h) After 2 days, the cell pellets will have consolidated sufficiently to permit direct handling.

Since the introduction of this in vitro assay, several variations have been developed. Although done for improvement of the in vitro assay as performed by a given laboratory, these alterations may be exploited for tissue engineering if they result in greater extracellular matrix production or a more appropriate matrix constitution for the desired tissue. Barry et al. [2001] noted that use of TGF-β3 provided them with greater matrix elaboration. Sekiya et al. [2001] produced extensive chondrogenesis when BMP-6 was also added. Although dexamethasone is needed for initiation of chondrogenesis in this system, withdrawal of dexamethasone from the culture after 7 days promotes greater extracellular matrix production [Roh et al., 2001].

Regardless of the exact composition of the inductive medium, the cartilage produced in this manner is probably best described as a fibrocartilage, because there is type I collagen produced at all times during the culture. Clearly, the center of a pellet is more hyaline and the outer layers are more fibrous (Fig. 4.5, See Color Plate 3), but this method does not produce a cartilage that is identical to any of the cartilages found in the body, which have individual features based on their extracellular composition and organization. One other feature of the system that should also be noted is the progression of the cells through chondrogenesis to become hypertrophic chondrocytes, expressing type X collagen, that will die by apoptosis [Johnstone et al., 2002]. Although a good feature of the system for exploring differentiation, the production of hypertrophic cartilage is not necessarily desired, depending on the type of cartilage that is the repair goal. This aspect is discussed further in Section 6.

6. TISSUE ENGINEERING CARTILAGE IMPLANTS FROM CELLS WITH CHONDROGENIC POTENTIAL

Although cartilaginous aggregates of up to 5.0×10^5 cells can be produced with this in vitro assay system, these are very small (Fig. 4.3) and not of obvious use for tissue engineering cartilage as is. Modification of the in vitro assay system of progenitor cells to produce tissue-engineered cartilage for implantation has been the subject of many recent studies [Angele et al., 1999, 2000; Martin et al., 2001; Noth et al., 2002], and the subject is covered in depth in Chapters 6 and 8. All the considerations associated with tissue engineering of any tissue are relevant: use of scaffolds, bioreactors, mechanical conditioning of the implant, etc. It is not clear what stage of cell differentiation, matrix production, and matrix organization will be most appropriate in implants for each type of cartilage. Furthermore, we do not yet know exactly how to achieve some of the desired types of cartilage. For example, one of the challenges for production of articular cartilage implants with progenitor cells is the facilitation of chondrogenic differentiation without hypertrophy. Although many of the factors that are involved in controlling chondrocyte differentiation are known, the appropriate stimuli that produce articular chondrocytes are still being defined. For fibrocartilage implants, such as those for meniscus, controlling the production of large amounts of collagen fibers in appropriate ratio to proteoglycans and other matrix molecules is still a challenge. Cartilage tissue engineering is a relatively new field, and it needs to become more sophisticated. It is expected that substantial progress will be made in this area in the next few years.

SOURCES OF MATERIALS

Material	*Catalog #*	*Supplier*
Alginate (Keltone LV)	N/A	ISP
L-Ascorbic acid-2-phosphate	AC358610250	Fisher Scientific
Bovine serum albumin	A 9647	Sigma
$CaCl_2$	C69-500	Fisher Scientific
Cell culture flasks, 75 cm^2	430641	Corning
Cell culture flasks, 25 cm^2	430639	Corning
Petri dishes, 10 cm	3160-100	Corning
Centrifuge tubes, 15 ml conical polypropylene	352095	Becton Dickinson
Centrifuge tubes, 50 ml conical polypropylene	352079	Becton Dickinson
Collagenase A	1088785	Roche Diagnostics
Collagenase type II	LS004176	Worthington Biochemical Corp
Cryotubes	3471	CLP
DMEM, 4.5 g/l glucose, containing pyruvate and glutamate	10569-010	GIBCO Invitrogen
DMEM, 1.0 g/l glucose, containing pyruvate and glutamate	11885-084	GIBCO Invitrogen
Dexamethasone	D-4902	Sigma
Fetal bovine serum	16000-044	GIBCO Invitrogen
Filter, 500 ml, 0.22 μm	SCGP UO5 RE	Millipore
Filter, 250 ml, 0.22 μm	SCGP UO2 RE	Millipore
Filter (Steriflip) 50 ml, 0.22 μm	SCGP 005 25	Millipore
Filters, 40-μm tube	352340	Becton Dickinson
Fungizone™ antimycotic liquid	15290-018	GIBCO Invitrogen
Glutamine	25-05-C1	Cellgro
HCl	A1445	Fisher
Ham's F-12 nutrient medium	31765-035	GIBCO Invitrogen
Hanks' balanced salt solution	24020-117	GIBCO Invitrogen
ITS-G Supplement (100×)	41400-045	GIBCO Invitrogen
ITS + Premix	40352	B-D Biosciences
Insulin-transferrin-sodium selenite supplement	1074547	Roche Diagnostics
Opti-MEM 1 reduced-serum medium	31985-070	GIBCO Invitrogen
Penicillin-streptomycin liquid	15140-122	GIBCO Invitrogen
PBS, Mg^{2+}, Ca^{2+}-free	14190-144	GIBCO Invitrogen
Pronase	0165921	Roche Diagnostics
NaCl	S671-500	Fisher Scientific
Sodium pyruvate	11360-070	GIBCO Invitrogen
$C_6H_5O_7Na_32H_2O$	S279-500	Fisher Scientific
TGF-β1	240-B	R&D Systems
Trypan Blue	15250-061	GIBCO Invitrogen
Trypsin-EDTA: 0.25% trypsin, 1 mM EDTA	25200-056	GIBCO Invitrogen
Ultra-low-attachment plates, 35 mm	3471	Costar

REFERENCES

Ahrens, P.B., Solursh, M., Reiters, R. (1977) Stage-related capacity for limb chondrogenesis in cell culture. *Dev. Biol.* 60: 69–82.

Allard, S.A., Maini, R.N., Muirden, K.D. (1988) Cells and matrix expressing cartilage components in fibroblastic tissue in rheumatoid pannus. *Scand. J. Rheum.* 76: 125–129.
Angele, P., Johnstone, B., Kujat, R., Nerlich, M., Goldberg, V., Yoo, J. (2000) Meniscus repair with mesenchymal progenitor cells in a biodegradable composite matrix. *Trans Orthop. Res. Soc.* 25: 605.
Angele, P., Kujat, R., Nerlich, M., Yoo, J., Goldberg, V., Johnstone, B. (1999) Engineering of osteochondral tissue with bone marrow mesenchymal progenitor cells in a derivatized hyaluronan-gelatin composite sponge. *Tissue Eng.* 5: 545–553.
Archer, C.W., McDowell, J., Bayliss, M.T., Stephens, M.D., Bentley, G. (1990) Phenotypic modulation in sub-populations of human articular chondrocytes in vitro. *J. Cell Sci.* 97: 361–371.
Ballock, R.T., Heydemann, A., Wakefield, L.M., Flanders, K.C., Roberts, A.B., Sporn, M.B. (1993) TGF-beta 1 prevents hypertrophy of epiphyseal chondrocytes: regulation of gene expression for cartilage matrix proteins and metalloproteases. *Dev. Biol.* 158: 414–429.
Ballock, R.T., and Reddi, A.H. (1994) Thyroxine is the serum factor that regulates morphogenesis of columnar cartilage from isolated chondrocytes in chemically defined medium. *J. Cell Biol.* 126: 1311–1318.
Barry, F., Boynton, R.E., Liu, B., Murphy, J.M. (2001) Chondrogenic differentiation of mesenchymal stem cells from bone marrow: differentiation-dependent gene expression of matrix components. *Exp. Cell Res.* 268: 189–200.
Benya, P.D., and Shaffer, J.D. (1982) Dedifferentiated chondrocytes reexpress the differentiated collagen phenotype when cultured in agarose gels. *Cell* 30: 215–224.
Benz, K., Breit, S., Lukoschek, M., Mau, H., Richter, W. (2002) Molecular analysis of expansion, differentiation, and growth factor treatment of human chondrocytes identifies differentiation markers and growth-related genes. *Biochem Biophys Res Commun.* 293(1): 284–292.
Bergman, R.J., Gazit, D., Kahn, A.J., Gruber, H., McDougall, S., Hahn, T.J. (1996) Age-related changes in osteogenic stem cells in mice. *J. Bone Miner, Res,* 11: 568–577.
Binette, F., McQuaid, D.P., Haudenschild, D.R., Yaeger, P.C., McPherson, J.M., Tubo, R. (1998) Expression of a stable articular cartilage phenotype without evidence of hypertrophy by adult human articular chondrocytes in vivo. *J. Orthop. Res.* 16: 207–216.
Bonaventure, J., Kadhom, N., Cohen-Solal, L., Hg, K.H., Bourguignon, J., Lasselin, C. (1994) Reexpression of cartilage-specific genes by dedifferentiated human articular chondrocytes cultured in alginate beads. *Exp. Cell Res.* 212: 97–104.
Bradham, D.M., and Horton, W.E. (1998) In vivo cartilage formation from growth factor modulated articular chondrocytes. *Clin. Orthop.* 352: 239–249.
Brittberg, M., Lindahl, A., Nilsson, A., Ohlsson, C., Isaksson, O., Peterson, L. (1994) Treatment of deep cartilage defects in the knee with autologous chondrocyte transplantation. *N. Engl. J. Med.* 331: 889–895.
Brittberg, M., Nilsson, A., Lindahl, A., Ohlsson, C., Peterson, L. (1996) Rabbit articular cartilage defects treated with autologous cultured chondrocytes. *Clin. Orthop.* 326: 270–283.
Brown, D.G., Johnson, N.F., Wagner, M.M. (1985) Multipotential behaviour of cloned rat mesothelioma cells with epithelial phenotype. *Br. J. Cancer* 51: 245–252.
Cao, Y., Vacanti, J.P., Paige, K.T., Upton, J., Vacanti, C.A. (1997) Transplantation of chondrocytes utilizing a polymer-cell construct to produce tissue-engineered cartilage in the shape of a human ear. *Plast. Reconstr. Surg.* 100: 297–302; discussion 303–304.
Chang, S.C., Rowley, J.A., Tobias, G., Genes, N.G., Roy, A.K., Mooney, D.J., Vacanti, C.A., Bonassar, L.J. (2001) Injection molding of chondrocyte/alginate constructs in the shape of facial implants. *J. Biomed. Mater. Res.* 55: 503–511.
Cassiede, P., Dennis, J.E., Ma, F., Caplan, A.I. (1996) Osteochondrogenic potential of marrow mesenchymal progenitor cells exposed to TGF-beta 1 or PDGF-BB as assayed in vivo and in vitro. *J. Bone Miner. Res.* 11: 1264–1273.
Chen, Q., Johnson, D.M., Haudenschild, D.R., Goetinck, P.F. (1995) Progression and recapitulation of the chondrocyte differentiation program: cartilage matrix protein is a marker for cartilage maturation. *Dev. Biol.* 172: 293–306.
Cottrill, C.P., Archer, C.W., Wolpert, L. (1987) Cell sorting and chondrogenic aggregate formation in micromass culture. *Dev. Biol.* 122: 503–515.
Crone, M.H., and Watt, I. (1988) Synovial chondromatosis. *J. Bone Joint Surg.* 70-B: 807–811.

De Bari, C., Dell'Accio, F., Tylzanowski, P., Luyten, F.P. (2001) Multipotent mesenchymal stem cells from adult human synovial membrane. *Arth. Rheum.* 44: 1928–1942.
De Bari, C., Dell'Accio, F., Vandenabeele, F., Vermeesch, J.R., Raymackers, J.M., Luyten, F.P. (2003) Skeletal muscle repair by adult human mesenchymal stem cells from synovial membrane. *J. Cell Biol.* 160: 909–918.
Dell'Accio, F., De Bari, C., Luyten, F.P. (2001) Molecular markers predictive of the capacity of expanded human articular chondrocytes to form stable cartilage in vivo. *Arthritis Rheum.* 44: 1608–1619.
Dennis, J.E., Merriam, A., Awadallah, A., Yoo, J.U., Johnstone, B., Caplan, A.I. (1999) A quadripotential mesenchymal progenitor cell isolated from the marrow of an adult mouse. *J. Bone Miner. Res.* 14: 700–709.
D'Ippolito, G., Schiller, P.C., Ricordi, C., Roos, B.A., Howard, G.A. (1999) Age-related osteogenic potential of mesenchymal stromal stem cells from human vertebral bone marrow. *J. Bone Miner. Res.* 14: 1115–1122.
Domm, C., Schunke, M., Christesen, K., Kurz, B. (2002) Redifferentiation of dedifferentiated bovine articular chondrocytes in alginate culture under low oxygen tension. *Osteoarthritis Cartilage* 10: 13–22.
Edwards, J.C.W. (1994) The nature and origin of synovium: experimental approaches to the study of synoviocyte differentiation. *J. Anat.* 184: 493–501.
Erickson, G.R., Gimble, J.M., Franklin, D.M., Rice, H.E., Awad, H., Guilak, F. (2002) Chondrogenic potential of adipose tissue-derived stromal cells in vitro and in vivo. *Biochem. Biophys. Res. Commun.* 290: 763–769.
Fortier, L.A., Mohammed, H.O., Lust, G., Nixon, A.J. (2002) Insulin-like growth factor-I enhances cell-based repair of articular cartilage *J. Bone Jt. Surg.* 84-B: 276–288.
Fukumoto, T., Sanyal, A., Fitzsimmons, J.S., O'Driscoll, S.W. (2002) Expression of beta1 integrins during periosteal chondrogenesis. *Osteoarthritis Cartilage* 10: 135–144.
Gagne, T.A., Chappell-Alfonso, K., Johnson, J.L., McPherson, M., Oldham, C.A., Tubo, R.A., Vaccaro, C., Vasios, G.W. (2000) Enhanced proliferation and differentiation of human articular chondrocytes when seeded at low cell densities in alginate in vitro. *J. Orthop. Res.* 18: 882–890.
Gluhak, J., Mais, A., Mina, M. (1996) Tenascin-C is associated with early stages of chondrogenesis by chick mandibular ectomesenchymal cells in vivo and in vitro. *Dev. Dyn.* 205: 24–40.
Goshima, J., Goldberg, V.M., Caplan, A.I. (1991) The osteogenic potential of culture-expanded rat marrow mesenchymal cells assayed in vivo in calcium phosphate ceramic. *Clin. Orthop.* 262: 298–311.
Gronthos, S., Franklin, D.M., Leddy, H.A., Robey, P.G., Storms, R.W., Gimble, J.M. (2001) Surface protein characterization of human adipose tissue-derived stromal cells. *J. Cell Physiol.* 189: 54–63.
Gronthos, S., Zannettino, A.C., Hay, S.J., Shi, S., Graves, S.E., Kortesidis, A., Simmons, P.J. (2003) Molecular and cellular characterisation of highly purified stromal stem cells derived from human bone marrow. *J. Cell Sci.* 116: 1827–1835.
Harada, K., Oida, S., Sasaki, S. (1988) Chondrogenesis and osteogenesis of bone marrow-derived cells by bone-inductive factor. *Bone* 9: 177–183.
Häuselmann, H.J., Fernandes, R.J., Mok, S.S., Schnid, T.M., Block, J.A., Aydelotte, M.B., Kuettner, K.E., Thonar, E.J.M.-A. (1994) Phenotypic stability of bovine articular chondrocytes after long-term culture in alginate beads. *J. Cell Sci.* 107: 17–27.
Hendrickson, D.A., Nixon, A.J., Erb, H.N., Lust, G. (1994) Phenotype and biological activity of neonatal equine chondrocytes cultured in three-dimensional fibrin matrix. *Am. J. Vet. Res.* 55: 410–414.
Holtzer, H., Abbott, J., Lash, J., Holtzer, S. (1960) The loss of phenotypic traits by differentiated cells In vitro. I. Dedifferentiation of cartilage cells. *Proc. Natl. Acad. Sci. USA* 46: 1533.
Homicz, M.R., Chia, S.H., Schumacher, B.L., Masuda, K., Thonar, E.J., Sah, R.L., Watson, D. (2003) Human septal chondrocyte redifferentiation in alginate, polyglycolic acid scaffold, and monolayer culture. *Laryngoscope* 113: 25–32.

Inoue, H., Hiasa, K., Samma, Y., Nakamura, O., Sakuda, M., Iwamoto, M., Suzuki, F., Kato, Y. (1990) Stimulation of proteoglycan and DNA syntheses in chondrocytes by centrifugation. *J. Dent. Res.* 69: 1560–1563.
Isogai, N., Landis, W., Kim, T.H., Gerstenfeld, L.C., Upton, J., Vacanti, J.P. (1999) Formation of phalanges and small joints by tissue-engineering. *J. Bone Joint Surg.* 81: 306–316.
Iwamoto, M., Shimazu, A., Nakashima, K., Suzuki, F., Kato, Y. (1991) Reduction of basic fibroblasts growth factor receptor is coupled with terminal differentiation of chondrocytes. *J. Biol. Chem.* 266: 461–467.
Iwata, H., Ono, S., Sato, K., Sato, T., Kawamura, M. (1993) Bone morphogenetic protein-induced muscle- and synovium-derived cartilage differentiation in vitro. *Clin. Orthop.* 296: 295–300.
Izumi, T., Scully, S.P., Heydemann, A., Bolander, M.E. (1992) Transforming growth factor beta 1 stimulates type II collagen expression in cultured periosteum-derived cells. *J. Bone Miner. Res.* 7: 115–121.
Jakob, M., Démarteau, O., Schäfer, D., Hintermann, B., Dick, W., Herberer, M., Martin, I. (2001) Specific growth factors during the expansion and redifferentiation of adult human articular chondrocytes enhance chondrogenesis and cartilaginous tissue formation in vitro. *J. Cell Biochem.* 81: 368–377.
Jaroma, H.J., and Ritsila, V.A. (1988) Differentiation of periosteal cells in muscle. An experimental study using the diffusion chamber method. *Scand. J. Plast. Reconstr. Surg. Hand Surg.* 22: 193–198.
Jiang, Y., Jahagirdar, B.N., Reinhardt, R.L., Schwartz, R.E., Keene, C.D., Ortiz-Gonzalez, X.R., Reyes, M., Lenvik, T., Lund, T., Blackstad, M., Du, J., Aldrich, S., Lisberg, A., Low, W.C., Largaespada, D.A., Verfaillie, C.M. (2002) Pluripotency of mesenchymal stem cells derived from adult marrow. *Nature* 418: 41–49.
Johnstone, B., Barthel, T., Yoo, J.U. (2002) In vitro chondrogenesis with mammalian progenitor cells. In Rosier, R., and Evans, C., eds., *Molecular Biology in Orthopaedics*. Park Ridge, American Academy of Orthopaedic Surgeons Symposium.
Johnstone, B., Hering, T.M., Caplan, A.I., Goldberg, V.M., Yoo, J.U. (1998) In vitro chondrogenesis of bone marrow-derived mesenchymal progenitor cells. *Exp. Cell Res.* 238: 265–272.
Jones, E.A., Kinsey, S.E., English, A., Jones, R.A., Straszynski, L., Meredith, D.M., Markham, A.F., Jack, A., Emery, P., McGonagle, D. (2002) Isolation and characterization of bone marrow multipotential mesenchymal progenitor cells. *Arthritis Rheum.* 46: 3349–3360.
Kafienah, W., Jakob, M., Demarteau, O., Frazer, A., Barker, M.D., Martin, I., Hollander, A.P. (2002) Three-dimensional tissue engineering of hyaline cartilage: comparison of adult nasal and articular chondrocytes. *Tissue Eng.* 8: 817–826.
Kato, Y., Iwamoto, M., Koike, T., Suzuki, F., Takano, Y. (1988) Terminal differentiation and calcification in rabbit chondrocyte cultures grown in centrifuge tubes: regulation by transforming growth factor beta and serum factors. *Proc. Natl. Acad. Sci. USA* 85: 9552–9556.
Kobayashi, N., Koshino, T., Uesugi, M., Yokoo, N., Xin, K.Q., Okuda, K., Mizukami, H., Ozawa, K., Saito, T. (2002) Gene marking in adeno-associated virus vector infected periosteum derived cells for cartilage repair. *J. Rheumatol.* 29: 2176–2180.
Kolettas, E., Buluwela, L., Bayliss, M.T., Muir, H.I. (1995) Expression of cartilage-specific molecules is retained on long-term culture of human articular chondrocytes. *J. Cell Sci.* 108: 1991–1999.
Kuriwaka, M., Ochi, M., Uchio, Y., Maniwa, S., Adachi, N., Mori, R., Kawasaki, K., Kataoka, H. (2003) Optimum combination of monolayer and three-dimensional cultures for cartilage-like tissue engineering. *Tissue Eng.* 9: 41–49.
Lee, C.R., Grodzinsky, A.J., Hsu, H.-P., Spector, M. (2003) Effects of a cultured autologous chondrocyte-seeded type II collagen scaffold on the healing of a chondral defect in a canine model. *J. Orthop. Res.* 21: 272–281.
Lee, D.A., Noguchi, T., Knight, M.M., O'Donnell, L., Bentley, G., Bader, D.L. (1998) Response of chondrocyte populations within unloaded and loaded agarose. *J. Orthop. Res.* 16: 726–733.
Lennon, D.P., Haynesworth, S.E., Arm, D.M., Baber, M.A., Caplan, A.I. (2000) Dilution of human mesenchymal stem cells with dermal fibroblasts and the effects on in vitro and in vivo osteochondrogenesis. *Dev. Dyn.* 219: 50–62.

Mandl, E.W., Van Der Veen, S., Verhaar, J.A.N., van Osch, G.J.V.M. (2002) Serum-free medium supplemented with high-concentration FGF2 for cell expansion culture of human ear chondrocytes promotes redifferentiation capacity. *Tissue Eng.* 8: 573–580.
Marijnissen, W.J.C.M., van Osch G,J.V.M., Aigner, J., van der Veen, S.W., Hollander, A.P., Verwoerd-Verhoef, H.L., Verhaar, J.A.N. (2002) Alginate as a chondrocyte-delivery substance in combination with a non-woven scaffold for cartilage tissue engineering. *Biomaterials* 23: 1511–1517.
Martin, I., Shastri, V.P., Padera, R.F., Yang, J., Mackay, A.J., Langer, R., Vunjak-Novakovic, G., Freed, L.E. (2001) Selective differentiation of mammalian bone marrow stromal cells cultured on three-dimensional polymer foams. *J. Biomed. Mater. Res.* 55: 229–235.
Martin, I., Vunjak-Novakovic, G., Yang, J., Langer, R., Freed, L.E. (1999) Mammalian chondrocytes expanded in the presence of fibroblast growth factor 2 maintain the ability to differentiate and regenerate three-dimensional cartilaginous tissue. *Exp. Cell. Res.* 253: 681–688.
Mason, J.M., Grande, D.A., Barcia, M., Grant, R., Pergolizzi, R.G., Breitbart, A.S. (1999) Expression of human bone morphogenic protein 7 in primary rabbit periosteal cells: potential utility in gene therapy for osteochondral repair. *Gene Ther.* 5: 1098–1104.
Masuda, K., Sah, R.L., Hejna, M.J., Thonar, E.J.M.-A. (2003) A novel two-step method for the formation of tissue-engineered cartilage by mature bovine chondrocytes: the alginate-recovered-chondrocyte (ARC) method. *J. Orthop. Res.* 21: 139–148.
Merker, H.J., Zimmermann, B., Barrach, H.J. (1984) The significance of cell contacts for the differentiation of the skeletal blastema. *Acta Biol. Hung.* 35: 195–203.
Minas, T., and Nehrer, S. (1997) Current concepts in the treatment of articular cartilage defects. *Orthopedics* 20: 525–538.
Miralles, G., Baudoin, R., Dumas, D., Baptiste, D., Hubert, P., Stoltz, J.F., Dellacherie, E., Mainard, D., Netter, P., Payan, E. (2001) Sodium alginate sponges with or without sodium hyaluronate; in vitro engineering of cartilage *J. Biomed. Mater. Res.* 57: 268–278.
Miura, Y., Fitzsimmons, J.S., Commisso, C.N., Gallay, S.H., O'Driscoll, S.W. (1994) Enhancement of periosteal chondrogenesis In vitro. Dose-response for transforming growth factor-beta 1 (TGF-beta 1). *Clin. Orthop.* 310: 271–280.
Miura, Y., Parvizi, J., Fitzsimmons, J.S., O'Driscoll, S.W. (2002) Brief exposure to high-dose transforming growth factor-beta1 enhances periosteal chondrogenesis In vitro: a preliminary report. *J. Bone Joint Surg.* 84-A: 793–799.
Mizuno, S., and Glowacki, J. (1996a) Chondroinduction of human dermal fibroblasts by demineralized bone in three-dimensional culture. *Exp. Cell Res.* 227: 89–97.
Mizuno, S., and Glowacki, J. (1996b) Three-dimensional composite of demineralized bone powder and collagen for in vitro analysis of chondroinduction of human dermal fibroblasts. *Biomaterials* 17: 1819–1825.
Mizuta, H., Sanyal, A., Fukumoto, T., Fitzsimmons, J.S., Matsui, N., Bolander, M.E., Oursler, M.J., O'Driscoll, S.W. (2002) The spatiotemporal expression of TGF-beta1 and its receptors during periosteal chondrogenesis in vitro. *J. Orthop. Res.* 20: 562–574.
Mukherjee, N., Saris, D.B., Schultz, F.M., Berglund, L.J., An, K.N., O'Driscoll, S.W. (2001) The enhancement of periosteal chondrogenesis in organ culture by dynamic fluid pressure. *J. Orthop. Res.* 19: 524–530.
Murphy, C.L., and Sambanis, A. (2001) Effect of oxygen tension and alginate encapsulation on restoration of the differentiated phenotype of passaged chondrocytes *Tissue Eng.* 7: 791–803.
Nakahara, H., Bruder, S.P., Haynesworth, S.E., Holecek, J.J., Baber, M.A., Goldberg, V.M., Caplan, A.I. (1990) Bone and cartilage formation in diffusion chambers by subcultured cells derived from the periosteum. *Bone* 11: 181–188.
Nakahara, H., Goldberg, V.M., Caplan, A.I. (1991) Culture-expanded human periosteal-derived cells exhibit osteochondral potential in vivo. *J. Orthop. Res.* 9: 465–476.
Nishida, S., Endo, N., Yamagiwa, H., Tanizawa, T., Takahashi, H.E. (1999) Number of osteoprogenitor cells in human bone marrow markedly decreases after skeletal maturation. *J. Bone Miner. Metab.* 17: 171–177.
Nishimura, K., Solchaga, L.A., Caplan, A.I., Yoo, J.U., Goldberg, V.M., Johnstone, B. (1999) Chondroprogenitor cells of synovial tissue. *Arthritis Rheum.* 42: 2631–2637.

Noth, U., Tuli, R., Osyczka, A.M., Danielson, K.G., Tuan, R.S. (2002) In vitro engineered cartilage constructs produced by press-coating biodegradable polymer with human mesenchymal stem cells. *Tissue Eng.* 8: 131–144.

O'Driscoll, S.W. (1999) Articular cartilage regeneration using periosteum. *Clin. Orthop. Relat. Res.* 367 Suppl.: S186–S203.

O'Driscoll, S.W., and Fitzsimmons, J.S. (2001) The role of periosteum in cartilage repair. *Clin. Orthop. Relat. Res.* 391 Suppl.: S190–S207.

O'Driscoll, S.W., Fitzsimmons, J.S., Commisso, C.N. (1997) Role of oxygen tension during cartilage formation by periosteum. *J. Orthop. Res.* 15: 682–687.

O'Driscoll, S.W., Fitzsimmons, J.S., Saris, D.B.F. (1998) Age-related decline in periosteal chondrogenesis. *Trans. Orthop. Res. Soc.* 23: 12.

O'Driscoll, S.W., Recklies, A.D., Poole, A.R. (1994) Chondrogenesis in periosteal explants. An organ culture model for in vitro study. *J. Bone Joint Surg.* 76: 1042–1051.

O'Driscoll, S.W., Saris, D.B., Ito, Y., Fitzimmons, J.S. (2001) The chondrogenic potential of periosteum decreases with age. *J. Orthop. Res.* 19: 95–103.

Paige, K.T., Cima, L.G., Yaremchuk, M.J., Schloo, B.L., Vacanti, J.P., Vacanti, C.A. (1996) De novo cartilage generation using calcium alginate-chondrocyte constructs. *Plast. Reconstr. Surg.* 97: 168–178; discussion 179–180.

Pei, M., Seidel, J., Vunjak-Novakovic, G., Freed, L.E. (2002) Growth factors for sequential cellular de- and re-differentiation in tissue engineering. *Biochem. Biophys. Res. Commun.* 294: 149–154.

Perka, C., Schultz, O., Spitzer, R.S., Lindenhayn, K. (2000) The influence of transforming growth factor beta1 on mesenchymal cell repair of full-thickness cartilage defects. *J. Biomed. Mater. Res.* 52: 543–552.

Reginato, A.M., Iozzo, R.V., Jiménez, S.A (1994) Formation of nodular structures resembling mature articular cartilage in long-term primary cultures of human fetal epiphyseal chondrocytes on a hydrogel substrate. *Arthritis Rheum.* 9: 1338–1349.

Reyes, M., Lund, T., Lenvik, T., Aguiar, D., Koodie, L., Verfaillie, C.M. (2001) Purification and ex vivo expansion of postnatal human marrow mesodermal progenitor cells. *Blood* 98: 2615–2625.

Roh, J., Xu, L., Hering, T.M., Yoo, J.U., Johnstone, B. (2001) Modulation of bone morphogenetic protein-2 expression during in vitro chondrogenesis. *Trans. Orthop. Res. Soc.* 26: 149.

Rotter, N., Bonassar, L.J., Tobias, G., Lebl, M., Roy, A.K., Vacanti, C.A (2002) Age dependence of biochemical and biomechanical properties of tissue-engineered human septal cartilage. *Biomaterials* 23: 3087–3094.

Rubak, J.M., Poussa, M., Ritsila, V. (1982) Chondrogenesis in repair of articular cartilage defects by free periosteal grafts in rabbits. *Acta Orthop. Scand.* 53: 181–186.

Sanyal, A., Oursler, M.J., Clemens, V.R., Fukumoto, T., Fitzsimmons, J.S., O'Driscoll, S.W. (2002) Temporal expression patterns of BMP receptors and collagen II (B) during periosteal chondrogenesis. *J. Orthop. Res.* 20: 58–65.

Saris, D.B., Sanyal, A., An, K.N., Fitzsimmons, J.S., O'Driscoll, S.W. (1999) Periosteum responds to dynamic fluid pressure by proliferating in vitro. *J. Orthop. Res.* 17: 668–677.

Schulze-Tanzil, G., de Souza, P., Castrejon, H.V., John, T., Merker, H.-J., Scheid, A., Shakibaei (2002) Redifferentiation of dedifferentiated human chondrocytes in high density cultures. *Cell Tissue Res.* 308: 371–379.

Sekiya, I., Colter, D.C., Prockop, D.J. (2001) BMP-6 enhances chondrogenesis in a subpopulation of human marrow stromal cells. *Biochem. Biophys. Res. Commun.* 284: 411–418.

Sekiya, I., Larson, B.L., Smith, J.R., Pochampally, R., Cui, J.G., Prockop, D.J. (2002) Expansion of human adult stem cells from bone marrow stroma: conditions that maximize the yields of early progenitors and evaluate their quality. *Stem Cells* 20: 530–541.

Sims, C.D., Butler, P.E., Casanova, R., Lee, B.T., Randolph, M.A., Lee, W.P., Vacanti, C.A., Yaremchuk, M.J. (1996) Injectable cartilage using polyethylene oxide polymer substrates. *Plast. Reconstr. Surg.* 98: 843–850.

Solursh, M. (1991) Formation of cartilage tissue in vitro. *J. Cell. Biochem.* 45: 258–260.

Stewart, M.C., Saunders, K.M., Burton-Wurster, N., MacLeod, J.N. (2000) Phenotypic stability of articular chondrocytes in vitro : the effects of culture models, bone morphogenetic protein 2, and serum supplementation. *J. Bone Miner. Res.* 15: 166–174.
Stokes, D.G., Liu, G., Coimbra, I.B., Piera-Velazquez, S., Crowl, R.M., Jiménez, S.A. (2002) Assessment of the gene expression profile of differentiated and dedifferentiated human fetal chondrocytes by microarray analysis *Arthritis Rheum.* 46: 404–419.
Tholpady, S.S., Katz, A.J., Ogle, R.C. (2003) Mesenchymal stem cells from rat visceral fat exhibit multipotential differentiation in vitro. *Anat. Rec.* 272A: 398–402.
Vacanti, M.P., Roy, A., Cortiella, J., Bonassar, L., Vacanti, C.A. (2001) Identification and initial characterization of spore-like cells in adult mammals. *J. Cell. Biochem.* 80: 455–460.
Walsh, C.J., Goodman, D., Caplan, A.I., Goldberg, V.M. (1999) Meniscus regeneration in a rabbit partial meniscectomy model. *Tissue Eng.* 5: 327–337.
Weisser, J., Rahfoth, B., Timmermann, A., Aigner, T., Bräuer, R., von der mark, K. (2001) Role of growth factors in rabbit articular cartilage repair by chondrocytes in agarose. *Osteoarthritis Cartilage* 9, *suppl.* A: S48–S54.
Winter, A., Breit, S., Parsch, D., Benz, K., Steck, E., Hauner, H., Weber, R.M., Ewerbeck, V., Richter, W. (2003) Cartilage-like gene expression in differentiated human stem cell spheroids: a comparison of bone marrow-derived and adipose tissue-derived stromal cells. *Arthritis Rheum.* 48: 418–429.
Xue, C., Takahashi, M., Hasunuma, T., Aono, H., Yamamoto, K., Yoshino, S., Sumida, T., Nishioka, K. (1997) Characterization of fibroblast-like cells in pannus lesions of patients with rheumatoid arthritis sharing properties of fibroblasts and chondrocytes. *Ann. Rheum. Dis.* 56: 262–267.
Yates, K.E., Mizuno, S., Glowacki, J. (2001) Early shifts in gene expression during chondroinduction of human dermal fibroblasts. *Exp. Cell Res.* 265: 203–211.
Yoo, J.U., Barthel, T.S., Nishimura, K., Solchaga, L., Caplan, A.I., Goldberg, V.M., Johnstone, B. (1998) The chondrogenic potential of human marrow-derived mesenchymal progenitor cells. *J. Bone Joint Surg.* 80: 1745–1757.
Young, H.E., Ceballos, E.M., Smith, J.C., Mancini, M.L., Wright, R.P., Ragan, B.L., Bushell, I., Lucas, P.A. (1993) Pluripotent mesenchymal stem cells reside within avian connective tissue matrices. *In vitro Cell Dev. Biol. Anim.* 29A: 723–736.
Young, H.E., Steele, T.A., Bray, R.A., Hudson, J., Floyd, J.A., Hawkins, K., Thomas, K., Austin, T., Edwards, C., Cuzzourt, J., Duenzl, M., Lucas, P.A., Black, A.C., Jr. (2001) Human reserve pluripotent mesenchymal stem cells are present in the connective tissues of skeletal muscle and dermis derived from fetal, adult, and geriatric donors. *Anat. Rec.* 264: 51–62.
Zuk, P.A., Zhu, M., Ashjian, P., De Ugarte, D.A., Huang, J.I., Mizuno, H., Alfonso, Z.C., Fraser, J.K., Benhaim, P., Hedrick, M.H. (2002) Human adipose tissue is a source of multipotent stem cells. *Mol. Biol. Cell* 13: 4279–4295.
Zuk, P.A., Zhu, M., Mizuno, H., Huang, J., Futrell, J.W., Katz, A.J., Benhaim, P., Lorenz, H.P., Hedrick, M.H. (2001) Multilineage cells from human adipose tissue: implications for cell-based therapies. *Tissue Eng.* 7: 211–228.

5

Lipid-Mediated Gene Transfer for Cartilage Tissue Engineering

Henning Madry

Laboratory for Experimental Orthopaedics, Department of Orthopaedic Surgery, Saarland University Medical Center, Saarland University, 66421 Homburg, Germany, hmad@hotmail.com

Culture of Cells for Tissue Engineering, edited by Gordana Vunjak-Novakovic and R. Ian Freshney

1. INTRODUCTION TO GENE TRANSFER

Recombinant DNA technology has provided tools to introduce exogenous DNA into cultured mammalian cells. In cartilage tissue engineering, these methods can be used to study the effect of single genes on the regulation of chondrogenesis or to enhance structural and functional properties of engineered cartilage [Madry et al., 2002]. The techniques of transferring DNA molecules to cells can be divided into nonviral methods (including chemical, mechanical, and electrical techniques) and viral methods. The first critical step is the delivery of a gene of interest to the chondrocyte. For the purposes of cartilage tissue engineering, this can be achieved by two different strategies. First, the DNA may be introduced into component chondrocytes before seeding them into a scaffold. The second approach would be to transfer the DNA at a later time point directly into tissue-engineered cartilaginous constructs. This step requires the transfection of chondrocytes within their native matrix in situ and can be performed efficiently by using recombinant adeno-associated virus (rAAV) vectors [Madry et al., 2003b]. In general, an adequate delivery system requires (1) a high efficiency of transmission, (2) a desired level and length of gene expression, and (3) biological safety during the course of the transgene expression.

Recombinant genes have been efficiently introduced into isolated articular chondrocytes by virus-based carriers [Madry et al., 2003b; Baragi et al., 1995; Doherty et al., 1998; Ikeda et al., 2000; Smith et al., 2000; Brower-Toland et al., 2001] and nonviral synthetic methods. Lipid-based vectors in particular offer many advantages [Madry et al., 2002; Kim et al., 1996; Madry and Trippel, 2000]. They are relatively easy to prepare and to handle, do not restrict the length of the DNA molecule of interest that is to be transferred, and very rarely elicit an immune response [Fogler et al., 1987]. The lipid-based gene transfer protocol described in this chapter has been optimized for overexpressing recombinant DNA in primary articular chondrocytes.

1.1. Principle of Lipid-Mediated Gene Transfer

Liposomes and other lipid-mediated gene shuttle systems are important tools to transfect primary cells because of their very high transfer efficiencies. Nonviral, cationic lipid-mediated gene transfer was first described by Felgner [Felgner et al., 1987]. Cationic liposomes are composed of an amphiphilic cationic lipid and a neutral phospholipid, the helper lipid [Felgner and Ringold, 1989; Felgner et al., 1995]. When brought into aqueous solution, these lipids form liposomes as hexagonal phases [Zhou F and Huang, 1994; Safinya, 2001]. After addition of plasmid DNA, the liposomes spontaneously assemble into liposome-DNA complexes. The cellular events that facilitate the entry of the transferred DNA into the nucleus are not completely known as yet [Escriou et al., 2001; Neves et al., 1999; Seisenberger et al., 2001]. The interaction of the lipid-DNA complexes with the lipid bilayer occurs either by nonspecific adsorption or by endocytosis. Once the DNA has passed the cellular membrane, it is trapped in endosomes. The destabilization of the endosomal membrane is thought to be facilitated by the helper lipid [Farhood et al., 1992], enabling the DNA to escape degradation [Zhou X and Huang, 1994; Crystal, 1995]. The DNA is thereafter released into the cytoplasm, where it enters the nucleus through the nuclear pores by an unidentified mechanism [Seisenberger et al., 2001].

Cationic liposomes that are currently used to transfect a variety of cell lines and primary cells are the monocationic cholesterol derivative DC-Chol [Gao and Huang, 1991], the polycationic lipid DMRIE [San et al., 1993], DOTMA (part of Lipofectin®) [Felgner et al., 1987], and DOSPA (part of Lipofectamine®) [Felgner et al., 1995]. Transfection of chondrocytes of avian origin has been successful with cationic liposomes like Lipofectin [Long and Linsenmayer, 1995] or Lipofectamine [Pallante et al., 1996]. However, transfection by these methods with bovine and human articular chondrocytes has not been forthcoming. Recently, nonliposomal lipid-mediated methods have been described to offer high transfection efficiencies in primary chondrocytes. Among them, FuGENE 6, a nonliposomal mixture of lipids and other components, has been particularly effective [Madry et al., 2001, 2002, 2003a; Madry and Trippel, 2000; Lefebvre et al., 1998; Dinser et al., 2001]. The protocol outlined below will focus on transfection with FuGENE 6.

1.2. Advantages and Shortcomings

Lipid-based transfection offers many advantages over other transfection systems to introduce genes of interest into primary articular chondrocytes. The high transfection efficiency avoids the need for cell selection that might result in undesired phenotypic drift, as observed after prolonged monolayer culture of articular chondrocytes [Benya et al., 1978]. The very low toxicity avoids the need to change the cell culture medium after transfection. In addition, lipid-based transfections can be carried out in the presence of serum.

The main shortcoming of lipid-based transfection is its relatively high cost, especially when large-scale transfections are required.

1.3. Critical Factors for Efficient Transfection

Several factors greatly influence the success of lipid-based transfection with FuGENE 6. The starting point is to identify the optimal quantity of DNA per cell. For chondrocytes cultured in a 24-well plate, 1 μg of DNA per well is commonly used. The ratio of FuGENE 6 (μl) to DNA (μg) must always be greater than 1. In our own investigations, ratios of 2–5 (v/w) FuGENE 6 to DNA have been successful. Another important consideration is the pericellular matrix that surrounds the chondrocytes, thereby acting as a physical barrier. It is thought to inhibit the uptake of the lipid-DNA complexes into the cell by limiting their interaction with the cell membrane. When this matrix is partially degraded by incubation with hyaluronidase, transfection efficiencies are significantly enhanced [Madry and Trippel, 2000]. Other factors for a successful transfection include the growth state of the cells [Corsaro and Pearson, 1981] and an optimal cell density at the time of transfection. It is therefore important to maintain a standard seeding protocol.

The protocol described in Section 3 has proven successful for the authors in transfecting neonatal bovine and human articular chondrocytes. It may serve as a starting point for further improvements, as transfection conditions vary among species, as well as with the anatomic location and the age of the donor. Although a list of reagents is included with the protocols, similar results may be obtained with reagents from different manufacturers. When chondrocytes from a different species (e.g., avian chondrocytes) are used, transfections with the commercially available Lipofectin or Lipofectamine may be considered. Alternatively, cationic liposomes may be synthesized from their individual components [Ravid and Freshney, 1998].

2. PREPARATION OF REAGENTS AND MEDIA

2.1. Expression Plasmid Vectors

Plasmid vectors that may be manipulated to contain a gene of interest are commercially available. These vectors contain an antibiotic resistance gene to allow for the selection of positive transformants. To express large quantities of DNA in vitro, a strong (or tissue specific) promoter and a stable polyadenylation sequence are desirable. The addition of a consensus Kozak sequence has been demonstrated to further increase translation. The efficiency of gene transfer is greatly affected by the purity of the plasmid DNA. Purification by cesium chloride centrifugation has been traditionally used to prepare DNA for transfections [Sambrook et al., 1989]. Alternatives based on commercially available column systems (e.g., Endofree Plasmid Maxi Kit, Qiagen) are less time consuming and less labor intensive and result in ultrapure DNA. Their additional advantage is the removal of endotoxins, in order to reduce the interference of bacterial endotoxins with the transfection processes [Cotten et al., 1994]. After purification, the DNA is usually resuspended in sterile Tris-EDTA (TE) buffer at a final concentration of 0.5–2.0 μg/μl. The vectors used in these protocols are as follows:

pCMVβgal

To express the *Escherichia coli* (*E. coli*) β-galactosidase (*lacZ*) marker gene, any vector containing the *lacZ* gene under the control of a strong promoter (e.g., the cytomegalovirus immediate-early (CMV-IE) promoter and enhancer) can be used [MacGregor and Caskey, 1989]. These vectors are also commercially available (e.g., pCMVβgal, Invitrogen, Carlsbad, CA). Expression of β-galactosidase can be monitored by staining the cells in situ or by applying a colorimetric assay to quantitatively determine the enzyme activity.

pcDNA3.1(–)

If expression of a therapeutic gene is required, pcDNA3.1/Zeo(+) (Invitrogen) containing the human CMV-IE promoter/enhancer and the bovine growth hormone polyadenylation signal may be used. This vector contains a multiple-cloning site.

2.2. Transfection Reagent

FuGENE 6 is a sterile-filtered mixture of different nonliposomal lipids dissolved in 80% ethanol. Any direct contact with plastic surfaces must be avoided because this interaction may greatly decrease transfection efficiency. We recommend the use of polypropylene tubes with a round bottom to prepare the transfection complexes as these have the lowest risk of binding, but direct contact with the FuGENE 6 reagent should still be avoided. Sometimes a precipitate may form in cold FuGENE 6 reagent that disappears once the reagent is warmed up to room temperature.

2.3. Tissue Culture Equipment and Preparation of Reagents

The use of regular tissue culture-treated plasticware for the cell cultures gives the best results. Special coatings (e.g., collagens) do not improve cell attachment. Transfection optimization experiments can be performed in 24-well plates. To transfect articular chondrocytes for tissue engineering, we found 10-cm cell culture dishes useful.

2.3.1. Basal Medium

Dulbecco's modified Eagle medium (DMEM). DMEM, correctly formulated, requires a gas phase of 10% CO_2.	
Penicillin G	50 U/ml
Streptomycin	50 μl/ml
Ascorbic acid	50 μg/ml

2.3.2. Growth Medium

Basal medium	
Fetal bovine serum (FBS)	10% (v/v)

Heat inactivate the FBS by incubation (in a 500-ml bottle) in a water bath for 1 h at 56 °C. Aliquot in 15-ml conical polypropylene centrifuge tubes (use 12 ml FBS per tube). Store at −20 °C. Thaw immediately before use.

2.3.3. Collagenase

Collagenase	0.08% (m/v)
In basal medium	

Vortex briefly to dissolve at room temperature. Sterilize the solution through a 0.22-μm filter. Use immediately.

2.3.4. Hyaluronidase

Hyaluronidase	200 U/ml
In UPW	

Vortex to dissolve at room temperature. Sterilize the solution through a 0.22-μm filter. Add 20 μl of this stock solution to 1 ml of growth medium, resulting in a 4 U/ml final concentration. Use immediately, do not store.

2.3.5. Tris-EDTA (TE) Buffer

Tris HCl	10 mM
EDTA	1 mM
In distilled deionized water	

Mix to dissolve with a stir bar at room temperature for 15–30 min. Adjust to pH 8.0. Sterilize the solution by autoclaving.

2.3.6. Fixative for X-Gal Staining

Paraformaldehyde (pH 7.0)	2% (m/v)
Glutaraldehyde	0.2% (v/v)
In phosphate-buffered saline without Ca^{2+} and Mg^{2+} (PBSA)	

2.3.7. X-Gal Stock Solutions

Solution A

Potassium ferricyanide crystalline	5 mM
Potassium ferrocyanide trihydrate	5 mM
Magnesium chloride	2 mM
In PBSA	

Store at 4 °C. Protect from light. Warm at 37 °C before use.

Solution B

X-Gal	40 mg/ml
In dimethyl sulfoxide (DMSO)	

Store at −20 °C in polypropylene tubes. Protect from light. Warm at 37 °C before use. Dilute solution B (X-Gal stock) 1:40 in solution A in a polypropylene tube. Combination of cold solutions may result in precipitation of X-Gal. Vortex, use immediately. Do not store.

3. PROTOCOLS

3.1. Tissue Harvest, Cell Isolation, and Primary Culture

Protocol 5.1. Primary Culture of Bovine Cartilage

Reagents and Materials

Sterile or aseptically prepared

- ❑ Radiocarpal joints of 1- to 2-week-old calves
- ❑ Basal medium (See Section 2.3.1)
- ❑ Growth medium (See Section 2.3.2)
- ❑ Collagenase (See Section 2.3.3)
- ❑ PBSA
- ❑ Scalpel
- ❑ Anatomic forceps
- ❑ Flat-end spatula
- ❑ Cell strainer with 100-μm nylon mesh
- ❑ Spinner flask (100 ml) with adjustable hanging bar
- ❑ Cell culture dishes, 10 cm, tissue culture treated
- ❑ Pipette (10 ml)

Nonsterile

- ❑ Hemocytometer
- ❑ Trypan Blue
- ❑ Centrifuge

Protocol

(a) Using bovine articular cartilage from radiocarpal joints of 1- to 2-week-old calves:
   i) Open the joint in a sterile fashion.
   ii) Harvest articular cartilage by removing cartilage chips from the articular surface with a sterile scalpel. Exclude the underlying vascularized cartilage.
   iii) Transfer the cartilage chips to a 10-ml dish filled with 10 ml PBSA.
   iv) Remove the PBSA.

(b) Wash the harvested cartilage 3× with PBSA.

(c) Dice the articular cartilage into 2-mm cubes with a scalpel.

(d) Remove the PBSA and measure the wet weight of the cartilage.

(e) Transfer cartilage chips with the spatula to a 100-ml spinner bottle.

(f) Add 100 ml basal medium containing 0.08% collagenase (use 100 ml enzyme solution for 1 g wet weight cartilage).

(g) Incubate at 37 °C in a humidified atmosphere with 10% CO_2 for 16 h (rotation speed: 50 rpm).

(h) Resuspend cells, using a 10-ml pipette to break down cell aggregates.

(i) Filter the solution containing isolated chondrocytes through a 100-μm cell strainer to remove undigested matrix.

(j) Wash once in 40 ml basal medium in a 50-ml Falcon tube. Centrifuge at 2000 *g* for 5 min.
(k) Determine the cell number by hemocytometer and viability by Trypan Blue exclusion. Do not use cells if their viability is below 90%.
(l) Place isolated chondrocytes in monolayer culture in 10-cm cell culture dishes at a density of 6.0×10^6 cells per dish in 10 ml growth medium. This cell number should be sufficient to obtain reliable results and will not overgrow the dish in 3 days. However, if cells are already near confluence after 1 day, decrease the initial seeding density, as a high cell density reduces transfection efficiency significantly.
(m) After 24 h check appearance of the cells. Not all cells may have settled on the plastic surface. Change the growth medium.
(n) Perform transfection experiments with primary cell cultures 1–2 days after seeding when cells reach 50–70% confluence.

3.2. Gene Transfer

Protocol 5.2. Gene Transfer to Bovine Cartilage Cells

Reagents and Materials

Sterile

- ❑ Basal medium (See Section 2.3.1)
- ❑ Growth medium (See Section 2.3.2)
- ❑ Opti-MEM
- ❑ PBSA
- ❑ Trypsin-EDTA (TE) buffer (See Section 2.3.5)
- ❑ Hyaluronidase (See Section 2.3.4)
- ❑ FuGENE 6
- ❑ Polypropylene tubes, round bottom, with cap
- ❑ Pipette tips, 10–1000 μl, aerosol resistant

Nonsterile

- ❑ Hemocytometer
- ❑ Trypan blue

Protocol

(a) When the cells are in exponential growth and at 50–70% confluence, add 200 μl hyaluronidase solution (4 U/ml) to the cells. It is important to prepare fresh hyaluronidase solution immediately before use.
(b) After 12 h, warm all remaining solutions to room temperature.
(c) For each 10-cm cell culture dish, prepare and label two polypropylene tubes. Add 177 μl Opti-MEM to the first polypropylene tube.
(d) Very gently add, in a dropwise fashion, 87 μl undiluted FuGENE 6 reagent directly into the Opti-MEM in this polypropylene tube with sterile, aerosol-resistant pipette tips. Gently tap the bottom of the tube. Be careful to avoid

direct contact of FuGENE 6 with the plastic surface of the tube, as a decrease in transfection efficiency may occur. Never add FuGENE 6 reagent to an empty tube.

(e) In the second polypropylene tube, add 29 μg plasmid DNA into a suitable volume of TE buffer to achieve a final concentration of 0.5–2.0 μg/μl to the bottom of the tube. Use sterile, aerosol-resistant pipette tips.

(f) Incubate the solutions in each of the two tubes for 5 min at room temperature.

(g) **Very carefully, dropwise, and slowly** add the diluted FuGENE 6 reagent from the first polypropylene tube directly onto the DNA drop in the second tube. Be careful to avoid direct contact of FuGENE 6 with the plastic surface of the tube. Gently tap the bottom of the tube to mix the two components. Close the tube with the cap. You may notice a precipitate as the lipid-DNA complexes form. It is very important to prepare this cocktail fresh, approximately 15–45 min before adding it to the cells.

(h) Incubate for 15–45 min at room temperature.

(i) In the meantime, remove the growth medium of the chondrocytes and add 10 ml fresh growth medium to the cells. A washing step is not necessary.

(j) Prepare fresh hyaluronidase solution.

(k) Approximately 15–45 min after combining the FuGENE 6 with the DNA, add the FuGENE 6-DNA complexes slowly and dropwise directly to the medium in the 10-cm cell culture dish. Swirl carefully to distribute the FuGENE 6-DNA complexes.

(l) Immediately after adding the FuGENE 6-DNA complexes, add 200 μl hyaluronidase solution (4 U/ml) directly to the medium in the 10-cm cell culture dish. It is important to prepare this solution fresh immediately before use. Swirl carefully.

(m) Check the appearance of the cells under the microscope. You may note the transfection complexes.

(n) Incubate the chondrocytes at 37 °C in a humidified atmosphere with 10% CO_2 for 6 h. As FuGENE 6-DNA complexes have not been reported to be harmful to the cells, it is usually not necessary to monitor the cells for signs of toxicity.

(o) After 6 h, add 5 ml growth medium to the cell culture dish containing 10 ml of medium and hyaluronidase.

(p) After 24 h, remove all medium.

(q) Wash the cell layer once with 5 ml PBSA.

(r) Trypsinize the cells.

  i) Apply 2 ml trypsin-EDTA solution.

  ii) Return the cells to the incubator for 15–45 min.

  iii) After dissociation from the culture vessel, wash the cells once in 40 ml basal medium in a 50-ml Falcon tube. Centrifuge at 2000 g for 5 min.

  iv) Determine the cell number by hemocytometer and cell viability by Trypan Blue exclusion.

  v) Proceed to Protocol 5.3 to determine transfection efficiency.

3.3. Cell Seeding in Scaffolds

Functional substitutes for native articular cartilage can be created in vitro from chondrogenic cells attached to polymer scaffolds in bioreactors. During this process, chondrocytes deposit extracellular matrix consisting of proteoglycans and type II collagen in parallel with the scaffold degradation [Vunjak-Novakovic et al., 1998, 1999; Freed et al., 1998]. For a detailed protocol on seeding scaffolds, See Chapter 6.

3.4. Reporter Gene Expression

Marker genes like the *E. coli* β-galactosidase (*lacZ*) gene, the chloramphenicol-acetyl transferase (CAT) or the *Photinus pyralis* luciferase gene are often used to detect transgene expression. A direct estimation of the efficiency of transfection is possible with *lacZ*. Transfection efficiency is expressed as the percentage of positive transfected cells. The sensitive CAT or luciferase assays are very useful to compare different transfection systems or to optimize transfection conditions. They do not allow a direct determination of transfection efficiency.

After a successful gene transfer, the exogenous bacterial enzyme β-galactosidase, which breaks β-galactoside into its component sugars, is produced inside the cell. It can be detected easily in the cytoplasm by a colorimetric reaction. The substrate 5-bromo-4-chloro-3-indolyl β-D-galactopyranoside (X-Gal) is cleaved and an insoluble indigo blue reaction product is generated in the presence of potassium ferricyanide and potassium ferrocyanide (Fig. 5.1). Thus all cells carrying the *lacZ* gene can be identified by their deep blue color. Some cells (e.g., synoviocytes) possess endogenous β-galactosidase-like enzymatic activity. The optimal pH for endogenous β-galactosidase is between 3.5 and 5.5, whereas that for the *E. coli* enzyme is 7.3 [Hatton and Lin, 1992]. Possible interference can therefore be minimized by adjusting the pH of the staining solution to 8.4 and by limiting the incubation to 2–4 h. The X-Gal staining solution should be prepared fresh every time. Alternatively, commercial β-galactosidase staining kits can be used, at somewhat higher total cost.

Protocol 5.3. Detection of Reporter Gene Expression in Transfected Cartilage

Reagents and Materials

Sterile

- ❑ Tissue culture-treated plates, 24 well
- ❑ PBSA

Nonsterile

- ❑ Fixative for X-Gal staining (See Section 2.3.6)
- ❑ X-Gal stock solutions, solutions A and B (See Section 2.3.7)

Protocol

(a) At 48 hours after transfection in a 24-well plate, remove medium, wash the cell layers twice in 10 ml PBSA, and fix the cells for 5 min in fixative.

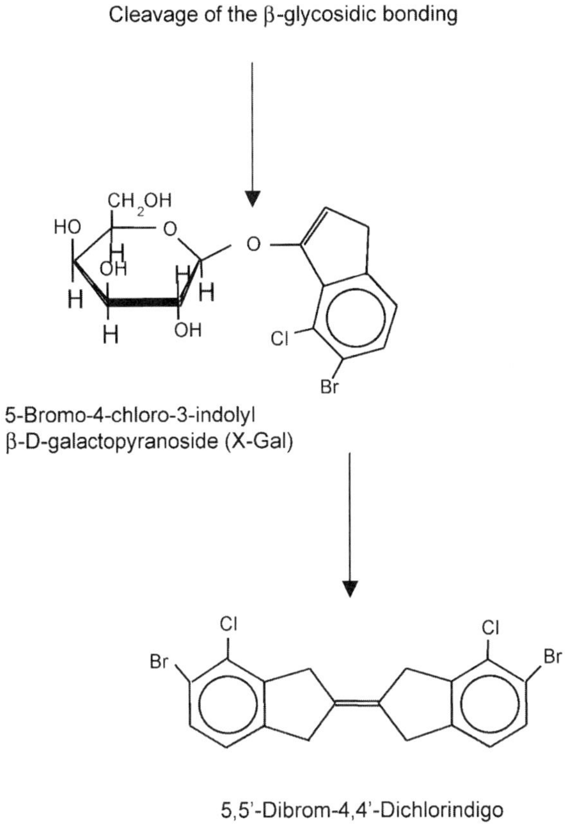

Cleavage of the β-glycosidic bonding

5-Bromo-4-chloro-3-indolyl
β-D-galactopyranoside (X-Gal)

5,5'-Dibrom-4,4'-Dichlorindigo

Figure 5.1. Cleavage of X-Gal by β-galactosidase. The colorless substrate 5-bromo-4-chloro-3-indolyl β-D-galactopyranoside (X-Gal) is cleaved by the bacterial enzyme β-galactosidase in galactose and in an indoxyl derivate. The indoxyl derivate later oxidizes to the blue dibrome-dichlor derivate that is insoluble in water. It can be easily identified as it stains the cytoplasm of the transfected cell blue.

(b) Rinse cells 3× with 10 ml PBSA; be careful not do dislodge the cell layer.

(c) Add 0.5 ml X-Gal staining solution to each well.

(d) Incubate for 2–4 h at 37 °C in a humidified incubator.

(e) Remove cells from the incubator after each 1 h to check for transgene expression. As there is no false-positive staining in chondrocytes when the solutions are properly prepared and its pH is adjusted, one may also stain overnight. The stain should develop after 1–2 h.

(f) Once the stain has developed, remove the X-Gal staining solution and add 10 ml PBSA.

(g) Express transfection efficiency as the percentage of positive to total cells by counting cells in 10–20 sequential microscope fields at 40× magnification along the horizontal and vertical diameters of each well of the 24-well plate. When transfection efficiencies are low (e.g., <1.0%), all stained cells per well may be counted.

Table 5.1. Structural and functional parameters of engineered cartilage after 4 weeks of bioreactor cultivation.

Construct parameters measured after 4 weeks of bioreactor cultivation	Constructs		
	Lac Z	IGF-I	Nontransfected
Construct structure			
Wet weight (mg/construct)	26.5 ± 2.1 ($n = 10$, $P < 0.001$)	67.7 ± 10.8 ($n = 10$)	29.0 ± 8.5 ($n = 10$, $P < 0.001$)
Dry weight (mg/construct)	2.2 ± 0.4 ($n = 10$, $P < 0.001$)	5.8 ± 1.1 ($n = 10$)	2.8 ± 0.4 ($n = 10$, $P < 0.001$)
Diameter (mm)	5.6 ± 0.1 ($n = 10$, $P < 0.001$)	6.5 ± 0.2 ($n = 10$)	5.3 ± 0.2 ($n = 10$, $P < 0.001$)
Water (% wet weight)	92.5 ± 1.0 ($n = 6$, $P > 0.05$)	91.3 ± 0.9 ($n = 5$)	88.7 ± 3.5 ($n = 6$, $P > 0.05$)
DNA (μg/construct)	48.8 ± 7.0 ($n = 6$, $P = 0.001$)	74.3 ± 11.5 ($n = 5$)	38.4 ± 7.0 ($n = 6$, $P = 0.001$)
Glycosaminoglycans (μg/construct)	96.3 ± 15.3 ($n = 6$, $P = 0.001$)	1068.6 ± 239.6 ($n = 5$)	133.3 ± 51.1 ($n = 6$, $P < 0.001$)
Glycosaminoglycans (μg/μg DNA)	2.0 ± 0.2 ($n = 6$, $P < 0.001$)	14.3 ± 1.2 ($n = 5$)	3.4 ± 0.9 ($n = 6$, $P < 0.001$)
Collagen (mg/construct)	0.6 ± 0.1 ($n = 6$, $P < 0.001$)	1.3 ± 0.2 ($n = 2$)	0.5 ± 0.1 ($n = 6$, $P < 0.001$)
Collagen (μg/μg DNA)	11.6 ± 1.3 ($n = 6$, $P < 0.001$)	17.0 ± 0.9 ($n = 5$)	14.1 ± 1.9 ($n = 6$, $P = 0.015$)
Construct function			
[^{35}S]sulfate incorporation (cpm/μg DNA/16 h)	2.3 ± 0.6 ($n = 6$, $P < 0.001$)	9.4 ± 1.0 ($n = 5$)	3.8 ± 0.9 ($n = 6$, $P < 0.001$)
[^{3}H]proline incorporation (cpm/μg DNA/16 h)	38.9 ± 8.3 ($n = 6$, $P > 0.05$)	45.4 ± 6.1 ($n = 5$)	43.6 ± 7.1 ($n = 6$, $P > 0.05$)
Glucose in medium (mg/cm^3)	3.22 ± 0.07 ($n = 2$, $P > 0.05$)	2.21 ± 0.52 ($n = 2$)	3.11 ± 0.20 ($n = 2$, $P > 0.05$)
Lactate in medium (mg/cm^3)	0.998 ± 0.07 ($n = 2$, $P > 0.05$)	1.80 ± 0.49 ($n = 2$)	1.04 ± 0.12 ($n = 2$, $P > 0.05$)
Equilibrium modulus (kPa)	35 ± 33 ($n = 3$, $P < 0.05$)	126 ± 52 ($n = 4$)	30 ± 4 ($n = 3$, $P < 0.03$)

Parentheses indicate the number of samples per group followed by the P value for comparing the respective control group with IGF-I constructs. Data represent average ± standard deviation.

4. APPLICATIONS

One focus of interest of our laboratory is the healing of articular cartilage defects by chondrocytes overexpressing therapeutic genes. We have reported previously that FuGENE 6-mediated transfection in the presence of hyaluronidase produced high transfection efficiencies in normal bovine and human articular chondrocytes [Madry and Trippel, 2000]. When articular chondrocytes were transfected with an expression plasmid vector based on pcDNA3.1/Zeo(+) containing the cDNA for human insulin-like growth factor-I (IGF-I), a polypeptide that is anabolic and mitogenic for cartilage, biologically relevant amounts of IGF-I protein were synthesized [Madry et al., 2001]. We have used Protocol 5.2 to modify bovine articular chondrocytes that were then used to engineer cartilaginous constructs. After seeding onto biodegradable polyglycolic acid scaffolds and culture in bioreactors, the structural and functional properties of tissue engineered cartilage were significantly enhanced when a human IGF-I cDNA was transfected (Table 5.1). Transgene expression was maintained over 28 days of in vitro cultivation followed by an additional 10 days either in vitro or in vivo [Madry et al., 2001]. In these experiments, the human IGF-I gene served as a model to demonstrate the potential benefits of gene transfer for tissue engineering. Genetically modified cartilaginous constructs overexpressing a growth or transcription factor gene may be used as a tissue substitute that simultaneously provides a stimulus for repair.

ACKNOWLEDGMENTS

Studies that generated data with these methods have been supported in part by a Leopoldina fellowship grant, provided by the German Academy for Natural Sciences Leopoldina (BMBF LPD 9801-10) and by the Deutsche Forschungsgemeinschaft (DFG MA 2 363 1-1). We thank Magali Cucchiarini for helpful discussions.

SOURCES OF MATERIALS

Item	*Supplier*
Ascorbic acid	Sigma
Bovine testicular hyaluronidase	Sigma
5-Bromo-4-chloro-3-indolyl β-D-galactopyranoside (X-Gal)	Roche
Cell culture dishes, standard tissue culture treated (100 × 20 mm)	Falcon (BD Biosciences)
Cell strainer with 100-μm nylon mesh	Falcon (BD Biosciences)
Collagenase (type I)	Worthington
DMEM (high glucose)	Invitrogen
Endofree Maxi Kit	Qiagen
FuGENE 6	Roche
Magnesium chloride	Fisher Scientific
N,*N*-dimethylethylformamide	Sigma
Opti-MEM	Invitrogen
PBSA	Invitrogen
pcDNA3.1/Zeo(+)	Invitrogen
pCMVβgal	Invitrogen
Penicillin	Invitrogen
Pipette tips, aerosol resistant	Fisher Scientific
Plasticware	Falcon (BD Biosciences)
Potassium ferricyanide, $K_3Fe(CN)_6$	Fisher Scientific
Potassium ferrocyanide trihydrate, $K_4Fe_3(CN)_6 \cdot 3H_2O$	Fisher Scientific
Round bottom test tubes with cap	Falcon (BD Biosciences)
Spatula, flat end	Fisher Scientific
Spinner flask, 100 ml, with adjustable hanging bar	Bellco Glass
Streptomycin	Invitrogen
Trypsin-EDTA (0.25% trypsin in 1.0 mM EDTA)	Invitrogen

REFERENCES

Baragi, V.M., Renkiewicz, R.R., Jordan, H., Bonadio, J., Hartman, J.W., Roessler, B.J. (1995) Transplantation of transduced chondrocytes protects articular cartilage from interleukin 1-induced extracellular matrix degradation. *J. Clin. Invest.* 96(5): 2454–2460.

Benya, P.D., Padilla, S.R., Nimni, M.E. (1978) Independent regulation of collagen types by chondrocytes during the loss of differentiated function in culture. *Cell* 15(4): 1313–1321.

Brower-Toland, B.D., Saxer, R.A., Goodrich, L.R., Mi, Z., Robbins, P.D., Evans, C.H,, Nixon, A.J. (2001) Direct adenovirus-mediated insulin-like growth factor I gene transfer enhances transplant chondrocyte function. *Hum. Gene Ther.* 12(2): 117–129.

Corsaro, C.M., and Pearson, M.L. (1981) Enhancing the efficiency of DNA-mediated gene transfer in mammalian cells. *Somatic Cell. Genet.* 7(5): 603–616.

Cotten, M., Baker, A., Saltik, M., Wagner, E., Buschle, M., (1994) Lipopolysaccharide is a frequent contaminant of plasmid DNA preparations and can be toxic to primary human cells in the presence of adenovirus. *Gene Ther.* 1(4): 239–246.

Crystal, R.G. (1995) Transfer of genes to humans: early lessons and obstacles to success. *Science* 270(5235): 404–410.

Dinser, R., Kreppel, F., Zaucke, F., Blank, C., Paulsson, M., Kochanek, S., Maurer, P. (2001) Comparison of long-term transgene expression after non-viral and adenoviral gene transfer into primary articular chondrocytes. *Histochem. Cell. Biol.* 116(1): 69–77.

Doherty, P.J., Zhang, H., Tremblay, L., Manolopoulos, V., Marshall, K.W. (1998) Resurfacing of articular cartilage explants with genetically-modified human chondrocytes in vitro. *Osteoarthritis Cartilage* 6(3): 153–159.

Escriou, V., Carriere, M., Bussone, F., Wils, P., Scherman, D. (2001) Critical assessment of the nuclear import of plasmid during cationic lipid-mediated gene transfer. *J. Gene Med.* 3(2): 179–187.

Farhood, H., Bottega, R., Epand, R.M., Huang, L. (1992) Effect of cationic cholesterol derivatives on gene transfer and protein kinase C activity. *Biochim. Biophys. Acta* 1111(2): 239–246.

Felgner, P.L., Gadek, T.R., Holm, M., Roman, R., Chan, H.W., Wenz, M., Northrop, J.P., Ringold, G.M., Danielsen, M. (1987) Lipofection: a highly efficient, lipid-mediated DNA-transfection procedure. *Proc. Natl. Acad. Sci. USA* 84(21): 7413–7417.

Felgner, P.L., and Ringold, G.M.(1989) Cationic liposome-mediated transfection. *Nature* 337 (6205): 387–388.

Felgner, P.L., Tsai, Y.J., Sukhu, L., Wheeler, C.J., Manthorpe, M., Marshall, J., Cheng, S.H. (1995) Improved cationic lipid formulations for in vivo gene therapy. *Ann. NY Acad. Sci.* 772: 126–139.

Fogler, W.E., Swartz, G.M., Jr., Alving, C.R. (1987) Antibodies to phospholipids and liposomes: binding of antibodies to cells. *Biochim. Biophys. Acta* 903(2): 265–272.

Freed, L.E., Hollander, A.P., Martin, I., Barry, J.R., Langer, R., Vunjak-Novakovic, G. (1998) Chondrogenesis in a cell-polymer-bioreactor system. *Exp. Cell Res.* 240(1): 58–65.

Gao, X., and Huang, L. (1991) A novel cationic liposome reagent for efficient transfection of mammalian cells. *Biochem. Biophys. Res. Commun.* 179(1): 280–285.

Hatton, J.D., and Lin, L. (1992) Demonstration of specific neuronal cell groups in rat brain by beta-galactosidase enzyme histochemistry. *J. Neurosci. Methods* 45(3): 147–153.

Ikeda, T., Kubo, T., Nakanishi, T., Arai, Y., Kobayashi, K., Mazda, O., Ohashi, S., Takahashi, K., Imanishi, J., Takigawa, M., Hirasawa, Y. (2000) Ex vivo gene delivery using an adenovirus vector in treatment for cartilage defects. *J. Rheumatol.* 27(4): 990–996.

Kim, H.J., Greenleaf, J.F., Kinnick, R.R., Bronk, J.T., Bolander, M.E. (1996) Ultrasound-mediated transfection of mammalian cells. *Hum. Gene Ther.* 7(11): 1339–1346.

Lefebvre, V., Li, P., de Crombrugghe, B. (1998) A new long form of Sox5 (L-Sox5), Sox6 and Sox9 are coexpressed in chondrogenesis and cooperatively activate the type II collagen gene. *EMBO J.* 17(19): 5718–5733.

Long, F., and Linsenmayer, T.F. (1995) Tissue-specific regulation of the type X collagen gene. Analyses by in vivo footprinting and transfection with a proximal promoter region. *J. Biol. Chem.* 270(52): 31310–31314.

MacGregor, G.R., and Caskey, C.T. (1989) Construction of plasmids that express *E. coli* beta-galactosidase in mammalian cells. *Nucleic Acids Res.* 17(6): 2365.

Madry, H., Cucchiarini, M., Stein, U., Remberger, K., Menger, M.D., Kohn, D., Trippel, S.B. (2003a) Sustained transgene expression in cartilage defects in vivo after transplantation of articular chondrocytes modified by lipid-mediated gene transfer in a gel suspension delivery system. *J. Gene Med.* 5(6): 502–509.
Madry, H., Cucchiarini, M., Terwilliger, E.F., Trippel, S.B. (2003b) Efficient and persistent gene transfer into articular cartilage using recombinant adeno-associated virus vectors in vitro and in vivo. *Hum. Gene Ther.* 14(4): 393–402.
Madry, H., Padera, B., Seidel, J., Freed, L., Langer, R., Trippel, S.B., Vunjak-Novakovic, G. (2002) Gene transfer of a human insulin-like growth factor I cDNA enhances tissue engineering of cartilage. *Hum. Gene Ther.* 13(13): 1621–1630.
Madry, H., and Trippel, S.B. (2000) Efficient lipid-mediated gene transfer to articular chondrocytes. *Gene Ther.* 7(4): 286–291.
Madry, H., Zurakowski, D., Trippel, S.B. (2001) Overexpression of human insulin-like growth factor-I promotes new tissue formation in an ex vivo model of articular chondrocyte transplantation. *Gene Ther.* 8(19): 1443–1449.
Neves, C., Escriou, V., Byk, G., Scherman, D., Wils, P. (1999) Intracellular fate and nuclear targeting of plasmid DNA. *Cell Biol. Toxicol.* 15(3): 193–202.
Pallante, K.M., Niu, Z., Zhao, Y., Cohen, A.J., Nah, H.D., Adams, S.L. (1996) The chick alpha2(I) collagen gene contains two functional promoters, and its expression in chondrocytes is regulated at both transcriptional and post-transcriptional levels. *J. Biol. Chem.* 271(41): 25233–25239.
Ravid, K., and Freshney, R.I. (1998) *DNA Transfer to Cultured Cells*. Wiley-Liss, New York.
Safinya, C.R. (2001) Structures of lipid-DNA complexes: supramolecular assembly and gene delivery. *Curr. Opin. Struct. Biol.* 11(4): 440–448.
Sambrook, J., Fritsch, E.F., Maniatis, T. (1989) In: Maniatis, T., ed., *Molecular Cloning: A Laboratory Manual*. Cold Spring Harbor, NY, Cold Spring Harbor Laboratory Press.
San, H., Yang, Z.Y., Pompili, V.J., Jaffe, M.L., Plautz, G.E., Xu, L., Felgner, J.H., Wheeler, C.J., Felgner, P.L., Gao, X., et al. (1993) Safety and short-term toxicity of a novel cationic lipid formulation for human gene therapy. *Hum. Gene Ther.* 4(6): 781–788.
Seisenberger, G., Ried, M.U., Endress, T., Buning, H., Hallek, M., Brauchle, C. (2001) Real-time single-molecule imaging of the infection pathway of an adeno-associated virus. *Science* 294(5548): 1929–1932.
Smith, P., Shuler, F.D., Georgescu, H.I., Ghivizzani, S.C., Johnstone, B., Niyibizi, C., Robbins, P.D., Evans, C.H. (2000) Genetic enhancement of matrix synthesis by articular chondrocytes. *J. Rheumatol.* 43(5): 1156–1164.
Vunjak-Novakovic, G., Martin, I., Obradovic, B., Treppo, S., Grodzinsky, A.J., Langer, R., Freed, L.E. (1999) Bioreactor cultivation conditions modulate the composition and mechanical properties of tissue-engineered cartilage. *J. Orthop. Res.* 17(1): 130–138.
Vunjak-Novakovic, G., Obradovic, B., Martin, I., Bursac, P.M., Langer, R., Freed, L.E. (1998) Dynamic cell seeding of polymer scaffolds for cartilage tissue engineering. *Biotechnol. Prog.* 14(2): 193–202.
Zhou, F., and Huang, L. (1994) Liposome-mediated cytoplasmic delivery of proteins: an effective means of accessing the MHC class I-restricted antigen presentation pathway. *Immunomethods* 4(3): 229–235.
Zhou, X., and Huang, L. (1994) DNA transfection mediated by cationic liposomes containing lipopolylysine: characterization and mechanism of action. *Biochim. Biophys. Acta* 1189: 195–203.

Part II

Tissue Engineering

6

Tissue Engineering: Basic Considerations

Gordana Vunjak-Novakovic

Department of Biomedical Engineering, Columbia University, 363G Engineering Terrace, Mail Code 8904, 1210 Amsterdam Avenue, New York, NY 10027, gv2131@columbia.edu

Culture of Cells for Tissue Engineering, edited by Gordana Vunjak-Novakovic and R. Ian Freshney

1. INTRODUCTION

Tissue engineering combines the principles of biology, engineering, and medicine to create biological substitutes for lost or defective native tissues. One approach to tissue engineering involves the generation of immature but functional tissue grafts in vitro and their maturation after implantation in vivo. Constructs engineered in vitro can also serve as high-fidelity models for quantitative studies of cell and tissue responses to genetic alterations, drugs, hypoxia, and mechanical stimuli. Extensive reviews of the methods and principles of tissue engineering can be found elsewhere [Lanza et al., 2000; Atala and Lanza, 2002]. In this book, we focus on selected examples of functional tissues engineered in vitro. Chapters 7–15 review the approaches used for successful tissue engineering of cartilage, bone, ligaments, blood vessels, cardiac muscle, skeletal muscle, and liver. In this introductory chapter to Part II, we review the principles and representative methods of tissue engineering, using cartilage and cardiac muscle as examples of two distinctly different tissues of substantial clinical interest that impose different requirements for the design and operation of tissue engineering systems.

2. IN VITRO CULTIVATION OF ENGINEERED TISSUES

2.1. Overall Approach

There is a serious lack of suitable donor tissues for transplantation, along with an increasing number of patients in need of transplantable tissues and organs. The disparity between the need and availability of donor tissues has motivated the tissue engineering approach, aimed at creating cell-based substitutes of native tissues. The overall concept of tissue engineering is depicted in Fig. 6.1. Living cells are

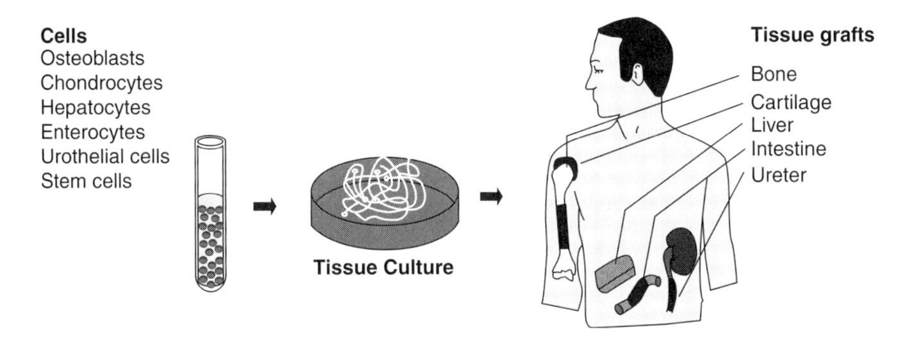

Figure 6.1. Overall approach to tissue engineering. Cells are cultured on a three-dimensional scaffold to engineer a graft that can replace a lost or damaged tissue. (Adapted from Langer and Vacanti, 1994).

obtained from a tissue harvest (either from the patient to engineer an *autograft*, or from a different person to engineer an *allograft*) and cultured in vitro on a three-dimensional scaffold to obtain a tissue construct suitable for transplantation. This approach, pioneered by Langer and Vacanti [1993] has been explored in numerous variations that differ from each other with respect to the cell source, the biomaterial scaffold, the conditions and duration of tissue culture, and the type and properties of the tissue being engineered.

2.2. Functional Tissue Engineering

In this book, we focus on *functional tissue engineering*, defined as the in vitro cultivation of tissue constructs with the structural and functional properties of the tissue being replaced. As compared to the transplantation of cells alone, engineered tissue constructs have the potential advantage of *immediate functionality*. As compared to transplantation of native tissues, engineered tissues can alleviate donor-recipient compatibility and disease transmission (for allografts), and donor site morbidity (for autografts). Engineered tissues can also serve as physiologically relevant models for *controlled studies of cells and tissues* under normal and pathological conditions [Vunjak-Novakovic and Goldstein, 2005].

Ideally, a lost or damaged tissue could be replaced by an engineered graft that can reestablish appropriate structure, composition, cell signaling, and function of the native tissue. In light of this paradigm, the clinical utility of tissue engineering will likely depend on our ability to replicate the site-specific properties of the particular tissue across different size scales. An engineered graft should provide regeneration, rather than repair, and undergo orderly remodeling in response to environmental factors in order to provide normal function in the long term (Table 6.1).

2.3. Model System

Tissue engineering generally attempts to recapitulate certain aspects of the environment present during normal development in order to stimulate functional assembly of the cells into specialized tissues. This involves the presence of

Table 6.1. Repair, regeneration, and remodeling.

1. *Repair* is rapid replacement of the damaged, defective, or lost tissue with functional new tissue that resembles, but does not replicate the structure, composition, and function of the native tissue.
2. *Regeneration* is slow restoration of all components of the repair tissue to their original condition such that the new tissue is indistinguishable from normal tissue with respect to structure, composition, and functional properties.
3. *Remodeling* is the change in tissue structure and composition in response to the local and systemic environmental factors that alter the functional tissue properties.

Buckwalter and Mankin, 1998; Einhorn, 1998; O'Driscoll, 2001.

reparative cells, the use of scaffolds (designed to provide a structural and logistic template for tissue development and to biodegrade at a controlled rate), and bioreactors (designed to control cellular microenvironment, facilitate mass transport to and from the cells, and provide the necessary biochemical and physical regulatory signals). Many different tissues have been engineered by utilizing variations of the "biomimetic" approach depicted in Fig. 6.2. In each case, the material properties of the native tissue provide the basis for establishing tissue engineering requirements and standards of success.

We will describe the components of this model system by referring to cartilage and myocardium, which perform structural and mechanical functions vital for health and survival and are of high clinical interest because of their inability for self-repair and distinctly different in many respects. Articular cartilage is an avascular tissue containing only one cell type, the chondrocyte, at a very low concentration. Chondrocytes maintain an extracellular matrix (ECM) consisting of a fibrous network of collagen type II and glycosaminoglycan (GAG)-rich proteoglycan [Buckwalter and Mankin, 1997]. Cartilage covers the surfaces of our joints, and its main function is to transfer compressive and shear forces during joint loading. The myocardium (cardiac muscle) is a highly vascularized muscular organ composed of cardiac myocytes, fibroblasts, and macrophages present at very high

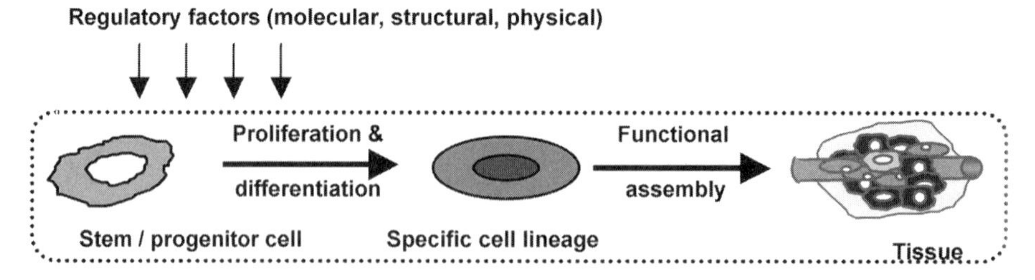

Figure 6.2. Environmental factors for functional tissue assembly. Cell function in vivo depends on a number of factors: a three-dimensional template; normal physicochemical milieu (temperature, pH, osmolality); exchange of nutrients, oxygen, and metabolites; and the presence of biochemical and physical regulatory factors. The design of tissue engineering systems is governed by the need to mimic these environmental factors in vitro and thereby direct the cells to assemble functional tissue structures.

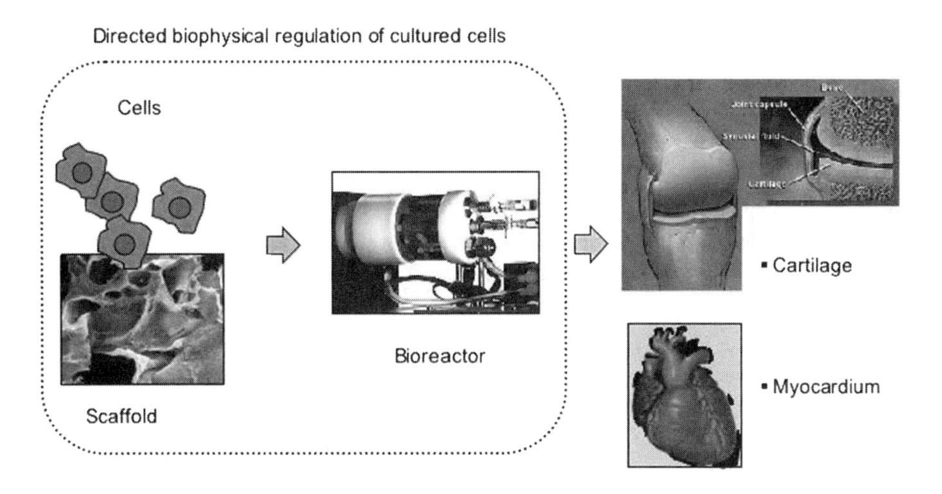

Figure 6.3. Model system. The cell–scaffold–bioreactor system for tissue engineering has been designed to modulate cell differentiation and functional assembly by providing environmental factors described in Fig. 6.2.

concentrations [MacKenna et al., 1994; Brilla et al., 1995]. The myocytes form a three-dimensional syncytium that enables propagation of electrical signals across specialized intracellular junctions to produce coordinated mechanical contractions that pump blood.

The cell–scaffold–bioreactor system (Fig. 6.3) was designed to utilize the factors thought to play regulatory roles during tissue development (See Fig. 6.2). This system has been extensively studied in vitro [See, e.g., Carrier et al., 1999, 2002a,b; Freed and Vunjak-Novakovic, 2000a,b] and in vivo [Schaefer et al., 2002]. It involves bioreactor cultivation of dissociated cells seeded at a high initial density on a three-dimensional scaffold.

Once implanted, an engineered tissue should (i) develop and integrate with adjacent host tissues and (ii) provide some minimal level of function immediately postimplantation that should improve progressively until normal function has been restored. All engineered tissues should possess a certain minimal size, thickness, and mechanical integrity to allow for handling and should permit construct survival under physiological conditions (e.g., in an articular joint for engineered cartilage or the myocardial wall for engineered cardiac tissue), as well as specific functional requirements (i.e., engineered cartilage should withstand and transmit loads, and engineered cardiac tissue should contract in a coordinated manner). We discuss here the general requirements for tissue engineering as well as the additional requirements for engineering cartilage and cardiac muscle.

3. CELLS

3.1. Basic Requirements

An ideal cell source for tissue engineering should have the capacity to proliferate and then differentiate in vitro, in a manner that can be reproducibly controlled. For cartilage tissue engineering, these criteria can be met by either articular chondrocytes or bone marrow-derived mesenchymal stem cells (MSC) [Freed et al., 1999; Meinel et al., 2004a], and in vivo studies have shown that both chondrocyte- and MSC-based grafts can be used to repair large, full-thickness cartilage defects in rabbit knee joints [Caplan et al., 1997; Kawamura et al., 1998; Schreiber et al., 1999]. For cardiac tissue engineering, the above criteria have not yet been met, because precursor cell sources have not been established and in vivo studies in immunocompetent animals have not yet been published.

3.2. Overview of Cell Sources

The cells used thus far to engineer cartilage have varied with respect to donor age (embryonic, neonatal, immature, or adult), differentiation state (precursor or phenotypically mature), and method of preparation (selection, expansion, gene transfer). The sources of chondrogenic cells have included bovine chondrocytes [See, e.g., Buschmann et al., 1992; Freed et al., 1998; Pei et al., 2002b], rabbit chondrocytes [See, e.g., Fedewa et al., 1998], equine chondrocytes [See, e.g., Heath, 2000; Litzke et al., 2004], embryonic chick limb bud cells [See, e.g., Elder et al., 2000], human chondrocytes [See, e.g., Sittinger et al., 1994] and mesenchymal stem cells derived from bone marrow [Meinel et al., 2004a]. The choice of cell type can affect in vitro culture requirements (e.g., medium supplements, structure, and degradation rate of scaffold) and in vivo function (e.g., potential for integration) of engineered cartilage constructs.

Articular chondrocytes are phenotypically stable if cultured under appropriate conditions (e.g., up to 7–8 months in vitro [Hauselmann et al., 1994; Freed et al., 1997]) and can be used to engineer mechanically functional cartilaginous constructs [Buschmann et al., 1995; Freed et al., 1997; Vunjak-Novakovic et al., 1999; Obradovic et al., 2001; Schaefer et al., 2002]. However, adult chondrocytes are not easily harvested; cells from younger donors tend to be more responsive to environmental stimuli [(See, e.g., Heath, 2000]), and cell expansion responds to growth factors [Martin et al., 1999, 2001b; Pei et al., 2002a]. Articular chondrocytes are obtained by enzymatic digestion of full-thickness articular cartilage harvested from 2- to 4-week-old bovine calves [Freed et al., 1993] or 2- to 8-month old New Zealand White rabbits [Freed et al., 1994a]. For both sources, the cell yield is $3–5 \times 10^7$ cells per gram of wet tissue. Primary chondrocytes can be directly seeded onto scaffolds, or first amplified by subculture in the chondrocyte medium (See below) supplemented with 5 ng/ml of fibroblast growth factor FGF-2 [Martin et al., 1999]. Expanding chondrocytes undergo approximately 10 doublings, in two passages. During monolayer culture, chondrocytes can be transfected, for example, by human insulin-like growth factor I [Madry and Trippel, 2000].

Precursor cells from the bone marrow are relatively easier to harvest and expand in culture, remain metabolically active in older donors [Haynesworth et al., 1998], and can recapitulate some aspects of skeletal tissue development [Caplan et al., 1997]. Bone marrow-derived MSC are obtained from 16-day embryonic chicks [Martin et al., 1998], 2- to 4-week old bovine calves [Martin et al., 2001a], or human bone marrow aspirates [Meinel et al., 2004b]. BMSC are selected from the mixed-cell population based on their ability to adhere to the dish, fully characterized (for the presence of surface receptors and ability for selective differentiation [Meinel et al., 2004b]), and passaged in monolayers (twice, to undergo 10–20 doublings).

In the case of engineered cardiac tissue, heart cells were obtained from embryonic chicks and fetal or neonatal rats [See, e.g. Carrier et al., 1999; Li et al., 1999; Zimmermann et al., 2000]. In vivo studies avoided immunorejection by using donor heart cells from inbred rats [Li et al., 1999]; in general, the younger the cell donor, the higher the proliferative capacity and metabolic activity of the harvested cells. Whereas all cell types used to engineer cartilage were able to proliferate in vitro, only embryonic and fetal heart cells proliferated in vitro [Li et al., 1999; Fink et al., 2000].

Primary cardiac myocytes are obtained by enzymatic digestion of heart ventricles harvested from 1- to 2-day neonatal rats [Bursac et al., 1999; Carrier et al., 1999] or 14- to 15-day embryonic chicks [Carrier et al., 1999]. Cell yields range from 1.5 to 7×10^6 cells/heart [Springhorn and Claycomb, 1989; Toraason et al., 1989; Barnett et al., 1993]. The fraction of cardiac myocytes in the cell preparation can be increased by preplating, which allows preferential attachment of fibroblasts [Maki et al., 1996]. After two 1-h periods, the cells that remain unattached are used to seed polymer scaffolds.

4. MEDIA

4.1. Chondrocytes

Chondrocytes are cultured in Dulbecco's modified Eagle's medium (DMEM) containing 4.5 g/l glucose and 4 mM glutamine supplemented with 10% fetal bovine serum (FBS), 10 mM *N*-2-hydroxyethylpiperazine-*N*′-2-ethanesulfonic acid (HEPES), 0.1 mM nonessential amino acids (NEAA), 0.4 mM proline, 0.2 mM (50 μg/ml) ascorbic acid 2-phosphate, 100 U/mL penicillin, 100 μg/ml streptomycin, and 0.5 μg/ml Fungizone (optional) [Sah et al., 1989].

4.2. MSCs

Bone marrow-derived precursor cells are cultured in DMEM containing 4.5 mg/ml glucose and 4 mM glutamine supplemented with 10% FBS, 0.1 mM NEAA, 0.2 mM (50 μg/ml) ascorbic acid 2-phosphate, 10 nM dexamethasone, 5 μg/ml insulin, 5 ng/ml TGF-β1, 100 U/ml penicillin, 100 μg/ml streptomycin, and 0.5 μg/ ml Fungizone (optional) [Meinel et al., 2004b].

4.3. Cardiac Myocytes

Neonatal rat cardiac myocytes are cultured in DMEM containing 5.5 mM (1 mg/ml) glucose, 1 mM pyruvate, 4 mM glutamine, and 25 mM HEPES and supplemented with 100 U/ml penicillin (optional) and 0.5 μg/ml Fungizone (optional) [Bursac et al., 1999; Carrier et al., 1999]. Medium can be further supplemented with 10% FBS or 2% adult horse serum (HyClone) [Papadaki et al., 2001]. Embryonic chick cardiac myocytes are cultured in medium supplemented with 6% FBS [Barnett et al., 1993].

5. BIOMATERIAL SCAFFOLDS

5.1. Basic Requirements

Most studies suggest that the scaffold is essential for promoting orderly regeneration of cartilage, in vivo and in vitro. A biomaterial scaffold provides a template for cell attachment and tissue development, and it biodegrades in parallel with the accumulation of tissue components. Scaffold structure determines the transport of nutrients, metabolites, and regulatory molecules to and from the cells, and the scaffold chemistry may have an important signaling role. Ideally, scaffolds are made of materials that are biocompatible and biodegradable, preferentially those already used in products approved by the Food and Drug Administration. To achieve isomorphous tissue replacement, the scaffold should biodegrade at a rate matching the rate of extracellular matrix deposition and without any toxic or inhibitory byproducts. The maintenance or the rate of decline of the mechanical properties of the scaffold may be critical for its efficacy, as well as for the modulation of the stress-strain environment at the cellular and tissue levels.

5.2. Overview of Scaffolds

Scaffolds used in tissue engineering vary with respect to material chemistry (e.g., collagen, hydrogels, or synthetic polymers), geometry (e.g., gels, fibrous meshes, porous sponges), structure (e.g., porosity, pore size, pore distribution, orientation, and connectivity), mechanical properties (e.g., tension, compression, resistance to shear, and permeability), and the sensitivity to and rate of degradation [See, e.g., Freed and Vunjak-Novakovic, 2000b for review].

A variety of scaffolds have been used to engineer cartilage, including gels of agarose [Buschmann et al., 1995; Lee and Bader, 1997; Mauck et al., 2000] collagen [Wakitani et al., 1994, 1998], and chitosan [Di Martino et al., 2005]; meshes of collagen [Grande et al., 1997] and polyglycolic acid (PGA) [Freed et al., 1997, 1998; Grande et al., 1997; Carver and Heath, 1999; Obradovic et al., 1999; Vunjak-Novakovic et al., 1999; Schaefer et al., 2000a]; and sponges of PLA [Chu et al., 1997]. Scaffolds used to engineer cardiac tissues include gels of collagen with or without Matrigel® [Eschenhagen et al., 1997; Fink et al., 2000; Zimmermann et al., 2000], meshes of PGA with or without laminin coating [Freed et al., 1997; Bursac et al., 1999; Carrier et al., 1999; Papadaki et al., 2001], sponges of native collagen (Ultrafoam®) [Radisic et al., 2003, 2004a,b], sponges of denatured collagen (Gelfoam®) [Li et al., 1999], and fibers of collagen and polystyrene beads [Akins et al., 1999] (See Fig. 6.4 for representative structures of a fibrous mesh and a sponge).

One representative fibrous scaffold is a nonwoven mesh made of PGA fibers (13 μm fiber diameter, >95% void volume) [Freed et al., 1994a]. In most cases, the mesh is punched into 1- to 5-mm-thick disks, 5–10 mm in diameter. Other shapes are used for other applications (e.g., tubular scaffolds made of 1-mm-thick mesh to engineer small-caliber arteries [Niklason et al., 1999]). In culture, fibrous PGA loses its mechanical integrity over 12 days [Niklason et al., 1999] and degrades to approximately 50% of the initial mass over 4 weeks [Freed et al., 1994c]. PGA mesh can be surface-hydrolyzed to increase hydrophilicity [Gao et al., 1998] and then coated with laminin to enhance cell attachment [Papadaki et al., 2001] (Fig. 6.4).

One representative porous scaffold is the Davol Ultrafoam™, a clinically used hemostatic sponge made of a water-insoluble hydrochloric acid salt of purified bovine corium collagen, in the form of 3-mm-thick sheets. For tissue engineering studies, this sheet is punched into 13-mm-diameter disks; after hydration in culture medium, disks reduce their diameters to approximately 10 mm. For rapid cell inoculation, Ultrafoam™ is used in conjunction with Matrigel®, a basement membrane preparation that is liquid at low temperatures (2–8 °C) and gels at 22–35 °C (Fig. 6.4).

6. BIOREACTORS

6.1. Basic Requirements

Ideally, a bioreactor should provide an in vitro environment for rapid and orderly development of functional tissue structures by isolated cells on three-dimensional

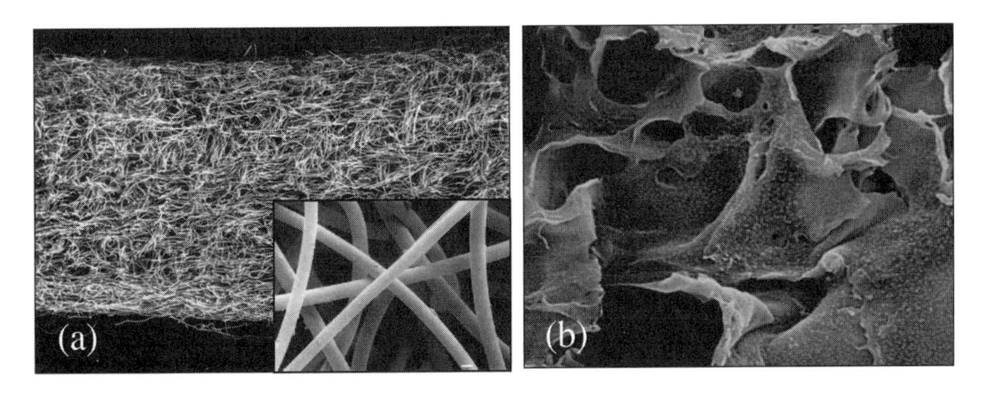

Figure 6.4. Biomaterial scaffolds. a) Fibrous mesh made of polyglycolic acid. b) Collagen sponge.

scaffolds. In a general case, bioreactors are designed to perform one or more of the following functions:

1. Establish spatially uniform concentrations of cells within biomaterial scaffolds.
2. Control conditions in culture medium (e.g., temperature, pH, osmolality, levels of oxygen, nutrients, metabolites, regulatory molecules).
3. Facilitate mass transfer between the cells and the culture environment.
4. Provide physiologically relevant physical signals (e.g., interstitial fluid flow; shear, and compression for cartilage; pulsatile pressure and stretch for cardiac muscle).

6.2. Overview of Bioreactor Types

Representative culture vessels that are frequently used for tissue engineering (static flasks, spinner flasks, rotating vessels) are compared in Fig. 6.5. All culture vessels are operated in incubators (to maintain the temperature and CO_2, and, thereby, control pH), with continuous gas exchange and periodic medium replacement.

Starter dishes are 6-well or 96-well plates containing one scaffold per well seeded with 0.06 ml (96-well plate) to 6 ml of cell suspension (6-well plate). Dishes are used static [Freed et al., 1993], on an orbital shaker mixed at 50–75 rpm [Freed et al., 1994b], or on an XYZ gyrator mixed at 60 rpm [Papadaki et al., 2001].

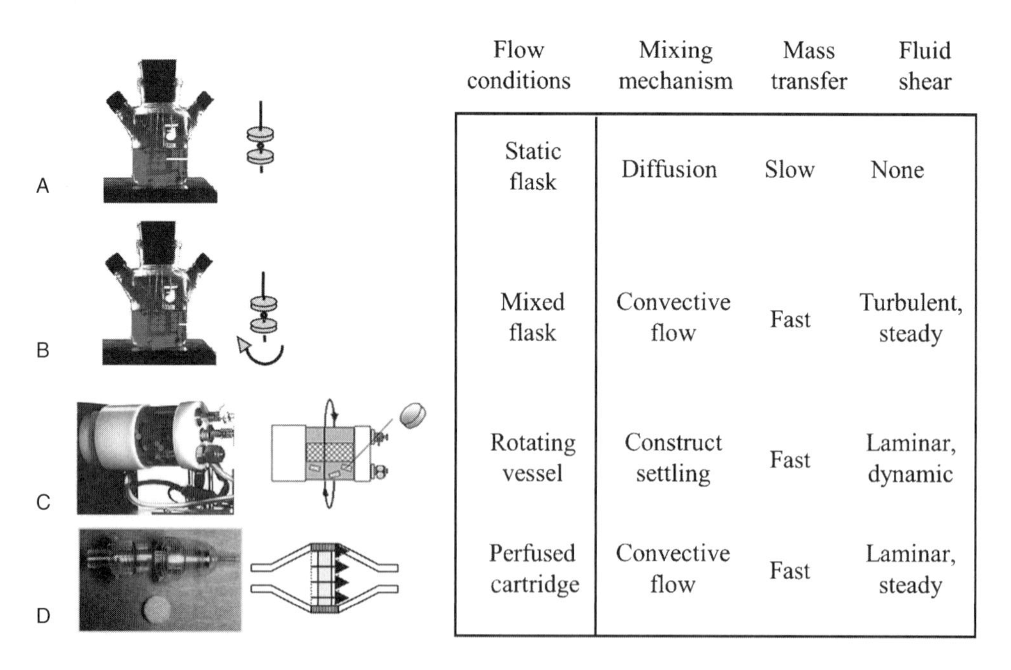

	Flow conditions	Mixing mechanism	Mass transfer	Fluid shear
A	Static flask	Diffusion	Slow	None
B	Mixed flask	Convective flow	Fast	Turbulent, steady
C	Rotating vessel	Construct settling	Fast	Laminar, dynamic
D	Perfused cartridge	Convective flow	Fast	Laminar, steady

Figure 6.5. Bioreactors. a) Static flask. b) Mixed flask. c) Rotating vessel. d) Perfused cartridge.

Spinner flasks (100-ml nominal capacity) are 6.5 cm in diameter × 12 cm high, filled with 120 ml of medium, and fitted with a stopper. Flasks can be operated statically or mixed at 50–75 rpm, using a nonsuspended 0.8-cm-diameter × 4-cm-length stir bar. Gas exchange is via surface aeration through the sidearms [Vunjak-Novakovic et al., 1996, 1998].

Rotating vessels. The Slow Turning Lateral Vessel (STLV) and the High Aspect Ratio Vessel (HARV) were developed at NASA. The STLV is configured as two concentric cylinders that have diameters of 5.75 cm and 2 cm. The annular space, approximately 110 ml in volume, is used for tissue culture, and the inner cylinder serves as a membrane gas exchanger. The HARV is configured as a cylinder 1.3 cm high and 10 cm in diameter, approximately 110 ml in capacity, with one base serving as a membrane gas exchanger. Both vessels are primed with medium to displace all air and mounted on a base that simultaneously rotates the vessel around its central axis at the desired rate (10–45 rpm) and pumps filter-sterilized incubator air over the gas exchange membrane at a rate of about 1 l/min [Freed and Vunjak-Novakovic, 1995].

Perfusion cartridges are small, 10-mm-diameter, 42-mm-length, 1.5-ml-volume polycarbonate vessels made by Advanced Tissue Sciences to culture one scaffold apiece [Dunkelman et al., 1995]. For cell seeding, each cartridge is fitted with two stainless steel screens (with 85% void area) supporting the scaffold during perfusion, each with a silicone gasket (1 mm thick, 10-mm OD, 5-mm ID) preventing the bypass of medium around the scaffold (See Fig. 6.2). Each cartridge is filled with medium to displace all air and connected to a recirculation loop containing two gas exchangers (each configured as a coil of platinum-cured silicone tubing, 80 cm long, 1.6-mm ID, 3.2-mm OD), one on each side of the cartridge. Medium is recirculated by a push/pull syringe pump that can operate two loops at a time (Fig. 6.6).

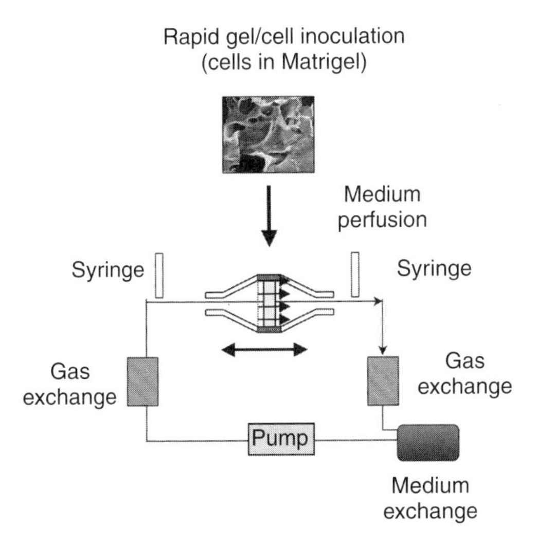

Figure 6.6. Perfused cartridges with interstitial flow of culture medium.

7. SCAFFOLD SEEDING

7.1. Basic Requirements

Cell seeding of 3D scaffolds is the first step of bioreactor cultivation of engineered tissues. Seeding requirements include (a) high seeding efficiency, to maximize cell utilization; (b) high kinetic rate of cell attachment, to minimize the time in suspension for anchorage-dependent cells; and (c) high and spatially uniform distribution of attached cells, for rapid and uniform tissue assembly [Vunjak-Novakovic and Radisic, 2004a].

Before cell seeding, scaffolds must be thoroughly prewetted with culture medium such that all air is displaced. Methods in use include the application of pressure (by a small forceps) or vacuum (by a pipette) on scaffolds immersed in culture medium. Synthetic scaffolds can be first exposed to 70% ethanol to increase wettability [Freed et al., 1993]. Scaffolds are incubated in culture medium for 2–24 h before cell seeding. Incubation in medium containing serum can further enhance cell attachment.

After cell inoculation in a laminar hood, scaffolds are incubated in a humidified, 37 °C incubator containing 5–10% CO_2 in air. The duration of attachment ranges from 1 h to 3 days, depending on the cells, scaffold, seeding vessel, and specific method used. We and others have studied a number of seeding methods for different cell types, initial cell concentrations ($5–50 \times 10^6$ cells per cm^3 of scaffold volume), and scaffolds (fibrous and porous; 5–10 mm in diameter $\times$ 1–5 mm thick). Thin (<2 mm) scaffolds can be seeded statically; spatially uniform cell seeding of thicker (≥ 2 mm) scaffolds requires mixing. A variety of mechanisms was utilized to provide mixing and flow of culture medium, in order to suspend the isolated cells and to generate convective motion of the cells into the scaffold interior. We present two well-characterized cell seeding protocols that give consistently good results, one for cartilage tissue engineering and one for cardiac tissue engineering. [See Vunjak-Novakovic et al., 1998; Radisic et al., 2003; Vunjak-Novakovic and Radisic, 2004a for more details.]

7.2. Dynamic Seeding in Spinner Flasks

Seeding in spinner flasks is the preferred system for seeding fibrous scaffolds. This is a well-characterized method that results in a relatively uniform spatial distribution of attached cells within a period of 24 hours at a yield of essentially 100% [Vunjak-Novakovic et al., 1998]. The main steps of scaffold seeding in spinner flasks are as follows.

Protocol 6.1. Dynamic Seeding in Spinner Flasks for Cartilage Tissue Engineering

Reagents and Materials

Sterile

- ❑ Growth medium (See Section 4)

❑ Spinner flask assembly (See Fig. 6.5):
  (i) Spinner flask, 100-ml nominal capacity, containing a 4-cm-long magnetic stir bar.
  (ii) Silicone stopper, 5-cm diameter, into which four 4-in.-long, 22-gauge needles have been symmetrically placed (See Fig. 6.5).
  (iii) Sterilize the flask by autoclaving with the stopper in place such that the needles extend into the flask.
❑ Polymer scaffold disks: 10-mm diameter × 5-mm thick, 10 mm × 2 mm thick, or 5 mm × 2 mm thick, sterilized with ethylene oxide (ETO, for 12 h), aerated (for at least 24 h), packaged in trilaminate aluminum foil pouches in a dry box to minimize hydrolytic degradation, and stored at room temperature
❑ Silicone tubing, #13, 3-mm lengths, sterilized by autoclaving
❑ Gloves

Nonsterile
❑ Magnetic stirrer

Protocol
(a) Wearing sterile gloves, thread prewetted polymer scaffolds onto the needles such that each needle contains 2 disks that are 10 mm in diameter × 2–5 mm thick, or 3 disks that are 5 mm in diameter × 2 mm thick.
(b) Position disks, using 3-mm-long segments of #13 silicone tubing.
(c) Place the stopper in the mouth of the flask.
(d) Add 120 ml medium via one of the sidearms. Remove any droplets and flame to ensure sterility.
(e) Leave the sidearm caps slack to allow gas exchange.
(f) Transfer the assembled stirrer flask onto a magnetic stirrer in a humidified 37 °C, 5% CO_2 incubator. Set stirring rate to 50–75 rpm and leave overnight.
(g) When a suspension of freshly isolated cells is prepared, remove 16 ml medium from the flask, through the sidearm, using a pipette. Flame the sidearm after removing the cap.
(h) Add cells (e.g., 16 ml of 4×10^6 cells/ml to seed 12 scaffolds that are 5 mm in diameter × 2 mm thick).
(i) Flame the sidearm and replace the cap (leave loosely capped, to allow gas exchange).
(j) Transfer flask onto the magnetic stirrer in a 37 °C, 5% humidified CO_2 incubator.
(k) Completely replace medium after 1–3 days.
(l) Continue to culture cell-polymer constructs in flasks, or transfer into a different culture vessel.

Dynamic seeding in flasks has been extensively used to seed fibrous PGA scaffolds with primary and passaged chondrocytes [Vunjak-Novakovic et al., 1999; Schaefer et al., 2000b; Gooch et al., 2001b; Obradovic et al., 2001], embryonic chick and bovine MSCs [Martin et al., 1998, 1999], and cardiac myocytes [Carrier et al., 1999; Papadaki et al., 2001]. Spinner flasks are also used to seed porous scaffolds made of poly(lactic-co-glycolic acid) (PLGA) and polyethylene glycol (PEG) with bovine MSC [Martin et al., 2001a] and periosteal cells [Schaefer et al., 2000b]. Seeding efficiency (percentage of total cells that attached to scaffolds) is essentially 100% for chondrocytes on PGA scaffolds [Vunjak-Novakovic et al., 1998], approximately 60% for cardiac myocytes on PGA scaffolds [Carrier et al., 1999], and 30–60% for periosteal cells on PLGA-PEG scaffolds [Schaefer et al., 2000b].

In spinner flasks, mixing during cell seeding maintains the cells in suspension and provides relative velocity between the cells and the scaffolds. The probable mechanism by which cells populate the scaffold interior is convective motion of suspended cells into the scaffold followed by inertial impacts between the cells and the fibers and cell attachment [Vunjak-Novakovic et al., 1998].

7.3. Gel-Cell Seeding of Porous Scaffolds

Rapid inoculation of metabolically active cells at high initial densities can be achieved by using gels as cell delivery vehicles. Collagen gel can be used to seed mesenchymal stem cells onto biodegradable sutures [Awad et al., 2000]. Fibrin gel can be used for rapid inoculation of PGA scaffolds with bovine articulate chondrocytes [Ameer et al., 2002]. We used Matrigel® to inoculate Ultrafoam™ scaffolds (10-mm diameter × 3-mm thickness) with cardiac myocytes at the approximate density of 70×10^6 cells/cm^3 according to the procedure described in Protocol 6.2 [Radisic et al., 2003, 2004a,b].

Protocol 6.2. Gel-Cell Seeding of Porous Scaffolds for Cardiac Tissue Engineering

Reagents and Materials

Sterile or aseptically prepared

- ❑ Counted cell suspension
- ❑ Growth medium (See Section 4)
- ❑ Matrigel®
- ❑ Scaffold disks (as in Protocol 6.1)
- ❑ Petri dishes, 9 cm
- ❑ Kimwipes

Protocol

(a) Thaw Matrigel overnight in a 4 °C refrigerator.

(b) Prewet the scaffolds with culture medium and place into a 37 °C, 5% CO_2 incubator for ≥2 h before inoculation.

(c) Blot-dry the scaffolds with sterile Kimwipes, and transfer them to empty Petri dishes (4–5 scaffolds/dish.)

(d) Centrifuge the cell suspension at 1000 rpm for 10 min in a 15-ml conical tube. Place the tube on ice in the laminar hood and aspirate off the supernatant.
(e) Using an automatic micropipette, resuspend a cell pellet in Matrigel. It is recommended that between 5 and 10 μl Matrigel per 10^6 cells be used.
(f) Load the homogeneous cell suspension onto a scaffold as uniformly as possible.
(g) Place the Petri dish with inoculated scaffolds for 10 min in the 37 °C, 5% CO_2 humidified incubator to allow gelation to occur.

To avoid diffusional limitations of mass transfer during scaffold seeding, the scaffolds are directly perfused with culture medium such that the transport of oxygen from the medium to the cells occurs via both diffusion and convection. For seeding in perfusion, the system shown in Fig. 6.6 is used with one scaffold per cartridge in an alternating flow regime according to Protocol 6.3 [See Radisic et al., 2003 for more details].

Protocol 6.3. Perfusion of Porous Scaffolds for Tissue Engineering

Reagents and Materials

Sterile

- ❑ Growth medium (See Section 4)
- ❑ Silicone tubing, screens, gaskets, and cartridges (See Protocol 6.2), sterilized by autoclaving (20 min autoclaving/20 min drying on a dry cycle, 121 °C)
- ❑ Syringes, 10 ml, 2

Protocol

(a) Assemble the seeding loops according to Fig. 6.6 in a laminar flow hood, using sterile technique.
(b) Using one syringe, prime one loop with total of 8 ml culture medium (5.5 ml in perfusion cartridge and tubing; 2.5 ml in one reservoir syringe; the other syringe is empty).
(c) Place the primed loop in the 37 °C, 5% CO_2 incubator for minimum 2 h before seeding.
(d) In the laminar flow hood, carefully open a perfusion cartridge and, using forceps, place the gel-cell inoculated scaffold between two silicone gaskets and stainless steel screens.
(e) Close the perfusion cartridge, and remove any air bubbles by injecting culture medium from syringe 2 into syringe 4 (See Fig. 6.6) while holding the cartridge in a vertical position so that bubbles can escape upward.
(f) Place the whole assembly in a 37 °C, 5% CO_2 incubator and insert the seeding loop into the Push/Pull pump. Program the pump to perfuse the medium at a desired flow rate (0.5–1.5 ml/min) with flow direction reversal after every 2.5 ml for a period of 1.5–4.5 h.

Perfusion of gel-cell inoculated scaffolds helps maintain viability, metabolic activity, and uniform cell distribution [Radisic et al., 2003]. It is also possible to seed cells directly from suspension (omitting the gel inoculation step) to PGA scaffolds in a perfusion loop [Kim et al., 2000]. The procedure yields constructs with relatively uniform distribution of metabolically active cells, but the cell density and seeding efficiency are significantly lower than in gel-cell inoculated scaffolds.

8. BIOREACTOR CULTIVATION

8.1. Hydrodynamic Environment

Convective mixing improves the kinetic rate, efficiency, and spatial uniformity of cell seeding on three-dimensional polymer scaffolds and improves tissue structure and composition by enhancing mass transport within the culture medium and at the tissue surfaces [Vunjak-Novakovic et al., 1998; Radisic et al., 2003]. Hydrodynamic factors present during culture can modulate cell function and tissue development in at least two ways: via associated effects on mass transport between the developing tissue and culture medium (e.g., oxygen, nutrients, growth factors) and by physical stimulation of the cells (e.g., shear, pressure). The composition, morphology, and mechanical properties of engineered tissues grown in hydrodynamically active environments were generally better than in static environments. For cultivations in vessels described in Fig. 6.5, construct compositions and mechanical properties were better in stirred flasks and rotating vessels than in static flasks.

Direct perfusion through cultured tissue constructs stimulated chondrogenesis, presumably because of the combined effects of enhanced mass transport, pH regulation, and fluid shear in the cell microenvironment [Sittinger et al., 1994; Dunkelman et al., 1995; Pazzano et al., 2000], in particular at physiological interstitial flow velocities [Maroudas, 1979; Mow et al., 1980, 1991; Vunjak-Novakovic et al., 1996]. However, physiologically thick, functional cartilage could be engineered with diffusional transport within the cultured tissue constructs (e.g., in rotating bioreactors, [Vunjak-Novakovic et al., 1999; Pei et al., 2002a]). In contrast, the same culture conditions yielded engineered cardiac tissue in which the functional layer was only approximately 0.1 mm thick [Carrier et al., 1999]. This finding could be attributed to diffusional limitations of oxygen transport to the cells, as supported by calculations that showed that the pO_2 would decrease to zero at a depth of approximately 0.1 mm [Carrier et al., 2002b]. The engineering of a thick layer of functional cardiac tissue requires the perfusion of oxygen-rich culture medium directly through the growing construct [Radisic et al., 2004a]. The combination of rapid cell inoculation and immediate establishment of medium perfusion enabled physiological densities of viable cells in engineered cardiac constructs, because of the maintenance of efficient oxygen supply to the cells at all times of cultivation.

8.2. Growth Factors

Growth factors (e.g., FGF-2, TGF-β1) are generally required to engineer cartilaginous tissues starting from bone marrow-derived mesenchymal stem cells

[Johnstone et al., 1998; Martin et al., 1998; Meinel et al., 2004a] and expanded chondrocytes [Martin et al., 1999, 2001b]. Specific combinations of bioactive factors were shown to promote chondrocytes cells to first dedifferentiate during expansion in monolayers and then redifferentiate and regenerate cartilaginous tissues during subsequent cultivation on biomaterial scaffolds [Pei et al., 2002a]. Growth factors supplemented sequentially to culture medium (TGF-β/FGF-2 early, IGF-I later) markedly and significantly improved the compositions and mechanical properties of engineered cartilage [Pei et al., 2002a]. Gene transfer of human IGF-I augmented the structural and functional properties of cartilaginous constructs grown in bioreactors, suggesting that spatially defined overexpression of growth factors may be advantageous for cartilage tissue engineering [Madry et al., 2001].

Importantly, the mechanical environment and supplemental growth factors independently modulate the growth and mechanical properties of engineered cartilage, interact to produce results not suggested by the independent responses, and in certain combinations can produce tissues superior to those obtained by utilizing these factors individually [Gooch et al., 2001a]. Beneficial effects of growth factors can be amplified by dynamic mechanical loading. TGF-β and IGF-I interacted with dynamic loading applied during culture in a synergetic manner and improved the compositions and mechanical properties of cultured constructs to an extent greater than the sum of effects of either stimulus applied alone [Mauck et al., 2000, 2002, 2003]. Likewise, the hydrodynamically active environment present in rotating bioreactors amplified the beneficial effects of polymer scaffolds on construct compositions and mechanical properties and yielded engineered cartilage that had equilibrium moduli of 400–540 kPa after only 4 weeks of bioreactor cultivation [Pei et al., 2002b].

8.3. Physical Signals

Chondrogenesis in vitro and remodeling in vivo of native and engineered cartilage has been studied with a variety of physical signals, including fluid flow [Wu et al., 1999; Pazzano et al., 2000], dynamic fluctuations in hydrodynamic shear and pressure [Freed et al., 1997; Vunjak-Novakovic et al., 1999], cyclic hydrostatic pressure [Carver and Heath, 1999], cyclic mechanical compression [Buschmann et al., 1995; Mauck et al., 2000; Seidel et al., 2004], and cyclic stretch [Wu and Chen, 2000]. In vitro, dynamic compression enhanced synthesis of proteoglycans in cartilage explants [Sah et al., 1989], improved the mechanical function of engineered cartilage [Buschmann et al., 1992, 1995; Lee and Bader, 1997; Carver and Heath, 1999; Mauck et al., 2000], and enhanced chondrogenesis of chick limb bud cells [Elder et al., 2000, 2001]. In vivo, cyclic loading caused the mesenchymal cells to differentiate into cartilage overlaying bone [Tagil and Aspenberg, 1999]. The application of mechanical strain during cultivation improved the properties of engineered cardiac constructs, in vitro and in vivo [Zimmermann et al., 2002a,b].

Mechanical stimulation can cause multiple changes to the extracellular environment (a) by direct effects on cell shape and interfibrillar spacing, (b) by increase in hydrostatic pressure, (c) by fluid flow that can enhance mass transport to and from the cells, or (d) by change in fluid volume that can cause changes in concentrations of chemical and ionic species [Mow et al., 1999]. All these effects can modulate the synthesis, breakdown, and structural adaptations of the ECM, which in turn serves as a transducer of mechanical and electrochemical signals, and thereby mediate the catabolic and anabolic changes in cell metabolism (See Vunjak-Novakovic and Goldstein, 2005 for detailed review).

Most recently, cardiac constructs prepared by seeding collagen sponges with neonatal rat ventricular cells were stimulated during cultivation with suprathreshold square biphasic pulses (2-ms duration, 1 Hz, 5 V) [Radisic et al., 2004b]. Over only 8 days of culture, stimulation resulted in significantly better contractile responses to pacing as compared to unstimulated controls, as evidenced by the sevenfold higher amplitude of contractions in response to pacing. Excitation-contraction coupling of cardiac myocytes in stimulated constructs was evidenced by transmembrane potentials that were similar to the action potentials reported previously for cells from mechanically stimulated constructs [Zimmermann et al., 2002b]. Stimulated constructs exhibited higher levels of cardiac markets and a remarkable level of ultrastructural organization. These studies suggest that electrical stimulation of construct contractions during cultivation enhanced the properties of engineered myocardium at the cellular, ultrastructural, and tissue levels [Radisic et al., 2004b].

8.4. Structural and Functional Assessment of Engineered Constructs

For histologic assessment, engineered constructs are fixed in neutral buffered formalin, embedded in paraffin, and sectioned (5–8 μm thick). Engineered cartilage sections are stained with hematoxylin and eosin (H&E) for cells, safranin-O for GAGs, and monoclonal antibodies for collagen types I, II, IX, and X [Freed et al., 1998; Martin et al., 1998; Riesle et al., 1998]. Engineered cardiac tissue sections are stained with H&E for cells and monoclonal antibodies for sarcomeric α-actin, cardiac troponin I, sarcomeric tropomyosin, and connexin 43 [Carrier et al., 1999; Papadaki et al., 2001; Radisic et al., 2004a,b]. Construct size and distributions of cells and tissue components are assessed by image analysis [Freed et al., 1998; Vunjak-Novakovic et al., 1998]. Ultrastructural analyses include scanning and transmission electron microscopy (SEM, TEM) [Riesle et al., 1998; Carrier et al., 2002a,b; Radisic et al., 2004b].

For biochemical evaluation, cartilage constructs are digested with papain or protease-K, and lyophilized [Freed et al., 1993, 1994c, 1998]. The amount of DNA is determined fluorometrically with Hoechst 33258 dye [Kim et al., 1988]. Sulfated GAG content is determined spectrophotometrically by Dimethylmethylene Blue dye binding [Farndale et al., 1986]. Total collagen content is determined spectrophotometrically from hydroxyproline content after acid hydrolysis and reaction with p-dimethylaminobenzaldehyde and chloramine-T [Woessner, 1961]. Type II collagen content is determined by inhibition ELISA [Hollander et al., 1994; Freed et al., 1998; Riesle et al., 1998]. The presence of other collagens (e.g., I, IX, X) is semiquantitatively measured by SDS-PAGE and Western blot [Riesle et al., 1998].

GAG distribution is determined by magnetic resonance imaging (MRI) [Bashir et al., 1996; Williams et al., 1998]. Synthesis rates of GAG, total protein, and collagen are measured by incorporation and release of radiolabeled tracers [Freed et al., 1998].

The molar ratio of lactate production to glucose consumption ($\sim$1 for aerobic, $\sim$2 for anaerobic cell metabolism) and the rate of ammonia production are determined from concentrations measured in culture medium at timed intervals with blood gas and lactate analyzers [Obradovic et al., 1997; Radisic et al., 2004a]. Cell number and cell viability are assessed with ethidium monoazide bromide (EMA) in conjunction with fluorescence-activated cell sorting (FACS [Radisic et al., 2003]. Cell cycle analysis was done by FACS after incubation with propidium iodide [Radisic et al., 2004a].

For engineered cartilage, mechanical construct properties (e.g., compressive modulus, dynamic stiffness, hydraulic permeability, streaming potential) are measured in static and dynamic radially confined compression [Frank and Grodzinsky, 1987; Vunjak-Novakovic et al., 1999, Vunjak-Novakovic and Goldstein, 2005]. The capacity for integration with native cartilage is evaluated in vitro in bioreactors [e.g., Obradovic et al., 2001]) and in vivo in an animal model [e.g., Schaefer et al., 2002].

For engineered cardiac tissue, electrophysiological studies are determined by measuring signal propagation over macroscopic distances with a linear array of extracellular microelectrodes [Burșac et al., 1999]. At the cellular level, transmembrane potentials (action potentials) are measured under physiologic conditions and in response to a potassium current blocker [Bursac et al., 2003]. The contractile function of engineered cardiac constructs is evaluated by measuring contraction amplitude and frequency in response to electrical stimulation, using video microscopy and image processing[Radisic et al., 2003, 2004a,b].

9. SUMMARY

The primary functions of cartilage and myocardium, biomechanical for cartilage and contractile for myocardium, drive tissue engineering toward the restoration of the functions inherent in the tissue being replaced. The current paradigm is that the restoration of normal tissue function can be best achieved by using in vitro or in vivo engineered constructs that can regenerate the exact site-specific properties (molecular, structural, functional) of native tissues across different size scales [Vunjak-Novakovic and Goldstein, 2005].

Cells, biomaterial scaffolds, and regulatory factors (biochemical and physical) have been utilized in a variety of ways, in vitro with bioreactors and in vivo by implantation, to engineer functional tissues. In a general sense, tissue engineering tends to recapitulate some aspects of the environment present in vivo during tissue development and thereby stimulate the cells to regenerate functional tissue structures. Cultivation of biosynthetically active cells on an appropriate scaffold, facilitated mass transport, and the provision of physical regulatory signals are among common requirements for rapid and orderly tissue regeneration.

ACKNOWLEDGMENTS

Research described in this chapter has been conducted by research teams supported by grants from NASA and NIH. The expert help of Sue Kangiser with manuscript preparation is greatly appreciated.

SOURCES OF MATERIALS

Item	*Supplier*
Ascorbic acid 2-phosphate	Sigma
Dexamethasone	Sigma
DMEM	Invitrogen (Gibco)
FBS	Invitrogen (Gibco)
Fungizone	Invitrogen (Gibco)
Gyratory mixer	Boeker Scientific
High Aspect Ratio Vessel (HARV)	Synthecon
Horse serum	HyClone
Insulin	Sigma
Matrigel®	B-D Biosciences
Multiwell plates or dishes, 6 well or 96 well	Corning Costar
Nonessential amino acids	Invitrogen (Gibco)
Orbital shaker	Belco
Penicillin	Invitrogen (Gibco)
Perfusion cartridges, polycarbonate	Advanced Tissue Sciences
PGA fibrous scaffold	Albany International
Platinum-cured silicone	Cole Parmer
Slow Turning Lateral Vessel (STLV)	Synthecon
Spinner flasks	Belco
Streptomycin	Invitrogen (Gibco)
Syringe pump, PHD 2000	Harvard Apparatus
Syringes	Becton Dickinson
TGF-β1	Invitrogen (Gibco)
Three-way stopcocks	Baxter Healthcare
Ultrafoam™	Davol

REFERENCES

Akins, R.E., Boyce, R.A., Madonna, M.L., Schroedl, N.A., Gonda, S.R., McLaughlin, T.A., Hartzell, C.R. (1999) Cardiac organogenesis in vitro: Reestablishment of three-dimensional tissue architecture by dissociated neonatal rat ventricular cells. *Tissue Eng.* 5: 103–118.

Ameer, G.A., Mahmood, T.A., Langer, R. (2002) A biodegradable composite scaffold for cell transplantation. *J. Orthop. Res.* 20: 16–19.

Atala, A., and Lanza, R.P., Eds. (2002) *Methods of Tissue Engineering*. San Diego, Academic Press.

Awad, H.A., Butler, D.L., Harris, M.T., Ibrahim, R.E., Wu, Y., Young, R.G., Kadiyala, S., Boivin, G.P. (2000) In vitro characterization of mesenchymal stem cell-seeded collagen scaffolds for tendon repair: Effects of initial seeding density on contraction kinetics. *J. Biomed. Mater. Res.* 51: 233–240.

Barnett, J.V., Taniuchi, M., Yang, M.B., Galper, J.B. (1993) Co-culture of embryonic chick heart cells and ciliary ganglia induces parasympathetic responsiveness in embryonic chick heart cells. *Biochem. J.* 292: 395–399.

Bashir, A., Gray, M.L., Burstein, D. (1996) Gd-DTPA2 as a measure of cartilage degradation. *Magn. Reson. Med.* 36: 665–673.

Brilla, C.G., Maisch, B., Rupp, H., Sunck, R., Zhou, G., Weber, K.T. (1995) Pharmacological modulation of cardiac fibroblast function. *Herz* 20: 127–135.

Buckwalter, J.A., and Mankin, H.J. (1997) Articular cartilage, part I: Tissue design and chondrocyte-matrix interactions. *J. Bone Joint Surg. Am.* 79A: 600–611.

Buckwalter, J.A., and Mankin, H.J. (1998) Articular cartilage: degeneration and osteoarthritis, repair, regeneration, and transplantation. *Instr. Course Lect.* 47: 487–504.

Bursac, N., Papadaki, M., Cohen, R.J., Schoen, F.J., Eisenberg, S.R., Carrier, R., Vunjak-Novakovic, G., Freed, L.E. (1999) Cardiac muscle tissue engineering: toward an in vitro model for electrophysiological studies. *Am. J. Physiol. Heart Circ. Physiol.* 277: H433–H444.

Bursac, N., Papadaki, M., White, J.A., Eisenberg, S.R., Vunjak-Novakovic, G., Freed, L.E. (2003) Cultivation in rotating bioreactors promotes maintenance of cardiac myocyte electrophysiology and molecular properties. *Tissue Eng.* 9: 1243–1253.

Buschmann, M.D., Gluzband, Y.A., Grodzinsky, A.J., Kimura, J.H., Hunziker, E.B. (1992) Chondrocytes in agarose culture synthesize a mechanically functional extracellular matrix. *J. Orthop. Res.* 10: 745–758.

Buschmann, M.D., Gluzband, Y.A., Grodzinsky, A.J., Hunziker, E.B. (1995) Mechanical compression modulates matrix biosynthesis in chondrocyte/agarose culture. *J. Cell Sci.* 108: 1497–1508.

Caplan, A.I., Elyaderani, M., Mochizuki, Y., Wakitani, S., Goldberg, V.M. (1997) Principles of cartilage repair and regeneration. *Clin. Orthop. Rel. Res.* 342: 254–269.

Carrier, R.L., Papadaki, M., Rupnick, M., Schoen, F.J., Bursac, N., Langer, R., Freed, L.E., Vunjak-Novakovic, G. (1999) Cardiac tissue engineering: cell seeding, cultivation parameters and tissue construct characterization. *Biotechnol. Bioeng.* 64: 580–589.

Carrier, R.L., Rupnick, M., Langer, R., Schoen, F.J., Freed, L.E., Vunjak-Novakovic, G. (2002a) Effects of oxygen on engineered cardiac muscle. *Biotechnol. Bioeng.* 78: 617–625.

Carrier, R.L., Rupnick, M., Langer, R., Schoen, F.J., Freed, L.E., Vunjak-Novakovic, G. (2002b) Perfusion improves tissue architecture of engineered cardiac muscle. *Tissue Eng.* 8: 175–188.

Carver, S.E., and Heath, C.A. (1999) Semi-continuous perfusion system for delivering intermittent physiological pressure to regenerating cartilage. *Tissue Eng.* 5: 1–11.

Chu, C., Dounchis, J.S., Yoshioka, M., Sah, R.L., Coutts, R.D., Amiel, D. (1997) Osteochondral repair using perichondrial cells: a 1 year study in rabbits. *Clin. Orthop. Rel. Res.* 340: 220–229.

Di Martino, A., Sittinger, M., Risbud, M.V. (2005) Chitosan: A versatile biopolymer for orthopaedic tissue-engineering. *Biomaterials* May 12; [Epub ahead of print].

Dunkelman, N.S., Zimber, M.P., Lebaron, R.G., Pavelec, R., Kwan, M., Purchio, A.F. (1995) Cartilage production by rabbit articular chondrocytes on polyglycolic acid scaffolds in a closed bioreactor system. *Biotechnol. Bioeng.* 46: 299–305.

Einhorn, T.A. (1998) The cell and molecular biology of fracture healing. *Clin. Orthop. Rel. Res.* 355S: S7–S21.

Elder, S.H., Kimura, J.H., Soslowsky, L.J., Lavagnino, M., Goldstein, S.A. (2000) Effect of compressive loading on chondrocyte differentiation in agarose cultures of chick limb-bud cells. *J. Orthop. Res.* 18: 78–86.

Elder, S.H., Goldstein, S.A., Kimura, J.H., Soslowsky, L.J., Spengler, D.M. (2001) Chondrocyte differentiation is modulated by frequency and duration of cyclic compressive loading. *Ann. Biomed. Eng.* 29: 476–482.

Eschenhagen, T., Fink, C., Remmers, U., Scholz, H., Wattchow, J., Woil, J., Zimmermann, W., Dohmen, H.H., Schafer, H., Bishopric, N., Wakatsuki, T., Elson, E. (1997) Three-dimensional reconstitution of embryonic cardiomyocytes in a collagen matrix: a new heart model system. *FASEB J.* 11: 683–694.

Farndale, R.W., Buttler, D.J., Barrett, A.J. (1986) Improved quantitation and discrimination of sulphated glycosaminoglycans by the use of dimethylmethylene blue. *Biochim. Biophys. Acta* 883: 173–177.

Fedewa, M.M., Oegema, T.R., Jr., Schwartz, M.H., MacLeod, A., Lewis, J.L. (1998) Chondrocytes in culture produce a mechanically functional tissue. *J. Orthop. Res.* 16: 227–236.

Fink, C., Ergun, S., Kralisch, D., Remmers, U., Weil, J., Eschenhagen, T. (2000) Chronic stretch of engineered heart tissue induces hypertrophy and functional improvement. *FASEB J.* 14: 669–679.

Frank, E.H., and Grodzinsky, A.J. (1987) Cartilage electromechanics II. A continuum model of cartilage electrokinetics and correlation with experiments. *J. Biomech.* 20: 629–639.

Freed, L.E., Marquis, J.C., Nohria, A., Emmanual, J., Mikos, A.G., Langer, R. (1993) Neocartilage formation in vitro and in vivo using cells cultured on synthetic biodegradable polymers. *J. Biomed. Mater. Res.* 27: 11–23.

Freed, L.E., Grande, D.A., Lingbin, Z., Emmanual, J., Marquis, J.C., Langer, R. (1994a) Joint resurfacing using allograft chondrocytes and synthetic biodegradable polymer scaffolds. *J. Biomed. Mater. Res.* 28: 891–899.

Freed, L.E., Marquis, J.C., Vunjak-Novakovic, G., Emmanual, J., Langer, R. (1994b) Composition of cell-polymer cartilage implants. *Biotechnol. Bioeng.* 43: 605–614.

Freed, L.E., Vunjak-Novakovic, G., Biron, R., Eagles, D., Lesnoy, D., Barlow, S., Langer, R. (1994c) Biodegradable polymer scaffolds for tissue engineering. *Bio/Technology* 12: 689–693.

Freed, L.E., and Vunjak-Novakovic, G. (1995) Cultivation of cell-polymer constructs in simulated microgravity. *Biotechnol. Bioeng.* 46: 306–313.

Freed, L.E., Langer, R., Martin, I., Pellis, N., Vunjak-Novakovic, G. (1997) Tissue engineering of cartilage in space. *Proc. Natl. Acad. Sci. USA* 94: 13885–13890.

Freed, L.E., Hollander, A.P., Martin, I., Barry, J.R., Langer, R., Vunjak-Novakovic, G. (1998) Chondrogenesis in a cell-polymer-bioreactor system. *Exp. Cell Res.* 240: 58–65.

Freed, L.E., Martin, I., Vunjak-Novakovic, G. (1999) Frontiers in tissue engineering: in vitro modulation of chondrogenesis. *Clin. Orthop. Rel. Res.* 367S: S46–S58.

Freed, L.E., and Vunjak-Novakovic, G. (2000a) Tissue engineering bioreactors. In Lanza, R.P., Langer, R., Vacanti, J., eds, *Principles of Tissue Engineering*. San Diego, Academic Press, pp. 143–156.

Freed, L.E., and Vunjak-Novakovic, G. (2000b) Tissue engineering of cartilage. In Bronzino, J.D., ed, *The Biomedical Engineering Handbook*. Boca Raton, CRC Press, pp. 124–126.

Gao, J., Niklason, L., Langer, R. (1998) Surface hydrolysis of poly(glycolic acid) meshes increases the seeding density of vascular smooth muscle cells. *J. Biomed. Mater. Res.* 42: 417–424.

Gooch, K.J., Blunk, T., Courter, D.L., Sieminski, A.L., Bursac, P.M., Vunjak-Novakovic, G., Freed, L.E. (2001a) IGF-I and mechanical environment interact to modulate engineered cartilage development. *Biochem. Biophys. Res. Commun.* 286: 909–915.

Gooch, K.J., Kwon, J.H., Blunk, T., Langer, R., Freed, L.E., Vunjak-Novakovic, G. (2001b) Effects of mixing intensity on tissue-engineered cartilage. *Biotechnol. Bioeng.* 72: 402–407.

Grande, D.A., Halberstadt, C., Naughton, G., Schwartz, R., Manji, R. (1997) Evaluation of matrix scaffolds for tissue engineering of articular cartilage grafts. *J. Biomed. Mater. Res.* 34: 211–220.

Hauselmann, H.J., Fernandes, R.J., Mok, S.S., Schmid, T.M., Block, J.A., Aydelotte, M.B., Kuettner, K.E., Thonar, E.J.M. (1994) Phenotypic stability of bovine articular chondrocytes after long-term culture in alginate beads. *J. Cell Sci.* 107: 17–27.

Haynesworth, S.E., Reuben, D., Caplan, A.I. (1998) Cell-based tissue engineering therapies: The influence of whole body physiology. *Adv. Drug Deliv. Res.* 33: 3–14.

Heath, C.A. (2000) The effects of physical forces on cartilage tissue engineering. *Biotechnol. Genet. Eng. Rev.* 17: 533–551.

Hoemann, C.D., Sun, J., Legare, A., McKee, M.D., Ranger, P., Buschmann, M.D. (2001) A thermosensitive polysaccharide gel for cell delivery in cartilage repair. *Trans. Orthop. Res. Soc.* 26.

Hollander, A.P., Heathfield, T.F., Webber, C., Iwata, Y., Bourne, R., Rorabeck, C., Poole, R.A. (1994) Increased damage to type II collagen in osteoarthritic articular cartilage detected by a new immunoassay. *J. Clin. Invest.* 93: 1722–1732.

Johnstone, B., Hering, T.M., Caplan, A.I., Goldberg, V.M., Yoo, J.U. (1998) In vitro chondrogenesis of bone marrow-derived mesenchymal progenitor cells. *Exp. Cell Res.* 238: 265–272.

Kawamura, S., Wakitani, S., Kimura, T., Maeda, A., Caplan, A.I., Shino, K., Ochi, T. (1998) Articular cartilage repair-rabbit experiments with a collagen gel-biomatrix and chondrocytes cultured in it. *Acta Orthop. Scand.* 69: 56–62.

Kim, S.S., Sundback, C.A., Kaihara, S., Benvenuto, M.S., Kim, B.S., Mooney, D.J., Vacanti, J.P. (2000) Dynamic seeding and in vitro culture of hepatocytes in a flow perfusion system. *Tissue Eng.* 6: 39–44.

Kim, Y.J., Sah, R.L., Doong, J.Y.H., Grodzinsky, A.J. (1988) Fluorometric assay of DNA in cartilage explants using Hoechst 33258. *Anal. Biochem.* 174: 168–176.

Langer, R., and Vacanti, J.P. (1993) Tissue engineering. *Science* 260: 920–926.

Lanza, R.P., Langer, R., Vacanti, J., Eds. (2000) *Principles of Tissue Engineering*. San Diego, Academic Press.

Lee, D.A., and Bader, D.L. (1997) Compressive strains at physiological frequencies influence the metabolism of chondrocytes seeded in agarose. *J. Orthop. Res.* 15: 181–188.

Li, R.-K., Jia, Z.Q., Weisel, R.D., Mickle, D.A.G., Choi, A., Yau, T.M. (1999) Survival and function of bioengineered cardiac grafts. *Circulation* 100: II63–II69.

Litzke, L.E., Wagner, E., Baumgaertner, W., Hetzel, U., Josimovic-Alasevic, O., Libera, J. (2004) Repair of extensive articular cartilage defects in horses by autologous chondrocyte transplantation. *Ann. Biomed. Eng.* 32: 57–69.

MacKenna, D.A., Omens, J.H., McCulloch, A.D., Covell, J.W. (1994) Contribution of collagen matrix to passive left ventricular mechanics in isolated rat heart. *Am. J. Physiol Heart Circ. Physiol.* 266: H1007–H1018.

Madry, H., and Trippel, S.B. (2000) Efficient lipid-mediated gene transfer to articular chondrocytes. *Gene Ther.* 7: 286–291.

Madry, H., Padera, R., Seidel, J., Freed, L.E., Langer, R., Trippel, S.B., Vunjak-Novakovic, G. (2001) Tissue engineering of cartilage enhanced by the transfer of a human insulin-like growth factor-I gene. *Trans. Orthop. Res. Soc.* 26: 289.

Maki, T., Gruver, E., Davidoff, A., Izzo, N., Touplin, D., Colucci, W., Marks, A. (1996) Regulation of calcium channel expression in neonatal myocytes by catecholamines. *J. Clin. Invest.* 97: 656–663.

Maroudas, A. (1979) Physiochemical properties of articular cartilage. In Freeman, M.A.R., ed, *Adult articular cartilage*, Pitman Medical, pp. 215–290.

Martin, I., Padera, R.F., Vunjak-Novakovic, G., Freed, L.E. (1998) In vitro differentiation of chick embryo bone marrow stromal cells into cartilaginous and bone-like tissues. *J. Orthop. Res.* 16: 181–189.

Martin, I., Vunjak-Novakovic, G., Yang, J., Langer, R., Freed, L.E. (1999) Mammalian chondrocytes expanded in the presence of fibroblast growth factor-2 maintain the ability to differentiate and regenerate three-dimensional cartilaginous tissue. *Exp. Cell Res.* 253: 681–688.

Martin, I., Shastri, V.P., Padera, R.F., Yang, J., Mackay, A.J., Langer, R., Vunjak-Novakovic, G., Freed, L.E. (2001a) Selective differentiation of mammalian bone marrow stromal cells cultured on three-dimensional polymer foams. *J. Biomed. Mater. Res.* 55: 229–235.

Martin, I., Suetterlin, R., Baschong, W., Heberer, M., Vunjak-Novakovic, G., Freed, L.E. (2001b) Enhanced cartilage tissue engineering by sequential exposure of chondrocytes to FGF-2 during 2D expansion and BMP-2 during 3D cultivation. *J. Cell. Biochem.* 83: 121–128.

Mauck, R.L., Soltz, M.A., Wang, C.C.B., Wong, D.D., Chao, P.G., Valhmu, W.B., Hung, C.T., Ateshian, G.A. (2000) Functional tissue engineering of articular cartilage through dynamic loading of chondrocyte-seeded agarose gels. *J. Biomech. Eng.* 122: 252–260.

Mauck, R.L., Seyhan, S.L., Ateshian, G.A., Hung, C.T. (2002) Influence of seeding density and dynamic deformational loading on the developing structure/function relationships of chondrocyte-seeded agarose hydrogels. *Ann. Biomed. Eng.* 30: 1046–1056.

Mauck, R.L., Nicoll, S.B., Seyhan, S.L., Ateshian, G.A., Hung, C.T. (2003) Synergistic action of growth factors and dynamic loading for articular cartilage tissue engineering. *Tissue Eng.* 9: 597–611.

Meinel, L., Hofmann, S., Karageorgiou, V., Zichner, L., Langer, R., Kaplan, D., Vunjak-Novakovic, G. (2004a) Engineering cartilage-like tissue using human mesenchymal stem cells and silk protein scaffolds. *Biotechnol. Bioeng.* 88: 379–391.

Meinel, L., Karageorgiou, V., Fajardo, R., Snyder, B., Shinde-Patil, V., Zichner, L., Kaplan, D., Langer, R., Vunjak-Novakovic, G. (2004b) Bone tissue engineering using human mesenchymal stem cells: Effects of scaffold material and medium flow. *Ann. Biomed. Eng.* 32: 112–122.

Mow, V.C., Kuei, S.C., Lai, W.M., Armstrong, C.G. (1980) Biphasic creep and stress relaxation of articular cartilage in compression: Theory and experiments. *J. Biomech. Eng.* 102: 73–84.

Mow, V.C., Ratcliffe, A., Rosenwasser, M.P., Buckwalter, J.A. (1991) Experimental studies on the repair of large osteochondral defects at a high weight bearing area of the knee joint: a tissue engineering study. *Trans. Am. Soc. Mech. Engr.* 113: 198–207.

Mow, V.C., Wang, C.B., Hung, C.T. (1999) The extracellular matrix, intersitial fluid and ions as a mechanical signal transducer in articular cartilage. *Osteoarthr. Cartilage* 7: 41–58.

Niklason, L.E., Gao, J., Abbott, W.M., Hirschi, K.K., Houser, S., Marini, R., Langer, R. (1999) Functional arteries grown in vitro. *Science* 284: 489–493.

Obradovic, B., Freed, L.E., Langer, R., Vunjak-Novakovic, G. (1997) Bioreactor studies of natural and engineered cartilage metabolism. In Peppas, N.A., Mooney, D.J., Mikos, A.G., Brannon-Peppas, L., eds, *Proceedings of the Topical Conference on Biomaterials, Carriers for Drug Delivery, and Scaffolds for Tissue Engineering*. AIChE, p. 335.

Obradovic, B., Carrier, R.L., Vunjak-Novakovic, G., Freed, L.E. (1999) Gas exchange is essential for bioreactor cultivation of tissue engineered cartilage. *Biotechnol. Bioeng.* 63: 197–205.

Obradovic, B., Martin, I., Padera, R.F., Treppo, S., Freed, L.E., Vunjak-Novakovic, G. (2001) Integration of engineered cartilage. *J. Orthop. Res.* 19: 1089–1097.

O'Driscoll, S.W. (2001) Preclinical cartilage repair: current status and future perspectives. *Clinical Orthopaedics* 391 Suppl: S397–S401.

Papadaki, M., Bursac, N., Langer, R., Merok, J., Vunjak-Novakovic, G., Freed, L.E. (2001) Tissue engineering of functional cardiac muscle: molecular, structural and electrophysiological studies. *Am. J. Physiol. Heart Circ. Physiol.* 280: H168–H178.

Pazzano, D., Mercier, K.A., Moran, J.M., Fong, S.S., DiBiasio, D.D., Rulfs, J.X., Kohles, S.S., Bonassar, L.J. (2000) Comparison of chondrogensis in static and perfused bioreactor culture. *Biotechnol. Prog.* 16: 893–896.

Pei, M., Seidel, J., Vunjak-Novakovic, G., Freed, L.E. (2002a) Growth factors for sequential cellular de- and re-differentiation in tissue engineering. *Biochem. Biophys. Res. Commun.* 294: 149–154.

Pei, M., Solchaga, L.A., Seidel, J., Zeng, L., Vunjak-Novakovic, G., Caplan, A.I., Freed, L.E. (2002b) Bioreactors mediate the effectiveness of tissue engineering scaffolds. *FASEB J.* 16: 1691–1694.

Radisic, M., Euloth, M., Yang, L., Langer, R., Freed, L.E., Vunjak-Novakovic, G. (2003) High density seeding of myocyte cells for tissue engineering. *Biotechnol. Bioeng.* 82: 403–414.

Radisic, M., Yang, L., Boublik, J., Cohen, R.J., Langer, R., Freed, L.E., Vunjak-Novakovic, G. (2004a) Medium perfusion enables engineering of compact and contractile cardiac tissue. *Am. J. Physiol. Heart Circ. Physiol.* 286: H507–H516.

Radisic, M., Park, H., Shing, H., Consi, T., Schoen, F.J., Langer, R., Freed, L.E., Vunjak-Novakovic, G. Functional assembly of engineered myocardium by electrical stimulation of cardiac myocytes cultured on scaffolds. *Proc. Natl. Acad. Sci. USA.* 101(52): 18129–18134 (2004b).

Riesle, J., Hollander, A.P., Langer, R., Freed, L.E., Vunjak-Novakovic, G. (1998) Collagen in tissue-engineered cartilage: Types, structure and crosslinks. *J. Cell. Biochem.* 71: 313–327.

Sah, R.L.Y., Kim, Y.J., Doong, J.Y.H., Grodzinsky, A.J., Plaas, A.H.K., Sandy, J.D. (1989) Biosynthetic response of cartilage explants to dynamic compression. *J. Orthop. Res.* 7: 619–636.

Schaefer, D., Martin, I., Heberer, M., Jundt, G., Seidel, J., Bergin, J., Grodzinsky, A.J., Vunjak-Novakovic, G., Freed, L.E. (2000a) Tissue engineered composites for the repair of large osteochondral defects. *Trans. Orthop. Res. Soc.* 25: 619.

Schaefer, D., Martin, I., Shastri, V.P., Padera, R.F., Langer, R., Freed, L.E., Vunjak-Novakovic, G. (2000b) In vitro generation of osteochondral composites. *Biomaterials* 21: 2599–2606.

Schaefer, D., Martin, I., Jundt, G., Seidel, J., Heberer, M., Grodzinsky, A.J., Bergin, I., Vunjak-Novakovic, G., Freed, L.E. (2002) Tissue engineered composites for the repair of large osteochondral defects. *Arthritis Rheum.* 46: 2524–2534.

Schreiber, R.E., Ilten-Kirby, B.M., Dunkelman, N.S., Symons, K.T., Rekettye, L.M., Willoughby, J., Ratcliffe, A. (1999) Repair of osteochondral defects with allogeneic tissue-engineered cartilage implants. *Clin. Orthop. Rel. Res.* 367S: S382–S395.

Seidel, J.O., Pei, M., Gray, M.L., Langer, R., Freed, L.E., Vunjak-Novakovic, G. (2004) Long-term culture of tissue engineered cartilage in a perfused chamber with mechanical stimulation. *Biorheology* 41: 445–458.

Sittinger, M., Bujia, J., Minuth, W.W., Hammer, C., Burmester, G.R. (1994) Engineering of cartilage tissue using bioresorbable polymer carriers in perfusion culture. *Biomaterials* 15: 451–456.

Springhorn, J.P., and Claycomb, W.C. (1989) Prepoenkephalin mRNA expression in developing rat heart and in cultured ventricular cardiac muscle cells. *Biochem. J.* 258: 73–78.

Tagil, M., and Aspenberg, P. (1999) Cartilage induction by controlled mechanical stimulation in vivo. *J. Orthop. Res.* 17: 200–204.

Toraason, M., Luken, M., Breitenstein, M., Krueger, J., Biagini, R. (1989) Comparative toxicity of allylamine and acrolein in cultured myocytes and fibroblasts from neonatal rat heart. *Toxicology* 56: 107–113.

Vunjak-Novakovic, G., Freed, L.E., Biron, R.J., Langer, R. (1996) Effects of mixing on the composition and morphology of tissue-engineered cartilage. *AIChE J.* 42: 850–860.

Vunjak-Novakovic, G., Obradovic, B., Bursac, P., Martin, I., Langer, R., Freed, L.E. (1998) Dynamic cell seeding of polymer scaffolds for cartilage tissue engineering. *Biotechnol. Prog.* 14: 193–202.

Vunjak-Novakovic, G., Martin, I., Obradovic, B., Treppo, S., Grodzinsky, A.J., Langer, R., Freed, L.E. (1999) Bioreactor cultivation conditions modulate the composition and mechanical properties of tissue engineered cartilage. *J. Orthop. Res.* 17: 130–138.

Vunjak-Novakovic, G., and Radisic, M. (2004) Cell seeding of polymer scaffolds. *Methods Mol. Biol.* 238: 131–146.

Vunjak-Novakovic, G., and Goldstein, S. (2005) Biomechanical principles of cartilage and bone tissue engineering. In: *Basic Orthopaedic Biomechanics and Mechanobiology*, Mow, V.C., and Huiskes, R. 3rd Ed., Lippincot-Williams and Wilkens, Chapter 8, pp. 343–408.

Wakitani, S., Goto, T., Pineda, S.J., Young, R.G., Mansour, J.M., Caplan, A.I., Goldberg, V.M. (1994) Mesenchymal cell-based repair of large, full-thickness defects of articular cartilage. *J. Bone Joint Surg. Am.* 76A: 579–592.

Wakitani, S., Goto, T., Young, R.G., Mansour, J.M., Goldberg, V.M., Caplan, A.I. (1998) Repair of large full-thickness articular cartilage defects with allograft articular chondrocytes embedded in a collagen gel. *Tissue Eng.* 4: 429–444.

Williams, S.N.O., Burstein, D., Freed, L.E., Gray, M.L., Langer, R., Vunjak-Novakovic, G. (1998) MRI measurements of fixed charge density as a measure of glycosaminoglycan content and distribution in tissue engineered cartilage. *Trans. Orthop. Res. Soc.* 23: 203.

Woessner, J.F. (1961) The determination of hydroxyproline in tissue and protein samples containing small proportions of this imino acid. *Arch. Biochem. Biophys.* 93: 440–447.

Wu, F., Dunkelman, N., Peterson, A., Davisson, T., De La Torre, R., Jain, D. (1999) Bioreactor development for tissue-engineered cartilage. *Ann. N. Y. Acad. Sci.* 875: 405–411.

Wu, Q.-q., and Chen, Q. (2000) Mechanoregulation of chondrocyte proliferation, maturation and hypertrophy: ion-channel dependent transduction of matrix deformation signals. *Exp. Cell Res.* 256: 383–391.

Zimmermann, W.H., Fink, C., Kralish, D., Remmers, U., Weil, J., Eschenhagen, T. (2000) Three-dimensional engineered heart tissue from neonatal rat cardiac myocytes. *Biotechnol. Bioeng.* 68: 106–114.

Zimmermann, W.H., Didie, M., Wasmeier, G.H., Nixdorff, U., Hess, A., Melnychenko, I., Boy, O., Nehuber, W.L., Weyand, M., Eschenhagen, T. (2002a) Cardiac grafting of engineered heart tissue in syngenic rats. *Circulation* 106: I151–I157.

Zimmermann, W.H., Schneiderbanger, K., Schubert, P., Didie, M., Munzel, F., Heubach, J.F., Kostin, S., Nehuber, W.L., Eschenhagen, T. (2002b) Tissue engineering of a differentiated cardiac muscle construct. *Circ. Res.* 90: 223–230.

7

Tissue Engineering of Articular Cartilage

Koichi Masuda[1] and Robert L. Sah[2]

Departments of Orthopedic Surgery and Biochemistry[1], Rush University Medical Center, Chicago, Illinois 60612; Department of Bioengineering[2], University of California-San Diego, La Jolla, California 92093-0412

Corresponding author: rsah@ucsd.edu

Culture of Cells for Tissue Engineering, edited by Gordana Vunjak-Novakovic and R. Ian Freshney

1. INTRODUCTION

Articular cartilage has a limited intrinsic capacity for repair [Mankin, 1982; Buckwalter and Mankin, 1997]. In experimental studies on adult animals, defects, extending through the full thickness of cartilage into the subchondral bone, heal with the formation of fibrocartilaginous repair tissue [Hjertquist and Lemperg, 1971; Shapiro et al., 1993; Wei and Messner, 1999]. Such repair tissue differs from normal cartilage in that it contains relatively low amounts of type II collagen and aggrecan, functionally important markers of the chondrocytic phenotype [Benya and Shaffer, 1982], and a relatively high amount of type I collagen, a fibrillar molecule not present in measurable amounts in normal adult articular cartilage [Eyre et al., 1992]. Associated with the abnormal composition of such repair tissue is a deficiency in biomechanical function [Athanasiou et al., 1995; Wei et al., 1998; Dounchis et al., 2000]. Because of the ineffectiveness of intrinsic repair, methods for stimulating cartilage repair are under active investigation [Sah, et al., 2005].

An underlying strategy of a variety of experimental and clinical methods to enhance repair of an articular cartilage defect is the introduction or recruitment of an appropriate cell population that can undergo chondrogenesis. Recruitment of endogenous chondroprogenitor cells from bone marrow has been achieved by penetration through the base of cartilage defects into the subchondral vasculature by subchondral drilling [Muller and Kohn, 1999], microfracture [Steadman et al., 2001], abrasion [Johnson, 2001], and spongialization [Ficat et al., 1979]. Alternatively, after multiplication of cells ex vivo, autologous chondrocytes have been introduced directly into a cartilage defect and secured with a periosteal flap [Brittberg et al., 1994]. The process of introducing and securing transplanted cells within the defect site may be simplified surgically, and accelerated temporally, by the use of preformed, cell-laden tissues. Although used infrequently, fresh osteochondral allografts, with fully formed and mature articular cartilage, have a long history of clinical success [Bugbee and Convery, 1999; Aubin et al., 2001], and, as a consequence, autologous osteochondral grafts have been introduced [Hangody et al., 2001]. Other chondrogenic tissues, such as periosteum [O'Driscoll and Fitzsimmons, 2001] and perichondrium [Bouwmeester et al., 1999], have also been implanted with the goal of promoting the differentiation of chondrogenic cells within the tissue into chondrocytes capable of producing a cartilage matrix. Likewise, cartilaginous tissues, engineered in vitro, may be suitable for repairing articular defects.

2. CULTURE CONDITIONS FOR CHONDROCYTES

Normally, chondrocytes within cartilage express a phenotype that is characterized by cellular secretion of components, including type II collagen and aggrecan, that are typically present in cartilage and comprise the cartilaginous extracellular matrix [Benya and Shaffer, 1982]. However, when cultured in monolayer under conditions that are considered standard for a variety of cell types, chondrocytes dedifferentiate to a fibroblastic phenotype. The dedifferentiated phenotype is characterized by the abnormal secretion of type I collagen and the reduced secretion of type II collagen and aggrecan. The propensity to dedifferentiate is especially marked when cultures are initiated at a low density and stimulated to proliferate. Compared to the culture of chondrocytes in monolayer, culture of chondrocytes in three-dimensional gel materials promotes the retention and restoration of the chondrocytic phenotype.

When cultured in alginate, chondrocytes can express stably, or be induced to redifferentiate from a dedifferentiated state, the chondrocytic phenotype. Alginate culture has such an effect on chondrocytes not only from animals but even from adult humans, with a stable phenotype having been maintained for as long as 8 months [Häuselmann et al., 1992]. Such cultures accumulate two distinct cartilage matrix components, termed the cell-associated matrix (CM) and the further-removed matrix (FRM). The presence of alginate gel promotes the retention of newly synthesized proteoglycan and collagen around cells, for example, compared to monolayer culture [Mok et al., 1994]. The CM forms rapidly around the cells, and, with time, the FRM forms further away from the cells. The cells with their CM can be recovered from alginate beads cultured for 1–2 weeks by dissolving the alginate polymer with agents that chelate divalent cations [Mok et al., 1994]. A number of in vitro studies have characterized the properties of matrix components and the regulation of their metabolism in this culture system [Häuselmann et al., 1992, 1994; Mok et al., 1994; Petit et al., 1996].

3. PREPARATION OF MEDIA AND REAGENTS

3.1. Media

3.1.1. Isolation Medium

Sterile DMEM-F-12, 50 μg/ml gentamicin, 360 μg/ml L-glutamine.

3.1.2. Culture Medium

Sterile DMEM-F-12, 50 μg/ml gentamicin, 360 μg/ml (2.5 mM) L-glutamine, 25 μg/ml ascorbic acid. Usually with 10% FBS, but sometimes 20% or serum-free culture medium as indicated in the text.

3.2. Saline Wash

Sterile NaCl, 0.9% (0.15 M), with 100 μg/ml gentamicin

3.3. Enzymes

3.3.1. Pronase Digestion Solution

(i) Prewarm the isolation medium to 37 °C in a sterile 50-ml centrifuge tube.

(ii) Mix the total amount of Pronase enzyme needed (0.10 g or 0.20 g) with 50 ml prewarmed isolation medium.

(iii) Dissolve for at least 1 hour.

(iv) Filter sterilize the Pronase enzyme digestion solution, using a 0.22-μm sterile Millipore Steriflip-GP 50-ml filter unit or Steritop-GV Filter Unit (150 ml).

Use 0.2% Pronase for bovine cartilage and 0.4% Pronase for adult human cartilage.

Total volume of Pronase enzyme digestion solution needed: ~50 ml/8 g of cartilage to be digested.

3.3.2. Collagenase-P Digestion Solution

Prepare a 2× collagenase-P enzyme digestion solution as follows:

(i) Prewarm 25 ml culture medium to 37 °C in a sterile 50-ml centrifuge tube.

(ii) Mix the total amount of collagenase-P enzyme needed (12.5 mg per 25 ml) with serum-free culture medium.

(iii) Dissolve for at least 1 h.

(iv) Filter sterilize the 2× collagenase-P enzyme digestion solution, using a 0.22-μm sterile Millipore Steriflip-GP 50-ml filter unit.

Total volume of the final 1× Collagenase-P enzyme digestion solution needed: ~50 ml/8 g of cartilage to be digested (the same as for the Pronase digestion).

3.3.3. Papain Enzyme Digestion Solution

Papain, 20 μg/ml in 0.1 M sodium acetate (NaAc), 0.05 M EDTA, pH 5.53. Prepare as follows:

(i) Prepare papain buffer solution—0.1 M NaAc, 0.05 M EDTA, pH 5.53

NaAc (MW 82.03)	4.102 g
EDTA (MW 372.2)	9.360 g
UPW	500 ml

Add UPW to less than 500 ml. Adjust the pH to 5.53 and then add UPW to a final volume of 500 ml. Store at 4 °C.

(ii) Prepare papain enzyme stock—23 mg protein/ml in UPW.

(iii) Just before use, activate the papain buffer solution by adding 5 mM L-cysteine hydrochloride hydrate (21.95 mg L-cysteine/25 ml of papain buffer).

(iv) Withdraw 5 ml for use with the controls.

(v) Add 17.4 μl 23 mg protein/ml papain enzyme stock to 20 ml activated (with L-cysteine) papain buffer to make a 20 μg/ml papain enzyme digestion solution.

Note: The concentration of protein per milliliter for each lot of papain enzyme changes. Therefore, the amount of papain enzyme to add must be calculated for each lot (bottle) of enzyme.

3.3.4. Pepsin Stock Solution

Pepsin, 1 mg/ml in 0.05 M acetic acid

3.3.5. Pancreatic Elastase

1 mg/ml stock solution of pancreatic elastase in high-salt Tris buffer (See Section 3.6.8.)

3.4. Alginate Encapsulation

3.4.1. Alginate Solution

Sodium alginate, 1.2% in 0.9% (0.15 M) NaCl; ***prepare fresh***

0.9 g	Alginate
75 ml	0.9% (0.15 M) NaCl, sterile solution

(i) Add 75 ml of 0.9% NaCl, sterile solution, to a sterile sample cup along with a sterile stir bar.

(ii) Add 0.9 g alginate.

(iii) Stir for at least 3 h to dissolve.

(iv) Once dissolved, filter the solution through a sterile 0.22-μm Millipore Steriflip-GP 50-ml filter unit. Do not autoclave the alginate solution.

3.4.2. Alginate Dissolving Buffer

Without EDTA: sodium citrate, 0.055 M, in 0.15 M NaCl, pH 6.8.

Na citrate (MW = 294.1)	16.2 g
NaCl (MW = 58.44)	8.8 g
UPW	1 L

With EDTA: sodium citrate, 0.055 M, in 0.03 M EDTA, 0.15 M NaCl, pH 6.8

(i) Add UPW to less than 1 L.

(ii) Adjust the pH to 6.80 and then add UPW to a final volume of 1 L.

(iii) Filter through a sterile 1-L bottle filter system.

(iv) Store at 4 °C.

3.4.3. Calcium Chloride Solution

$CaCl_2$, 102 mM

$CaCl_2$ (MW = 147.0)	15 g
UPW	1 L

(i) Dissolve and then filter through a sterile 1-L bottle filter system.

(ii) Store at 4 °C

3.5. Proteoglycan Assays

3.5.1. Proteoglycan Standard Stock

Proteoglycan, 2 mg/ml in 0.05 M NaAc and 0.05% Tween 20 in UPW, pH 6.8
Aliquot and freeze.

3.5.2. Proteoglycan Dilution Buffer

NaAc, 0.05 M, with 0.05% Tween 20, at pH 6.8

3.5.3. Dimethylmethylene Blue (DMMB) Dye Reagent

DMMB, 16 μg/ml, in 0.03 M sodium formate, 0.2% V/V formic acid, pH 6.8.

To prolong the activity of the DMMB dye, it is desirable to first solubilize the reagent in organic solvents before diluting it into a mostly aqueous environment.

(i) Dissolve 16 mg DMMB dye in 5.0 ml absolute ethanol. Solubilize for 15 min at room temperature (RT).

(ii) Add 2.0 ml formic acid and 2.0 g sodium formate to approximately 20 ml UPW. Dissolve and add to the ethanol-DMMB solution.

(iii) Stir for 20 min at RT.

(iv) Add UPW to less than 1 L.

(v) Adjust the pH to 6.80.

(vi) Add UPW to a final volume of 1 L.

Note: By adjusting the pH before making the final volume 1 liter, the dye solution should be more reproducible.

Do not filter or refrigerate the DMMB reagent!

DMMB activity is transient, so shelf life is only 1 month.

3.5.4. Guanidine Hydrochloride (GuHCl)

Prepare a 2.88 M GuHCl stock solution in 0.05 M NaAc pH 6.8 as follows:

(i) Prepare a 0.078 M NaAc pH 7.0 solution

NaAc (MW = 82.03), 0.078 M	6.40 g/L

(ii) For 1 L, mix the 0.078 M NaAc pH 7.0 solution with an 8 M GuHCl solution as described below:

GuHCl, 8 M	360 ml
NaAc, 0.078 M, pH 7.0	640 ml

(iii) The pH should be measured after mixing to validate that it is 6.8.

3.6. Collagen Assays

3.6.1. Hydroxyproline Drying Solution

TEA-methanol: methanol:dH_2O:triethylamine (TEA) (2:2:1)

Methanol	500 μl
Deionized water	500 μl
TEA	250 μl

Store at 4 °C
This will yield enough drying solution to dry 25 samples twice.

3.6.2. Hydroxyproline Derivatizing Solution

Methanol:dH_2O:TEA: phenylisothiocyanate (PITC)—7: 1:1:1; ***prepare fresh***.

(i) For 25 samples prepare 500 μl derivatizing solution. Always make in excess, so for 25 samples, prepare 600 μl total.

Methanol	420 μl
TEA	60 μl
Deionized water	60 μl

(ii) Finally, add 60 μl PITC to the above solution.

△ ***Safety note.*** **PITC IS VERY HARMFUL. WEAR GLOVES AND DO THIS STEP IN A FUME-HOOD**. Any excess derivatizing solution that is not used with samples should not be used again and should be stored at −20 °C until disposal. This should be treated as hazardous waste for disposal purposes.

Note: Unopened vials of PITC are stored at RT. Once opened, any unused PITC should first be "bubbled" under nitrogen, to remove all air, and then the vial should be stored at $-20\,^{\circ}C$. PITC treated in this manner is good for use for no longer than 3 weeks. For disposal, expired, unused PITC should be treated as hazardous waste.

3.6.3. Hydroxyproline HPLC Standard

Hydroxy-L-proline, 0.2 mg/ml:

Hydroxy-L-proline	10 mg
HPLC-grade water	50 ml

(i) Prepare 100-μl aliquots and lyophilize.

Standard concentration: 20 mg/tube; 20 ng/μl.

Lyophilized standard is derivatized simultaneously with samples and then reconstituted in 1 ml of reconstitution buffer.

3.6.4. Collagen Reconstitution Buffer

Na_2HPO_4, 4.75 mM in 5% V/V acetonitrile, pH 7.4.

(i) Prepare buffer: sodium phosphate, dibasic, anhydrous (Na_2HPO_4), 5 mM

Na_2HPO_4 (MW = 142.0)	710 mg
HPLC-grade water	1 L

(ii) Titrate to pH 7.4 with 10% V/V phosphoric acid (H_3PO_4)

(iii) Add acetonitrile:

Acetonitrile	50 ml
Na_2HPO_4 buffer pH 7.4, 5 mM	950 ml

(iv) Store at $4\,^{\circ}C$

3.6.5. Hydroxyproline HPLC Eluent A

Acetonitrile:dH_2O:140 mM sodium acetate trihydrate buffer (6:4:90), pH 6.4 with 0.5 ml/l TEA.

(i) Prepare 140 mM sodium acetate trihydrate-TEA solution

Sodium acetate trihydrate (MW = 136.08)	38 g
HPLC-grade water	1800 ml

(ii) Once dissolved, add 1 ml TEA

(iii) Titrate to pH 6.4 with glacial acetic acid

(iv) Bring resulting solution to 2 L in a 2-L volumetric flask with HPLC-grade water

(v) In a separate 2-L glass bottle add:

Acetonitrile (CH_3CN)	120 ml
Sodium acetate trihydrate, 140 mM-TEA solution	1880 ml

(vi) Store at 4 °C

(vii) Before use, de-gas in a sonicator.

3.6.6. Hydroxyproline HPLC Eluent B

60% Acetonitrile:

Acetonitrile	600 ml
HPLC-grade water	400 ml

(i) Prepare in a glass bottle.

(ii) Store at 4 °C

(iii) Before use, de-gas in a sonicator.

3.6.7. Collagen Extraction Buffer

Guanidine hydrochloride (GuHCl), 6 M, in 50 mM Tris-HCl, pH 7.4

3.6.8. High-salt Tris Buffer

Tris, 1.0 M, in 2.0 M NaCl, 50 mM $CaCl_2$, pH 8.0

3.7. DNA Assay

3.7.1. Proteinase K Buffer

Tris-HCl, 0.05 M, $CaCl_2$, 0.005 M, pH 7.0

Tris-HCl (MW = 121.4)	6.07 g
$CaCl_2$ (MW = 147.0)	0.735 g
UPW	1 L

(i) Add UPW to less than 1 L.

(ii) Adjust the pH to 7.0 with concentrated HCl.

(iii) Add UPW to a final volume of 1 L.

(iv) Store at 4 °C.

3.7.2. Proteinase K Solution

(i) Prepare proteinase K Stock Solution, 5 mg/ml

Proteinase K	100 mg
Proteinase K buffer (See Section 3.7.1)	20 ml

(ii) Dissolve and then freeze 1.0-ml aliquots in 1.5-ml tubes.

(iii) Store at −20 °C

(iv) Prepare a working solution of proteinase K: 125 μg/ml proteinase K in proteinase K buffer (See Section 3.7.1) by adding 500 μl 5 mg/ml proteinase K stock solution in proteinase K buffer to 19.5 ml proteinase K buffer.

Final working solution: 20 ml 125 μg/ml proteinase K (This is a 1/40 dilution).

3.7.3. DNA Standard Solution

Stock Solution: 1 mg/ml calf thymus DNA in UPW.

Prepare approximately 20 ml solution; stir overnight at room temperature, aliquot, and freeze.

Working Solution: 40 μg/ml calf thymus DNA in proteinase K digestion buffer.

Add 40 μl Stock Solution to 960 μl proteinase K digestion buffer.

Prepare Working Solution fresh, as needed.

3.7.4. Hoechst 33258 Dye Buffer

Tris, 0.1 M, in 0.1 M NaCl, pH 8.0

Tris (MW = 121.14)	12.114 g
NaCl (MW = 58.44)	5.844 g

(i) Add UPW to less than 1 L.

(ii) Adjust the pH with concentrated HCl and then add UPW to a final volume of 1 L.

(iii) Filter solution with a 0.22-μm filter.

A higher concentration of Tris and a higher pH than the typical optimum for the Hoechst dye are used in this dye buffer to compensate for the low pH in the papain digestion buffer and the reduced volumes used in the microtiter plate. If the above-listed papain digestion buffer (See Section 3.3.3) is not used, the following dye buffer should be used: 10 mM Tris, 0.1 M NaCl, pH 7.4.

(iv) Store at 4 °C.

Note: This dye buffer is for samples that already contain EDTA in solution. For samples that do not contain EDTA, add 1 mM EDTA to the dye buffer.

3.7.5. Hoechst 33258 Dye Solution

△ ***Safety note.*** HOECHST 33258 DYE IS A MUTAGEN. GLOVES SHOULD BE WORN AT ALL TIMES WHEN HANDLING!

Stock Solution: 1 mg/ml in UPW; stable at 4 °C for 6 months in the dark (wrap the tube in aluminum foil).

Working Solution: 1 μg/ml in Hoechst dye buffer (15 ml/plate; add 15 μl Stock Solution to 15 ml Hoechst dye buffer).

Prepare immediately before use and store in a dark tube/bottle.

4. HARVESTING CARTILAGE AND ISOLATION OF CHONDROCYTES

Protocol 7.1. Harvest of Articular Cartilage Tissue (Day 1)

Reagents and Materials

Sterile

- ❑ Isolation medium (See Section 3.1.1)
- ❑ Saline wash (See Section 3.2)
- ❑ Alcohol, 70%
- ❑ Sample cups, 1 for every 3 hooves
- ❑ Scalpels, #10, #15 blades

Nonsterile

- ❑ Bovine lower front limbs, 3 or more, distal to the carpus, isolated by transecting at the carpus.
- ❑ Paper towels
- ❑ Clamp for holding leg
- ❑ Balance, 100 mg–100 g range

Protocol

A. Bovine Articular Cartilage Tissue–Metacarpophalangeal Joint

One experiment should pool cartilage from 3 or more hooves.

(a) Using sterile technique, fill a sterile sample cup with 50 ml isolation medium.
(b) Cap the cup with the lid before removing it from the tissue culture hood. Prepare 1 sample cup for every 3 hooves being cut.
(c) Preweigh each container with medium alone (without tissue).
(d) Thoroughly clean the leg by running under hot water while scrubbing with a brush.
(e) Use a scalpel with a new #10 blade to remove the skin from the leg:
  i) Cut around the leg just above the hoof.
  ii) Cut lengthwise from the top of the leg to just above the first incision.
  iii) Remove skin.
(f) Thoroughly wash the skinned leg with running water and remove all contaminated material.
(g) Dry the leg with paper towels and place clean dry paper towels on the cut end of the leg to prevent blood from running down the leg.
(h) Place the leg in an appropriate holder.

(i) Place paper towels around the hoof.
(j) Clean the outside of the metacarpophalangeal joint with 70% EtOH.
(k) Using sterile technique (mask and gloves recommended), open the joint with a scalpel (new #10 blade). TO MAINTAIN STERILITY, DO NOT TOUCH THE CARTILAGE INSIDE THE JOINT OR THE SYNOVIAL FLUID WITH THE BLADE THAT HAS TOUCHED THE OUTSIDE OF THE JOINT.
(l) Cut the cartilage from the lower joint, using a scalpel with a #15 blade.
(m) Transfer the cartilage pieces into the sample cup containing medium with sterile forceps.
(n) Cut the cartilage of the upper joint, using a scalpel with a #10 blade, and transfer to sterile cup as in Step (m).
(o) After all possible usable cartilage in the joint is removed, weigh the container with the cartilage.
(p) In the tissue culture hood, remove the medium from each sample cup, using a sterile pipette.
(q) Wash twice, 5 min each wash, with sterile 0.9% NaCl, 100 μg/ml gentamicin (50 ml/wash).
(r) Combine the cartilage from up to 4 hooves in 1 sample cup in preparation for enzyme digestion.

B. Human Articular Cartilage Tissue
(a) If the whole joint is provided, follow a similar procedure to isolate cartilage as *A*.

△ *Safety note.* To minimize the chance for transmission of infectious material, use extra care to avoid injury and splashing. Also, use proper protective gear such as a surgical gown and a facemask with a plastic shield.

(b) In the tissue culture hood, remove the medium from each sample cup with a sterile pipette.
(c) Wash twice, 5 min each wash, with sterile 0.9% NaCl, 100 μg/ml gentamicin (50 ml/wash).
(d) Cut the cartilage into $5 \times 5\text{-mm}^2$ pieces to obtain optimum digestion, keeping the tissue moist at all times.

Protocol 7.2. Isolation of Chondrocytes from Bovine or Human Articular Cartilage (Day 1)

Reagents and Materials

Sterile

- ❑ Culture medium (See Section 3.1.2)
- ❑ Isolation medium (See Section 3.1.1)
- ❑ Sample cup with cartilage pieces
- ❑ Pronase solution (See Section 3.3.1)

- ❑ Collagenase-P solution (See Section 3.3.2)
- ❑ Stirrer bar
- ❑ Cell strainer: 40-μm mesh sieve
- ❑ Centrifuge tube, 50 ml

Nonsterile

- ❑ Trypan Blue, 0.4%
- ❑ Electronic cell counter, or hemocytometer

Pronase Digestion

Protocol

(a) Add the sterile filtered Pronase enzyme digestion solution to the sample cup containing the cartilage pieces.

(b) Insert the stirrer bar and incubate with the cap of the sample cup attached loosely for 1 h, with gentle stirring, in a 37 °C, 5% CO_2 tissue culture incubator.

(c) At the end of the Pronase enzyme digestion incubation period, wash the cartilage 3 times, 2 min each, with isolation medium (30–50 ml per wash).

Collagenase-P Digestion

(d) Add 25 ml 2× Collagenase-P enzyme digestion solution and 25 ml culture medium to the sample cup containing the cartilage pieces, along with a sterile stir bar. The final Collagenase-P concentration is 0.0025%.

(e) Loosely cap the specimen cup and incubate for a maximum of 16 h, with gentle stirring, in a 37 °C, 5% CO_2 tissue culture incubator.

(f) At the end of the Collagenase-P enzyme digestion incubation period, transfer the suspension of released chondrocytes through a 40-μm cell strainer sieve into a 50-ml centrifuge tube.

(g) Centrifuge at 110 *g* for 10 min at 4 °C.

(h) Wash the chondrocytes 3 times as follows:

   i) Discard the supernatant and resuspend the pellet in 25 ml isolation medium.

   ii) Centrifuge at 110 *g* for 10 min at 4 °C.

(i) After the last centrifugation, discard the supernate and resuspend the chondrocytes in 10 ml isolation medium.

Determination of Cell Number

(j) Count the number of chondrocytes in suspension with an electronic cell counter (e.g. Coulter Z1 Particle Counter, upper threshold: 21.54 μm, lower threshold: 8 μm) or hemocytometer.

Determination of Cell Viability

(k) Add 100 μl total chondrocyte cell suspension to 100 μl Trypan Blue stain 0.4%.

(l) Mix and immediately add to a hemocytometer for viewing under the microscope. Dead cells will appear blue because Trypan Blue stain only penetrates the membranes of dead cells. A photographic record should be made.

---

5. CULTURE OF CHONDROCYTES

5.1. Proliferating Monolayer

Protocol 7.3. Monolayer Culture of Articular Chondrocytes Under Proliferating Conditions

Reagents and Materials

- ❑ Culture medium (See Section 3.1.2)
- ❑ Isolation medium (See Section 3.1.1)
- ❑ Pronase solution, 0.2% (See Section 3.3.1)
- ❑ Collagenase-P solution (See Section 3.3.2)
- ❑ Culture flasks, 150 cm^2
- ❑ Centrifuge tube, 50 ml

Protocol

Initial Seeding

(a) Plate cells at a concentration of 2.5×10^6 cells per 150-cm^2 flask (17,000 cells/cm^2) with a total volume of 20 ml/flask.
(b) Change the medium every other day until cells reach confluence ($\sim$1.5 weeks for first passage, $\sim$1 week for second and subsequent passages).

Passage of Cells after Monolayer Culture

(c) Aspirate media from flasks.
(d) Add equal volumes of standard media and 0.2% Pronase solution (a minimum of 5 ml standard media and 5 ml 0.2% Pronase solution should be used) to each flask. Place in the incubator for 10–30 min or until cells are released from the flask.
(e) Transfer cells from all flasks from the same donor to one 50-ml tube. Rinse each flask with 5–10 ml culture medium, pipetting off any remaining cells. Combine the cells from the rinse with the cells in the 50-ml tube.
(f) Centrifuge at 110 g for 10 min at 4 °C.
(g) Aspirate the medium, add 25 ml isolation medium, and centrifuge at 110 g for 10 min at 4 °C.
(h) Aspirate the media again, add 12.5 ml culture medium, and transfer cells to a specimen cup containing a sterile stir bar.
(i) Prepare 12.5 ml 2× sterile Collagenase-P solution (0.05% Collagenase-P) in a 50-ml tube and then add to the specimen cups.
(j) Loosely cap the specimen cups and place in the incubator at 37 °C and 5% CO_2 for 1.5 h.
(k) At the end of the Collagenase-P enzyme digestion incubation period, transfer the suspension of released chondrocytes through a 40-μm cell strainer sieve filter into a sterile 50-ml centrifuge tube.
(l) Centrifuge at 110 g for 10 min at 4 °C.

(m) Resuspend in growth medium and reseed at 2.5×10^6 cells per 150-cm^2 flask (17,000 cells/cm^2) with a total volume of 20 ml/flask, or proceed to alginate encapsulation (See Protocol 7.4).

5.2. Alginate Beads

Protocol 7.4. Encapsulation of Chondrocytes in Alginate Beads

Reagents and Materials

Sterile

- ❑ Isolation medium (See Section 3.1.1)
- ❑ Culture medium (See Section 3.1.2)
- ❑ Sodium alginate solution (See Section 3.4.1)
- ❑ NaCl solution, 0.9%
- ❑ $CaCl_2$ solution, 102 mM (See Section 3.4.3)
- ❑ Stirrer bar
- ❑ Pipettes, including 50 ml
- ❑ Petri dishes, 10 cm, deep well

Protocol

(a) Prepare a 1.2% alginate solution in sterile 0.9% NaCl; start early, to allow sufficient time for dissolving the alginate. Prepare 25 ml 1.2% alginate solution for cartilage from each hoof (e.g., 150 ml for cartilage from 6 hooves).

(b) Centrifuge the original total chondrocyte cell suspension at 110 *g* for 10 min at 4°C.

(c) Remove the supernate and add sufficient sterile 1.2% alginate solution to have 4×10^6 chondrocytes/ml of 1.2% alginate solution. Determine the volume of 1.2% alginate solution needed by dividing the total cell number by 4×10^6.

(d) When resuspending the chondrocytes in the 1.2% alginate solution, do not add more than 5 ml alginate solution initially. Mix well and then add the remaining alginate solution needed to obtain the desired chondrocyte concentration.

(e) Add 50 ml sterile 102 mM $CaCl_2$ solution to a sterile sample cup.

(f) Add 7 ml chondrocyte-alginate mixture to 50 ml sterile 102 mM $CaCl_2$ solution. This step is particularly important for human chondrocyte preparations. Up to 12 ml chondrocyte-alginate mixture may be added to 50 ml sterile 102 mM $CaCl_2$ solution:

  i) The $CaCl_2$ solution should be stirred constantly with a sterile stirrer bar.
  ii) Use either a sterile syringe that has a sterile 22-gauge needle attached or a disposable sterile pipette (5 ml or 10 ml) attached to a peristaltic pump.
  iii) Suspend the syringe/pipette above the sample cup at a 45° angle, making sure that the end of the needle/pipette is at a 90° angle with the surface of the $CaCl_2$ solution. A distance of 5 cm from the end of the needle/pipette to the surface of the $CaCl_2$ solution is desired.

iv) Discard the first few drops of the chondrocyte/alginate mixture and then express the chondrocyte/alginate mixture into the $CaCl_2$ solution. The speed should be adjusted so that all of the solution is added in less than 2 min.

(g) Equilibrate the newly formed chondrocyte-containing alginate beads in the $CaCl_2$ solution for 8–10 min, but not longer than 10 min.

(h) Wash the beads 3 times with 25–30 ml sterile 0.9% NaCl, 10 min each wash. Use a Pasteur pipette connected to the vacuum line in the hood to remove the wash solutions.

(i) Wash the beads once for 5 min with 25 ml sterile isolation medium.

(j) Collect the beads with a 50-ml sterile pipette slowly (avoid breaking beads) and transfer to a sterile 10-cm non-tissue culture-grade Petri dish (deep well), where they can be counted and then:

i) Batch culture, 500 beads/one or two 10-cm dishes

or

ii) Dispense into 6-well (150–200 beads/well), 12-well (27 beads/well), or 24-well (9 beads/well) tissue culture plates according to the design of the experiment.

(k) Add culture medium: 22 ml/500 beads; 1.2 ml/27 beads; 400 μl/9 beads.

(l) Incubate in a tissue culture incubator at 37 °C/5% CO_2.

Maintenance of Chondrocytes in Alginate Culture

(m) Change medium daily. Use a 10-ml pipette to remove the medium, with caution to avoid sucking the beads into the pipette. Use 22 ml culture medium per 500 beads.

---

6. FABRICATION OF CARTILAGINOUS TISSUE BY THE SCAFFOLD-FREE ALGINATE-RECOVERED CHONDROCYTE (ARC) METHOD

Cell-laden cartilaginous tissues can be engineered in vitro and may provide a suitable graft material for cartilage repair. Such tissues are typically formed by multiplication of cells from an appropriate source (e.g., by monolayer culture starting at a low seeding density), differentiation of the cells to the chondrocytic phenotype, and then growth in vitro under controlled conditions. Compared to direct implantation of cells, such engineered tissues would be predicted to shorten the postoperative rehabilitation time because the processes of differentiation and matrix formation have already occurred or begun in vitro. Such tissues have been synthesized primarily with immature chondrocytes or chondroprogenitor cells, in combination with various types of scaffolds.

Chondrogenic cells have often been mixed with or infiltrated into a number of scaffold materials to form cartilaginous tissue. Such materials are typically degradable and synthetic, such as polylactic acid [Dounchis et al., 2000] or polyglycolic acid [Dunkelman et al., 1995; Vunjak-Novakovic et al., 1999], or natural, such as those derived from or analogous to collagen [Wakitani et al., 1989], hyaluronan (HA) [Robinson et al., 1990; Solchaga et al., 2001], fibrin [Itay et al., 1987; Hendrickson et al., 1994], and alginate-fibrin mixture [Perka et al., 2000]. Other materials, such as agarose, have been used primarily as model systems for analyzing the regulation of tissue formation [Buschmann et al., 1992, 1995; Mauck et al., 2003]

Other methods have been developed to form implants that are composed only of cells and their products. Immature chondrocytes can form cartilaginous tissue when cultured as a monolayer at high density in the absence of a scaffold [Adkisson et al., 2001] or as a multilayer [Yu et al. 1997], a configuration that recapitulates the high density of cells in fetal cartilage [Li et al., 2001]. On the other hand, chondrocytes from adult articular cartilage form only a very small amount of matrix when cultured under the same conditions [Adkisson et al., 2001]. For the production of cartilaginous tissue in vitro, we developed a two-step culture method, termed the alginate-recovered chondrocyte (ARC) method [Masuda et al., 2003].

The ARC method allows formation of cartilaginous tissue from animal or human cells with culture in alginate gel as an intermediate step. The methodology is robust enough to use cells from adults, which are typically difficult to stabilize in a chondrogenic phenotype. The first step of the ARC method consists of culturing chondrocytes encapsulated in alginate gel under conditions that maintain the normal phenotype of chondrocytes and modulate the formation of a CM rich in aggrecan molecules. After 1 week, the alginate gel is solubilized and the cells with their attached CM are recovered and cultured on a porous membrane. Within 1 further week the alginate-recovered chondrocytes become integrated into a cohesive cartilaginous tissue mass containing abundant amounts of aggrecan and type II collagen and minimal amounts of type I collagen [Masuda et al., 2003].

Once there is a sufficient quantity of cell-associated matrix in the cultured alginate beads, the beads can be dissolved and the cells with their cell-associated matrix recovered. For bovine cells, an appropriate preculture period is 7–10 days when cultured in the presence of 20% FBS. For human cells, approximately 2 weeks is effective, when incubation is done in medium supplemented with 200 ng/ml osteogenic protein 1 (OP-1; BMP-7) and 20% FBS.

Protocol 7.5. Release of Chondrocytes from Alginate Beads

Reagents and Materials

Sterile

- ❑ Alginate dissolving buffer (See Section 3.4.2)
- ❑ Saline wash (See Section 3.2)
- ❑ Isolation medium (See Section 3.1.1)
- ❑ Culture medium (See Section 3.1.2)
- ❑ Spatula
- ❑ Centrifuge tube, 50 ml

Protocol

(a) Collect beads in a sterile 50-ml tube with a sterile spatula as follows:
   i) Bovine, collect 300 beads/insert (See Protocol 7.6)
   ii) Human, collect 350 beads/insert
   iii) Porcine, collect 300–350 beads/insert
(b) Add 20 ml alginate dissolving buffer to each tube.
(c) Mix the tube gently. **Do Not Vortex.** Let it sit at room temperature for 30 min.
(d) After the beads have dissolved, centrifuge at 110 *g* for 10 min at 4 °C.
(e) Remove the supernate and wash the cell pellet 2× with saline wash. Centrifuge at 110 *g* for 10 min at 4 °C each time, removing the wash solutions with a Pasteur pipette connected to the vacuum line in the hood. Leave approximately 2 ml after each aspiration.
(f) Wash the cell pellet once with 20 ml isolation medium.
(g) Centrifuge at 110 *g* for 10 min at 4 °C.
(h) Resuspend in culture medium as required.

Protocol 7.6. Culture of ARC Chondrocytes and Associated Matrix in Filter Well Inserts

Reagents and Materials

Sterile

- ❑ Culture medium (See Section 3.1.2)
- ❑ Filter well inserts, 23 mm, 4-μm porosity, polyethylene terephthalate (PET)
- ❑ Multiwell plate, 6 well, deep well
- ❑ Scalpel

Protocol

Seeding

(a) Add 15 ml culture medium to each well of the 6-well deep-well companion plate before positioning the cell culture insert.
(b) Resuspend the cells in 5 ml culture medium.
(c) Pipette the 5 ml cell suspension onto the 0.4-μm inserts in the 6-well plates. (***Note***: inspect inserts for holes before adding the cell suspension.)
(d) The cell suspension should be allowed to settle on the insert for 1 full day without movement before changing the medium.
(e) Change the medium every other day after the first 2 days (15 ml outer well, followed by 5 ml to the insert).
(f) Collect inserts after 2–3 weeks.

Maintaining Cultures of ARC Tissue

(g) Remove medium from both the bottom of the deep-well plate and the filter well insert.

(h) Carefully add complete medium onto the insert. (***Note:*** do not feed dropwise at the beginning of culture; touch the pipette to the wall of the insert and add slowly.)

Recovering ARC Tissue

(i) Once culture is complete, remove each insert from the 6-well plate and place in a Petri dish.

(j) Cut the PET membrane along its circumference with a scalpel and peel away the membrane carefully to release the de novo ARC tissue.

7. CRITERIA FOR EVALUATION OF CARTILAGINOUS TISSUE FORMATION

There are a variety of ways to test whether engineered cartilaginous tissue has a composition, structure, function, and metabolism similar to cartilage and indicative of cells expressing the normal chondrocytic phenotype. Biochemical analyses include the determination of proteoglycan (PG) content, collagen content and type, and cellularity assessed as DNA content. Biomechanical analyses include the determination of compressive and tensile functions. The results of these analyses can be compared to those of normal cartilage, with consideration of the natural changes in the tissue that occurs with growth, aging, and degeneration as well as the heterogeneity of cartilage with depth from the articular surface and location in various joints [Sah, 2002].

7.1. Proteoglycan

Protocol 7.7. Determination of Proteoglycan Content of Chondrocyte Cultures from Filter Well Inserts

Reagents and Materials

Nonsterile

- ❑ Excised insert cultures (See Protocol 7.6 Step (j))
- ❑ Papain enzyme digestion solution, 20 μg/ml (See Section 3.3.3)
- ❑ PG standard stock, 2 mg/ml (See Section 3.5.1)
- ❑ PG dilution buffer (See Section 3.5.2)
- ❑ Boiling water bath
- ❑ Guanidine hydrochloride (See Section 3.5.4)
- ❑ Dimethylmethylene blue (DMMB) dye reagent (See Section 3.5.3)

Protocol

A. Papain Digestion of Sample

Digestion can be performed on all of the samples from one experiment at the same time.

(a) Add an appropriate amount of the 20 μg/ml papain enzyme digestion solution to each sample. The volume required for a 23-mm diameter piece of ARC tissue is at least 3 ml.

(b) Incubate the samples (and blank controls; See Section 3.3.3) overnight at 56–60 °C.
(c) At the completion of digestion, all samples should be boiled for 5 min to inactivate the enzyme.

B. Assay for PG Content

(d) Determine the content of sulfated PG molecules in the papain-digested sample with a modified dimethylmethylene blue (DMMB) dye binding method [Chandrasekhar et al., 1987]. The dynamic range of the DMMB assay is very narrow. Obtaining a pinkish color on PG-dye binding indicates that the dye is already saturated and that dilution to a lower concentration is necessary. For that reason, it is recommended to perform at least 6 dilutions for each sample in a 96-well plate to accurately measure the amount of glycosaminoglycan by the DMMB method.

Standard Dilution Scheme

i) From the 2 mg/ml PG stock solution, prepare a working stock solution of 16 μg/ml PG. This is a dilution factor of 1/125.
ii) A total of 7 dilutions should be performed for the standard, in duplicate, for each plate.
iii) Place standards on the top and bottom of the plate (A2-A9, H2-H9).
iv) Wells A1 and H1 should be left empty for the blank.
v) Pipette 263 μl 16 μg/ml PG standard working stock solution to the first well (A2, H2).
vi) To the remaining wells pipette 75 μl dilution buffer.
vii) Perform serial dilutions of 1/1.4 for the remaining wells by transferring 188 μl with a multichannel pipette.
viii) 188 μl of A2 is transferred to A3 and mixed, then 188 μl of A3 is transferred to A4, etc. A dilution of 188 μl up to 263 μl is a 1/1.4 dilution factor

Final volume in all standard wells after dilution = 75 μl.
Concentration Range of Standard (μg/ml): 16; 11.43; 8.16; 5.83; 4.16; 2.97; 2.12

Note: The working range of the standard is only from 11.43 μg/ml to 2.12 μg/ml. The concentration of 16 μg/ml is a convenient starting dilution by which to obtain the working range and should not be used in calculations because it is off scale.

Sample Dilution Scheme

i) Pipet 150 μl diluted sample into the first well.
ii) To the remaining wells pipet 75 μl dilution buffer.
iii) Perform serial dilutions of 1/2 by transferring 75 μl; 75 μl of B1 is transferred to B2 and mixed, then 75 μl of B2 is transferred to B3, etc. A dilution of 75 μl up to 150 μl is a 1/2 dilution factor, and the final volume in all sample wells after dilution is 75 μl.

Addition of Guanidine Hydrochloride (GuHCl)

It is important to maintain a concentration of 0.24 M GuHCl/well to prevent PG-dye complex precipitation as well as the interference of HA and/or DNA.

i) After dilution of standards and samples is complete, add 25 μl 2.88 M GuHCl to each well.
ii) The final concentration per well is 0.24 M after the addition of DMMB reagent.

Reaction with DMMB

iii) Add 200 μl DMMB reagent to each well with a multichannel pipette.

Blank: (Microtitration wells A1, H1) 275 μl dilution buffer + 25 μl 2.88 M GuHCl

Background: (Microtitration wells A12, H12) 75 μl dilution buffer + 25 μl 2.88 M GuHCl +200 μl DMMB reagent

iv) Put plates on a shaker for 30 s to ensure thorough mixing.
v) Measure the absorbance for each plate at 530 nm and 595 nm

Note: It is important to read the plates as soon as possible after the addition of DMMB because the color intensity decreases over time. If assaying more than one plate, reagent should not be added to all plates at once. Each plate should have its absorbance read immediately after the addition of DMMB.

Once the 530-nm and 595-nm absorbance data are measured, the ratio should be calculated by using software provided with the plate reader or an Excel spreadsheet.

Calculation of Amount of PG in Samples

A plot of absorption against PG standard concentrations yields a linear response between roughly 2.97 μg/ml and 11.43 μg/ml. The equation of this line can be used to determine the amount of PG present in the unknown samples. It is important to correct for the dilution factor when necessary.

7.2. Collagen Content

The content of hydroxyproline can be measured by phenylisothiocyanate (PITC) derivatization and isocratic reverse-phase high-performance liquid chromatography (HPLC) [Chiba et al., 1997] as a measure of collagen molecules. To calculate the amount of collagen content in each sample, the measured hydroxyproline content is multiplied by a conversion factor of 7.25 [Herbage et al., 1977; Pal et al., 1981].

Protocol 7.8. Determination of Collagen Content of Chondrocyte Cultures from Filter Well Inserts

Reagents and Materials

Nonsterile

❑ HCl, 12 N

- ❑ TEA drying solution (See Section 3.6.1)
- ❑ Derivatizing solution (See Section 3.6.2)
- ❑ Collagen assay standard tubes: lyophilized L-proline (20 mg/tube)
- ❑ Reconstitution buffer (See Section 3.6.4)
- ❑ Hydroxyproline HPLC eluent A (See Section 3.6.5)
- ❑ Hydroxyproline HPLC eluent B (See Section 3.6.6)
- ❑ Hydroxyproline HPLC standard (See Section 3.6.3)
- ❑ Screw cap tubes, glass (hydrolysis tubes) with caps with PTFE liners
- ❑ Microcentrifuge filter tubes, 0.22-μm pore size, nylon membrane (Corning Costar Spin-X HPLC tubes)
- ❑ Heating block to fit hydrolysis tubes
- ❑ Lyophilizer
- ❑ HPLC
- ❑ HPLC columns: 5-μm C18 precolumn (45 × 4.6 mm) and a 5-μm C18 column (250 × 4.6 mm)

Protocol

A. Sample Preparation

(a) Estimate the volume of sample required that gives 2 μg hydroxyproline (1 or 0.5 μg may be used if necessary). Make sure that the total volume is under 200 μl and then bring up to a final volume of 200 μl by adding dH_2O.

(b) Pipette this into hydrolysis tubes.

B. Sample Hydrolysis

(c) Add 200 μl 12 N HCl/200 μl sample.

(d) Firmly tighten the screw caps and vortex.

(e) Place samples in the heating block at 120 °C for 16–20 h (overnight).

(f) Centrifuge the tubes before uncapping and then lyophilize to dry.

C. Sample Drying

(g) Once the samples are dry, add 20 μl drying solution to each sample. The drying solution neutralizes any excess HCl.

(h) Lyophilize or concentrate the samples to dry.

(i) Repeat this procedure once.

D. Sample Derivitization

(j) Add a 20-μl aliquot of freshly prepared derivatizing solution to each sample, as well as to the proline standards, and vortex.

(k) Leave for 10 min at room temperature in a fume hood for the reaction to occur before lyophilizing or concentrating the samples and standards to dryness.

△ *Safety note.* PITC is very harmful—wear gloves and use in a fume hood.

E. Sample Reconstitution

(l) Dissolve samples in 1 ml reconstitution buffer and centrifuge for 5 min in microcentrifuge filter tubes, 0.22-μm pore size, nylon membrane (Corning Costar Spin-X HPLC tubes).

(m) Dissolve standard in 1 ml reconstitution buffer and centrifuge as above. Final standard concentration = 20 mg/ml.

F. Hydroxyproline Determination with HPLC

Hydroxyproline content is determined with isocratic separation and fluorescence detection at 254λ.

(n) Utilizing a 5-μm C18 precolumn (45 × 4.6 mm) and a 5-μm C18 column (250 × 4.6 mm), elute samples in the following manner.

   i) At start up, equilibrate columns with 100% eluant A (See Section 3.6.5) for 15 min at a flow rate of 1 ml/min.
   ii) Column pressure is typically 2000 psi at 1 ml/min.
   iii) Load samples onto the column (10–100 μl) with an autosampler.
   iv) Elution of the hydroxyproline peak will occur at approximately 9 min with 100% eluant A at a flow rate of 1 ml/min.
   v) At 13 minutes, switch to 100% eluant B at a flow rate of 1.5 ml/min for a duration of 3 min. This wash step serves to strip the column of previous sample residues.
   vi) Finally, reequilibrate the columns with 100% eluant A for 13 min at a flow rate of 1 ml/min.
   vii) Once the initial sample run is complete, proceed to the next sample immediately after the 13-min reequilibration.
   viii) Samples are monitored on an absorbance detector at 254 nm, yielding a hydroxyproline peak at approximately 9 min.
   ix) Hydroxyproline sample content is quantified relative to a known amount of derivatized hydroxyproline standard.
   x) Standards are run after every 10 samples to ensure appropriate elution times and run conditions. A reproducible and linear profile of standard concentrations is achieved by running 20-, 50-, and 100-μl injections alternately. 20 μl injected standard (20 ng/μl) equals 400 ng hydroxyproline.

Identification and content of hydroxyproline are determined by the standard elution time and standard peak area ratios.

Protocol 7.9. Determination of Collagen Type

Reagents and Materials

- ❑ Collagen extraction buffer (See Section 3.6.7)
- ❑ High-salt Tris buffer (See Section 3.6.8)
- ❑ Phosphate-buffered saline (PBS) or Tris-buffered saline (TBS)
- ❑ Laemmli sample buffer (Bio-Rad)
- ❑ Acetic acid, 0.05 M
- ❑ NaOH, 1 N
- ❑ NaCl, crystalline

- ❑ Ammonium sulfate, crystalline
- ❑ Phenol red indicator solution
- ❑ Pepsin stock solution (See Section 3.3.4)
- ❑ Elastase stock solution (See Section 3.3.5)
- ❑ Materials for 7.5% SDS-PAGE
- ❑ Coomassie brilliant blue R-250 stain (CB)
- ❑ 10% acetic acid, 10% methanol solution or Bio-Rad destaining solution (overnight) or 10% acetic acid, 40% methanol solution (1 h) for destaining gel
- ❑ Liquid nitrogen
- ❑ Mortar and pestle or glass homogenizer
- ❑ Centrifuge tubes, 5 ml, high *g*
- ❑ Shaker
- ❑ Electrophoresis gel imager and software

Protocol

A. Collagen Extraction from Native and ARC Cartilage

(a) All of the procedures described here should be performed at 4 °C unless otherwise specifically indicated.
(b) The native cartilage should be pulverized after freezing in liquid nitrogen. The ARC tissue can be minced and homogenized in a glass homogenizer.
(c) Add 10 volumes collagen extraction buffer and extract overnight. This helps to enhance the pepsin digestion by removing the PG.
(d) Centrifuge at 10,600 *g* for 10 min at 4 °C. Remove the supernatant and save the pellet.
(e) Wash the pellet with 3 ml 0.05 M acetic acid by spinning and removing the supernate as described above.
(f) To the pellet, add 10 volumes 0.05 M acetic acid, and then adjust the pH to 2.8–3.0 with 0.05 M acetic acid.
(g) Add 1/10 volume pepsin stock solution.
(h) Digest overnight at 4 °C on a rocker/shaker.
(i) After overnight digestion, add 1/10 volume high-salt Tris buffer.
(j) Adjust pH to 8.0 with 1 N NaOH. Adding a drop of phenol red helps to determine the approximate pH.
(k) Add 1/10 volume pancreatic elastase.
(l) Digest for 30 min at 35 °C.
(m) Centrifuge at 10,600 *g* for 10 min at 4 °C.
(n) Collect the supernate.
(o) Save the pellet for further extraction if needed.
(p) Add crystalline NaCl to 3 M and crystalline ammonium sulfate to 30% and incubate for 30 min at 4 °C.
(q) Centrifuge at 10,600 *g* for 10 min at 4 °C and remove the supernate. The pellet contains the precipitated collagen.
(r) Resuspend the pellet in a low-salt buffer, such as 1 × PBS or 1 × TBS.

B. SDS-PAGE of Sample

(a) Dilute samples 1:2 with Laemmli sample buffer according to the supplier's instructions.
(b) Load the samples on a 7.5% SDS-PAGE gel.
(c) Run at room temperature: 60 V for run through the stacking gel, approximately 15 min; 200 V for run through the separating gel, approximately 30–40 min.
(d) Stain gel for protein
  i) Immediately after the gel run is complete, place the gel in 1× Coomassie brilliant blue R-250 stain (CB).
  ii) Incubate for 30 min to 1 h at room temperature with shaking.
(e) Destain gel
  i) *Overnight:* Use either a 10% acetic acid, 10% methanol solution or the Bio-Rad destaining solution.
  ii) *One hour:* Use a 10% acetic acid, 40% methanol solution.
  iii) Multiple solution changes, at room temperature with shaking, are necessary depending on the length of staining time.
(f) Gel storage: Gels may be stored indefinitely in a 10% acetic acid solution before drying. Acetic acid maintains gel integrity.
(g) Gel imaging:
  i) While wet, the gels can be digitized with a commercially available imager. To keep as a record, dry the gels and scan, using a regular scanner with a transparency unit.
  ii) Each $\alpha 1$ and $\alpha 2$ band can be scanned and quantified with NIH Image software to calculate the ratio of types I and II.

Protocol 7.10. Determination of DNA Content

Reagents and Materials

- ❑ Sodium dodecylsulfate (SDS) 0.02%
- ❑ Alginate dissolving buffer with EDTA (See Section 3.4.2)
- ❑ DNA standard solution, 40 μg/ml (See Section 3.7.3)
- ❑ Hoechst dye solution: 1 μg/ml Hoechst 33258 (See Section 3.7.5)
- ❑ Proteinase K solution (See Section 3.7.2)
- ❑ Proteinase K working solution (See Section 3.7.2)
- ❑ Microplate reader

A. Proteinase K Digestion of Alginate Beads for DNA Analysis

Digestion should be performed on all of the samples from one experiment at the same time.

(a) Add 100 μl 0.02% SDS to 9 alginate beads (do not dissolve beads beforehand!).
(b) Heat at 100 °C for 5 min.
(c) Add 400 μl 125 μg/ml proteinase K working solution (See Section 3.7.2) to each sample.

(d) Incubate overnight at 56–60 °C.
(e) Add 400 μl alginate dissolving buffer with EDTA to each sample.
(f) Shake on the shaker for 3 h at room temperature.
(g) The final volume for all samples is 1 ml.

B. Fluorometric Assay of DNA

(h) Prepare standard dilutions. A total of 10 dilutions should be performed for the standard, in duplicate, for each plate. Place standards on the top and bottom of the plate (A2–A7, H2–H7).
  i) Pipette 200 μl DNA Standard Working Solution (40 μg/ml) to wells A2 and H2.
  ii) Pipette 100 μl papain digestion buffer to wells A3–A7 and H3–H7.
  iii) Perform serial dilutions of $\frac{1}{2}$ by transferring 100 μl from A2 to A3 and mixing, then 100 μl from A3 to A4, etc.

Final Standard Concentrations (μg/ml) are: 40; 20; 10; 5; 2.5; 1.25.

(i) Add sample preparation: 100 μl proteinase K-digested sample/well (rows B–G).
(j) Add 100 μl Hoechst 33258 working-strength dye solution/well.
  i) The dye/sample complex is stable for 2 h, so the fluorescence should be measured within this time.
  ii) Blank: (A1, H1); 100 μl papain buffer + 100 μl Hoechst dye buffer.
  iii) Background: (A12, H12); 100 μl papain buffer + 100 μl 1 μg/ml Hoechst 33258 dye solution.

Read fluorescence on microplate reader at 360 nm excitation and 460 nm emission. Sensitivity: 36% and 40%.

C. Calculation of DNA in Samples

A plot of fluorescence against DNA concentration of the standards yields a linear response between 1.25 μg/ml and 40 μg/ml. The equation of this line can be used to determine the amount of DNA present in the unknown samples.

7.3. Outcome of Tissue Analyses

The addition of certain growth factors, such as osteogenic protein-1 (OP-1; BMP-7), to both steps of culture can significantly enhance the formation of tissue in vitro [Masuda et al., 2003]. With the ARC method and the inclusion of OP-1, cells obtained from a small biopsy of knee cartilage and expanded in monolayer culture can be induced to form cartilaginous tissue within 6 weeks [Masuda et al., 2002].

Variations of the ARC method involve the use of cells with different phenotypic characteristics [Klein et al., 2003]. The use of such cell populations allows formation of tissue with stratification and depth-varying properties, resembling native cartilage. In particular, cells from the superficial zone of cartilage have been used as the source of cells forming the superficial region of engineered tissue. Such cells normally secrete superficial zone protein (SZP) [Schumacher et al., 1994, 1999; Flannery et al., 1999]. SZP appears to have an important role for the mechanical function of articular cartilage, as it is identical or closely related to the molecule termed lubricin, which imparts lubrication properties to the articular surface [Swann et al., 1985; Jay et al., 1998, 2001; Schmid et al., 2002]. Cartilaginous tissue, engineered to be stratified, also secretes SZP from cells in its surface layer.

7.4. Assessment of Biomechanical Properties

The biomechanical properties of the formed tissue can be determined with compressive [Williamson et al., 2001] and tensile [Williamson et al., 2003] test methods.

Protocol 7.11. Measurement of Compressive Properties of Filter Well Insert Constructs

Reagents and Materials

- ❑ Phosphate-buffered saline (PBS)
- ❑ Calipers
- ❑ Disk punch, 9.6 mm
- ❑ Compression test instrument

(a) Removing the cultured tissue from the insert, as previously described (See Protocol 7.6, Step (j)).

(b) Measure the thickness of the tissue disk in several locations, and take the average to determine the thickness for mechanical testing.

(c) Measure compressive strength:

  i) Punch out a disk that is 9.6 mm in diameter.

  ii) Place the disk into a radially confining well of the mechanical test instrument, filled with PBS, and apply to the top a fluid-saturated porous compression platen.

  iii) Apply compressive displacements to 15%, 30%, and then 45% amplitude, at a strain rate of 0.02%/s. After each compression, allow stress-relaxation to occur until equilibrium is reached (typically ~30 min for a 1-mm-thick sample of engineered cartilage tissue). Record the equilibrium loads.

  iv) Fit the load-displacement data to determine the compressive modulus extrapolated to the free-swelling state, H_{A0} [Kwan et al., 1990; Chen et al., 2001; Williamson et al., 2001].

(d) Measure tensile properties

  i) After removing the cultured tissue from the insert, as described previously (See Protocol 7.6, Step (j)), measure the thickness of the tissue disk in several locations near the center, and take the average to determine the thickness for mechanical testing.

  ii) From the disk, punch out a tapered specimen, using a custom punch.

  iii) Place the specimen into the clamps of a mechanical test instrument, separated by 5 mm, with PBS recirculating over the sample. Limit the clamps so that the tissue is compressed by 30–50%, to enable secure gripping while not tearing the sample.

iv) Apply tensile displacements until a tare load of ~0.02 N is attained. Take this position to be the "zero" state. From there, elongate the sample to 10% and then 20% amplitude, at a strain rate of 0.2%/second. After each extension, allow stress-relaxation to occur until equilibrium is reached (typically ~10 min for a 1-mm-thick sample of engineered cartilage tissue). Record the equilibrium loads.

v) Then apply tensile displacement at a constant extension rate of 2%/s, recording the load data every 0.1 s, until the sample breaks.

vi) Fit the equilibrium load-displacement data to determine the tensile modulus, E_t, and the dynamic load-displacement data to determine the ramp stiffness and strength (maximum stress) [Williamson et al., 2003].

SOURCES OF MATERIALS

Item	*Comment*	*Catalog #*	*Supplier*
Articular cartilage, human adult	More than 5 g for primary cultures, more than 20 mg for monolayer expansion		Clinical collaboration
L-Ascorbic acid	Nonsterile	A4544	Sigma-Aldrich
Bovine lower legs with hooves	Each experiment should use more than two hooves		Obtained from a local slaughterhouse
Calcium chloride		C-7902	Sigma-Aldrich
Calf thymus DNA Type I	Store at 4 °C	D1501	Sigma-Aldrich
Cell culture flask, 150 cm²,	Canted neck, plug-seal cap	355000	BD Biosciences
Cell strainer, 40 μm		2340	BD Biosciences
Centrifuge tubes, 50 ml		2070	BD Biosciences
Collagenase-P, *Clostridium histolyticum*,	Nonsterile; store at −20 °C	1213873	Boehringer Mannheim (Roche Diagnostics)
Costar Spin-X HPLC tubes		8161	Corning
L-Cysteine hydrochloride hydrate	For papain buffer activation; store at 4 °C	C-7880	Sigma-Aldrich
1,9 Dimethylmethylene blue		03610	Polysciences, Inc
DMEM-F-12, 50-50 mix without L-glutamine	Sterile	Fisher Cat. # MT-15-090-CM	Mediatech, Inc
Elastase	Pancreatic, from porcine pancreas	E6883	Sigma-Aldrich
Fetal bovine serum (FBS) defined	Sterile	SH30070.03	Hyclone

Item	Comment	Catalog #	Supplier
Filter well inserts (polyethylene terephthalate (PET) membrane)	Transparent, 0.4-μm pore size, 23-mm effective diameter, 4.2-cm^2 effective growth area of membrane	353090	BD Biosciences
Filter well inserts, 12 well	0.4-μm pore size, 10.5-mm effective diameter, 0.9-cm^2 effective growth area of membrane	353180	BD Biosciences
Formic acid		F0507	Sigma-Aldrich
Gentamicin (50 mg/ml)	Sterile	15750-029	Invitrogen
L-Glutamine (29.2 mg/ml)	Sterile	Fisher Cat. # MT-25005-CI	Mediatech, Inc
Guanidine hydrochloride, 8 M		24115	Pierce Biotechnology, Inc
Hoechst 33258 dye	Store at 4 °C	09460	Polysciences
Human articular cartilage, adult	More than 5 g for primary cultures, more than 20 mg for monolayer expansion		Clinical collaboration
Laemmli sample buffer		161-0737	Bio-Rad
Microtitration plates, 96 well	Cytoplates	CFCPN9650	Applied Biosystems
Multiwell plates, 6 well,	For cell culture inserts	353502	BD Biosciences
Multiwell plates, 12 well	For cell culture inserts	353503	BD Biosciences
Multiwell plates, 6 well,	Deep-well	355467	BD Biosciences
NaCl, 0.9%	Sterile	2F7124	Baxter
Osteogenic protein 1 (OP-1; BMP-7)		354-BP-010	R&D systems
Papain	Store at 4 °C	P-3125	Sigma-Aldrich
Pepsin	Porcine gastric mucosa	P1143	Sigma-Aldrich
Phenylisothiocyanate	(PITC: Edman's Reagent)	26922	Pierce
Polyacrylamide gels:	Ready gel Tris-HCl Gel, 7.5% resolving gel, 4% stacking gel, 10 well	161-1154	BioRad
Pronase protease, *Streptomyces griseus,*	Nonsterile; store at 4 °C	53702	Calbiochem
Proteinase K		P-6556	Sigma-Aldrich
Scalpel holders and #10 and #15 scalpel blades			VWR
Sodium alginate	NE/EP grade for pharmaceutical use	Keltone® LV-(HM)	ISP Alginate
Sodium citrate	Tissue culture grade	BP327500	Fisher Scientific
Sodium formate		S648	Fisher Scientific

Item	Comment	Catalog #	Supplier
Spatula		3004	Corning
Specimen container, 100 ml		25384-078	VWR
Steriflip-GP 50-ml tube, 0.22 μm		SCGP00525	Millipore
Steritop-GV filter unit		SCGV S01 RE	Millipore
Trypan Blue 0.4%		15250-061	Invitrogen
Triethylamine		04884-100	Fisher Scientific

REFERENCES

Adkisson, H.D., Gillis, M.P., et al. (2001) In vitro generation of scaffold independent neocartilage. *Clin. Orthop.* 391S: 280–294.

Athanasiou, K.A., Fischer, R., et al. (1995) Effects of excimer laser on healing of articular cartilage in rabbits. *J. Orthop. Res.* 13: 483–494.

Aubin, P.P., Cheah, H.K., et al. (2001) Long-term followup of fresh femoral osteochondral allografts for posttraumatic knee defects. *Clin. Orthop.* 391S: 318–327.

Benya, P.D., and Shaffer, J.D. (1982) Dedifferentiated chondrocytes reexpress the differentiated collagen phenotype when cultured in agarose gels. *Cell* 30: 215–224.

Bouwmeester, P., Kuijer, R., et al. (1999) Histological and biochemical evaluation of perichondrial transplants in human articular cartilage defects. *J. Orthop. Res.* 17: 843–849.

Brittberg, M., Lindahl, A., et al. (1994) Treatment of deep cartilage defects in the knee with autologous chondrocyte transplantation. *N. Engl. J. Med.* 331: 889–895.

Buckwalter, J.A., and Mankin, H.J. (1997). Articular cartilage. Part II: degeneration and osteoarthrosis, repair, regeneration, and transplantation. *J. Bone Joint Surg. Am.* 79-A: 612–632.

Bugbee, W.D., and Convery, F.R. (1999) Osteochondral allograft transplantation. *Clin. Sports Med.* 18: 67–75.

Buschmann, M.D., Gluzband, Y.A., et al. (1995) Mechanical compression modulates matrix biosynthesis in chondrocyte/agarose culture. *J. Cell Sci.* 108: 1497–1508.

Buschmann, M.D., Gluzband, Y.A., et al. (1992) Chondrocytes in agarose culture synthesize a mechanically functional extracellular matrix. *J. Orthop. Res.* 10: 745–758.

Chandrasekhar, S., Esterman, M.A., et al. (1987) Microdetermination of proteoglycans and glycosaminoglycans in the presence of guanidine hydrochloride. *Anal. Biochem.* 161: 103–108.

Chen, A.C., Bae, W.C., et al. (2001) Depth- and strain-dependent mechanical and electromechanical properties of full-thickness bovine articular cartilage in confined compression. *J. Biomech.* 34: 1–12.

Chiba, K., Andersson, G.B., et al. (1997) Metabolism of the extracellular matrix formed by intervertebral disc cells cultured in alginate. *Spine* 22(24): 2885–2893.

Dounchis, J., Harwood, F.L., et al. (2000) Cartilage repair with autogenic perichondrium cell and polylactic acid grafts. *Clin. Orthop.* 377: 248–264.

Dunkelman, N.S., Zimber, M.P., et al. (1995) Cartilage production by rabbit articular chondrocytes on polyglycolic acid scaffolds in a closed bioreactor system. *Biotechnol. Bioeng.* 46: 299–305.

Eyre, D.R., Wu, J.J., et al. (1992) Cartilage-specific collagens: structural studies. In Kuettner, K.E., Schleyerbach, R., Peyron, J.G., Hascall, V.C., eds., *Articular Cartilage and Osteoarthritis*. New York, Raven Press, pp. 119–131.

Ficat, R.P., Ficat, C., et al. (1979) Spongialization: a new treatment for diseased patellae. *Clin. Orthop.* 144: 74–83.

Flannery, C.R., Hughes, C.E., et al. (1999). Articular cartilage superficial zone protein (SZP) is homologous to megakaryocyte stimulating factor precursor and is a multifunctional proteoglycan with potential growth-promoting, cytoprotective, and lubricating properties in cartilage metabolism. *Biochem. Biophys. Res. Commun.* 254(3): 535–541.

Hangody, L., Feczko, P., et al. (2001) Mosaicplasty for the treatment of articular defects of the knee and ankle. *Clin. Orthop.* 391S: 328–336.
Häuselmann, H.J., Aydelotte, M.B., et al. (1992) Synthesis and turnover of proteoglycans by human and bovine adult articular chondrocytes cultured in alginate beads. *Matrix* 12: 116–129.
Häuselmann, H.J., Fernandes, R.J., et al. (1994) Phenotypic stability of bovine articular chondrocytes after long-term culture in alginate beads. *J. Cell Sci.* 107: 17–27.
Hendrickson, D.A., Nixon, A.J., et al. (1994) Chondrocyte-fibrin matrix transplants for resurfacing extensive articular cartilage defects. *J. Orthop. Res.* 12: 485–497.
Herbage, D., Bouillet, J., et al. (1977) Biochemical and physicochemical characterization of pepsin-solubilized type-II collagen from bovine articular cartilage. *Biochem, J* 161: 303–312.
Hjertquist, S.O., and Lemperg, R. (1971) Histological, autoradiographic and microchemical studies of spontaneously healing osteochondral articular defects in adult rabbits. *Calcif. Tiss. Res.* 8: 54–72.
Itay, S., Abramovici, A., et al. (1987) Use of cultured embryonal chick epiphyseal chondrocytes as grafts for defects in chick articular cartilage. *Clin. Orthop.* 220: 284–303.
Jay, G.D., Haberstroh, K., et al. (1998) Comparison of the boundary-lubricating ability of bovine synovial fluid, lubricin, and Healon. *J. Biomed. Mater. Res.* 40: 414–418.
Jay, G.D., Tantravahi, U., et al. (2001) Homology of lubricin and superficial zone protein (SZP): products of megakaryocyte stimulating factor (MSF) gene expression by human synovial fibroblasts and articular chondrocytes localized to chromosome 1q25. *J. Orthop. Res.* 19: 677–687.
Johnson, L.L. (2001) Arthroscopic abrasion arthroplasty: a review. *Clin. Orthop.* 391S: 306–317.
Klein, T.J., Schumacher, B.L., et al. (2003) Tissue engineering of articular cartilage with stratification using chondrocyte subpopulations. *Osteoarthritis Cartilage* 11: 595–602.
Kwan, M.K., Lai, W.M., et al. (1990) A finite deformation theory for cartilage and other soft hydrated connective tissues—I. equilibrium results. *J. Biomech.* 23: 145–155.
Li, K.W., Williamson, A.K., et al. (2001) Growth responses of cartilage to static and dynamic compression. *Clin. Orthop.* 391S: 34–48.
Mankin, H.J. (1982) The response of articular cartilage to mechanical injury. *J. Bone Joint Surg. Am.* 64-A: 460–466.
Masuda, K., Miyazaki, T., et al. (2002) Human tissue engineered cartilage by the alginate-recovered-chondrocyte method after an expansion in monolayer. *Trans. Orthop. Res. Soc.* 27: 467.
Masuda, K., Sah, R.L., et al. (2003) A novel two-step method for the formation of tissue engineered cartilage: the alginate-recovered-chondrocyte (ARC) method. *J. Orthop. Res.* 21: 139–148.
Mauck, R.L., Nicoll, S.B., et al. (2003) Synergistic action of growth factors and dynamic loading for articular cartilage tissue engineering. *Tissue Eng.* 9(4): 597–611.
Mok, S.S., Masuda, K., et al. (1994) Aggrecan synthesized by mature bovine chondrocytes suspended in alginate. Identification of two distinct metabolic matrix pools. *J. Biol. Chem.* 269(52): 33021–33027.
Muller, B., and Kohn, D. (1999) Indikation und durchführung der knorpel-knochen-anbohrung nach Pridie. *Orthopade* 28(1): 4–10.
O'Driscoll, S.W., and Fitzsimmons, J.S. (2001) The role of periosteum in cartilage repair. *Clin. Orthop.* 391S: 190–207.
Pal, S., Tang, L.-H., et al. (1981) Structural changes during development in bovine fetal epiphyseal cartilage. *Collagen Relat. Res.* 1: 151–76.
Perka, C., Spitzer, R.S., et al. (2000) Matrix-mixed culture: new methodology for chondrocyte culture and preparation of cartilage transplants. *J. Biomed. Mater. Res.* 49(3): 305–411.
Petit, B., Masuda, K., et al. (1996) Characterization of crosslinked collagens synthesized by mature articular chondrocytes cultured in alginate beads—comparison of two distinct matrix compartments. *Exp. Cell Res.* 225: 151–161.
Robinson, D., Halperin, N., et al. (1990) Regenerating hyaline cartilage in articular defects of old chickens using implants of embryonal chick chondrocytes embedded in a new natural delivery substance. *Calcif. Tissue Int.* 46: 246–253.
Sah, R.L. (2002) The biomechanical faces of articular cartilage. in Kuettner, K.E., and Hascall, V.C., eds., *The Many Faces of Osteoarthritis*. New York, Raven Press pp. 409–422.

Sah, R.L., Klein, T.J., Schmidt, T.A., Albrecht, D.R., Bae, W.C., Nugent, G.E., McGowan, K.B., Temple, M.M., Jadin, K.D., Schumacher, B.L., Chen, A.C., Sandy, J.D. (2005): Articular cartilage repair, regeneration, and replacement. In: *Arthritis and Allied Conditions: A Textbook of Rheumatology*, ed by WJ Koopman, Lippincott Williams & Wilkins, Philadelphia, pp. 2277–2301.

Schmid, T.M., Su, J.-L., et al. (2002) Superficial zone protein (SZP) is an Abundant glycoprotein in human synovial fluid with lubricating properties. In Kuettner, K.E., and Hascall, V.C., eds., *The Many Faces of Osteoarthritis*. New York, Raven Press, pp. 159–161.

Schumacher, B.L., Block, J.A., et al. (1994) A novel proteoglycan synthesized and secreted by chondrocytes of the superficial zone of articular cartilage. *Arch. Biochem. Biophys.* 311: 144–152.

Schumacher, B.L., Hughes, C.E., et al. (1999). Immunodetection and partial cDNA sequence of the proteoglycan, superficial zone protein, synthesized by cells lining synovial joints. *J. Orthop. Res.* 17: 110–120.

Shapiro, F., Koido, S., et al. (1993) Cell origin and differentiation in the repair of full-thickness defects of articular cartilage. *J. Bone Joint Surg. Am.* 75-A: 532–553.

Solchaga, L.A., Goldberg, V.M., et al. (2001) Cartilage regeneration using principles of tissue engineering. *Clin. Orthop.* 391S: 161–170.

Steadman, J.R., Rodkey, W.G., et al. (2001) Microfracture: surgical technique and rehabilitation to treat chondral defects. *Clin. Orthop.* 391S: 362–369.

Swann, D.A., Silver, F.H., et al. (1985) The molecular structure and lubricating activity of lubricin isolated from bovine and human synovial fluids. *Biochem. J.* 225: 195–201.

Vunjak-Novakovic, G., Martin, I., et al. (1999) Bioreactor cultivation conditions modulate the composition and mechanical properties of tissue-engineered cartilage. *J. Orthop. Res.* 17: 130–139.

Wakitani, S., Kimura, T., et al. (1989) Repair of rabbit articular surfaces with allograft chondrocytes embedded in collagen gel. *J. Bone Joint Surg. Br.* 71-B: 74–80.

Wei, X., and Messner, K. (1999) Maturation-dependent durability of spontaneous cartilage repair in rabbit knee joint. *J. Biomed. Mater. Res.* 46: 539–548.

Wei, X., Reaseanen, T., et al. (1998) Maturation-related compressive properties of rabbit knee articular cartilage and volume fraction of subchondral tissue. *Osteoarthritis Cartilage* 6: 400–409.

Williamson, A.K., Chen, A.C., et al. (2003) Tensile mechanical properties of bovine articular cartilage: variations with growth and relationships to collagen network components. *J. Orthop. Res.* 21: 872–880.

Williamson, A.K., Chen, A.C., et al. (2001) Compressive properties and function-composition relationships of developing bovine articular cartilage. *J. Orthop. Res.* 19: 1113–1121.

Yu, H., Grynpas, M., et al. (1997) Composition of cartilaginous tissue with mineralized and non-mineralized zones formed in vitro. *Biomaterials* 18: 1425–1431.

8

Ligament Tissue Engineering

Jingsong Chen[1], Jodie Moreau[1], Rebecca Horan[1], Adam Collette[1], Diah Bramano[1], Vladimir Volloch[1], John Richmond[2], Gordana Vunjak-Novakovic[3], David L. Kaplan[1], and Gregory H. Altman[1]

Department of Biomedical Engineering[1], Bioengineering Center, Tufts University, Medford, Massachusetts 02155; Department of Orthopaedics[2], New England Baptist Hospital, Boston, Massachusetts; Department of Biomedical Engineering, Columbia University, 363G Engineering Terrace, Mail Code 8904, 1210 Amsterdam Avenue, New York, NY 10027

Corresponding author: Gregory.Altman@tufts.edu

Culture of Cells for Tissue Engineering, edited by Gordana Vunjak-Novakovic and R. Ian Freshney

1. BACKGROUND

1.1. Context of Tissue Engineering

In any tissue engineering effort, the preparation of cells is a critical system component along with the biomaterial matrix and bioreactor environment. Cell preparation typically involves the appropriate choice of cell type (e.g., differentiated or progenitor, human or animal), expansion, seeding, and genotypic and phenotypic characterization of cells within the context of the specific tissue engineering goal (tissue-specific outcomes).

1.2. ACL Injury and Limitations of Prevalent Treatments

The need to explore tissue engineering options for the anterior cruciate ligament (ACL) has arisen from the inability of currently available clinical options to fully restore knee function. Over the last twenty years, orthopedic sports medicine has been faced with continued repair needs associated with tears and ruptures of the ACL, a major cause of athletic disability. Presently, more than 200,000 ACL ruptures occur in the United States each year. Injury to the ACL and the high frequency of subsequent knee instability often result in further damage to the joint, motivating improvements in ACL reconstructive techniques. Currently, autologous tendon grafts harvested at the time of reconstruction are the most widely utilized ACL replacement tissues, but this practice typically results in tendon donor site morbidity. Allograft tendons have gained acceptance because they alleviate donor site morbidity problems associated with autograft harvest, but cost, risk of infection, and disease transmission remain problematic. Synthetic polymers have also been widely used as ligament replacements, including polytetrafluorethylene (Gore-tex®), polyester (Dacron®), carbon fiber, and polypropylene ligament augmentation devices (LAD). Complications due to stress shielding, particulate debris, and early mechanical failures are commonly associated with these prostheses.

1.3. Tissue Engineering May Be the Solution for ACL Replacement

Tissue engineering may potentially provide improved clinical options in orthopedic medicine through the generation of biologically based functional tissues in vitro for transplantation at the time of injury or disease. A tissue-engineered

ACL with the appropriate biological and mechanical properties would eliminate many of the deleterious effects associated with current clinical options for restoring knee function. In addition, model systems available in vitro which better represent tissues in vivo may provide new opportunities for the study of disease onset, disease prevention, pharmaceuticals, and fundamentals of tissue structure and function.

1.4. Criteria for the Tissue-Engineered ACL

The failure of the ruptured ACL to heal is related to the lack of vascular supply, deficits in intrinsic cell migration, impaired growth factor availability, and environmental effects of the synovial fluid on cell morphology. These problems have led to research on the biology of the ACL, particularly its response to injury and wound healing, providing the backdrop of issues important to consider in any tissue engineering strategy, such as marker profiling [Bramono et al., 2004; Frank et al., 1999; Lo et al., 1998, 2003]. Accumulated data from clinical studies combined with fundamental insight into the biological response to ACL rupture provide criteria to consider for any tissue engineering strategy to achieve success for ACL replacement. These criteria include (1) minimal patient morbidity, (2) surgically simple insertion with reliable methods of fixation that will withstand aggressive rehabilitation, (3) generation and maintenance of immediate knee stability without tissue-fixation device creep to allow rapid return to preinjury function, (4) minimal risk to the patient for infection or disease transmission, (5) biocompatibility and minimal host immune response, (6) support of host tissue ingrowth without causing stress shielding (i.e., adequate communication of environmental signals including mechanical, biochemical, and transport) to the developing host tissue such that ingrowth is properly directed and organized, and (7) biodegradation at a rate that provides adequate mechanical stability during replacement by new extracellular matrix (ECM).

2. PRINCIPLES OF METHODOLOGY

Through the incorporation of an appropriate biomaterial matrix, judiciously selected cells, and appropriate differentiation signaling, tissue engineering can offer options for generating an unlimited amount of autologous ligament tissue in vitro.

2.1. Background of ACL Tissue Engineering

A tissue engineering approach to ACL replacements has been under investigation since the early 1990s. A ligament prosthesis combining the advantages of synthetic materials (high strength, simple fabrication, and storage) and biological materials (biocompatibility and ingrowth promotion) was initially reported using a collagenous composite consisting of reconstituted type I collagen fibers in a collagen I matrix with polymethylmethacrylate bone fixation [Dunn et al., 1992, 1994]. However, inconsistent neoligament formation and significant weakening of the prosthesis in a rabbit model were observed. Collagen fiber-poly(L-lactic acid) (PLA) composites were explored to improve mechanical integrity and allow for neoligament tissue ingrowth; however, mechanical integrity could not be maintained for the rigorous ACL rehabilitation protocols.

2.2. Silk As a Candidate for ACL Tissue Engineering

Native silkworm silk is being explored as a scaffold for ACL tissue engineering because of silk's superior mechanical and biological properties. Historically, two common misconceptions have limited silk's broader use as a biomaterial in tissue engineering: first, that silk is immunogenic and second, that it is nondegradable. We have recently shown that silk fibroin, when contaminating sericin proteins secreted by the silkworm have been properly extracted, is nonantigenic, biocompatible, and capable of supporting BMSC attachment, spreading, growth, and differentiation (Fig. 8.1) [Altman et al., 2002a]. In vivo studies indicate that sericin-extracted silk induces a foreign body response comparable to most common degradable synthetic and natural polymers such as poly(glycolic acid) (PGA)-PLA copolymers and collagen [Altman et al., 2003]. With regard to biodegradability, silk maintains its mechanical integrity in tissue culture conditions, but in vivo is susceptible to proteolytic degradation resulting from a foreign body response. The slow rate of degradation in comparison to other degradable natural and synthetic polymers (e.g., collagen, polyesters) is viewed as a benefit, meeting criteria established for ACL tissue engineering [Weitzel et al., 2002]. The slow rate of silk degradation allows for the gradual transfer of stabilizing properties from the matrix to the new tissue without exposing the patient to periods of joint destabilization.

2.3. Choice of Cells for ACL Tissue Engineering

The cells incorporated into tissue-engineered constructs can provide signals needed for tissue regeneration. Two types of autologous cells were explored for reaching the goal of tissue engineering ACL: ACL fibroblasts and bone marrow stromal cells (BMSC). The first approach required a surgical procedure (e.g., arthroscopy) to harvest ACL tissue from the patient's knee and a relatively long period for cell culture expansion because of the limited proliferation capability of adult ACL fibroblasts. BMSCs attained via a small bone marrow aspirate (5–10 ml harvested from the patient's iliac crest in the physician's office) proved a better cell source because of their potential for differentiation into ligament lineage and superior growth ability compared to ACL fibroblasts [Chen et al., 2003]. In addition, this process does not involve the knee, thus eliminating the risk of local infection.

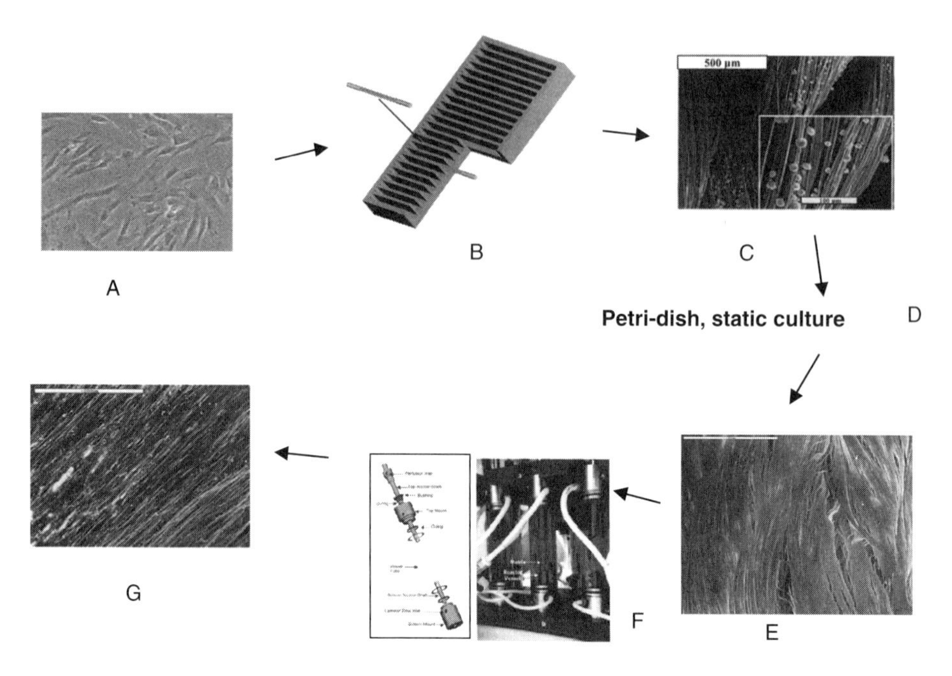

Figure 8.1. Cell culture for ligament tissue engineering. A. Step 1—Cell expansion: Human BMSCs were plated on tissue culture plastic at a density of 5000/cm^2 in DMEM-10% FCS containing 1 ng/ml bFGF (functions as the mitogen to enhance BMSC proliferation and as differentiation inhibitor) and culture-expanded for 6 days (Chen et al, 2003). B. Step 2—Cell seeding: Before confluence, BMSCs were trypsinized, centrifuged, and resuspended at a concentration of 2 million/ml in DMEM-10% FCS. BMSCs were seeded onto RGD-modified silk matrices in a custom-designed seeding vessel (Chen et al, 2003) to increase cell-matrix contact. C. As early as 5 min after seeding, BMSCs attached to RGD-modified silk matrices and started to spread (Chen et al, 2003). D. Step 3—Cell cultivation on silk matrices: Two hours after seeding, matrices were transferred from seeding vessel into tissue culture flasks, followed by a 2-week cultivation in DMEM-10% FCS at 37°C/5% CO_2. E. After the 2-week static culture, BMSCs reached confluence and fully covered the silk matrix. Cell/ECM was not aligned (Chen et al, 2003). F. Step 4—Bioreactor and mechanical stimulation: Cell-seeded silk matrices were then transferred into bioreactor and cultured under dynamic conditions with direct cyclic mechanical stimulation for 6 days (Altman et al, J Biomech Eng. 2002 Dec;124(6):742–9). Culture medium was reflushed for 1 h/day. G. The mechanically stimulated matrices showed highly organized cell/ECM alignment (Chen et al, 2003).

2.4. Rationale for Using Mechanical Stimulation as ACL Differentiation Inducer

BMSCs have the potential to differentiate into multiple lineages such as bone, cartilage, muscle, ligament, and tendon. Therefore, ACL-specific differentiation cues must be developed in order to direct BMSC differentiation toward ligament tissue. Recent research has focused on the impact of mechanical forces on progenitor cell differentiation toward ligament-like tissues. The motivation for this approach was based on two major factors: the finding that most fibroblastic cells, including those of the ACL, are load responsive and the absence of defined biochemical/chemical signaling factors to direct progenitor cells to differentiate into ligament cells. The focus on mechanical forces imparted through the bioreactor environment and transmitted by the matrix to the cells is an important principle to consider in tissue engineering ligaments.

3. PREPARATION OF REAGENTS

3.1. Bone Marrow Stromal Cell Culture Medium (BMSC Medium)

BMSC medium is prepared from Dulbecco's modified Eagle's medium (DMEM) supplemented with 10% fetal bovine serum (FBS), 0.1 mM nonessential amino acids, 40 U/ml penicillin and 40 μg/l streptomycin (P/S), and 1 ng/ml basic fibroblast growth factor (bFGF) (Table 1A).

3.2. Engineered Ligament Culture Medium (Ligament Medium)

Ligament medium is prepared from DMEM supplemented with 10% FBS, 0.1 mM nonessential amino acids, P/S, and 100 μM L-ascorbic acid 2-phosphate (Table 1B).

3.3. Freezing Medium

Freezing medium is prepared from DMEM supplemented with 10% FBS and 8% DMSO. Prepare fresh solution before use.

3.4. Trypsin-EDTA

Trypsin-EDTA is prepared from 0.15 M NaCl supplemented with 0.25% trypsin and 1mM EDTA.

3.5. 3-[4,5-dimethylthiazol-2-y]-2,5-dipheynyl tetrazolium bromide (MTT) Solution

MTT solution is prepared from ligament medium supplemented with 5 mg/ml MTT to make a 10× stock solution. Sterilize by filtration. Dilute into 1× with ligament medium for use. Avoid light and store at 4 °C.

Table 1. Preparation of cell culture media.

Reagent	Stock solution concentration	Volume of reagent needed (ml)	Final concentration
A. Media Preparation for Bone Marrow Stromal Cell Culture			
DMEM	1×	438	1×
FBS	—	50	10%
Nonessential amino acids	10 mM	5	0.1 mM
P/S antibiotics	P: 10,000 U/ml, S: 10 mg/ml	2	40 U/ml
BFGF	250 μg/ml	5	1 ng/ml
Total volume of reagents		500	
B. Media Preparation for Engineered Ligament Culture			
DMEM	1×	442	1×
FBS	—	50	10%
Nonessential amino acids	10 mM	5	0.1 mM
P/S antibiotics	P: 10,000 U/ml, S: 10 mg/ml	2	40 U/ml
L-Ascorbic acid 2-phosphate	50 mM (0.128 g in 10 ml H_2O)	1	100 μM
Total volume of reagents		500	

3.6. Buffers for Scanning Electron Microscopy (SEM)

Sodium Cacodylate Buffer (0.2 M)

Sodium cacodylate	4.28 g
Calcium chloride	25 g
Hydrochloric acid, 0.2N	2.5 ml
Dilute with distilled water, pH 7.4, to	200 ml

Karnovsky Fixative Stock Solution

Paraformaldehyde	2.0 g
Sodium hydroxide, 1 M	2–4 drops
Glutaraldehyde, 50%	5.0 ml
Dilute with 0.2 M cacodylate buffer, pH 7.4, to	25 ml

Mix the paraformaldehyde with 25 ml of distilled water. Heat to 60 °C on a stir plate. When moisture condenses on the sides of flask, add sodium hydroxide drop-wise and stir the solution until it clears. Cool solution in cold water bath. Filter and then add glutaraldehyde and 0.2 M buffer, pH range 7.2–7.4.

△ ***Safety note.*** Use caution when preparing the solutions. Work in a well-ventilated area; wear gloves and a lab coat.

3.7. Collagen ELISA Buffers

Acetic acid, 0.05 M, Containing 0.5 M NaCl, pH 2.9

Acetic acid (17.4 M stock solution)	1.436 ml
NaCl (MW: 58.44)	14.61 g
dH_2O to	500 ml
Adjust to pH 2.9 with formic acid	

Tris-Base, 1 M

Tris-Base (MW: 121.14)	6.057 g
dH_2O to	50 ml

Tris-HCl, 0.1 M, Containing 0.2 M NaCl and 0.05 M $CaCl_2$, pH 7.8

Tris-HCl (1 M stock solution)	5 ml
NaCl (MW: 58.44)	0.584 g
$CaCl_2$ (MW: 111)	0.276 g
dH_2O to	50 ml

3.8. Pepsin

Pepsin solution is made by dissolving 10 mg/ml in 0.05 M acetic acid.

3.9. Elastase

Elastase solution is made by dissolving 1 mg/ml pancreatic elastase in Tris-HCl buffer containing 0.2 M NaCl and 0.05 M $CaCl_2$ at a pH of 7.8.

4. TISSUE HARVEST AND CELL ISOLATION

A variety of different cell types have been considered and compared for developing tissue-engineered ligaments. Human BMSC (hBMSC) and differentiated human ACL fibroblasts were culture-expanded and seeded onto silk matrices. Rat bone marrow stem cells were isolated, culture-expanded, and assayed by flow cytometry to clarify the mesenchymal stem cell phenotype.

4.1. Human Bone Marrow Stromal Cells (hBMSCs)

Stromal cells were isolated from unprocessed whole bone marrow aspirates obtained from donors <25 years of age (See Sources of Materials).

Protocol 8.1. Primary Culture of Human Bone Marrow Stromal Cells

Reagents and Materials

Sterile or aseptically collected

- ❑ Human bone marrow (See Sources of Materials)
- ❑ BMSC medium (See Section 3.1)
- ❑ Culture flasks or Petri dishes
- ❑ Trypsin-EDTA (See Section 3.4)
- ❑ Freezing medium (See Section 3.3)

Protocol

(a) Suspend the marrow in BMSC medium (25 ml whole bone marrow aspirate in 100 ml medium).
(b) Plate at 8 μl aspirate/cm^2 in flasks or Petri dishes.
(c) Change medium after 4 days, removing nonadherent hematopoietic cells with the culture medium.
(d) Change the medium twice per week thereafter.
(e) Detach primary hBMSCs just before confluence, using trypsin-EDTA.
(f) Replate at 5×10^3 cells/cm^2.
(g) Trypsinize first-passage (P_1) BMSCs just before they reach confluence; resuspend in freezing medium, and store in liquid nitrogen for future use.

4.2. Human ACL Fibroblasts (ACLFs)

Fibroblasts were obtained by an explant culture method. ACL tissue was harvested from patients undergoing total ACL reconstruction and transferred to the laboratory in saline.

Protocol 8.2. Primary Explant Culture of Human Ligament

Reagents and Materials

- ❑ ACL tissue
- ❑ Culture flasks, 25 cm^2
- ❑ Scalpels, #11 blade
- ❑ Ligament medium (See Section 3.2)

Protocol

(a) Clean ligament of all synovial tissue and cut into ~1-mm^3 pieces.
(b) Transfer explants into a culture flask at ~1 piece per cm^2.
(c) Add 1 ml ligament medium and culture at 37 °C under 5% CO_2.
(d) Add additional ligament medium for a total of 5 ml once explants have attached.
(e) Change medium twice per week until the fibroblastic outgrowth reaches near-confluence, usually within 1–2 weeks.
(f) Detach primary ACLFs with trypsin-EDTA and replate at 5×10^3 cells/cm^2.
(g) Trypsinize first-passage (P_1) ACLFs; resuspend in freezing medium (See Section 3.3), and store in liquid nitrogen for future use.

---

5. SILK MATRICES AND RGD SURFACE MODIFICATION

5.1. Silk Matrices

Matrices were prepared from raw *Bombyx mori* silkworm silk as previously described [Altman et al., 2002a]. Bundles of 12 twisted fibers were generated and complete sericin removal was achieved through batch extraction in an aqueous solution of 0.02 M Na_2CO_3 and 0.3% (w/v) detergent at 90 °C for 1 h. Polyolefin heat shrink tubing was used to cuff the first and fifth centimeters (creating a 3-cm exposed sample length in 5-cm total matrix length) of 24 parallel bundles anchored under equal tension.

5.2. Surface Modification

We recently demonstrated that arginine-glycine-aspartic acid (RGD) surface modification of silk fibers enhanced BMSC adhesion and collagen matrix production [Chen et al., 2003]. RGD peptides were covalently coupled to the matrices as previously described [Chen et al., 2003; Sofia et al., 2001].

Protocol 8.3. Surface Modification of Silk Fibers with RGD for Bone Marrow Stromal Cell Adhesion

Reagents and Materials

- ❑ PBSA: Dulbecco's phosphate-buffered saline without Ca^{2+} and Mg^{2+}, pH 6.5
- ❑ EDC and NHS solution: 0.5 mg/ml 1-ethyl-3-(3-dimethylaminopropyl)carbodiimide hydrochloride and 0.7 mg/ml N-hydroxysuccinimide in PBSA
- ❑ GRGDS peptide: glycine-arginine-glycine-aspartate-serine, 0.4 mg/ml in PBSA
- ❑ Silk matrices (See Section 5.1)

Protocol

(a) Hydrate in PBSA for 1 h.
(b) Activate exposed carboxyl groups of aspartate and glutamate on silk with 20 ml EDC and NHS solution for 15 min at room temperature.
(c) Rinse matrices extensively with PBSA to remove excess EDC/NHS and react with 1 ml GRGDS peptide in PBSA for 2 h at room temperature with gentle shaking.
(d) Rinse matrices thoroughly in PBSA and then in UPW, and dry in air.

6. CELL SEEDING AND CULTIVATION

6.1. For Study of Silk Matrix Design (See Section 9.1.1)

A custom seeding chamber was designed with 24 wells, each 3.2 mm wide by 8 mm deep by 40 mm long (1 ml total volume) machined into a Teflon block designed to fit within a 15-cm Petri dish. Frozen first passage (P_1) BMSCs were used for matrix seeding within the Teflon wells. Sterilized silk cords and bundles were seeded with cells in the customized seeding chambers to minimize the cell to medium volume and increase cell-matrix contact. Cords were seeded with 3.3×10^6 cells in 1 ml of cell culture medium without bFGF as described in Protocol 8.4.

Protocol 8.4. Seeding Silk Matrix Cords with BMSCs

Reagents and Materials

Sterile

- ❑ Frozen P_1 BMSC cells (See Protocol 8.1)
- ❑ BMSC medium without bFGF (See Section 3.1)
- ❑ Silk cords sterilized in ethylene oxide
- ❑ Teflon seeding block, steam sterilized (See Section 6.1)

Protocol

(a) Thaw frozen P_1 BMSCs.
(b) Plate at 5×10^3 cells/cm^2 (P_2).

(c) Trypsinize when near ~85% confluence and resuspend at 1×10^6 cells/ml in BMSC medium (See Section 3.1)
(d) Place sterilized cords within the wells of the steam-sterilized Teflon block.
(e) Add 250 μl cell suspension (1×10^6cells/ml) to the wells containing cords.
(f) Incubate at 37 °C under 5% CO_2 for 30 min.
(g) Rotate the cords through 90° and add a second 250 μl of the cell suspension.
(h) Incubate for a further 30 min.
(i) Repeat the procedure twice more for a total of 270° of rotation to uniformly seed the cords.

After seeding, the cords were cultured independently in an appropriate amount of cell culture medium for 0 (immediately postseeding), 1, 7, and 14 days ($n = 2$ per time point for SEM and 2 for DNA), at which times cell morphology, growth, and marker expression ($n = 3$ at day 14) were assessed.

6.2. Modification for Study of Silk Surface Modification (See Section 9.1.2)

Frozen P_1 BMSCs or ACLFs were defrosted and replated at 5×10^3 cells/cm^2 (P_2), trypsinized when near confluence, and used for matrix seeding. The seeding procedure was dependent on matrix morphology. For single-side seeding, fibrous matrices were inoculated with 1 ml of cell suspension at a concentration of 2×10^6 cells/ml by direct pipetting, incubated for 2 h at 37 °C/5%CO_2, and transferred to tissue culture flasks for experiments in an appropriate amount of cell culture medium without bFGF.

6.3. Modification for Study of Mechanical Stimulation as the Ligament Differentiation Inducer (See Section 9.2)

For multiple (4-time) seeding, matrices were loaded into a custom Teflon seeding vessel designed to allow circumferential cell seeding while minimizing the volume of cell suspension (0.8 ml total) required for full coverage. Each RGD-coupled silk matrix was suspended horizontally (~4mm above the well surface) and anchored between two stainless steel shafts. Once matrices were loaded, the complete vessel was sterilized with ethylene oxide. After aeration, the matrices were wet with PBSA in their independent wells. P_3 BMSCs were delivered via four inoculations of 0.75 ml of cell suspension (3×10^6 cells/ml) at 0°, 180°, 90°, and 270° of rotation. Each inoculation was followed by an incubation period of 30 min at 37 °C under 5%CO_2. After seeding, matrices were incubated in the seeding vessel with culture medium overnight and subsequently transferred to Petri dishes for 2D culture.

7. BIOREACTOR

We have developed a specialized bioreactor to establish conditions in vitro that mimic the knee environment in vivo in terms of mechanical and biochemical conditions. This computer-controlled bench-top bioreactor system can accommodate up to 12 reactor vessels (2.5 cm in diameter × 8 cm long). Each vessel provides an environment for the growth of one vertically oriented BMSC-seeded silk matrix attached longitudinally between two anchors positioned 3 cm apart. Tube dimensions were determined by culture volumes previously used to grow ligaments (20 ml per ligament). Anchor placement was selected based on a human ACL length of 28 mm. Silk matrices were seeded four times with BMSCs in the Teflon chamber as described above (See Protocol 8.4), followed by 6 days of incubation in a static environment, then loaded into bioreactors, and cultured in DMEM with 10% FBS at 37 °C for 6 days. Gas exchange was controlled by recirculating growth medium through a custom-designed environmental chamber (pO_2 21%, pCO_2 10%, pN_2 69%, and pH 7.4). Mechanical stimulation (linear displacement: 0.5 mm; rotational displacement: 45° from the neutral position) was applied with a frequency of 1.39e-4 Hz.

8. ANALYTICAL ASSAYS

8.1. Mechanical Analysis

Mechanical testing was performed with a servohydraulic Instron 8511 tension/compression system with Fast-Track software. Fatigue analysis was performed on single ACL matrix cords. The data for a single cord were extrapolated to represent the six-cord ACL matrix because the cords were designed to be loaded in parallel. Single pull-to-failure testing was performed at a strain rate of 100%/s, and data were analyzed with Instron Series IX software. Fatigue testing to determine cycles to failure at ultimate tensile strength (UTS), 1680 N, and 1200 N ($n = 5$ for each load) was conducted with an H-sine wave function at 1 Hz generated by Wavemaker32 version 6.6 (Instron). Fatigue testing was performed in neutral phosphate-buffered saline (PBSA) at 23 °C. Samples were stored after preparation for no more than 7 days at room temperature before testing.

8.2. Cell Viability

Cell viability was determined by MTT staining. Seeded matrices were incubated in MTT solution (0.5 mg/ml MTT in cell culture medium) at 37 °C under 5%CO_2 for 2 h. The intense purple-colored formazan derivative formed via cell metabolism was eluted and dissolved in 0.04 M HCl in 95% isopropanol, and the absorbance was measured at 570 nm with a reference wavelength of 690 nm. Cell number was correlated to optical density (OD).

8.3. Morphology

Phase-contrast light microscopy was used to observe the morphology of hBMSCs with a Zeiss Axiovert S100 light microscope. Digital image analysis incorporated a CCD color video camera (DXC-390, Sony), a frame grabber card (CG-7 RGB, Scion), and Scion Image software version 1.9.1.

SEM was used to determine cell morphology on matrices. After harvest, seeded matrices were immediately rinsed in 0.2 M sodium cacodylate buffer and fixed in Karnovsky fixative overnight at 4 °C. Samples were then dehydrated through exposure to a gradient of alcohol followed by 1,1,2-trichlorotrifluoroethane and allowed to air dry in a fume hood. Specimens were sputter coated with gold, using a Polaron SC502 Sputter Coater, and examined with a JEOL JSM-840A scanning electron microscope at 15 kV.

8.4. Transcript Levels

RNA Isolation and Real-Time Reverse Transcriptase-Polymerase Chain Reaction (RT-PCR)

Cells were detached with 0.25% trypsin solution and collected after brief centrifugation (1000 *g*, 10 min). RNA was isolated from the collected cells, using the QuiAmp DNA mini kit according to the manufacturer's protocol (Quiagen). The RNA samples were reverse transcribed with oligo(dT) selection (Superscript Preamplification System). Transcripts were assessed with suitable primer sets to target specific markers, using an ABI Prism 7000 real-time system. PCR reaction conditions were 2 min at 50 °C, 10 min at 95 °C, and then 50 cycles at 95 °C for 15 s, and 1 min at 60 °C. The expression data were normalized to the expression of the housekeeping gene glyceraldehyde-3-phosphate dehydrogenase (GAPDH). Primer sequences for the human GAPDH gene were (5′-3′) : forward primer ATG GGG AAG GTG AAG GTC G, reverse primer TAA AAG CCC TGG TGA CC, probe CGC CCA ATA CGA CCA AAT CCG TTG AC. Primers and probes for specific markers of ligament formation can be purchased from Applied Biosciences (Assay on Demand).

8.5. Protein Formation—Extracellular Matrix

Collagen type I protein content was assessed with a commercially available ELISA kit against human collagen type I according to the manufacturer's protocol (Chondrex) and measured at 405 nm with a Spectra Max 250 microplate spectrophotometer; collagen type I concentration was determined by generating a standard curve of known human collagen type I (from ELISA kit) concentrations correlated to optical density.

Protocol 8.5. Isolation of Collagen from Cell-Seeded Silk Matrices

Reagents and Materials

- ❑ PBSA
- ❑ Acetic acid, 0.05 M in 0.5 M NaCl, pH 2.9–3.0 (adjust the pH with formic acid)
- ❑ Pepsin (10 mg/ml) dissolved in 0.05 M acetic acid (See Section 3.8)

- ❑ Pancreatic elastase (1 mg/ml) (See Section 3.9)
- ❑ 1 M Tris-base

Protocol

(a) Rinse the silk matrices with PBSA.

(b) Transfer the matrices to tubes that contain 1.0 ml 0.05 M acetic acid containing 0.5 M NaCl, pH 2.9–3.0.

(c) Add 0.1 ml of pepsin (10 mg/ml) and shake at 4 °C for 48 h

(Pepsin digests telopeptides located on both N- and C-terminals of the collagen molecule and solubilizes the collagen from collagen fibrils; however, pepsin cannot digest the intra- and intermolecular cross-linkages)

(d) Add 0.1 ml 1 M Tris-base to neutralize the pH to 7.5–7.8.

(e) Add 0.1 ml pancreatic elastase (1 mg/ml) and shake at 4 °C overnight.

(Elastase digests the collagen molecule at the N-terminal region, which contains the intra- and intermolecular cross-linkages. As a result of N-terminal cleavage by elastase, the dimer and trimer of collagen are converted into monomeric collagen. ***Note:*** **Elastase digests the denatured collagen into small fragments, so avoid increasing the sample temperature**.)

(f) Centrifuge at 10,000 rpm in a microfuge for 5 min and collect supernate.

(***Note***: Tissue should be completely solubilized; however, a trace of insoluble materials might remain.)

(g) Dilute the supernate at desired times with the Sample Dilution Buffer provided in Collagen Capture ELISA Kits.

8.6. Statistical Analysis

Data for cell attachment, viability, and collagen matrix production were analyzed by one-way ANOVA to evaluate differences between groups. Post hoc comparison of means is accomplished with the Student-Newman-Keuls test to determine significance between groups in which a p value <0.05 is typically defined as significant. All data are reported as means ± the standard error of the mean.

9. REPRESENTATIVE STUDIES

9.1. Silk Fibroin as the Scaffold for ACL Tissue Engineering

9.1.1. Silk Matrix Design

The geometry of the knee and kinematics related to ACL structure must be incorporated into any design strategy if a tissue-engineered ACL generated in vitro is to stabilize the knee successfully and function correctly in vivo. A mismatch in the ACL structure-function relationship would result in graft failure. Our hypothesis is that mechanical signals applied in vitro to a growing ACL will induce the formation of the structural and functional features required to meet performance requirements in vivo. The objective of the study was to engineer a mechanically and biologically functional matrix for ACL tissue engineering. Proper matrix design is essential to a successful tissue-engineered ACL and provides a foundation upon which to explore mechanics and biological function.

A silk fiber-based matrix was used in combination with a wire rope design to mirror the ACL in ultimate tensile strength (UTS) and linear stiffness. Having already been used clinically as suture material for decades [Inouye et al., 1998; Minoura et al., 1990; Santin et al., 1999; Sofia et al., 2001], silk was selected from the array of alternative synthetic and natural fibers for this application because it was the only protein-based fiber that: (1) could match the required mechanical properties of an ACL, (2) was biocompatible when properly prepared [Inouye et al., 1998], (3) avoided bioburdens associated with mammalian-derived materials, (4) maintained mechanical tensile integrity in vitro in tissue culture conditions, and (5) exhibited slow degradation in vivo [Greenwald et al., 1994], allowing adequate time for host tissue infiltration and eventual stabilization.

The number of fibers and geometry of the silk matrix were designed to mirror human ACL in UTS, linear stiffness, yield point, and percent elongation at break, as well as to support cell seeding and tissue ingrowth. Sericin removal, UTS measured in newtons (N), and linear stiffness (N/mm) were characterized as a function of extraction temperature up to 90 °C for a 60-minute processing period by SEM and single-pull-to-failure mechanical analysis.

The geometric hierarchy of the matrix that best mirrored the ACL was as follows: 1 ACL prosthesis = 6 parallel cords; 1 cord = 3 twisted strands (3 twists/cm counterclockwise); 1 strand = 6 twisted bundles (3 twists/cm clockwise); 1 bundle = 30 parallel extracted fibers. The resulting ACL matrix contained 3240 preextracted fibers with a combined cross-sectional area of 3.67 mm^2 and could tightly fit within a 4-mm-diameter bone tunnel. The selected silk matrix exhibited mechanical properties comparable to those of the native human ACL: UTS of ~2100 N, stiffness of ~240 N/mm, yield point of ~1200 N, and 33% elongation at break. Regression analysis of matrix fatigue data, when extrapolated to physiological load levels (400 N) [Chen and Black, 1980] to predict number of cycles to failure in vivo, indicated a matrix life of 3.3 million cycles.

To test the mechanical integrity of silk fibroin in physiologic cell culture conditions, bundles of 30 extracted silk fibers 3 cm between sutures were prepared. Bundles were seeded as described above (See Protocol 8.4), using a 1×10^6 cells/ml cell suspension and cultured for 1, 7, 14, and 21 days ($n = .8$ per time point); otherwise identical nonseeded matrices were generated for use as controls. Single pull-to-failure mechanical analysis was performed as described above by clamping directly on the knotted sutures, eliminating the need for molded epoxy ends; also, matrices were saturated with cell culture medium throughout the testing procedure. The silk fibers retained mechanical tensile strength over 21 days in tissue culture conditions, and no statistically significant changes in UTS were observed when comparing seeded and nonseeded matrices.

Human BMSCs readily adhered, spread, and grew on the silk fiber matrix after 1 day in culture and formed cellular extensions to bridge neighboring fibers. A uniform cell sheet covering the fibrous construct was observed by 14 days of culture. Measures of total DNA, from 2 cords per time point, confirmed BMSC proliferation on the silk, with the highest amount of DNA measured after 14 days in culture. Real-time RT-PCR assessment of 3 seeded cords cultured for 14 days indicated ligament-specific marker expression (e.g., collagen types I and III, tenascin-C) by the cultured BMSCs. Collagen types II and bone sialoprotein, as markers of cartilage- and bone-specific differentiation, respectively, were negligibly expressed. Furthermore, the ratio of collagen type I expression to collagen type III was 8.9:1, consistent with that of cruciate ligaments [Amiel et al., 1984] and not indicative of a wound-healing response characterized by excessive collagen type III expression. In comparison to human BMSCs grown in collagen gels for 14 days in a static environment [Altman et al., 2002b], baseline levels of collagen types I and III and tenascin-C expression on the silk matrix were 6.1-fold, 8.2-fold, and 7.6-fold greater, respectively.

9.1.2. Silk Matrix Surface Modification

In ligament tissue engineering the appropriate matrix, when combined with cells and environmental stimuli, must support cell attachment, spreading, growth and collagen matrix production. We have shown that a silk matrix can support BMSC attachment and differentiation. As a first step in in vitro ligament optimization, surface modification of the silk fiber matrix with RGD peptide was selected to enhance early stages of cell-matrix interactions. RGD-containing peptide is the main integrin-mediated cell attachment domain found in many extracellular matrix proteins; however, RGD binding domains are not present in the native *B. mori* silk. We hypothesize that enhanced initial cell attachment, spreading, and migration resulting from RGD coupling will increase cell density and ECM production, leading to higher rates of ligament development in vitro. This hypothesis is derived from related studies in which high initial cell densities led to better gap junction communication. The effects of silk matrix-RGD peptide coupling on BMSC and ACL fibroblast (as a positive control) behavior over 14 days of culture were explored.

RGD modification increased the rates of both BMSC matrix adhesion and spreading within 1 h of seeding; BMSC and ACLF matrix density was significantly increased by 1.6- to 2.1-fold (MTT assay) 1 day after seeding. It is known that RGD surface modification can improve cell mobility and, in turn, dramatically increase proliferation [Sottile et al., 1998]; however RGD modification of the fibroin matrix did not affect cell growth rates from 7 to 14 days of culture. Thus the effect of RGD coupling was only evident at initial seeding, resulting in an increase in cell attachment and spreading. SEMs of both cell types on modified and nonmodified matrices indicated the development of a cell sheet and possible ECM formation over 14 days in culture; however, the cell/ECM on RGD-modified matrices was thicker and more continuous than on nonmodified matrices to such an extent that silk fibroin fibers could not be seen in the micrographs at 14 days.

Collagen type I, the predominant protein of most connective tissues and the main constituent of the ACL, occupying >86% of its dry tissue weight, was chosen as a representative marker for cell matrix production. RGD coupling, through the enhancement of initial cell density, induced changes in collagen type I production over time as determined by mRNA transcript (monitored via semiquantitative RT-PCR) levels at day 7 and protein concentration at day 14. The significant increases observed in collagen type I mRNA levels from BMSCs and ACLFs grown on RGD-modified matrices are likely the result of an increase in cell density; enhanced cell-cell interactions likely induced a switch from cell proliferation to matrix production (differentiation). The increase in collagen type I transcript levels from BMSCs was fivefold greater than the increase observed from ACLFs (i.e., 180% compared with 30%).

ELISA against collagen type I at day 14 confirmed the trends observed in transcription levels at day 7. By 14 days of culture, a 410% increase in BMSC collagen type I production on RGD-modified compared with nonmodified matrices was observed, supporting the notion of a switch from a proliferative nondifferentiated state at low cell density (e.g., 7 days) to a matrix-producing state at high cell densities (14 days). Furthermore, although the effects on collagen type I matrix production may simply be the result of an increase in cell density, the introduction of additional cell binding sites through RGD modification may also induce subsequent intracellular signaling events to promote cell matrix production and differentiation [Grzesiak et al., 1997].

For tissue engineering, BMSCs should attach firmly on silk matrices, and, appropriately, most BMSCs remained on the RGD-modified silk film after trypsinization for 15 min, whereas most of the cells seeded on tissue culture plastic and EDC/NHS-treated and nonmodified silk films detached. These data are consistent with several studies that indicated strong adhesion of cells to RGD peptides on surfaces [Massia and Hubbell, 1990; Neff et al., 1999].

The present study has begun to address these issues by demonstrating a role for RGD-modified silk fibers in accelerating the attachment, coverage, and differentiation of BMSCs. These studies extend our prior work [Altman et al., 2002a, b] and suggest a path forward for continued optimization of the process.

9.2. Mechanical Stimulation as the Ligament Differentiation Inducer

Mechanical stimuli are known to be able to trigger a number of signaling pathways (protein kinase: FAK, ILK, Src; adaptor proteins: Shc, Grb-2, Crk; small GTPases: Rho, Ras) through cell surface receptors, $\alpha_5\beta_1$-integrin for example, [Giancotti, 1997]. The receptors, in turn, may activate the mitogen-activated protein kinase (MAPK) pathway, which counteracts apoptosis and promotes cell survival, thus increasing cell density. Mechanical stimulation could also activate MAPK via autocrine release of growth factors [Chiquet, 1999; Kim et al., 1999; MacKenna et al., 2000; Ruoslahti, 1997]. The identification of the signal transduction pathways involved in the present BMSC-silk-bioreactor system may lead future research in interesting directions.

Previously we proved that mechanical stimulation alone, without the addition of growth factors, can induce BMSC differentiation into ligament lineage [Altman et al., 2002b]. BMSCs seeded in a collagen gel system aligned along the direction of mechanical stimulation and produced collagen types I and III and tenascin-C, the typical markers of ligament differentiation.

To investigate the impact of mechanical stimulation upon the cells cultured on silk fibers, matrices were exposed to multidimensional cyclic stresses (torsion and tension). SEM analysis showed that BMSCs seeded on the RGD-modified silk matrices remained attached after 6 days of stimulation and aligned along the direction of mechanical stimuli, whereas most of the cells seeded on the nontreated silk fibers detached [Chen et al., 2003]. Ongoing experiments focus on optimizing the parameters of mechanical regime and dynamic medium flow to support BMSC growth and differentiation into ligament fibroblasts.

ACKNOWLEDGMENTS

Support from Tissue Regeneration, Inc., the Dolores Zorhab Liebmann National Foundation, the American Orthopaedic Society for Sports Medicine, New England Medical Center, the NIH (R01-DE13405-01), the NSF (DBI), and NASA (NCC8-174) are greatly appreciated. We thank Annette Shephard-Barry (New England Medical Center Pathology Laboratory) for her technical assistance.

SOURCES OF MATERIALS

Item	*Supplier*	*Storage Condition*
ABI Prism 7000 real time system	Applied Biosystems	
Axiovert S100 light microscope	Zeiss	
Basic fibroblast growth factor (bFGF)	Invitrogen	−20 °C
Bombyx mori silkworm silk	Rudolph-Desco	
CCD color video camera (DXC-390)	Sony	
3-[4,5-dimethylthiazol-2-yl]-2,5-diphenyl tetrazolium bromide (MTT)	Sigma	4 °C
DMSO	Sigma	Room temperature
Donkey anti-rabbit IgG	Jackson ImmunoResearch	4 °C
Dulbecco's modified Eagle's medium (DMEM)	Invitrogen	4 °C
ELISA kit for collagen type I	Chondrex	−20 °C
ELISA kit, human collagen type I	Chondrex	
1-ethyl-3-(3-dimethylaminopropyl)carbodiimide hydrochloride (EDC)	Pierce	
Fast-Track software Wavemaker32 version 6.6	Instron	
Fetal bovine serum (FBS)	Invitrogen	−20 °C
Frame grabber card (CG-7 RGB)	Scion	
Glycine-arginine-glycine-aspartate-serine (GRGDS) peptide	Sigma	
Human bone marrow stromal cells	Clonetic-Poietics	

Item	*Supplier*	*Storage Condition*
N-hydroxysuccinimide (NHS)	Pierce	
Instron 8511 servohydraulic tension/compression system	Instron	
L-Ascorbic acid 2-phosphate	Sigma	Room temperature
Microplate reader	Molecular Devices	
MTT	Sigma	4 °C
Pancreatic elastase	Sigma	−20 °C
Penicillin-streptomycin (P/S)	Invitrogen	−20 °C
Pepsin	Sigma	−20 °C
Phosphate-buffered solution (PBS)	Invitrogen	Room temperature
PicoGreen	Molecular Probes	−20 °C
Polaron SC502 Sputter Coater	Fison	
Polyolefin heat shrink tubing	Appleton Electronics	
Primers and probes, markers of ligament formation (Assay on Demand)	Applied Biosciences	
QuiAmp DNA mini kit	Quiagen	−20 °C
Rabbit anti-human collagen type I antibody	Biodesign	−20 °C
Scanning Electron Microscope JEOL JSM-840A	JEOL	
Scion Image software version 1.9.1	Scion	
Spectra Max 250 microplate spectrophotometer	Molecular Devices	
Sputter coater	Fison	
Superscript Preamplification System	Invitrogen	
1,1,2-Trichlorotrifluoroethane	Aldrich	Room temperature
Trypsin-EDTA	Invitrogen	−20 °C
Wavemaker32 version 6.6	Instron	

REFERENCES

Altman, G.H., Diaz, F., Jakuba, C., Calabro, T., Horan, R.L., Chen, J., Lu, H., Richmond, J., Kaplan, D.L. (2003) Silk-based biomaterials. *Biomaterials* 24: 401–416.

Altman, G.H., Horan, R.L., Lu, H.H., Moreau, J., Martin, I., Richmond, J.C., Kaplan, D.L. (2002a) Silk matrix for tissue engineered anterior cruciate ligaments. *Biomaterials* 23: 4131–4141.

Altman, G.H., Horan, R.L., Martin, I., Farhadi, J., Stark, P.R., Volloch, V., Richmond, J.C., Vunjak-Novakovic, G., Kaplan, D.L. (2002b) Cell differentiation by mechanical stress. *FASEB J.* 16: 270–272.

Amiel, D., Frank, C., Harwood, F., Fronek, J., Akeson, W. (1984) Tendons and ligaments: a morphological and biochemical comparison. *J. Orthop. Res.* 1: 257–265.

Bramono, D.S., Richmond, J.C., Weitzel, P.P., Kaplan DL, Altman GH. (2004) Matrix metalloproteinases and their clinical applications in orthopaedics. *Clin Orthop Relat Res.* 2004 Nov;(428): 272–85. Chen, E.H., Black, J. (1980) Materials design analysis of the prosthetic anterior cruciate ligament. *J. Biomed. Mater. Res.* 14: 567–586.

Chen, J., Altman, G.H., Karageorgiou, V., Horan, R., Collette, A., Volloch, V., Colabro, T., Kaplan, D.L. (2003) Human bone marrow stromal cell and ligament fibroblast responses on RGD-modified silk fibers. *J. Biomed. Mater. Res.* 67A: 559–570.

Chiquet, M. (1999) Regulation of extracellular matrix gene expression by mechanical stress. *Matrix Biol.* 18: 417–426.

Dunn, M.G., Maxian, S.H., Zawadsky, J.P. (1994) Intraosseous incorporation of composite collagen prostheses designed for ligament reconstruction. *J. Orthop. Res.* 12: 128–137.
Dunn, M.G., Tria, A.J., Kato, Y.P., Bechler, J.R., Ochner, R.S., Zawadsky, J.P., Silver, F.H. (1992) Anterior cruciate ligament reconstruction using a composite collagenous prosthesis. A biomechanical and histologic study in rabbits. *Am. J. Sports. Med.* 20: 507–515.
Frank, C.B., Hart, D.A., Shrive, N.G. (1999) Molecular biology and biomechanics of normal and healing ligaments—a review. *Osteoarthritis Cartilage* 7: 130–140.
Giancotti, F.G. (1997) Integrin signaling: specificity and control of cell survival and cell cycle progression. *Curr. Opin. Cell. Biol.* 9: 691–700.
Greenwald, D., Shumway, S., Albear, P., Gottlieb, L. (1994) Mechanical comparison of 10 suture materials before and after in vivo incubation. *J. Surg. Res.* 56: 372–377.
Grzesiak, J.J., Pierschbacher, M.D., Amodeo, M.F., Malaney, T.I., Glass, J.R. (1997) Enhancement of cell interactions with collagen/glycosaminoglycan matrices by RGD derivatization. *Biomaterials* 18: 1625–1632.
Inouye, K., Kurokawa, M., Nishikawa, S., Tsukada, M. (1998) Use of *Bombyx mori* silk fibroin as a substratum for cultivation of animal cells. *J. Biochem. Biophys. Methods* 37: 159–164.
Kim, B.S., Nikolovski, J., Bonadio, J., Mooney, D.J. (1999) Cyclic mechanical strain regulates the development of engineered smooth muscle tissue. *Nat. Biotechnol.* 17: 979–983.
Lo, I.K., Marchuk, L., Hart, D.A., Frank, C.B. (2003) Messenger ribonucleic acid levels in disrupted human anterior cruciate ligaments. *Clin. Orthop. Relat. Res.* 407: 249–258.
Lo, I.K., Marchuk, L.L., Hart, D.A., Frank, C.B. (1998) Comparison of mRNA levels for matrix molecules in normal and disrupted human anterior cruciate ligaments using reverse transcription-polymerase chain reaction. *J. Orthop. Res.* 16: 421–428.
MacKenna, D., Summerour, S.R., Villarreal, F.J. (2000) Role of mechanical factors in modulating cardiac fibroblast function and extracellular matrix synthesis. *Cardiovasc. Res.* 46: 257–263.
Massia, S.P., and Hubbell, J.A. (1990) Covalent surface immobilization of Arg-Gly-Asp- and Tyr-Ile-Gly-Ser-Arg-containing peptides to obtain well-defined cell-adhesive substrates. *Anal. Biochem.* 187: 292–301.
Minoura, N., Tsukada, M., Nagura, M. (1990) Physico-chemical properties of silk fibroin membrane as a biomaterial. *Biomaterials* 11: 430–434.
Neff, J.A., Tresco, P.A., Caldwell, K.D. (1999) Surface modification for controlled studies of cell-ligand interactions. *Biomaterials* 20: 2377–2393.
Ruoslahti, E. (1997) Stretching is good for a cell. *Science* 276: 1345–1346.
Santin, M., Motta, A., Freddi, G., Cannas, M. (1999) In vitro evaluation of the inflammatory potential of the silk fibroin. *J. Biomed. Mater. Res.* 46: 382–389.
Sofia, S., McCarthy, M.B., Gronowicz, G., Kaplan, D.L. (2001) Functionalized silk-based biomaterials for bone formation. *J. Biomed. Mater. Res.* 54: 139–148.
Sottile, J., Hocking, D.C., Swiatek, P.J. (1998) Fibronectin matrix assembly enhances adhesion-dependent cell growth. *J. Cell Sci.* 111 (19): 2933–2943.
Weitzel, P.P., Richmond, J.C., Altman, G.H., Calabro, T., Kaplan, D.L. (2002) Future direction of the treatment of ACL ruptures. *Orthop. Clin. North. Am.* 33: 653–661.

9

Cellular Photoencapsulation in Hydrogels

Jennifer Elisseeff[1], Melanie Ruffner[1], Tae-Gyun Kim[3], and Christopher Williams[2]

Department of Biomedical Engineering[1], and Department of Plastic Surgery[2], Johns Hopkins University, Baltimore, Maryland 21218; Seoul National University, Seoul, Korea[3]

Corresponding author: jhe@bme.jhu.edu

Culture of Cells for Tissue Engineering, edited by Gordana Vunjak-Novakovic and R. Ian Freshney

1. BACKGROUND

The goal of tissue engineering is to regenerate tissue and organ structures to replace those lost from trauma, congenital abnormalities, or disease [Mooney and Mikos, 1999]. One strategy for tissue engineering encompasses seeding cells by photoencapsulation within a hydrogel scaffold. The scaffold is designed to promote desired cell function and tissue development while physically protecting the nascent tissue after its implantation in vivo. Hydrogels are a class of biomaterial scaffolds that have shown potential for numerous tissue engineering applications.

1.1. Hydrogels

Hydrogels are formed by cross-linking water-soluble polymer chains to form a water-insoluble polymer network [Brannon-Peppas, 1994]. Cells may be encapsulated during the cross-linking process to create cell-hydrogel constructs for drug delivery and tissue engineering applications. Hydrogels have unique properties that make them potentially useful for tissue engineering, such as high water content for nutrient and waste transport, elasticity, and the ability to encapsulate or immobilize cells in a three-dimensional environment in situ. The properties of a hydrogel can be altered by manipulating polymer chemistry and cross-linking density. The distance between cross-links, or cross-linking density, directly influences the pore size of a hydrogel and related physical properties such as water content and mechanical strength. A scaffold with a high cross-linking density and a smaller pore size will imbibe less water and exhibit stronger mechanical properties compared to a hydrogel with a lower cross-linking density and a larger pore size.

Cross-linking density and pore size also influence cell behavior and tissue development. Researchers have shown that chondrocytes have increased extracellular matrix production in hydrogels with larger pore sizes. Therefore a balance must be found in the hydrogel formulation for optimal scaffold physical properties for a desired application with optimal cell function and tissue development. Hydrogels may be designed to remain stable or degrade over time, and their degradation properties can be controlled to match the rate of tissue development or matrix formation. Polymers that form the hydrogel can be chemically altered to incorporate growth factors or cell binding sites in order to promote cell proliferation, extracellular matrix production, or cell differentiation. Aside from cell encapsulation and tissue engineering technologies, numerous applications for hydrogels exist, including biosensors, dentistry, surgery, and drug-delivery systems. This chapter will focus on methods for isolating cells and encapsulating them in photopolymerizing hydrogels, with an emphasis on cartilage tissue engineering.

1.2. Methods for Forming a Hydrogel

There are numerous chemical options and methods for forming hydrogels under mild conditions that are compatible with cell encapsulation and tissue engineering applications. Both synthetic and naturally derived hydrogels have been applied to cell encapsulation. Naturally derived polymers that can form hydrogels include collagen, fibrin, agarose, hyaluronic acid, chondroitin sulfate, and chitosan [Buschmann et al., 1992; Ye et al., 2000; Silverman et al., 1999; Madihally and Matthew, 1999]. These natural hydrogels often have interesting biological properties that help promote tissue development. The physical properties and cross-linking of naturally derived hydrogels are often more difficult to control, compared to synthetic materials, leading in some cases to mechanically weak scaffolds. Mechanical properties of hydrogels pose a challenge for musculoskeletal tissue engineering because of their inability to withstand physiological loading. Synthetic polymers can form hydrogels with highly controlled cross-linked structures. For example, polyethylene glycol (PEG), polyvinyl alcohol, and polypropylene fumarate are synthetic polymers that have been modified to create cross-linked gels with controlled porosity [Drumheller and Hubbell, 1994; Behravesh et al., 2002; He et al., 2000].

Numerous methods may be used to form cross-links between polymer chains and create a hydrogel by covalent, ionic, or physical cross-links (van der Waals forces, hydrogen bonds). The polymerization or cross-linking process can be triggered by radiation, temperature changes, addition of a chemical cross-linker, or ionic agents. For many of these methods, once cross-linking is induced, the process cannot be stopped or accelerated. This lack of control of the cross-linking process and the subsequent difficulties in clinical application led to the development of photopolymerization to encapsulate cells in hydrogels for cartilage tissue engineering.

2. PRINCIPLES OF METHODOLOGY: PHOTOPOLYMERIZATION FOR CELL ENCAPSULATION

Photopolymerization is a method to covalently cross-link polymer chains to form a hydrogel. A photopolymerization occurs when a photoinitiator and polymer (with groups sensitive to the initiating species) are exposed to a light source specific to the photoinitiator species. The reaction is rapid, allowing for fast curing of a liquid to a cross-linked, water-swollen polymer network at room or body temperature. This technique allows for enhanced control over the gelation process compared to gel formation by physical or ionic interactions. Photoinduced gelation provides spatial and temporal control during scaffold formation, even permitting shape manipulation after injection and during gelation in vivo. Photopolymerizing hydrogels have been used in a wide variety of biomedical applications and have the potential to create a significant impact in tissue engineering [Burdick et al., 2002]. Photopolymerizations are utilized in the field of dentistry for applications ranging from sealants for caries prevention to root canal procedures and in the fields of drug delivery and tissue engineering [Tarle et al., 1998; Anseth et al., 1994].

Photopolymerizing polymer networks with a range of physical and mechanical properties have been developed, and their function as tissue engineering scaffolds has been studied. For example, Anseth and colleagues have examined photopolymerizing polyanhydrides for bone tissue engineering and polyvinyl alcohols and polyethylene oxide (PEO) for cartilage tissue engineering [Young et al., 2000; Poshusta and Anseth, 2001; Bryant et al., 1999; Burkoth and Anseth, 2000]. Hubbell and colleagues developed novel degradable photopolymerizing hydrogels based on PEO [Sawhney et al., 1996]. These polymers have been applied to drug delivery and cell encapsulation and have been studied as lung sealants and for the prevention of postoperative adhesions [Sawhney et al., 1996; Lyman et al., 1996; Hill-West et al., 1994]. Our research has demonstrated the feasibility of cell photoencapsulation for cartilage tissue engineering, and the key methods are presented in this chapter [Ye et al., 2000; Lyman et al., 1996; Elisseeff et al., 1999].

3. PREPARATION OF MEDIA AND REAGENTS

3.1. Transport Medium

Either cold phosphate-buffered saline (PBSA) supplemented with antibiotics or DMEM (Dulbecco's modified Eagle's medium with high glucose) with antibiotics.

3.2. Collagenase

DMEM containing 0.2% collagenase and 5% fetal bovine serum.

3.3. DMEM-FB-PS

DMEM supplemented with 10% fetal bovine serum, 100 U/ml penicillin. and 100 μg/ml streptomycin.

3.4. Papain

Papain 125 μg/ml in 0.1 M phosphate buffer, 10 mM cysteine, 10 mM EDTA, pH 6.3.

4. TISSUE HARVEST AND CELL ISOLATION

When designing a tissue engineering system, the source of the cells that are seeded on the biomaterial scaffold will significantly impact on the quality of engineered tissue and must therefore be chosen carefully. As we are studying tissue engineering in the musculoskeletal system, chondrocytes (the cells that comprise cartilage) and bone marrow-derived mesenchymal stem cells (capable of differentiating into cartilage) are two potential cell options for incorporation in the scaffold. Isolation of chondrocytes is essentially a two-step procedure: harvesting the tissue and isolating the cells from the tissue. Previously, most authors performed sequential digestion of cartilage with selected enzymes, including collagenase, hyaluronidase, and trypsin [Sah et al., 1991]. More recently, collagenase alone has been found to be adequate for cartilage digestion.

4.1. Chondrocytes

In current autologous chondrocyte transplantations, isolated chondrocytes are expanded by serial passage in monolayer culture because of the limited quantity of cartilage tissue that can be harvested from a patient. However, the expansion in monolayer culture causes the dedifferentiation or loss of chondrocyte-specific gene expression. Chondrocytes propagated in monolayers are marked by a decrease in their ability to produce cartilage-specific proteins such as collagen type II and aggrecan. Loss of the chondrocytic phenotype during monolayer expansion may be linked to the variable results of chondrocyte transplantation for cartilage injury.

Protocol 9.1. Chondrocyte Isolation from Bovine Knee Joint

Reagents and Materials

Sterile

- ❑ Transport medium, refrigerated (See Section 3.1)

- ❑ Dissection instruments
- ❑ Collagenase, 0.2% (See Section 3.2)
- ❑ Plastic tube, 50 ml
- ❑ Nylon filter, 70-μm mesh

Nonsterile

- ❑ Hemocytometer
- ❑ Trypan Blue

Protocol

(a) Harvest cartilage from the knee joint of 5- to 8-week-old bovine calves under aseptic conditions.

(b) Transfer to the tissue culture laboratory in transport medium.

(c) Dissect the cartilage free from connective tissue or bone under sterile conditions.

(d) Mince into small pieces (1–3 mm^3) with a scalpel.

(e) Rinse the tissue several times with cold PBSA.

(f) Transfer to a preweighed digestion vessel and weigh the tissue. A 50-ml tube works well for tissue samples less than 300 mg.

(g) Incubate the tissue in 0.2% collagenase for 14–16 h at 37 °C under 5% CO_2 on an orbital shaker rotating at approximately 75 rpm.

(h) Resuspend, using a pipette. Filter the resulting cell suspensions through a 70-μm nylon filter.

(i) Wash the cells three times with PBSA to remove collagenase, any matrix debris, or undigested particles.

(j) Count the cells with a hemocytometer, assessing cell viability by Trypan Blue dye exclusion.

4.2. Mesenchymal Stem Cells

Bone marrow-derived mesenchymal stem cells (MSCs) are another cell option for musculoskeletal tissue engineering. They are isolated based on their ability to adhere to the culture dish while nonadherent hematopoietic cells are removed by media change. MSCs are capable of differentiating into multiple cell types to form cartilage, bone, muscle, and fat. Methods to isolate these cells are presented in Protocol 9.2.

Protocol 9.2. Bone Marrow-Derived Stem Cell Isolation and Expansion

Reagents and Materials

Sterile or Aseptically Prepared

- ❑ Femurs or iliac crest of three- to three-and-a-half-year old castrated male goats
- ❑ DMEM-FB-PS (See Section 3.3)
- ❑ MSCGM: mesenchymal stem cell growth medium (See Sources of Materials)
- ❑ FBS: fetal bovine serum

- ❑ Trypsin-EDTA: 0.025% trypsin in 0.01% EDTA
- ❑ DMSO
- ❑ Syringe, 10 ml, with heparin, 6000 U
- ❑ Syringe needles, 16 and 21 gauge
- ❑ Tissue culture flasks, 75 cm^2, 175 cm^2

Protocol

(a) Aspirate bone marrow from the femurs or iliac crests of three- to three-and-a-half-year-old castrated male goats into 10-ml syringes with 6000 U heparin.

(b) Make single-cell suspensions by passing the marrow through 16- and 21-gauge needles three times.

(c) Resuspend the cells in DMEM-FB-PS.

(d) Wash the marrow samples twice in mesenchymal stem cell growth medium before suspension in fresh MSCGM.

(e) Count the number of mononuclear cells with a hemocytometer.

(f) Plate in 75-cm^2 tissue culture plastic flasks at a density of approximately 1.2×10^5 mononuclear cells/cm^2.

(g) Change the culture medium after 4 days and then every 2–3 days thereafter until confluence (12–14 days).

(h) When cells are near confluent, passage the cells with trypsin-EDTA for 5 min at 37 °C and replate in 75-cm^2 or 175-cm^2 flasks at 5000 MSCs/cm^2.

(i) Freeze MSCs in liquid nitrogen at 5×10^6–1×10^7 cells/ml in 50% MSCGM, 40% FBS until needed.

5. CELL PHOTOENCAPSULATION

Now that the cells are isolated and, if necessary, expanded, they can be encapsulated in the hydrogel scaffold. As discussed above, photopolymerization is a fast and efficient method for encapsulating cells, but care must be taken to ensure cell survival in the hydrogel (Fig. 9.1). In particular, a cytocompatible photoinitiator, which produces the radical that is responsible for polymerization and hydrogel formation, must be chosen. A concentration of photoinitiator may be determined that is not toxic for the cells yet allows the polymerization reaction to proceed efficiently. For each new cell type that we photoencapsulate, the toxicity of the photoinitiator is evaluated and if cytotoxicity is observed, the concentration is modified or a new initiator is chosen. There are numerous methods to monitor cell viability including Trypan Blue dye exclusion, fluorescent live-dead cell assay (e.g., with diacetylfluorescein and propidium iodide), MTT [Plumb et al., 1989], and WST-1 [Ukeda et al., 2002; Huhtala et al., 2003]. MTT and WST-1 assays are based on the use of a chemical compound that is converted by mitochondrial enzymes to a dye that can be monitored by a spectrophotometer. As augmentation of enzymatic activity leads to an increase in dye production, there is a correlation with the dye absorbance and the number of metabolically active cells. An example of a protocol for WST-1 analysis to determine photoinitiator toxicity is provided in Protocol 9.3.

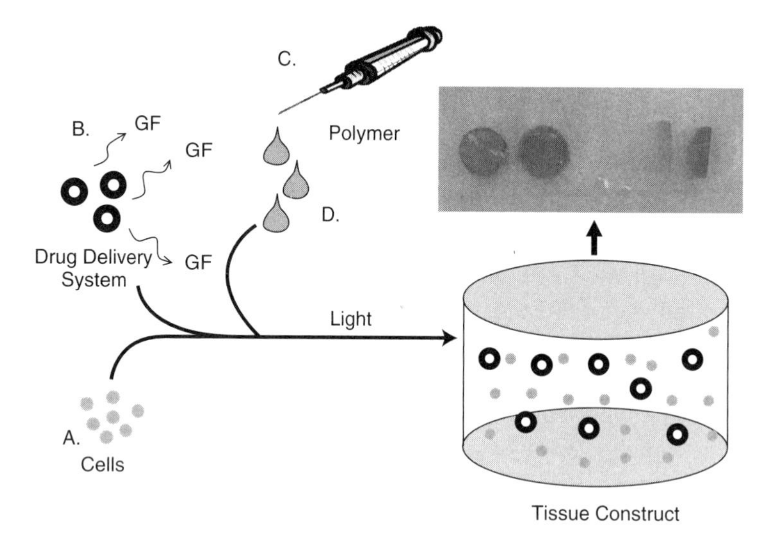

Figure 9.1. Schematic diagram of the photoencapsulation process. Cells are isolated (A) and combined with a drug delivery vehicle (B), if desired, and polymer (C). The cell-polymer liquid is placed in a mold or injected into an animal. The mixture is then exposed to light (D) to cause a photopolymerization and form a cell-laden hydrogel as pictured.

Protocol 9.3. WST-1 Analysis of Photoinitiator Toxicity

Reagents and Materials

Sterile

- ❑ Cells under study plus materials for trypsinization (See Protocol 9.2)
- ❑ Growth medium, e.g., Section 3.3
- ❑ PBSA
- ❑ Initiator stock solution (i.e., 100 mg/ml Irgacure D2959)
- ❑ WST-1 solution
- ❑ Tissue culture plate, 24 well

Nonsterile

- ❑ Light source
- ❑ Multiwell spectrophotometer (ELISA plate reader)

Protocol

(a) Prepare desired cell type to be studied at a concentration of 5×10^5 cells/ml medium.
(b) Add 400 μl medium to each well of a 24-well tissue culture plate.
(c) Add 100 μl cell suspension to each well.
(d) Add 5 μl initiator stock solution.
(e) Incubate for 30 min.
(f) Expose plates to light for photoinitiator activation (i.e., for Irgacure D2959 365 nm 4 mW/cm^2 for 0, 3, 6, 10 min).

(g) Incubate for 48 h.
(h) Aspirate medium from each well and rinse with 1–2 ml phosphate-buffered saline (PBSA)
(i) Add 500 μl medium to each well (including three extra cell-free wells for controls).
(j) Add 50 μl WST-1 solution to each well, incubate at 37 °C, 5% CO_2 for 4 h, and then measure absorbance at 440 nm on a multiwell spectrophotometer (ELISA reader).

Protocol 9.4. Polymer-Chondrocyte Preparation with Photoinitiated Hydrogels

Reagents and Materials

Sterile

- ❑ PEGDA: polyethylene glycol diacrylate
- ❑ PBSA with antibiotics (See Section 3.1)
- ❑ Photoinitiator: Irgacure D2959
- ❑ Cylindrical molds

Nonsterile

- ❑ UV light, 365 nm

Protocol

(a) Prepare the hydrogel solution by mixing 10% w/v PEGDA in sterile PBSA with antibiotics.
(b) Add the photoinitiator to the PEGDA solution and mix thoroughly to make a final concentration of 0.05% w/v.
(c) Immediately before photoencapsulation, resuspend chondrocytes in the solution to make a concentration of 2.0×10^7 cells/ml and gently mix to make a homogeneous suspension.
(d) Transfer 100 μl cell-polymer-photoinitiator suspension into cylindrical molds with a 6-mm internal diameter and expose the suspension for 5 min to long-wave, 365-nm UV light at 4 mW/cm^2.
(e) Remove the hydrogels from the molds and incubate them in separate wells of 12-well plates with the appropriate medium.
(f) Change the culture medium twice a week.

---

6. ENGINEERED TISSUE ANALYSIS

After cellular photoencapsulation and in vitro or in vivo incubation, the quality of tissue that develops in the material must be evaluated. In the case of cartilage, two major matrix components, type II collagen and proteoglycan, should be analyzed. In addition, the cellular content should be monitored to determine whether cell death or proliferation occurred and the gene expression (mRNA levels) of cartilage-specific markers may be evaluated. Protocols 9.5 through 9.7 describe procedures to evaluate cartilage tissue production in photopolymerizing hydrogels.

Protocol 9.5. Biochemical Analysis of Cartilage Hydrogels

Reagents and Materials

Nonsterile

- ❑ Papain, 125 μg/ml (See Section 3.4)
- ❑ Hoechst 33258
- ❑ Dimethylmethylene Blue dye
- ❑ *p*-Dimethylaminobenzaldehyde
- ❑ Chloramine-T
- ❑ Balance, mg range
- ❑ Tissue grinder

Protocol

(a) Remove construct from culture medium, lightly blot dry, and obtain the wet weight. Lyophilize for 48 h under vacuum and obtain dry weight.
(b) Crush the dried construct with a tissue grinder.
(c) Digest each specimen in 1 ml papain solution for 18 h at 60 °C. Digested construct may be stored in the freezer for future analysis.
(d) Determine the DNA content (ng of DNA/mg dry weight) with Hoechst 33258 [Kim et al., 1988].
(e) Estimate glycosaminoglycan (GAG) content by measuring the amount of chondroitin sulfate using dimethylmethylene blue dye [Farndale et al., 1986].
(f) Determine total collagen content by measuring the hydroxyproline content of the specimens after acid hydrolysis and reaction with *p*-dimethylaminobenzaldehyde and chloramine-T, using 0.1 as the ratio of hydroxyproline to collagen [Woessner, 1961].

Protocol 9.6. Gene Expression Analysis by RT-PCR of Cartilage Hydrogels

Reagents and Materials

Nonsterile

- ❑ TRIzol reagent
- ❑ Chloroform
- ❑ RNeasy mini kit
- ❑ Superscript amplification system
- ❑ Gel electrophoresis system with 2% agarose gel in TAE buffer

- ❑ Microcentrifuge tubes
- ❑ Tissue grinder
- ❑ Vortex mixer

Protocol

(a) Harvest constructs or explanted tissues and immediately grind with a tissue grinder that is RNase free in 1.5-ml microcentrifuge tube containing 200 μl TRIzol reagent.

(b) Add 800 μl more TRIzol reagent to the microcentrifuge tubes and leave for 10 min at room temperature.

(c) Add 200 μl chloroform, vortex, and incubate the tube for 5 min.

(d) Centrifuge the tube for 15 min at 4 °C at 12,000 g.

(e) After centrifugation, transfer upper transparent aqueous layer to a collecting tube of RNeasy mini kit (Qiagen) and add equal volume of 70% ethanol.

(f) Perform the next steps of the manufacturer's protocol of the RNeasy mini kit.

(g) Make cDNA, using random hexamers with the Superscript amplification system per the manufacturer's instructions.

(h) Amplify 1-μl aliquots of the resulting cDNA in a total 50-μl volume at annealing temperatures optimized for cartilage-specific phenotypic markers (type II collagen, aggrecan, link protein, COMP, type IX collagen, etc).

(i) Analyze each PCR product by separating 4 μl amplicon and 1 μl loading buffer in a 2% agarose gel in TAE buffer. Compare the relative levels of band intensity of the gene of interest to those of the internal control of housekeeping gene.

Protocol 9.7. Histologic Analysis of Cartilage Hydrogels

Reagents and Materials

Nonsterile

- ❑ Paraformaldehyde, 4%
- ❑ Ethanol, 70%
- ❑ Reagents for embedding
- ❑ Safranin-O/Fast Green stain
- ❑ Histostain-SP kit

Protocol

(a) Observe the hydrogels by inverted light microscopy during incubation. In general, the cells may be observed and their distribution in the gel can be monitored. Also, opacity that develops in the gel, indicative of matrix formation, can be observed over time.

(b) After the required culture period, harvest the constructs and fix overnight in 4% paraformaldehyde at 4 °C.

(c) After the overnight incubation, transfer the samples to 70% ethanol until embedded in paraffin. Section to 5 μm.

(d) Stain with Safranin-O/Fast Green to assess the presence of proteoglycans.
(e) Immunostain with the antibodies of interest (type II collagen, type I collagen, aggrecan, link protein, etc) with the Histostain-SP kit, following the manufacturer's protocol.

---

7. PHOTOENCAPSULATION OF BOVINE CHONDROCYTES FOR CARTILAGE TISSUE ENGINEERING

Our previous research has focused on encapsulating chondrocytes in PEG-based hydrogels both in vitro and in vivo to engineer cartilage-like tissue [Elisseeff et al., 1999, 2000]. Recent studies have demonstrated the importance of cartilage-specific tissue architecture. In particular, the superficial, middle, and deep zones of cartilage each have unique genetic and biochemical characteristics that contribute to the structural and functional properties of the tissue. We studied the relevance of depth variation in tissue-engineered cartilage with a bovine chondrocyte model system. We were interested in recreating the zones of articular cartilage in a photopolymerizing hydrogel system capable of forming complex, multilayered structures for cartilage tissue engineering. Chondrocytes from varying depths (superficial, middle, and deep) differ in proliferation and expression of matrix markers on plating for amplification and matrix production after encapsulation in the hydrogel. The goal of this study was to prove that chondrocytes isolated from the three layers would differ in gene expression patterns and matrix formation after being encapsulated in a photopolymerizing hydrogel.

7.1. Design

Cartilage slices were removed from three (upper, middle, and lower) zones of articular cartilage of young bovine legs. Histology and biochemical composition of the cartilage slices were analyzed to confirm that they had been obtained from the proper zone. Gene expression of chondrocytes in monolayer culture and matrix formation in photopolymerizing hydrogels were evaluated. Cell viability and maintenance of cell viability from each respective layer were evaluated with the Live/Dead viability kit. After 3 weeks, the constructs were harvested for gene expression, biochemical, and histologic examination including immunohistochemistry for type II collagen.

7.2. Methods

Cartilage Layers

Articular cartilage was isolated from the patellofemoral groove and distal femoral chondyles of 5- to 8-week-old bovine legs (Research 87, Marlboro, MA). The top 10%, central 20%, and lower 10% of the excised tissue were removed to isolate the superficial (S), middle (M), and deep (D) layers (Fig. 9.2, See Color Plate 4A).

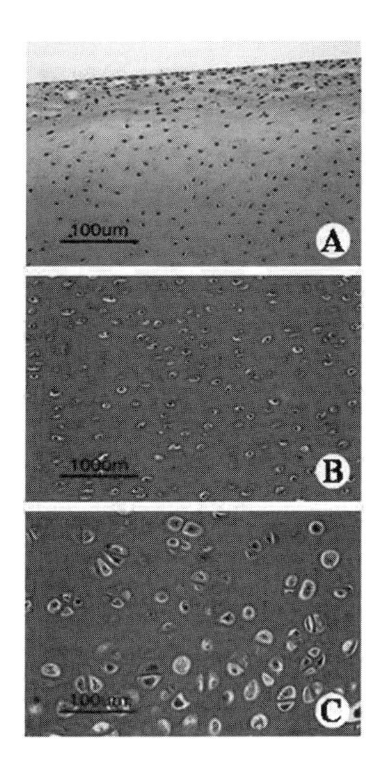

Figure 9.2. Histologic sections of the superficial, middle, and deep zones of juvenile bovine cartilage. The intensity of extracellular matrix staining and cell size increase with depth of zone: superficial (A), middle (B), and deep (C) zones. [Adapted from Kim et al., 2003] (See Color Plate 4A).

Cell Isolation

Chondrocytes from the individual layers were isolated by digestion of the cartilage chips in 0.2% collagenase overnight (See Protocol 9.1). Number and size of isolated cells was determined with a Z2 Coulter Particle Size and Number Analyzer.

RT-PCR

RNA was isolated from chondrocytes with the RNeasy Mini Kit. cDNA was synthesized with random hexamers. Cartilage-specific primers included type II collagen and aggrecan with β-actin as a housekeeping gene (See Protocol 9.6).

Cell Encapsulation

Polyethylene oxide-diacrylate was dissolved in PBSA to make a 10% w/v solution to which photoinitiator (0.05% Irgacure D2959) was added. The polymer solution was combined with a cell pellet to make a final concentration of 2×10^7 cells/ml. Approximately 100 μl of thoroughly mixed cell suspension in polymer solution was placed in an 8-mm cylindrical mold under a UVA lamp (365 nm $\sim$4 mW/cm^2) for 5 minutes. The resulting polymerized gel was removed from the mold, placed in complete DMEM (See Section 3.3), and incubated for 3 weeks.

Biochemical Characterization

Wet weights (ww) and dry weights (dw, after 48 h of lyophilization under vacuum) were obtained from constructs from each group ($n = 3$–4). The dried constructs were crushed with a tissue grinder and digested in 1 ml of papain, and the amounts of DNA, GAG, and total collagen were measured (See Protocol 9.5).

7.3. Results

Analysis of histology and biochemical composition confirmed that the cartilage slices had been obtained from the specific zone (superficial, middle, and deep; See Fig. 9.2, Color Plate 4A). The superficial layer exhibited minimal staining for GAG and type II collagen and smaller cells compared to the middle and deep zones. Chondrocytes from each zone differed in gene expression in monolayer and in matrix synthesis in three-dimensional culture (Figs. 9.3, 9.4). The gene expression of the cartilage-specific markers differed among the cells from different zones (See Fig. 9.3). Type II collagen expression of the superficial-zone chondrocytes was notably lower than the middle- and deep-zone chondrocytes. The aggrecan expression in freshly isolated cells had no remarkable differences among the zones. A slight decrease in aggrecan expression was observed in all groups on plating.

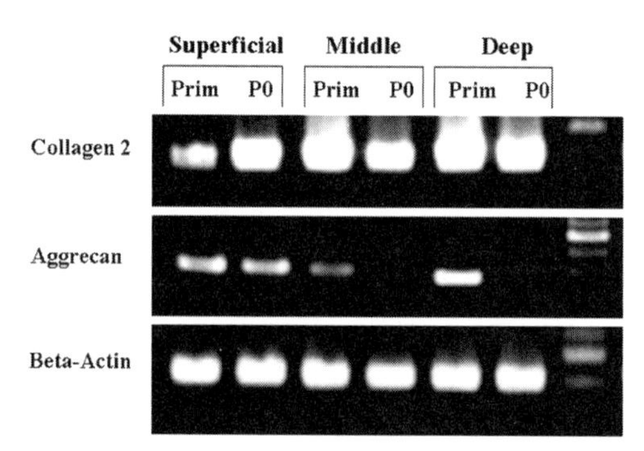

Figure 9.3. Gene expression profiles for cartilage-specific proteins in cells isolated from the different zones of cartilage.

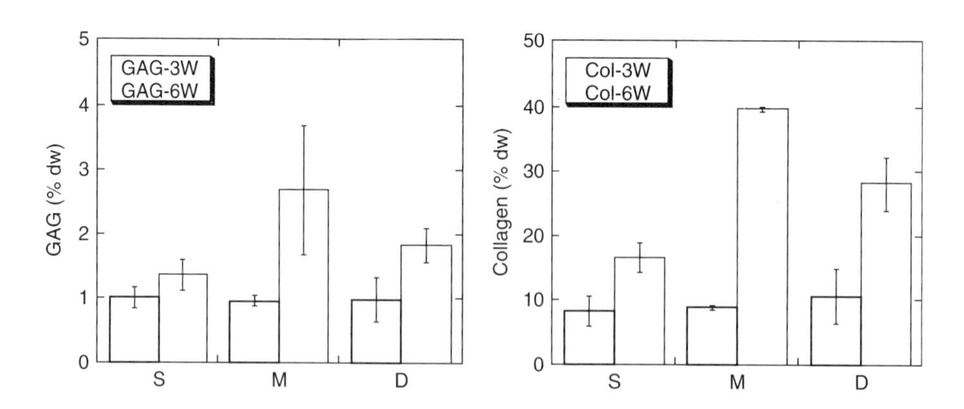

Figure 9.4. Biochemical analysis (GAG and total collagen) of zone-specific constructs incubated for 2 and 6 weeks.

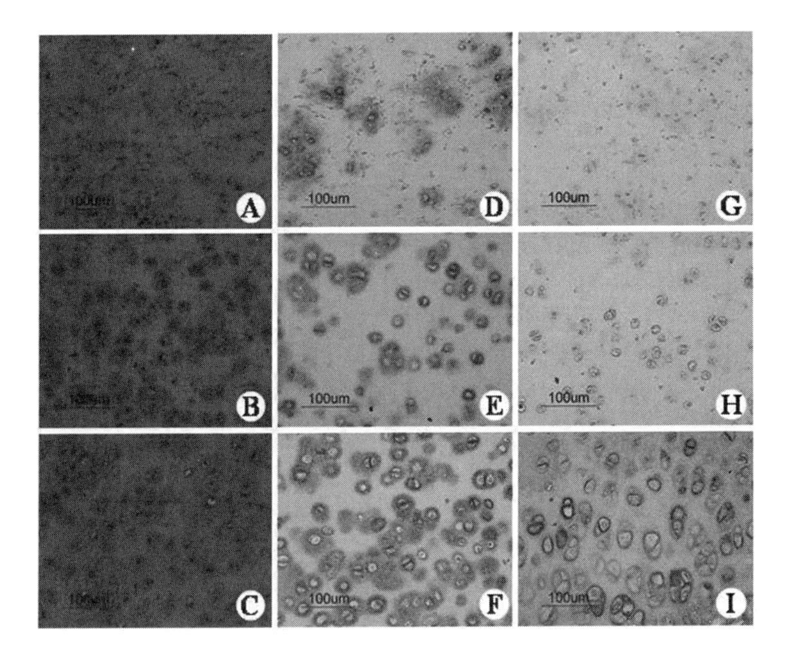

Figure 9.5. Histology of a multilayered PEGDA hydrogel encapsulated with different zone chondrocytes for different layers: Safranin-O staining (A–C) and immunohistochemistry for type II collagen (D–I) (200×, meter bar = 100 μm). Each layer of the constructs showed findings similar to those in native cartilage (A, D—upper; B, E—middle; and C, F—lower zones). Negative controls without primary antibody had no positive signals (G—upper zone, H—lower zone). I) Positive control with a cartilage slice of lower zone. [Adapted from Kim et al., 2003] (See Color Plate 4C.).

After 3 weeks in culture, the histologic differences between hydrogel constructs prepared with chondrocytes from the three zones were similar to the respective differences between the three zones in native articular cartilage. Superficial-zone cells maintained their relatively small size after the encapsulation in hydrogel, and accumulated less extracellular matrix staining, compared to the hydrogel-encapsulated deep cells. (See Fig. 9.5 and Color Plate 4C).

In native bovine cartilage, there is a 22% increase in GAG (%dw) from the S to the D layer and a 10% decrease in collagen content (graph not shown). The compositions of hydrogels cultured for 2 and 6 weeks with chondrocytes from the S, M, and D layers were significantly different from each other, and the changes were consistent with the zonal origin of chondrocytes (See Fig. 9.4). GAG content (%dw) in constructs containing chondrocytes from the D zone was 45% higher than in constructs containing S chondrocytes. Collagen production showed a similar trend, and the amount of collagen (%dw) of D hydrogels was 56% higher compared to S hydrogels.

Previous studies showed that articular chondrocytes retain the metabolic features characteristic of their zones of origin even after they are isolated from cartilage and cultured in suspension [Aydelotte et al., 1988; Aydelotte and Kuettner, 1988]. Biochemical assays showed that these metabolic differences between chondrocytes from different zones were maintained when chondrocytes were photoencapsulated and cultured in a hydrogel (See Fig. 9.4). The environment of the hydrogel is similar to that of the native tissue in that the chondrocytes are isolated with little cell-cell contact, which encourages them to synthesize extracellular matrix [Elisseeff et al., 1999., 2000, 2002; Anseth et al., 2002]. The differences in biosynthetic activity, with the deep-zone cells significantly exceeding the superficial-zone cells in terms of matrix synthesis, are in line with the study by Wong and his colleagues [Wong et al., 1996]. They investigated the zone-specific biosynthetic activity in mature bovine articular cartilage and found significant differences between the biosynthetic activities of chondrocytes in deep and superficial zones. The photopolymerization system can therefore efficiently encapsulate cells, and the differences in chondrocyte biology and engineering can be maintained and may potentially be used to create more complex tissue-engineered cartilage structures.

8. PHOTOENCAPSULATION OF GOAT BONE MARROW-DERIVED MESENCHYMAL STEM CELLS FOR CARTILAGE TISSUE ENGINEERING

Bone marrow-derived mesenchymal stem cells are another cell source for cartilage tissue engineering. MSCs have numerous advantages including a relatively easy clinical harvest, large potential for expansion, and possible allogeneic cell application in orthopedics. We studied the ability of MSCs to survive photoencapsulation and differentiate into chondrocytes and produce cartilage-like tissue.

8.1. Design

MSCs were photoencapsulated in hydrogels and cultured in three experimental groups: (1) 3 weeks with TGF-β1 (3wk + TGF), (2) 6 weeks with TGF-β1 (6wk + TGF), and (3) 6 weeks without TGF-β1 (6wk − TGF) ($n = 6$–7/group). Histologic, biochemical, and RNA analyses were performed to evaluate both the differentiation of MSCs into a chondrogenic phenotype and the accumulation of ECM products in the hydrogels.

8.2. Methods

Cell Isolation and Expansion

Bone marrow from the femurs of three- to three-and-a-half-year-old castrated male goats being sacrificed for other reasons was aspirated into 10-ml syringes with 6000 U of heparin and processed within 4 hours of harvest. The marrow samples were washed and centrifuged twice (1000 rpm, 1500 g for 10 min) in MSCGM before suspension in fresh MSCGM. The mononuclear cells were counted with a hemocytometer and plated in 75-cm^2 tissue culture plastic flasks at a density of approximately 1.2×10^5 mononuclear cells/cm^2. Culture medium was changed after 4 days and then every 2–3 days thereafter until confluence (12–14 days). Cells were passaged with 0.025% trypsin-EDTA for 5 min at 37 °C and replated in 75-cm^2 or 175-cm^2 flasks at 5×10^3 MSCs/cm^2. MSCs were frozen in liquid nitrogen in 50% MSCGM, 40% FBS, 10% DMSO until needed. Cell viability after thawing was consistently above 92%. When needed, the frozen cells were thawed, plated in 75-cm^2 or 175-cm^2 flasks in MSCGM, and grown until confluent. Passage 3 cells were trypsinized, centrifuged, and resuspended in the hydrogel solution as described below.

MSC Photoencapsulation

The hydrogel solution was prepared by mixing 10% weight/volume (w/v) of PEGDA in sterile PBSA with 100 U/ml penicillin and 100 μg/ml streptomycin. The photoinitiator, Irgacure D2959, was added to the PEGDA solution and mixed thoroughly to make a final concentration of 0.05% w/v. In experimental groups (1) and (2), TGF-β1, 10 ng/ml, was added to the hydrogel solution. Immediately before photoencapsulation, MSCs were resuspended in the hydrogel solution to make a concentration of 2×10^7 cells/ml and gently mixed to make a homogeneous suspension. Seventy-five microliters l of cell-polymer-photoinitiator suspension was transferred into cylindrical molds with a 6-mm internal diameter and exposed for 5 min to long-wave, 365-nm UV light at 4 mW/cm^2. The hydrogels were then removed from their molds, washed once with sterile PBSA containing penicillin-streptomycin, and incubated in separate wells of 12-well plates.

In Vitro Cultivation

The hydrogels were incubated at 37 °C in 5% CO_2 on an orbital rocker at 70 rpm in 2 ml of chondrogenic medium with or without TGF-β1. Chondrogenic medium consisted of high-glucose DMEM, 100 nM dexamethasone, 50 μg/ml ascorbic acid 2-phosphate, 100 μg/ml sodium pyruvate, 40 μg/ml proline, 100 U penicillin, 100 μg/ml streptomycin, and 5 ml of ITS + premix in 500 ml of medium (insulin (6.25 μg/ml), transferrin (6.25 μg/ml), selenous acid (6.25 μg/ml), linoleic acid (5.35 μg/ml), and bovine serum albumin (1.25 μg/ml)) with or without 10 ng/ml of TGF-β1. Medium was changed every 2–3 days.

Histology and Immunohistochemistry

Throughout the experiment, the hydrogels were observed by inverted light microscopy at least twice a week and digitally photographed at the beginning and end of the culture period. Particular attention was given to observing encapsulated cells for signs of cell division. At the end of the culture period, two constructs per group were harvested for histologic and immunohistochemical studies. The hydrogels were fixed overnight in 2% paraformaldehyde at 4 °C and transferred to 70% ethanol until being embedded in paraffin according to standard histologic technique. Sections were stained with hematoxylin and eosin and Safranin-O/Fast Green. Immunohistochemistry was performed with the Histostain-SP kit (Zymed

#95–9743), following the manufacturer's protocol. Rabbit polyclonal antibodies to collagen I and collagen II (Research Diagnostics Inc.) and mouse monoclonal antibodies to aggrecan and link protein (Hybridoma Bank, University of Iowa) were used as the primary antibodies.

RNA Extraction and Reverse Transcription-Polymerase Chain Reaction (RT-PCR)

Total RNA was isolated from three constructs per group and from goat MSCs of the same passage cultured in monolayer with the RNeasy Mini Kit. To extract the total RNA the constructs were homogenized in 1.5-ml microcentrifuge tubes containing 200 μl of RLT buffer from the RNeasy Mini Kit, using a tissue grinder (See Protocol 9.6). After complete homogenization, 400 μl more of the RLT buffer was added to the microcentrifuge tubes and the suspension was further homogenized with the QIAshredder column. The homogenates were transferred to columns from the RNeasy Mini Kit after an equal volume of 70% ethanol had been added. RNA was isolated, following the manufacturer's protocol. The RNA was reverse transcribed into cDNA, using random hexamers with the Superscript amplification system per the manufacturer's instructions. One-μl aliquots of the resulting cDNA were amplified in a total 50-μl volume at an annealing temperature of 58 °C (collagen type II was annealed at 60 °C) for 35 cycles, using the Ex Taq™ DNA Polymerase Premix. PCR primers (forward and backward, 5′ to 3′) were as follows: collagen I, 5′-TGACGAGACCAAGAACTG-3′ and 5′-CCATCCAAACCACTGAAACC-3′; collagen II, 5′-GTGGAGCAGCAAGAGCA-AGGA-3′ and 5′-CTTGCCCCACTTACCAGTGTG-3′; aggrecan, 5′-CACGCTAC ACCCTGGACT TG-3′ and 5′-CCATCTCCTCAGCGAAGCAGT-3′; β-actin, 5′-TGGCACCACACCTTCTACAATGAGC-3′ and 5′-GCACAGCTTCTCCTTAAT GTCACGC-3′. Each PCR product was analyzed by separating 4 μl of the amplicon and 1 μl of loading buffer in a 2% agarose gel in TAE buffer. The relative levels of band intensity of the gene of interest were compared to those of the internal control of housekeeping gene.

Biochemical Characterization

Wet weights (ww) and dry weights (dw, after 48 h of lyophilization under vacuum) were obtained from constructs from each group ($n = 3$–4). The dried constructs were crushed with a tissue grinder and digested in 1 ml of papain, and the amounts of DNA, GAG, and total collagen were measured (See Protocol 9.5).

8.3. Results

Histology

Observations of the hydrogels immediately after photoencapsulation under inverted light microscopy showed rounded cells evenly dispersed throughout the constructs (data not shown). As the culture period extended for the groups cultivated with TGF-β1, many single cells divided and produced small, multicellular aggregations with approximately two to five cells. Cell clusters were not seen in hydrogels cultured without TGF-β1.

Histologic study of fixed slides of the hydrogels showed that the experimental groups cultured with TGF-β1 were strongly positive for GAG compared to the other groups. Figure 9.6 shows histologic sections of the four groups stained with Safranin-O/Fast Green, which stains negatively charged glycosaminoglycans red and nuclei green. At Day 0 (Fig. 9.6A) only the light blue-green counterstain was present. The sections from the 3wk + TGF group (Fig. 9.6B) revealed an intense positive staining for GAG, particularly around the pericellular regions. The positive staining was enhanced in the 6wk + TGF group (Fig. 9.6C) and was strongly present in the intercellular matrix as well, indicating that the GAG had diffused throughout the PEGDA gels [Bryant and Anseth, 2002]. Conversely, in the 6wk − TGF section (Fig. 9.6D), only a small amount of GAG produced by spontaneous chondrogenic differentiation was seen in a few of the pericellular regions.

Immunohistochemical staining for aggrecan and link protein showed strong positive staining in the 6wk + TGF group but revealed negative or sporadic, weakly positive cells in the 6wk − TGF group (not shown). Staining for type I collagen was positive on both the 6wk + TGF and 6wk − TGF sections. Interestingly, type II collagen staining was also noted on sections from both 6wk + TGF and 6wk − TGF sections, which is in agreement with the gene expression results described below.

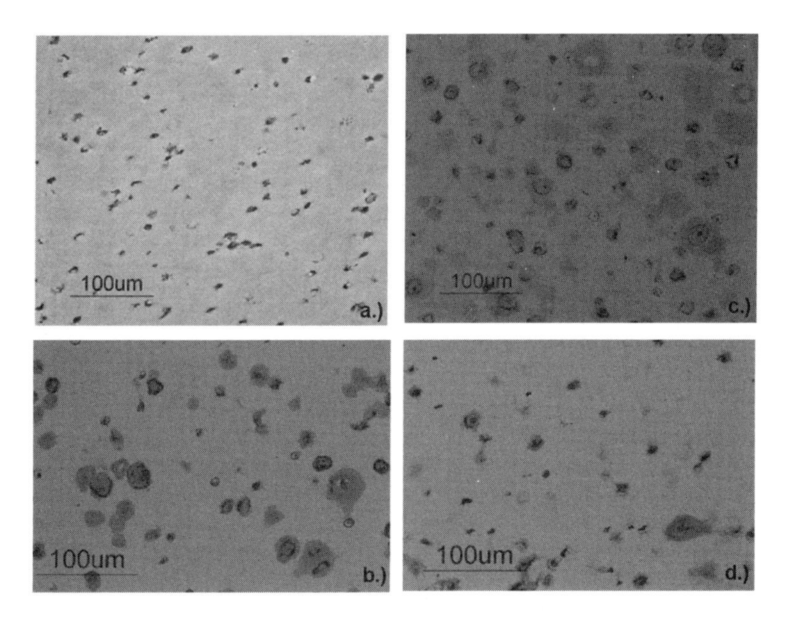

Figure 9.6. Paraffin embedded histologic sections of PEGDA-MSC hydrogels for Day 0 controls (A), group 1: 3wk + TGF (B), group 2: 6wk + TGF (C), and group 3: 6wk − TGF (D) stained with Safranin-O/Fast Green. Originally acquired at 200×. This dye combination stains GAG red and nuclei green. The scale bars are 100 µm. [Adapted from William et al., 2003] (See Color Plate 4B).

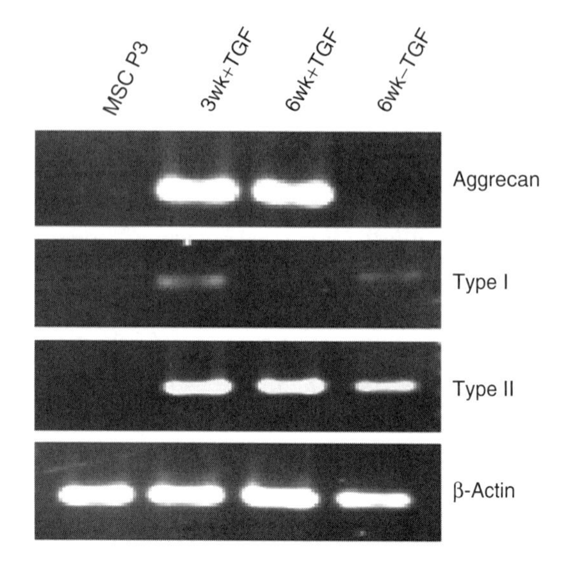

Figure 9.7. RT-PCR products for MSC passage 3 monolayer culture as control, 3wk + TGF, 6wk + TGF, and 6wk − TGF (left to right on gels). Primers used include aggrecan, type I collagen, type II collagen, and β-actin (from top to bottom of the gels). (Adapted from Williams et al., 2003).

RT-PCR

RT-PCR supported the histologic findings by demonstrating that the previously undifferentiated MSCs shifted their genetic expression during the culture period (Fig. 9.7). The expression of the aggrecan gene was almost entirely absent in monolayer controls, absent in the 6wk − TGF constructs, but strongly present in the 3wk + TGF and 6wk + TGF constructs. Type I collagen was not expressed in monolayer culture, almost absent in 6wk + TGF constructs, but weakly present in 3wk + TGF and 6wk − TGF constructs. Type II collagen was not expressed in monolayer culture of the MSCs, present in low quantities in 6wk − TGF, but strongly present in the 3wk + TGF and 6wk + TGF constructs.

Biochemical Analysis

The DNA content of the MSC constructs revealed a statistically significant increase to 1101 ng (±170 ng) of DNA/mg dw in the 6wk + TGF group from an initial value of 882 ng (±94 ng)/mg dw ($p = 0.036$) and a significant decrease to 681 ng (±43 ng)/mg dw in the 6wk − TGF group ($p = 0.028$).

Correlating with the RT-PCR and histologic findings, the hydrogels showed a significant increase in GAG and total collagen content (%dw) by detection of chondroitin sulfate and hydroxyproline at each time point when the constructs were cultured with TGF-β1 (Fig. 9.8B, C). The amount of GAG increased from 0 at Day 0 constructs to 1.4%dw in 3wk + TGF constructs ($p = 0.020$) and to 3.5%dw in the 6wk + TGF constructs ($p = 0.001$). The amounts of GAG in 6wk − TGF hydrogels (0.9%dw) were comparable to those in Day 0 controls ($p > 0.05$). Collagen contents (%dw) increased with time of culture in all experimental groups: to 2.3%dw in 3wk + TGF ($p = 0.001$); to 5.0%dw in 6wk + TGF ($p = 0.001$); and to 1.4%dw in 6wk − TGF ($p = 0.029$). Amounts of GAG and collagen in the cartilage-like tissue produced in this study are comparable to those

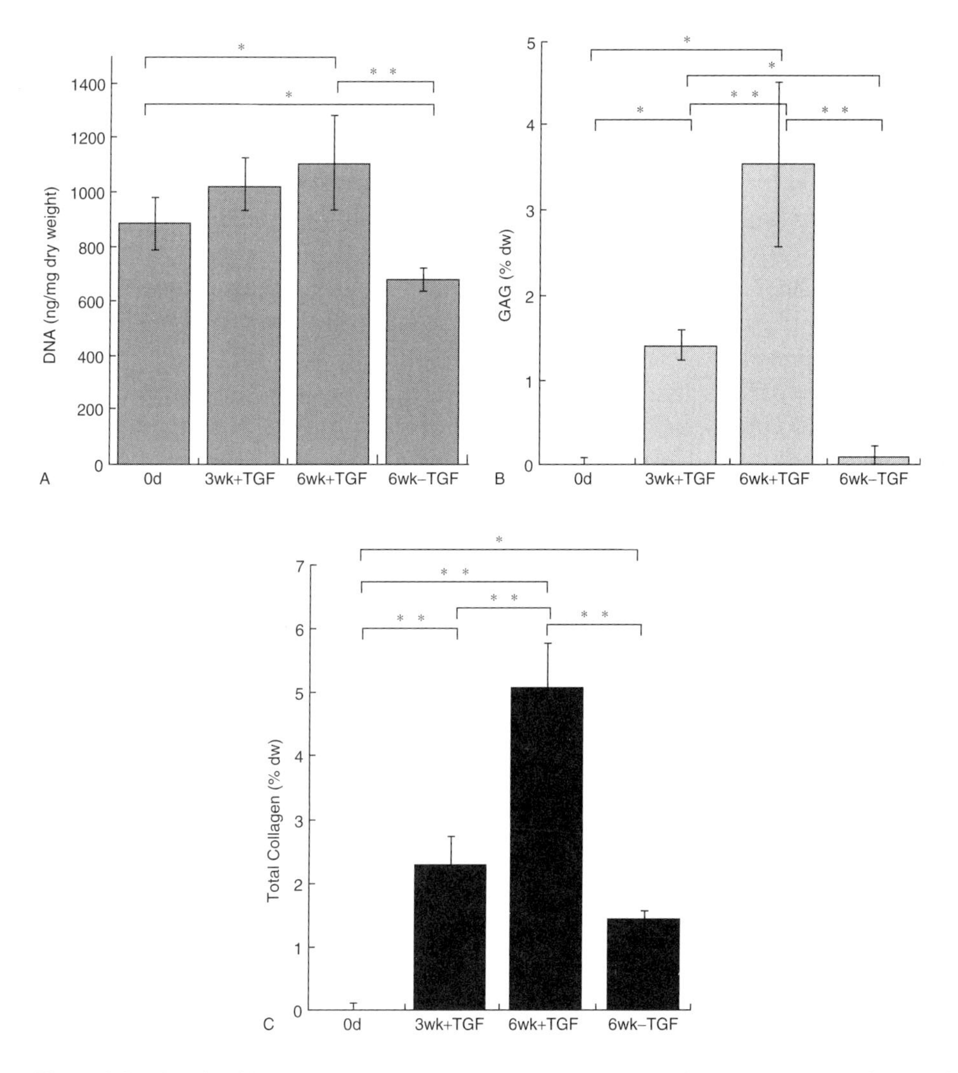

Figure 9.8. Results of biochemical assays for the Day 0 control, 3wk + TGF, 6wk + TGF, and 6wk − TGF hydrogels are depicted: DNA content (ng DNA/mg dry weight) (A), GAG content (% dry weight of construct) (B), and total collagen content (% dry weight of construct) (C). $*p < 0.05$, $**p < 0.001$. (Adapted from Williams et al., 2003).

in previous studies using cultured chondrocytes in PEGDM-based hydrogels and PGA mesh scaffolds [Bryant and Anseth, 2002; Freed and Vunjak-Novakovic, 1995]. Bryant and colleagues reported 0.4 ng of GAG/chondrocyte and 0.3 ng of collagen/chondrocyte produced after 4-week cultures of bovine chondrocytes in similar PEGDM-based hydrogel constructs. Our MSC constructs cultured in chondrogenic conditions for 6 weeks differentiate and produce 0.24 ng of GAG/cell and 0.34 ng of collagen/cell. Freed and Vunjak-Novakovic [1995] reported that a 6-week culture of approximately 1.0×10^7 bovine chondrocytes/mesh PGA construct produced 10% GAG/dw and 11–12% collagen/dw. Our 6-week PEGDA hydrogel constructs, which contain approximately 1×10^6 MSCs/construct, produce about 3.5% GAG/dw and 5% collagen/dw. These rough comparisons suggest that MSCs can differentiate and produce similar amounts of ECM as native chondrocytes when photoencapsulated in a polyethylene glycol-based hydrogel. Interestingly, Barry and colleagues suggested that MSCs in an alginate gel culture system were capable of producing more GAG, as determined by ^{35}S-sodium sulfate incorporation, than native, isolated, and dedifferentiated articular chondrocytes [Barry et al., 2001]. The findings of our study indicate the robust biochemical productivity of MSCs in a photopolymerizing hydrogel and may have powerful implications for the development of a cartilage replacement therapy using MSCs.

Despite these encouraging results, our cartilage-like tissue is still approximately three- to fourfold lower in GAG (%dw) and 10 to 12-fold lower in collagen content (%dw) than native cartilage (10–15% dw and 55–85% dw, respectively) [Mow et al., 1992]. We also noted that chondrogenic differentiation and extracellular matrix production from the MSCs was not evenly distributed throughout the gels. The central area of the hydrogels showed less production of GAG and collagen on histologic sections. This contrasts with bovine chondrocytes photoencapsulated in similar hydrogels, which demonstrate an even distribution of GAG throughout the entire gels [Bryant et al., 1999].

A polyethylene glycol-diacrylate hydrogel provides a three-dimensional, non-adhesive environment for encapsulated cells. The DNA assay and microscopic observations in this study demonstrate that the MSCs could not only survive but could also divide in the presence of TGF-β1. Conversely, in the absence of TGF-β1, the number of encapsulated cells decreased with time in culture. Even though the findings of this study indicate that the PEGDA photopolymerizing hydrogel supports MSC survival, phenotypic differentiation, and accumulation of chondrogenic extracellular matrix, it seems clear, as evidenced by the results for the 6wk + TGF group, that cellular signaling driven by TGF-β1 is important for enhanced tissue development.

Interestingly, in the absence of TGF-β1, there is still a small degree of differentiation in the 6wk – TGF group, as indicated by several assays. Histology reveals sporadic cells producing small amounts of matrix staining with Safranin-O. The immunohistochemistry, RT-PCR, and collagen assays for the 6wk – TGF hydrogels demonstrate the presence of type I and II collagen proteins and RNA in this group. Indeed, the amount of total collagen produced in the 6wk – TGF constructs is similar to the amount of total collagen produced by the 3wk + TGF constructs ($p > 0.05$) but is lower than in the 6wk – TGF constructs ($p = 0.001$). However, little to no GAG, aggrecan, or link protein was produced in the 6wk – TGF group. The mechanisms that induce this partial differentiation are unclear. Previous studies have shown that cellular morphology might be associated with cell differentiation [Benya and Shaffer, 1982; Johnstone et al., 1998]. One possible explanation is that the rounded cellular morphology in our hydrogels, which is much different from the flattened morphology typical of a monolayer culture, aided the spontaneous chondrogenic differentiation of the MSCs as opposed to cellular division.

In summary, bone marrow-derived mesenchymal stem cells are attractive cells for tissue engineers and biologists. Hydrogels are a class of polymers used in tissue engineering that have many advantages including high, tissuelike water content. The creation of three-dimensional hydrogels by photopolymerization gives a great deal of spatial and temporal control to the engineer and can be adapted to a number of minimally invasive surgical techniques currently in clinical use. These in vitro studies suggest the potential use of MSCs with hydrogels for cartilage tissue engineering, and this technology could potentially expand the plastic and orthopedic surgeons' armamentarium for cartilage repair or augmentation in the future.

ACKNOWLEDGMENTS

The authors would like to acknowledge the Arthritis Foundation and Johns Hopkins University for funding.

SOURCES OF MATERIALS

Item	*Supplier*
Ascorbic acid 2-phosphate	Sigma
Collagenase	Worthington
Dexamethasone	Sigma
Digital camera: DMX1200	Nikon
Dimethylene blue dye	
DMEM (Dulbecco's modified Eagle's medium with high glucose)	Invitrogen (GIBCO)
DMSO	Sigma
Ex Taq™ DNA Polymerase Premix	Takara Bio
FBS	Hyclone
Histostain-SP kit (#95–9743)	Zymed Laboratories
Inverted microscope: Eclipse TE200	Nikon
ITS^+ premix	BD Biosciences
Live/Dead viability kit	Molecular Probes
Monoclonal antibodies to aggrecan and link protein	Hybridoma Bank, University of Iowa
MSCGM	Clonetics, Cambrex
Nylon filter, 70 μm	Tekmar-Dohrmann
Papain	Worthington Biomedical Corporation
Penicillin-100 μg/ml streptomycin	Invitrogen (GIBCO)
Photoinitiator, Igracure D2959	Ciba Specialty Chemicals
Polyethylene glycol-diacrylate (PEGDA)	Shearwater Corp.
Proline	Sigma
QIAshredder	Qiagen
Random hexamers	Invitrogen (GIBCO)
RNeasy mini kit	Qiagen

Item	*Supplier*
Safranin-O/Fast Green	Sigma
Sodium pyruvate	Invitrogen (GIBCO)
SPSS (version 10.0)	SPSS
Superscript amplification system	Invitrogen (GIBCO)
TGF-β1	Research Diagnostics
Tissue grinder: Pellet Pestle Mixer	Kimble/Kontes
TRIzol reagent	Invitrogen (GIBCO)
Trypsin-EDTA	Clonetics, Cambrex
UV light	Glowmark Systems or VWR
WST-1 solution	Sigma

REFERENCES

Anseth, K., Wang, C., Bowman, C. (1994) Reaction behavior and kinetic constants for photopolymerizations of multi(meth)acrylate monomers. *Polymer* 35: 3243.

Anseth, K.S., Metters, A.T., Bryant, S.J., Martens, P.J., Elisseeff, J.H., Bowman, C.N. (2002) In situ forming degradable networks and their application in tissue engineering and drug delivery. *J. Control Release* 78: 199–209.

Aydelotte, M.B., Greenhill, R.R., Kuettner, K.E. (1988) Differences between sub-populations of cultured bovine articular chondrocytes. II. Proteoglycan metabolism. *Connect. Tissue Res.* 18: 223–234.

Aydelotte, M.B., and Kuettner, K. (1988) Differences between sub-populations of cultured bovine articular chondrocytes. I. Morphology and cartilage matrix production. *Connect. Tissue Res.* 18: 205–222.

Barry, F., Boynton, R.E., Liu, B., Murphy, J.M. (2001) Chondrogenic differentiation of mesenchymal stem cells from bone marrow: differentiation-dependent gene expression of matrix components. *Exp. Cell Res.* 268: 189–200.

Behravesh, E., Jo, S., Zygourakis, K., Mikos, A.G. (2002) Synthesis of in situ cross-linkable macroporous biodegradable poly(propylene fumarate-co-ethylene glycol) hydrogels. *Biomacromolecules* 3(2): 374–381.

Benya, P.D., Shaffer, J.D. (1982) Dedifferentiated chondrocytes reexpress the differentiated collagen phenotype when cultured in agarose gels. *Cell* 30: 215–224.

Brannon-Peppas, L. (1994) *Preparation and Characterization of Crosslinked Hydrophilic Networks*. Washington, DC, ACS.

Bryant, S.J., and Anseth, K.S. (2002) Hydrogel properties influence ECM production by chondrocytes photoencapsulated in poly(ethylene glycol) hydrogels. *J. Biomed. Mater. Res.* 59: 63–72.

Bryant, S.J., Nuttelman, C.R., Anseth, K.S. (1999) The effects of crosslinking density on cartilage formation in photocrosslinkable hydrogels. *Biomed. Sci. Instrum.* 35: 309–314.

Burdick, J., Mason, M., Hinman, A., Thorne, K., Anseth, K. (2002) Delivery of osteoinductive growth factors from degradable PEG hydrogels influences osteoblast differentiation and mineralization. *J. Control Release* 83(1): 53.

Burkoth, A.K., and Anseth, K.S. (2000) A review of photocrosslinked polyanhydrides: in situ forming degradable networks. *Biomaterials* 21: 2395–2404.

Buschmann, M.D., Gluzband, Y.A., Grodzinsky, A.J., Kimaru, J.H., Hunziker, E.B. (1992) Chondrocytes in agarose culture synthesize a mechanically functional matrix. *J. Orthop. Res.* 10: 745–758.

Drumheller, P.D., and Hubbell, J.A. (1994) Polymer networks with grafted cell adhesion peptides for highly biospecific cell adhesive substrates. *Anal. Biochem.* 222(2): 380–388.

Elisseeff, J., Anseth, K., Sims, D., McIntosh, W., Randolph, M., Langer, R. (1999) Transdermal photopolymerization for minimally invasive implantation. *Proc. Natl. Acad/Sci. USA* 96: 3104–3107.

Elisseeff, J., Anseth, K., Sims, D., McIntosh, W., Randolph, M., Yaremchuk, M., Langer, R. (1999) Transdermal photopolymerization of poly(ethylene oxide)-based injectable hydrogels for tissue-engineered cartilage. *Plast. Reconstr. Surg.* 104: 1014–1022.

Elisseeff, J., McIntosh, W., Anseth, K., Riley, S., Ragan, P., Langer, R. (2000) Photoencapsulation of chondrocytes in poly(ethylene oxide)-based semi-interpenetrating networks. *J. Biomed. Mater. Res.* 51: 164–171.

Elisseeff, J.H., Lee, A., Kleinman, H.K., Yamada, Y. (2002) Biological response of chondrocytes to hydrogels. *Ann. NYAcad. Sci.* 961: 118–122.

Farndale, R., Buttle, D., Barrett, A. (1986) Improved quantitation and discrimination of sulphated glycosaminoglycans by the use of dimethylmethylene blue. *Biochim. Biophys. Acta* 883: 173–177.

Freed, L., and Vunjak-Novakovic, G. (1995) Tissue engineering of cartilage. In Bronzind, J., ed., *The Biomedical Engineering Handbook*. Boca Raton, CRC, pp. 1778–1796.

He, S., Yaszemski, M.J., Yasko, A.W., Engel, P.S., Mikos, A.G. (2000) Injectable biodegradable polymer composites based on poly(propylene fumarate) crosslinked with poly(ethylene glycol)-dimethacrylate. *Biomaterials* 21(23): 2389–2394.

Hill-West, J., Chowdhury, S., Sawhney, A., Pathak, C., Dunn, R., Hubbell, J. (1994) Prevention of postoperative adhesions in the rat by in situ photopolymerization of bioresorbable hydrogel barriers. *Obstet. Gynecol.* 83: 59–64.

Huhtala, A., Alajuuma, P., Burgalassi, S., Chetoni, P., Diehl, H., Engelke, M., Marselos, M., Monti, D., Pappas, P., Saettone, M.F., Salminen, L., Sotiropoulou, M., Tahti, H., Uusitalo, H., Zorn-Kruppa, M. (2003) A Collaborative evaluation of the cytotoxicity of two surfactants by using the human corneal epithelial cell line and the WST-1 test. *J. Ocul. Pharmacol. Ther.* 19: 11–21.

Johnstone, B., Hering, T.M., Caplan, A.I., Goldberg, V.M., Yoo, J.U. (1998) In vitro chondrogenesis of bone marrow-derived mesenchymal progenitor cells. *Exp. Cell Res.* 238: 265–272.

Kim, Y., Sah, R., Doong, J., et al. (1988) Fluorometric assay of DNA in cartilage explants using Hoechst 33258. *Anal. Biochem.* 174: 168.

Kim, T.K., Sharma, B., Williams, C.G., Ruffner, M.A., Malik, A., McFarland, E.G., Elisseeff, J.H., (2003). Experimental model for cartilage tissue engineering to regenerate the zonal organization of articular cartilage. *Osteoarthritis Cartilage*, 11: 653–664.

Lyman, M., Melanson, D., Sawhney, A. (1996) Characterization of the formation of interfacially photopolymerized thin hydrogels in contact with arterial tissue. *Biomaterials* 17: 359–364.

Madihally, S.V., and Matthew, H.W. (1999) Porous chitosan scaffolds for tissue engineering. *Biomaterials* 20(12): 1133–1142.

Mooney, D.J., and Mikos, A.G. (1999) Growing new organs. *Sci. Am.* 280(4): 60–65.

Mow, V.C., Radcliffe, A., Poole, A.R. (1992) Review: Cartilage and diarthrodial joints as paradigms for hierarchical materials and structures. *Biomaterials* 13: 67–97.

Plumb, J.A., Milroy, R., Kaye, S.B. (1989) Effects of the pH dependence of 3-(4,5-dimethylthiazol-2-yl)-2,5-diphenyl-tetrazolium bromide-formazan absorption on chemosensitivity determined by a novel tetrazolium-based assay. *Cancer Res.* 49: 4435–4440.

Poshusta, A.K., and Anseth, K.S. (2001) Photopolymerized biomaterials for application in the temporomandibular joint. *Cells Tissues Organs* 169: 272–278.

Sah, R., Doong, J.-Y., Grodzinsky, A., Plaas, A., Sandy, J. (1991) Effects of compression on the loss of newly synthesized proteoglycans and proteins from cartilage explants. *Arch. Biochem. Biophys.* 286: 20–29.

Sawhney, A., Lyman, F., Yao, F., Levine, M., Jarrett, P. (1996) A novel in situ formed hydrogel for use as a surgical sealent or barrier. 23rd International Symposium of Controlled Release of Bioactive Materials. Kyoto, Japan, Controlled Release Society, pp. 236–237.

Silverman, R., Passaretti, D., Huang, W., Randolph, M., Yaremchuk, M. (1999) Injectable tissue-engineered cartilage using a fibrin glue polymer. *Plast. Reconstr. Surg.* 103: 1809–1818.

Tarle, Z., Meniga, A., Ristic, M., Sutalo, J., Pichler, G., Davidson, C. (1998) The effect of the photopolymerization method on the quality of composite resin samples. *J. Oral. Rehabil.* 25: 436–442.

Ukeda, H., Shimamura, T., Tsubouchi, M., Harada, Y., Nakai, Y., Sawamura, M. (2002) Spectrophotometric assay of superoxide anion formed in Maillard reaction based on highly water-soluble tetrazolium salt. *Anal. Sci.* 18: 1151–1154.

Williams, C.G., Kim, T.K., Taboas, A., Malik, A., Manson, P., Elisseeff, J. (2003) In vitro chondrogenesis of bone marrow-derived mesenchymal stem cells in a photopolymerizing hydrogel. *Tissue Eng.*, 9: 679–688.
Woessner, J.F. (1961) The determination of hydroxyproline in tissue and protein samples containing small proportions of this imino acid. *Arch. Biochem. Biophys.* 93: 440–447.
Wong, M., Wuethrich, P., Eggli, P., Hunziker, E. (1996) Zone-specific cell biosynthetic activity in mature bovine articular cartilage: a new method using confocal microscopic stereology and quantitative autoradiography. *J. Orthop. Res.* 14: 424–432.
Ye, Q., Zund, G., Benedikt, P., Jockenhoevel, S., Hoerstrup, S.P., Sakyama, S., Hubbell, J.A., Turina, M. (2000) Fibrin gel as a three dimensional matrix in cardiovascular tissue engineering. *Eur. J. Cardiothorac. Surg.* 17(5): 587–691.
Young, J.S., Gonzales, K.D., Anseth, K.S. (2000) Photopolymers in orthopedics: characterization of novel crosslinked polyanhydrides. *Biomaterials* 21: 1181–1188.

10

Tissue Engineering Human Skeletal Muscle for Clinical Applications

Janet Shansky, Paulette Ferland, Sharon McGuire, Courtney Powell, Michael DelTatto, Martin Nackman, James Hennessey, and Herman H. Vandenburgh

Cell Based Delivery Inc., and Brown University School of Medicine, Miriam Hospital, Providence, Rhode Island 02906

Corresponding author: herman_vandenburgh@brown.edu

Culture of Cells for Tissue Engineering, edited by Gordana Vunjak-Novakovic and R. Ian Freshney

1. INTRODUCTION

Skeletal muscle-derived cells isolated from a variety of mammalian as well as non-mammalian species can be tissue-engineered to generate three-dimensional muscle-like structures (alternately called bioartificial muscles (BAMs), organoids, or myoids) that, when cultured in vitro under passive or active tension, form parallel arrays of postmitotic myofibers and express sarcomeric contractile proteins [Strohman et al., 1990; Vandenburgh et al., 1991; Swasdison and Mayne, 1992; Okano et al., 1997; Dennis and Kosnik, 2000, reviewed in Kosnik et al., 2003]. BAMs formed from animal-derived muscle cell cultures have many applications for research, including performing muscle physiology studies, elucidating the mechanics of the development of organized, multinucleated, functional myofibers from proliferating myoblasts, and as a model for muscle wasting induced by disuse or decreased tension. For example, avian BAMs flown on several Space Shuttle missions have shown significant myofiber atrophy due to a pronounced decrease in protein synthesis, indicating that skeletal muscle fibers in BAMs are directly affected by microgravity in the absence of innervation or circulating hormones [Vandenburgh et al., 1999]. Tissue-engineered skeletal muscle BAMs can therefore provide a unique tool to develop countermeasures to muscle atrophy resulting from space travel, which may have applications to earth-based wasting muscle disorders.

BAMs formed from cells isolated from human skeletal muscle biopsy tissue have the potential to be used clinically for both structural and functional skeletal muscle repair or replacement, and for the treatment of a variety of diseases when used as a delivery vehicle for recombinant protein therapeutics from genetically engineered muscle cells. Human skeletal myoblasts have been transduced to secrete recombinant proteins and tissue engineered ex vivo into human BAMs (HBAMs), which continue to secrete the foreign gene products in vitro [Powell et al., 1999] and in vivo [unpublished data]. Autologous, genetically modified HBAMs could potentially be implanted into patients for delivery of recombinant proteins for therapeutic treatment of diseases such as growth hormone deficiency (human growth hormone), hemophilia A (factor VIII) or B (factor IX), and heart disease (vascular endothelial growth factor or insulin-like growth factor-I). The implanted HBAMs would serve as in vivo protein factories capable of delivering predictable levels of therapeutic gene products, and would offer the advantage over other gene therapy protocols of reversible implantation of postfused, nonmigrating cells with high, long-term protein synthesis levels. With additional structural/functional HBAM engineering improvements to more closely resemble in vivo muscle (including higher myofiber density, larger diameter fibers, reduced extracellular matrix materials, and innervation), HBAMs offer the additional promise of structural repair or replacement of skeletal muscle in the future [Powell et al., 2002].

Following protocols for rodent muscle cells, Powell et al. [1999] engineered human BAMs, which had morphological characteristics similar to those of other mammalian BAMs [Okano et al., 1997]. With the goal of developing a tissue-engineered product that could be implanted for clinical applications, our laboratory has extended these initial human muscle cell isolation and tissue-engineering techniques by optimizing the muscle cell isolation procedure, cell expansion protocol, and engineering methods, using clinically approved matrix materials.

2. PREPARATION OF REAGENTS AND MEDIA

2.1. Culture Media

2.1.1. DMEM/pen/strep

Dulbecco's modified Eagle's medium, supplemented with penicillin, 50 U/ml, streptomycin, 50 μg/ml (1% v/v of stock pen/strep)

2.1.2. SKGM/15, SKGM/2

Skeletal muscle growth medium (SKGM) supplemented with 15% or 2% fetal bovine serum

2.1.3. CMF

Ca^{2+}- and Mg^{2+}-free Earle's balanced salt solution (EBSS)

2.1.4. Differentiation Medium

DMEM supplemented with:

Insulin, 10 μg/ml
Bovine serum albumin (BSA), 0.5 mg/ml
hEGF, 10 ng/ml
Gentamycin, 0.05 mg/ml

2.2. Digestive Enzyme Solutions

2.2.1. Collagenase-Dispase Solution

Collagenase type II, 0.1% (w/v), dispase, 4 mg/ml in DMEM. Sterilize by filtration.

2.2.2. Trypsin Stock

Trypsin, 0.5%, EDTA, 5.4 mM (0.2%). Dilute 1:10 in CMF for use.

2.3. Staining Solutions

2.3.1. Coomassie Blue Staining Solution

Coomassie Brilliant Blue, 1 g/l in 40% (v/v) methanol, 10% (v/v) acetic acid

2.3.2. Blocking Buffer

Bovine serum albumin, 1% (w/v), 0.2% (v/v) Triton-X in PBS

3. SKELETAL MUSCLE NEEDLE BIOPSY

A complete description of the biopsy procedure has been previously described [Hennessey et al., 1997]. Biopsies are taken from the vastus lateralis, a mixed fast/slow muscle fiber type. Briefly, the biopsy site is shaved, sterilized, and anesthetized, and an incision is made through the skin. A muscle sample is removed with a 6-mm biopsy needle apparatus by a suction-enhanced technique (Fig. 10.1), and immediately thereafter, a second muscle sample is removed from the same incision site. The muscle tissues are transferred to sterile tubes containing chilled DMEM supplemented with antibiotics (See materials list), and transported on ice to the cell culture facility.

Protocol 10.1. Skeletal Muscle Needle Biopsy

Reagents and Materials

Sterile

- ❑ DMEM/pen/strep: Dulbecco's modified Eagle's medium, supplemented with penicillin, 50 U/ml, streptomycin, 50 μg/ml (See Section 2.1.1)
- ❑ Iodophor solution
- ❑ Novocain solution, 1%, without epinephrine
- ❑ Sterile saline
- ❑ Surgilube
- ❑ Tissue culture Petri dish, 10 cm
- ❑ Conical centrifuge tubes, 50 ml
- ❑ Biopsy needles (4- and 6-mm Bergstrom cutting trocars)
- ❑ Cutting cannula adapters with argyle female Luer lock connector
- ❑ Adapter to syringe tubing: Interlink catheter extension kit with male Luer lock adapter 15.2 cm long, 0.5-ml volume

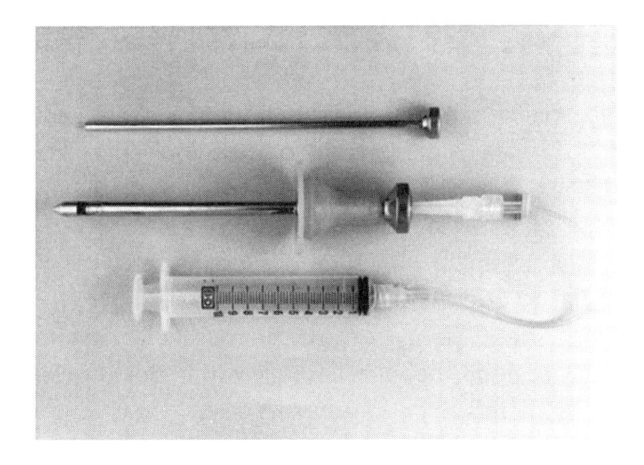

Figure 10.1. Human skeletal muscle biopsy instruments assembled and ready for use. From top to bottom: clearing rod used to expel samples; cutting trocar inserted through nipple (used to enhance suction) into outer needle with cutting chamber on left; syringe connected to adapter tubing, which attaches to the proximal end of the cutting cannula. The closed system generates enhanced suction, resulting in increased sample size compared to previous percutaneous muscle biopsy techniques [Hennessey et al., 1997]. Reprinted from Hennessey et al., 1997 by permission of the American Physiological Society.

- ❑ Suction-enhancing nipples: infant nipples, 4 mm
- ❑ Syringe, 10 ml
- ❑ No. 11 scalpel
- ❑ Fenestrated drape
- ❑ Sterile gauze sponges, 2 × 2 in. and 4 × 4 in.
- ❑ Steristrips, $\frac{1}{4}$ in.
- ❑ Foam or elastic tape
- ❑ Curved forceps, 12 cm

Nonsterile

- ❑ Ice bath

Protocol

(a) Set up sterile instruments on a sterile field.

(b) Assemble the Bergstrom biopsy needle apparatus:

- (i) Insert fully the clearing rod through the cutting cannula, easing the blunt end through the outside opening of the infant nipple.
- (ii) Place the assembly on the trocar so the nipple can slide firmly over the open end, creating a tight seal.
- (iii) Fully insert the cutting cannula into the trocar so that there is full closure of the cutting chamber, using surgical lubricant to maintain the seal.
- (iv) Insert the male end of the syringe-tubing adapter into the proximal end of the cutting cannula and attach to a 10-ml syringe (See Fig. 10.1).

(c) Withdraw syringe to 3-ml position to a create buffer of air.

(d) Soak a sterile 2 × 2-in. gauze sponge with sterile saline in 10-cm dish, replace lid, and place on ice.

(e) Place a 50-ml sterile conical centrifuge tube containing 25 ml DMEM/pen/strep on ice.
(f) Place patient in a comfortable reclining position and locate biopsy site 25 cm proximal from the tuberositas tibiae and 5 cm lateral from the midline of the femoral course.
(g) Shave the area, and sterilize with Iodophor solution.
(h) Anesthetize the skin and subcutaneous tissue with 10 ml 1% Novocain solution.
(i) Using the No.11 scalpel, make a 4- to 6-mm incision through the skin.
(j) Insert the trocar into the incision site, pushing through the muscle fascia, and advance to ensure that the cutting chamber opening lies fully within the muscle.
(k) Slide the nipple down against the patient's thigh to ensure a seal.
(l) Withdraw the trocar by 2.5 cm.
(m) Apply suction (an assistant is required) by withdrawing the plunger of the syringe, which will draw muscle tissue into the cutting chamber; advance the trocar forward, cutting off a sample as the assistant releases the suction.
(n) Rotate the trocar clockwise 90° to maximize the amount of muscle sample obtained.
(o) Withdraw the trocar, depress the syringe to expel the 3 ml of air, remove the muscle tissue sample from the chamber, and place on the 2 × 2-in. sterile gauze on ice.
(p) Repeat the procedure beginning at step (j) to obtain a second muscle sample.
(q) Place a clean, sterile 2 × 2-in. gauze sponge over the wound and fix into place with 3 Steristrips. Apply pressure on the wound for 10 min.
(r) Transfer the biopsy tissue with forceps to the chilled tube of DMEM/pen/strep.
(s) Transport the biopsy tissue on ice to the cell culture facility.

4. ISOLATION AND CULTURE OF HUMAN SKELETAL MYOBLASTS

Biopsies are performed on volunteers according to procedures approved by the Institutional Clinical Review Board. Skeletal muscle-derived cells have been isolated successfully from healthy volunteers as well as frail elderly, growth hormone-deficient, and heart failure patients. No significant differences in myoblast purity and cell growth characteristics were found among any of the patient populations when compared with healthy volunteers [Powell et al., 1999]. Wet weight of muscle tissue samples ranged from 140 to 1070 mg, with a mean weight of 398 ± 54 mg in 25 muscle biopsies from healthy volunteers.

Skeletal muscle biopsy tissue from each volunteer or patient is digested separately. The current procedure used for enzymatic digestion of the muscle tissue is modified from published protocols [Powell et al., 2002, Kosnik et al., 2001]. All muscle samples are handled with universal safety precautions because volunteers are not pretested for viral/bacterial disorders. Biopsies are given an identifying ID number at the time of biopsy to maintain privacy of the volunteer.

Protocol 10.2. Isolation and Culture of Human Skeletal Myoblasts

Reagents and Materials

Sterile

- ❑ DMEM/pen/strep (See Section 2.1.1)
- ❑ Collagenase-dispase solution (See Section 2.2.1)
- ❑ SKGM/15 (See Section 2.1.2)
- ❑ CMF: Ca^{2+}- and Mg^{2+}-free EBSS (See Section 2.1.3)
- ❑ Trypsin-EDTA solution: dilute stock 1:10 in CMF to give 0.05% trypsin in 5 mM EDTA (See Section 2.2.2)
- ❑ Fetal bovine serum at 4 °C
- ❑ Conical centrifuge tubes, 15 ml, 50 ml
- ❑ Dissecting forceps
- ❑ Dissecting scissors
- ❑ Petri dishes, 6 cm
- ❑ Tissue culture Petri dishes, 10 cm
- ❑ Polypropylene jar, 30 ml, autoclaved

Nonsterile

- ❑ Orbital platform shaker
- ❑ Balance
- ❑ Water bath at 37 °C
- ❑ Table top centrifuge

Protocol

A. Isolation of human skeletal myoblasts

(a) Obtain isolated muscle biopsy tissue in chilled DMEM/pen/strep solution.
(b) Working under sterile conditions, transfer muscle tissue to a 6-cm Petri dish containing 2 ml DMEM/pen/strep.
(c) Using sterile forceps, carefully remove excess connective tissue and fat from biopsy sample.
(d) Weigh tissue in a covered, sterile 6-cm Petri dish.
(e) Add 1–2 ml sterile DMEM/pen/strep to cover tissue.
(f) Using sterile scissors, cut tissue into strips approximately 1 mm × 10 mm.
(g) Place strips into sterile polypropylene jar containing 10 ml DMEM/pen/strep.
(h) Cap jar loosely and place in 37 °C CO_2 incubator.
(i) After 48 h of tissue incubation, prepare collagenase-dispase solution.
(j) Remove incubation medium from tissue strips and add 10 ml collagenase-dispase solution.
(k) Cap jar tightly, secure to orbital shaker in 37 °C incubator, and incubate for 60 min at approximately 60 rpm.
(l) Remove digested tissue along with enzyme solution and transfer to sterile centrifuge tube.
(m) Add 10% (v/v) chilled FBS to digested tissue solution.

(n) Pellet cells at 200 *g* for 5 min.
(o) Suspend cell pellet in 10.5 ml SKGM/15 and plate 10 ml tissue suspension into a 10-cm tissue culture dish.
(p) Plate a second 10-m tissue culture dish with 0.1 ml cell suspension in 10 ml SKGM/15 for cell yield determination (See Section 5).

B. Maintenance of human skeletal muscle cultures

(a) Feed muscle cultures 4 days after plating cells with 10 ml fresh prewarmed SKGM/15 for each 10-cm dish.
(b) Repeat feedings every 2–3 days until cultures are approximately 75–80% confluent.
(c) Subculture cells at 75–80% confluence.
(d) Aspirate medium from cells.
(e) Rinse plates with CMF, leaving rinse on dish for 2–5 min.
(f) Remove rinse and add 4 ml 1 × trypsin-EDTA solution to each 10-cm dish.
(g) Place cells in 37 °C incubator for 5 min to enhance detachment, monitoring process every 2–3 min under a microscope.
(h) Once cells are detached, remove to sterile 15-ml centrifuge tube.
(i) Rinse plate with 4 ml SKGM/15 and add rinse to tube.
(j) Centrifuge at 200 g for 5 min.
(k) Carefully aspirate supernate.
(l) Suspend cell pellet in SKGM/15, and count an aliquot on a hemocytometer.
(m) At this step, cells can be either expanded further, plated for characterization, plated for retroviral transduction, cryopreserved, or tissue engineered.
(n) For further expansion, plate cells at a density of not less than 1×10^5/10-cm dish and subculture after reaching 80% confluence.

---

5. CHARACTERIZATION OF HUMAN SKELETAL MYOBLASTS

Muscle-derived cells from each biopsy are characterized for total cell yield (number of cells isolated/100 mg of biopsy tissue), myoblast purity (percent desmin-positive cells) and cell population doubling time. Cell yield is determined by counting colonies in the cell yield plate (Protocol 10.2A, Step (p)) and extrapolating to the number of colonies that would be present in a 100-mg tissue sample. The cell isolation method currently in use, in which the muscle tissue is incubated for 48 h before being digested, consistently yields more cells/mg wet weight when compared to tissue samples digested immediately after biopsy [Powell et al., 1999, but also using a different enzyme solution]. Although the current protocol adds 2 days of incubation to the procedure outlined previously [Powell et al., 1999], significantly more cells are isolated. As a result, a clinical procedure requiring 1.0×10^8 myogenic cells would be achieved an average of 4 days earlier by this method (14 days) than when the tissue is enzymatically digested immediately after the biopsy (18 days).

Figure 10.2. Characterization of myoblast purity of skeletal muscle cultures. Human skeletal muscle-derived cells are immunostained with an antibody to desmin [Powell et al., 1999] and developed with AlexaFluor 488-conjugated goat anti-mouse IgG. Nuclei are visualized by DAPI staining. (See Color Plate 5A.)

Myoblast purity is determined for cells isolated from each volunteer by immunostaining with an antibody to desmin, an intermediate filament protein located in proliferating skeletal myoblasts [Powell et al., 1999]. Cultures are counterstained with DAPI (4′,6-diamidino-2-phenylindole) to identify all nuclei, and random fields were counted to determine average myoblast percentage (Fig. 10.2, See Color Plate 5A). Myoblast purity is routinely determined from cells plated after the first harvest from the tissue digest (P_1), although the percentage of myoblasts in a particular biopsy sample remains constant with time in culture and can therefore be assayed at any passage number (data not shown). As a minimum tissue-engineering requirement for use of a cell population isolated from a particular biopsy, we selected those with greater than 50% myoblasts. Approximately 80% of skeletal muscle tissue samples exceeded this requirement.

Typical characteristics of skeletal muscle derived cultures are as follows: yield of 1700–20,000 cells/100 mg wet weight; myoblast purity of $77.5 \pm 4.9\%$; and doubling time of 24.5 ± 0.8 h (Table 10.1).

Table 10.1. Characteristics of skeletal muscle biopsies from 9 representative donors.

Biopsy #	Tissue wet weight (mg)	Cell yield (no. of cells isolated/100 mg tissue)	Myoblast purity (% desmin-positive cells)	Cell doubling time (h)
1B	333	20090	64.6	28
2B	413	9010	65.3	22
3B	1070	5400	71.7	24
4B	1013	1710	91.8	24
5B	444	9275	89.6	24
6B	210	17765	82.0	25
7	285	4600	67.9	26
*8	443	4970	20.8	23
9	377	3310	79.5	26

Characteristics are determined as described in text. *Biopsy 8 was rejected for further expansion for tissue engineering, as the myoblast purity was <50%.

Protocol 10.3. Characterization of Human Skeletal Myoblast Cultures

Reagents and Materials

Sterile

- ❑ SKGM/15 (See Section 2.1.2)
- ❑ PBS (phosphate-buffered saline), containing calcium and magnesium
- ❑ SonicSeal slides, 4 well

Nonsterile

- ❑ Coomassie Blue staining solution (See Section 2.3.1)
- ❑ Methanol-acetone, 1:1 (v/v)
- ❑ Blocking buffer (See Section 2.3.2)
- ❑ Mouse anti-desmin
- ❑ Alexa Fluor 488 goat anti-mouse IgG
- ❑ Microscope with fluorescent light source; fluorescein and DAPI filters
- ❑ Vectashield® Mounting Medium with DAPI
- ❑ Circular coverslips, 12 mm

Protocol

A. Cell yield

Cell yield plate is ready for counting when macroscopic colonies are visible.

(a) Remove medium from dish, and rinse with 10 ml PBS.
(b) Remove rinse, and add 5 ml Coomassie Blue staining solution.
(c) Incubate 15 min.
(d) Remove staining solution.
(e) Count all colonies in plate.
(f) Calculate yield:

Cells/100 mg of biopsy tissue = (Number of colonies × 100) ÷ (wt of biopsy in mg/100)

B. Myoblast purity

(a) Plate 2 wells of a 4-well chamber slide with 3000 cells/well in 1 ml SKGM/15 and place in 37 °C incubator for 3 days.
(b) Aspirate medium, rinse wells 2 × quickly with PBS at room temperature.
(c) Fix for 10 min with 1:1 (v/v) methanol-acetone.
(d) Remove fixative, and rinse 3 × 5 min with PBS. Fixed cultures may be stored at 4° in PBS for several weeks before immunostaining.
(e) Incubate cultures with blocking buffer for 30 min at room temperature.
(f) Incubate with anti-desmin (1:200 dilution in blocking buffer) for 2 h at room temperature.
(g) Wash with 3 × 5-min rinses of PBS at room temperature.
(h) Incubate with Alexa Fluor 488 goat anti-mouse IgG (1:200 in blocking buffer) 30 min at room temperature.

(i) Wash with 3 × 5-min rinses of PBS at room temperature.
(j) Mount coverslip with Vectashield® DAPI.
(k) Count desmin-positive cells and total nuclei in 10 random fields for each isolated cell population under fluorescence microscope.
(l) Quantify myoblast percentage for each field, and calculate mean percentage of myoblasts.

---

6. TISSUE ENGINEERING OF HUMAN SKELETAL MYOBLASTS WITH CLINICALLY APPROVED EXTRACELLULAR MATRIX MATERIALS

Skeletal myoblasts can be tissue-engineered into 3-dimensional organ-like HBAM structures containing parallel arrays of postmitotic myofibers [Powell et al., 1999]. For casting BAMs, silicone rubber molds with internal dimensions of 6.5 mm wide × 30 mm long × 6 mm deep are formed in preconstructed aluminum molds (Fig. 10.3). Wells are washed and glued into six-well tissue culture dishes, with one rubber mold/well. Attachment points are constructed from syringe needles and inserted at each end of the well, approximately 20 mm apart (Fig. 10.4) (See *note (i)*.)

Myoblasts are suspended in an ice-cold extracellular matrix solution, and the cell suspension is pipetted into the silicone rubber molds. Within 2–3 days after casting, the cell-gel mixture contracts, detaching from the mold but remaining secured in place by the end attachment sites (See Fig. 10.4). BAM morphology is assessed by immunostaining with an antibody to sarcomeric tropomyosin, a contractile protein found only in fused muscle fibers. Cross sections of 10-day in vitro HBAMs show arrays of muscle fibers aligned primarily perpendicular to the long axis of the BAM and distributed uniformly throughout the entire cross-sectional area (Fig. 10.5, See Color Plate 5B) when the BAM diameter is small (<250 μm).

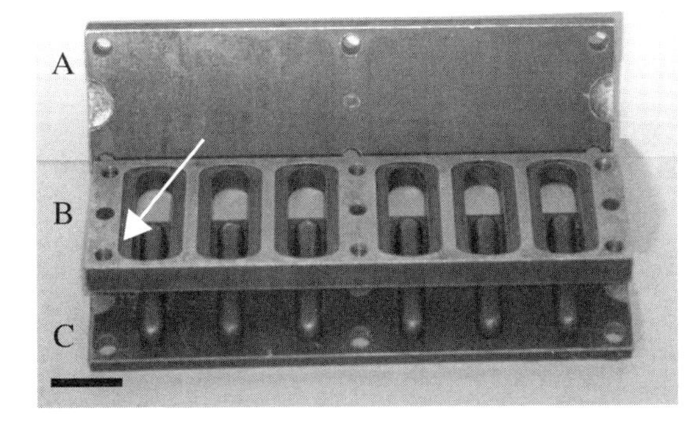

Figure 10.3. Aluminum casting mold before assembly. The aluminum mold used to cast silicone rubber wells for BAM formation is assembled from the three parts shown. Part B, containing a lip (See arrow) at the top of the cut-out region, provides a base for the cast silicone rubber well, and is placed on top of part C. Screws are inserted up through the holes in part C and part B to secure the assembly in position. The liquid silicone solution is poured into each well of the mold, and part A is set in place over the screws. Wing nuts are tightened on the screws to secure the assembly. Bar = 15 mm.

Figure 10.4. Fourteen day in vitro HBAM cast in a silicone rubber mold. Human skeletal muscle-derived cells are suspended in a 2 mg/ml collagen solution, and 2×10^6 cells in 1 ml collagen solution are poured into the silicone rubber mold. Arrow points to a pin attachment site. The contracted HBAM has detached from the mold sides and bottom and is held in place only by the pins. The HBAM is under passive tension because detachment from one pin results in a 60–70% retraction in length within 12 h. The passive tension in the construct aligns the fusing myogenic cells into parallel arrays of postmitotic muscle fibers [Powell et al., 1999]. Bar = 10 mm.

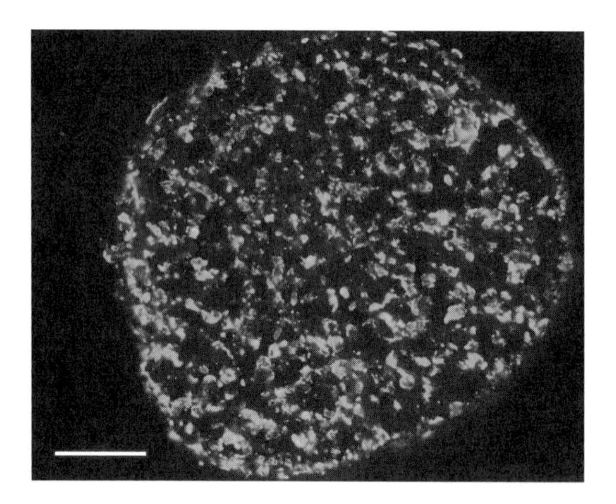

Figure 10.5. Cross section of 10-day in vitro HBAM. After 10 days in vitro, an HBAM is fixed for 1 h in 2% formaldehyde, and 10-μm cryostat cross sections are cut. Sections are stained with an antibody to sarcomeric tropomyosin, followed by incubation with AlexaFluor 488-conjugated goat anti-mouse IgG to identify muscle fibers. Bar = 50 μm. (See Color Plate 5B.)

Protocol 10.4. Formation of Silicone Rubber Casting Molds

Reagents and Materials

Nonsterile

- ❑ Ultra-pure water (UPW)

- ❑ Lemon Joy liquid dishwashing detergent (nontoxic mold release agent), 50% in ultra-pure water
- ❑ Tissue culture plates, 6 well, 35 mm diameter
- ❑ Centrifuge tubes, 50 ml
- ❑ Silicone rubber (Med6015 Silicone Elastomer)
- ❑ Preformed aluminum casting molds, machined to user specifications (See Fig. 10.3)
- ❑ Silicone rubber adhesive sealant (RTV108)
- ❑ Syringe needles, 22 g, cut into 8- to 10-mm lengths with beveled end removed
- ❑ Vacuum pump
- ❑ Vacuum chambers
- ❑ Standard bench-top lab heating oven, capable of 150 °C
- ❑ Air brush
- ❑ Ethylene oxide source

Protocol

(a) Mix silicone elastomer components, following the manufacturer's instructions.

(b) Pour mixture into 50-ml centrifuge tubes, place open tubes in vacuum chamber, and run vacuum until solution is clear (i.e., bubble free, approximately 25 min).

(c) Heat clean aluminum casting mold plates at 150 °C for 20 min.

(d) Using an air brush and working in a fume hood, spray hot disassembled mold plates with 50% Lemon Joy, being sure to cover all corners and angles (rotate mold as necessary).

(e) Assemble mold plates when soap has dried (See Fig. 10.3).

(f) Pour silicone mixture into assembled aluminum casting mold plates.

(g) Place mold plates containing silicone in 150 °C oven for 30 min.

(h) Remove from oven and rinse in cold running tap water for 1 min.

(i) Disassemble aluminum casting mold plates, and remove hardened silicone rubber molds.

(j) Rinse molds well with tap water to remove soap, and soak in UPW overnight.

(k) Assemble silicone molds in 6-well plates, gluing 1 BAM mold/well with a small amount of RTV sealant.

(l) Insert pins vertically at ends of each BAM mold at desired separation distance.

(m) Once the glue has dried, wash entire plate 3 times with distilled water, and then fill wells again with UPW and soak for 6 h minimum.

(n) Remove water and air-dry plates in a laminar flow hood.

(o) Sterilize with ethylene oxide.

Protocol 10.5. Tissue-Engineering and Maintenance of Human Skeletal Muscle

Outline

HBAMs are engineered by suspending isolated muscle-derived cells in an ice-cold collagen solution and casting the cell-gel solution into the silicone rubber molds. Each

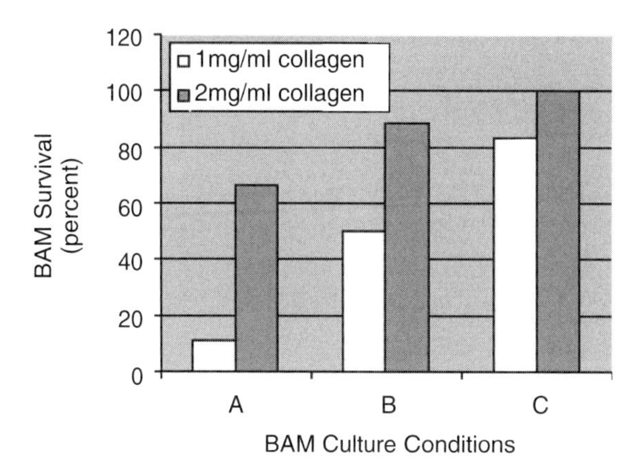

Figure 10.6. Survival of HBAMs engineered with varying collagen concentrations and cell numbers. HBAMs are engineered (See Protocol 10.5) with either 1 mg/ml or 2 mg/ml collagen, and with 1×10^6 cells in 500 μl of cell-matrix suspension (A), 1.5×10^6 cells in 750 μl (B), or 2.0×10^6cells in 1000 μl (C). BAMs are evaluated for survival (18 BAMs/group) on days 11–12 after casting. Surviving BAMs are identified as being intact and secured to both end attachment sites.

HBAM contains 2×10^6 cells suspended in 1 ml of a 2 mg/ml collagen solution. Cell number and collagen concentration may be varied depending on the application, but these conditions are optimal for cell and HBAM survival (Fig. 10.6).

Reagents and Materials

Sterile

- ❑ Zyderm
- ❑ Sterile silicone rubber BAM casting molds
- ❑ SKGM/15 (See Section 2.1.2)
- ❑ Acetic acid, 0.1 N
- ❑ NaOH, 1.0 N
- ❑ Tisseel stock solution, 7.5 mg/ml
- ❑ Differentiation medium (See Section 2.1.4)
- ❑ SKGM/2 maintenance medium (See Section 2.1.2)

Protocol

(a) Calculate the total volume of cell/Zyderm solution required.

(b) Prepare a working solution of Zyderm (3.25 mg/ml) by diluting stock Zyderm (65 mg/ml) 1:20 with 0.1 N sterile acetic acid.

(c) Suspend the Zyderm working solution well by vortexing gently, and incubate for at least 1 h at room temperature to solubilize. The working solution can be stored for at least several weeks at 4 °C.

(d) Calculate the volume of each component required for HBAM formation. (See *note (ii)*.)

i) Final concentration of Zyderm = 2 mg/ml. Calculate volume of working solution needed: final volume of cell-gel solution × 2 ÷ 3.25.
ii) Final concentration of tisseel = 0.125 mg/ml. Volume of stock tisseel required = 0.125 mg/ml × volume of Zyderm working solution ÷ 7.5 mg/ml
iii) 1 N NaOH volume: 0.065 × volume of Zyderm working solution
iv) SKGM/15 volume: final volume of Zyderm/cell solution−(Zyderm working solution volume + Tisseel volume + NaOH volume).

(e) Chill SKGM/15, Zyderm working solution, 1 N NaOH, and Tisseel on ice.
(f) Prepare collagen solution by adding Zyderm to SKGM/15, suspending slowly with sterile pipette and taking care not to create bubbles.
(g) Add Tisseel and NaOH, and suspend very well with sterile pipette, keeping solution chilled on ice. (See *note (iii)*.)
(h) Harvest and pellet required number of cells.
(i) Suspend cells in Zyderm solution at a concentration of 2×10^6 cells/ml.
(j) Cast Zyderm-cell mixture into molds with a sterile, chilled pipette tip, adding 1 ml/silicone rubber mold. Distribute solution evenly along length of mold and around pins with the pipette tip.
(k) Carefully transfer the mold plate to a 37 °C incubator for 1 h.
(l) Flood wells with SKGM/15, pipetting slowly so as not to disturb the partially solidified cell-gel mix, adding 8–9 ml medium/well so that entire mold and HBAMs are covered.
(m) Three days after casting, change medium to differentiation medium (8–9 ml/well).
(n) After HBAMs have been in differentiation medium for 3 days, change to SKGM/2 maintenance medium (8–9 ml/well).
(o) Continue to change SKGM/2 medium every 2–3 days for duration of experiment.

Notes:

1. Molds can also be formed from silicone rubber tubing (0.187-in. ID × 0.312-in. OD) cut into appropriate lengths. A wedge of 1/3 of the circumference of the tubing is removed longitudinally. Tubing ends are blocked by either gluing a small piece of a silicone rubber sheet, 0.01 in. thick, at each end or, alternatively, gluing Velcro (loop side only; available at fabric stores) pieces 2 mm × 2 mm as attachment points in place of pins. The cell solution will integrate well into the Velcro because of its high surface tension and will therefore not flow out through the ends of the tubing [Vandenburgh et al., 1996].
2. It is essential to make up enough cell-matrix solution (Steps (d)–(i)) for several extra BAMs as volume is lost while pipetting and casting BAMs.
3. Once Tisseel and NaOH have been added to the Zyderm-SKGM/15 mixture, the solution must be kept chilled on ice and used within 1 hour of preparation.

7. APPLICATIONS OF TISSUE-ENGINEERED HUMAN SKELETAL MUSCLE

7.1. Gene Therapy

BAMs made from nonhuman myoblasts and implanted subcutaneously have been shown to deliver biologically active, therapeutic proteins long-term in several animal models [Vandenburgh et al., 1996, 1998; Lu et al., 2001, 2002]. Human growth hormone (hGH) constantly secreted from implanted mouse BAMs attenuated skeletal muscle wasting in a muscle atrophy model in mice more effectively than daily hGH injections [Vandenburgh et al., 1998]. In a large-animal model, BAMs engineered from ovine skeletal muscle genetically modified to secrete recombinant vascular endothelial growth factor (rhVEGF) were implanted onto the surface of the sheep heart, resulting in increased capillary ingrowth in the neighboring host myocardium [Lu et al., 2002]. These studies suggest the potential of tissue-engineered HBAMs formed from human skeletal muscle-derived cells genetically modified to secrete therapeutic proteins as an implantable protein delivery device.

Human skeletal myoblasts can be genetically engineered to secrete sustained levels of therapeutic proteins [Powell et al., 1999]. When tissue engineered into HBAMs they continue to secrete biologically active foreign gene products. Implantation of genetically engineered HBAMs for patients with a variety of disorders would thus present an attractive alternative to daily injections, which often result in serious side effects. Further improvements to this model will require scaling up of protein output from the HBAMs to reach efficacious levels of the recombinant protein. Modifications to the cell packing density, size of implant, and genetic engineering methods will help to increase delivery dosage levels.

7.2. Muscle Repair/Replacement

In addition to their value as a platform for gene therapy, HBAMs also have the potential to be used for structural and functional skeletal muscle repair or replacement. To achieve this goal, HBAMs will need to be engineered to more closely resemble in vivo human skeletal muscle and generate meaningful force. Our laboratory has designed a novel computer-controlled electromechanical tissue stimulator device for this purpose, which mechanically loads HBAMs during both formation and later development, while recording real-time passive and active forces. To date these active forces are only 1–2% of the force generated by adult muscle [unpublished data] and thus are of little clinical relevance.

By mechanically stimulating BAMs, we have measured structural and functional improvements, including increased elasticity, mean myofiber diameter, and myofiber area percent [Powell et al., 2002]. The mechanically loaded BAMs, although closer to in vivo muscle, will need to be improved further before they can be used as functional human muscle analogs. Improved design of mechanical conditioning programs may allow for advanced dynamic engineering techniques that will enable the engineering of more in vivo-like BAMs that will be more applicable for structural repair.

Finally, in addition to use as a platform for gene therapeutics and muscle repair, tissue-engineered human skeletal muscle offers other potential applications, including in vitro drug screening, and as a model system for the study of muscle wasting in the microgravity of space [Vandenburgh et al., 1999]. We anticipate that additional improvements in tissue engineering of HBAMs will expand research and clinical opportunities in the future.

ACKNOWLEDGMENTS

Work was supported by the following grants: NIH R44-AG-14958, NIH R01-HL-60502, NASA NAG2-1205, and NIST ATP 70NANB9011.

SOURCES OF MATERIALS

Item	*Catalog No.*	*Supplier*
Air brush	Model 200, Deluxe Set	Badger Air Brush Co.
Alexa Fluor 488 goat anti-mouse IgG	11029	Molecular Probes (See also Invitrogen)
Argyle female Luer lock connector	8888-275008	Sherwood Medical
Biopsy needles (4- and 6-mm Bergstrom cutting trocars)	n/a	Stille
Bovine serum albumin (blocking buffer)	A4161	Sigma
Bovine serum albumin (differentiation medium)	CC4160	Clonetics (See Cambrex)
CMF: Ca^{2+}- and Mg^{2+}-free EBSS (Earle's balanced salt solution)	14155-063	GIBCO
Collagenase type II	C6885	Sigma
Conical centrifuge tubes, 15 ml	430052	Corning
Conical centrifuge tubes, 50 ml	430290	Corning
Coomassie Blue R	B0149	Sigma
Curved forceps 12 cm	11002-12	Fine Scientific Tools
Cutting cannula adapters (B-D interlink vial access)	303367	Becton Dickinson
Dispase	165859	Roche
DMEM/pen/strep (Dulbecco's modified Eagle's medium)	11995-040	GIBCO (See Invitrogen)
Fetal bovine serum (FBS)	F2442	Sigma
Gentamycin	G1272	Sigma
Human epidermal growth factor (hEGF)	CC4017	Clonetics (See Cambrex)
Insulin	CC4025	Clonetics (See Cambrex)
Interlink catheter extension kit	2N3374	Baxter HealthCare
Iodophor	n/a	Hospital pharmacy
Mouse anti-desmin	D1033	Sigma
Novocain solution, 1%, without epinephrine	n/a	Hospital pharmacy
PBS with Ca^{2+} and Mg^{2+}	14040	GIBCO (See Invitrogen)
Penicillin/streptomycin	15140-122	GIBCO (See Invitrogen)

Item	*Catalog No.*	*Supplier*
Polypropylene jar, 30 ml	2118-0001	Nalge Nunc
Silicone elastomer, silicone rubber	Med6015	Nusil Silicone Technology
Silicone rubber adhesive sealant	RTV108	GE Silicones
Silicone rubber sheet (0.01 in. thick)	Custom order	Silicone Specialty Fabricators
Silicone rubber tubing	8060-0040	Nalgene
Skeletal muscle growth medium (SKGM)	CC3160	Clonetics (See Cambrex)
SonicSeal slides, 4 well	138121	Nalge Nunc
Steristrips, $\frac{1}{4}$ in.	3 M #R-1542	Seaway
Suction-enhancing nipples: infant nipples, 4 mm	00079	Ross Laboratories
Surgilube	0168-0205-06	Fougera
Syringe needles, 22 g	305155 BD	BD Biosciences
Tisseel VH fibrin sealant	921029	Baxter
Triton X-100	T9284	Sigma
Trypsin-EDTA solution	T4174	Sigma
Vectashield® Mounting Medium with DAPI	H1200	Vector Labs
Zyderm® 2 Collagen Implant	5024A	McGhan Medical

REFERENCES

Dennis, R.G., and Kosnik, P.E. (2000) Excitability and isometric contractile properties of mammalian skeletal muscle constructs engineered in vitro. *In Vitro Cell Dev. Biol. Anim.* 36: 327–335.

Hennessey, J.V., Chromiak, J.A., DellaVentura, S., Guertin, J., MacLean, D.B. (1997) Increase in percutaneous muscle biopsy yield with a suction-enhancement technique. *J. Appl. Physiol.* 82: 1739–1742.

Kosnik, P.E., Dennis, R.G., Vandenburgh, H.H. (2003) Tissue engineering skeletal muscle. In Guilak, F., et al, eds., *Functional Tissue Engineering: The Role of Biomechanics*. Springer-Verlag, New York, N.Y., pp. 377–392.

Kosnik, P.E., Faulkner, J.A., Dennis, R.G. (2001) Functional development of engineered skeletal muscle from adult and neonatal rats. *Tissue Eng.* 7: 573–584.

Lu, Y., Shansky, J., Del Tatto, M., Ferland, P., et al. (2002) Therapeutic potential of implanted tissue-engineered bioartificial muscles delivering recombinant proteins to the sheep heart. *Ann. NY Acad. Sci.* 961: 1–5.

Lu, Y., Shansky, J., Del Tatto, M., Ferland, P., Wang, X., Vandenburgh, H.H. (2001) Recombinant vascular endothelial growth factor secreted from tissue-engineered bioartificial muscles promotes localized angiogenesis. *Circulation* 104(5): 594–599.

Okano, T., Satoh, S., Oka, T., Matsuda, T. (1997) Tissue engineering of skeletal muscle. Highly dense, highly oriented hybrid muscular tissues biomimicking native tissues. *ASAIO J.* 43: M749–M753.

Powell, C.A., Smiley, B.L., Mills, J., Vandenburgh, H.H. (2002) Mechanical stimulation improves tissue-engineered human skeletal muscle. *Am. J. Physiol. Cell Physiol.* 283: C1557–C1565.

Powell, C.P., Shansky, J., Del Tatto, M., Forman, D., Hennessey, J., Sullivan, K., Zielinski, B.A., Vandenburgh, H.H. (1999) Tissue-engineered human bioartificial muscles expressing a foreign recombinant protein for gene therapy. *Hum. Gene Ther.* 10(4): 565–577.

Strohman, R.C., Byne, E., Spector, D., Obinata, T., Micou-Eastwood, J., Maniotis, A. (1990) Myogenesis and histogenesis of skeletal muscle on flexible membranes in vitro. *In Vitro Cell Dev. Biol.* 26: 201–208.

Swasdison, S., and Mayne, R. (1992) Formation of highly organized skeletal muscle fibers in vitro. *J. Cell Sci.* 10: 643–652.
Vandenburgh, H.H., Chromiak, J., Shansky, J., Del Tatto, M., Lemaire, J. (1999) Space travel directly induces skeletal muscle atrophy. *FASEB J.* 13(9): 1031–1038.
Vandenburgh, H.H., Del Tatto, M., Shansky, J., Goldstein, L., et al. (1998) Attenuation of skeletal muscle wasting with recombinant human growth hormone secreted from a tissue-engineered bioartificial muscle. *Hum. Gene Ther.* 9: 2555–2564.
Vandenburgh, H.H., Del Tatto, M., Shansky, J., Lemaire, J., et al. (1996) Tissue-engineered skeletal muscle organoids for reversible gene therapy. *Hum. Gene Ther.* 7: 2195–2200.
Vandenburgh, H.H., Swasdison, S., Karlisch, P. (1991) Computer aided mechanogenesis of skeletal muscle organs from single cells in vitro. *FASEB J.* 5: 2860–2867.

Engineered Heart Tissue

Thomas Eschenhagen and Wolfgang H. Zimmermann

Institute of Experimental and Clinical Pharmacology, University Hospital, Hamburg Eppendorf 20246 Hamburg, Germany

Corresponding author: t.eschenhagen@uke.uni-hamburg.de

Culture of Cells for Tissue Engineering, edited by Gordana Vunjak-Novakovic and R. Ian Freshney

1. BACKGROUND

Cell-based therapies have been suggested as a novel and potentially curative approach for replacement of impaired myocardium [Reinlib and Field, 2000]. Presently, two principally different approaches have been developed: (1) implantation of isolated cells by direct injection into the myocardium or by percutaneous applications [Koh et al., 1993; Soonpaa et al., 1994; Li et al., 1996; Scorsin et al., 1997; Taylor et al., 1998; Reinecke et al., 1999; Sakai et al., 1999; Tomita et al., 1999; Condorelli et al., 2001; Etzion et al., 2001; Menasche, 2001; Orlic et al., 2001; Müller-Ehmsen et al., 2002a,b; Roell et al., 2002] and (2) construction of cardiac muscle constructs in vitro that can be surgically attached to the myocardium [Bursac et al., 1999; Carrier et al., 1999; Leor et al., 2000; Li et al., 2000]. The present article describes our own approach aimed at developing a functional cardiac tissue construct suitable for cardiac replacement therapy.

In 1994, our idea to generate 3-dimensional cardiac tissue constructs came from the observation that it is relatively easy to genetically manipulate cultured immature cardiac myocytes in the classic 2-dimensional culture format (and to observe biochemical or molecular consequences) but difficult to obtain reliable information from these cells in terms of contractile function. Evaluation of contractile consequences of any genetic manipulation would be essential, however, for the validation of potentially relevant genes identified by gene expression arrays that we and others performed at that time.

The description of a method to culture embryonic chick fibroblasts in collagen I in a 3-D format that allowed measurement of (tonic) contractile forces in the organ bath by the group of E. L. Elson [Kolodney and Elson, 1993] stimulated us to adapt the method to embryonic chick cardiac myocytes. The experiments quickly yielded success, and cardiac myocytes spontaneously generated a coherently and regularly beating cardiac tissue-like structure (cardiac myocyte-populated matrix, CMPM) that developed measurable forces when suspended in organ baths and connected to a force transducer [Eschenhagen et al., 1997]. The success was surprising because others had described that cardiac myocytes from neonatal rats do not grow *inside* a collagen I matrix but can only be cultured *on top* of a preformed collagen I matrix [Souren et al., 1992]. We later learned that, indeed, cardiac myocytes from neonatal rats, in contrast to embryonic chick cardiac myocytes, did not spontaneously form a tissuelike structure when cultured in collagen I, but needed supplementation of Matrigel and a modification of the cell-matrix composition in the original reconstitution mixture [Zimmermann et al., 2000].

The initial experimental setup followed the one proposed by E. L. Elson's group in St. Louis and consisted of two glass tubes held at a defined distance by a metal spacer in a rectangular well (Fig. 11.1). When cardiac myocytes were mixed with neutralized liquid collagen I and pipetted between the two glass tubes they yielded biconcave lattices spanning between the glass tubes. Contractile force could easily be measured by transferring CMPMs in an organ bath and suspending the glass tube-anchored lattices between a fixed holder and a force transducer.

What we learned in this system was the importance of mechanical strain. At the free edges, the region of increased strain, cells formed a dense network of muscle bundles that were oriented along the edges, that is, in the direction of the strain (See Fig. 11.1). In contrast, cells formed only a loose and randomly oriented network in the central parts of the lattices (See Fig. 11.1). Thus it appeared as if the orientation of cells reflected the biomechanical strain throughout the lattice. The importance of strain was substantiated by experiments in which CMPMs were subjected to a 5-day cyclic stretch regimen imposed by a motorized device [Fink et al., 2000]. This resulted in a three- to fivefold increase in contractile force development, hypertrophy of cardiac myocytes in CMPMs, and improved tissue formation at the free edges. A systematic analysis of the relation between the degree of stretch and force development showed that the strain had a threshold of approximately 5–7% that was necessary to induce CMPM formation with improved

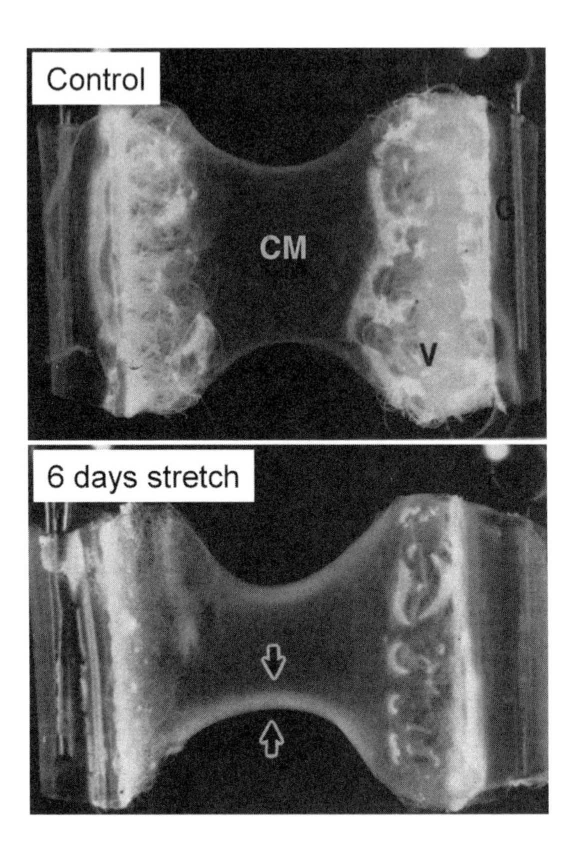

Figure 11.1. Morphology of cardiac myocyte-populated matrix (CMPM) with and without phasic stretch. Note the dense cell population at the free edges and sparse population in the center. After 6-day stretch the cell-rich area increased. Size of CMPM is 15 × 10 mm. From Fink et al. [2000].

tissue properties [Fink et al., 2000]. Larger strains did not further improve force, and strains higher than 20% tended to disrupt the CMPM structure.

CMPMs proved to be a valuable system to measure pharmacological responses to β-adrenergic and muscarinic agonists and extracellular calcium [Eschenhagen et al., 1997; Fink et al., 2000; Zimmermann et al., 2000] as well to transfer cardiac genes [Most et al., 2001; Remppis et al., 2003]. However, the inhomogeneity of cell orientation and tissue formation and the need for the production of casting devices with Velcro-covered glass tubes led to variable results and motivated us to think about an easier setup. The solution was a ring-shaped casting mold into which the cell-collagen-Matrigel mixture was pipetted. These casting molds are reproducibly produced from silicone in large quantities and can be used indefinitely [Zimmermann et al., 2000, 2002a]. In addition to their practical advantages, the ring-shaped engineered heart tissues (EHTs) had much better tissue homogeneity, likely due to the more homogeneous distribution of biomechanical strain. Circular EHTs have also been utilized to study gene function [El-Armouche et al., 2003; Zolk et al., 2003]. Hence, all our subsequent studies have been carried out with the improved circular EHT system.

With the growing interest in cardiac tissue engineering and the progress made in using embryonic and adult stem cells to derive cardiac myocytes, it became obvious that EHTs may prove valuable as tissue replacement material. Several questions have been addressed or are under current evaluation.

1.1. Can EHT Contractile Function and Tissue Formation be Improved as Compared to Our Standard Protocol?

Cell density, collagen I content, chick embryo extract, and horse serum requirements have been tested. Cell concentration had a biphasic effect on force development. When cell number per EHT was increased from 0.5 to 2.5×10^6, the force of contraction increased; for cell numbers above 2.5×10^6 cells/EHT the contractions ceased, probably because of metabolic problems. The contractile force was highest for 0.5 and 0.7 mg collagen/EHT and decreased both above and below this range, whereas the mechanical integrity increased with collagen content. The standard collagen content was set to 0.8 mg/EHT [Eschenhagen et al., 2002]. Chick embryo extract appeared not to be essential throughout the duration of culture. In contrast, EHTs depended on the continuous presence of horse serum at 10% v/v (Fig. 11.2). Any reduction of the serum content resulted in a rapid reduction of the contractile activity [Eschenhagen et al., 2002]. Higher serum concentrations have not been tested systematically.

1.2. Do EHTs Survive When Implanted into the Peritoneum of Rats?

Experiments in syngeneic rats showed that EHTs became quickly and strongly vascularized and survived for at least 4 weeks when implanted under the peritoneum [Eschenhagen et al., 2002]. Histology, however, showed infiltration of the EHT by mononuclear cells and multinucleated giant cells, predominantly at the implant periphery. Despite the apparent immune response, cardiac myocytes retained a cross-striated morphology for at least 4 weeks after implantation.

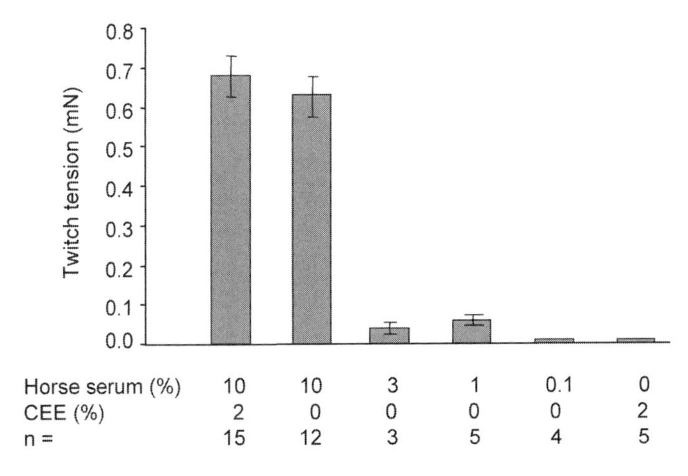

Figure 11.2. Serum dependence of EHT. EHTs were cultured for 7 days in the presence of 10% horse serum (HS) and 2% chick embryo extract (CEE) and for another 5 days in the presence of the indicated concentrations of supplements. The ordinate indicates twitch tension at maximally effective extracellular calcium concentrations in the organ bath. Note that reduction of horse serum reduced twitch tension to almost zero, whereas depletion of chick embryo extract had no effect. From Eschenhagen et al. [2002].

1.3. Do EHTs Survive When Implanted onto the Heart of Rats?

In contrast to implantation into the peritoneum, EHTs did not survive when implanted onto the heart of syngeneic rats [Zimmermann et al., 2002b]. After 2 weeks, we observed a strong immune response with destruction of implanted cardiomyocytes and replacement fibrosis. The difference in the immune responses following EHT implantation in the peritoneum and on the heart is presently unclear. The immune response in general was no surprise. Although collagen and heart cells were carefully isolated from syngeneic rats (Fischer 344) we utilized various xenogeneic components in the original EHT reconstitution mixture and during in vitro maturation: horse serum, Matrigel, and chick embryo extract. Any of these factors might have been the cause of immune activation. A likely explanation is an impregnation of otherwise syngeneic cells with xenogeneic proteins. When the experiments were repeated under immunosuppression with cyclosporine, azathioprine, and methylprednisolone, EHTs survived for up to 8 weeks, continued to contract, and became heavily vascularized and even innervated [Zimmermann et al., 2002b]. Importantly, the examination of myocyte ultrastructure revealed sarcomere development with clear signs of maturation and a regular formation of M-bands that were only rarely seen during cultivation of EHTs in vitro. These data support the hypothesis that the growth factor milieu and/or the cellular environment of adult host myocardium drives cardiac myocytes into terminal differentiation.

1.4. Do EHTs Become Vascularized and Integrate into the Recipient Heart?

As noted above, EHTs quickly become vascularized, both under the peritoneum and on the heart. Blood vessel ingrowth is visible as early as 3 days. After 4 weeks, the entire EHT is penetrated by functional blood vessels, as confirmed by the ultrastructural demonstration of erythrocytes in the blood vessels [Zimmermann et al., 2002b]. We are currently evaluating whether the preformed capillaries and primitive vascular structures that are regularly seen in EHTs during in vitro cultivation [Zimmermann et al., 2002a] participate in vascularization after implantation onto the heart. The relevance of large myelinated and nonmyelinated nerve fibers that have been found in EHTs 4 weeks after implantation remains unknown.

An important question is whether EHTs electrically and functionally integrate into the host myocardium. First series of electrical mapping experiments indicate that electrical coupling of EHT to host myocardium has occurred 4 weeks after implantation onto infarcted rat hearts [Zimmermann et al., 2003]. Mechanical coupling is more difficult to prove because mechanical activity of the EHT, when integrated well in the host myocardium, cannot be differentiated by eye from the beating of the heart. More sophisticated evaluation methods are needed to answer this question.

1.5. Do EHTs Alter the Contractile Function of Uninjured Hearts?

Contractile function of the uninjured hearts did not change after EHT implantation, as demonstrated by echocardiographic analysis [Zimmermann et al., 2002b].

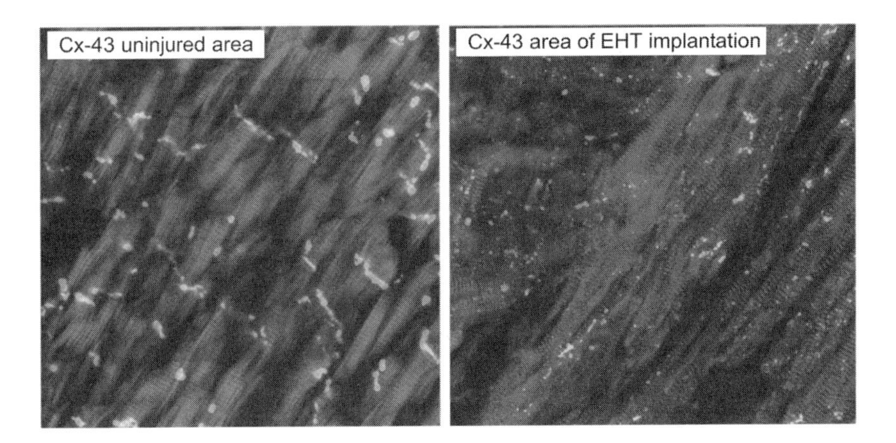

Figure 11.3. Effect of EHT implantation of the spatial organization of connexin 43 (Cx-43) in rat hearts. The left panel shows an immunoconfocal image from the middle of a rat heart, the right panel from the area of EHT implantation (EHT upper left corner, host myocardium right lower part). Cx-43 is indicated in green, phalloidin-stained actin in red. Note the disorganization of Cx-43 at the implantation site. (See Color Plate 6A.)

This may seem trivial, but one could argue that there is some reaction of host myocardium to the implanted EHT that may result in deterioration of contractile function. This could include impairment of diastolic function, which is typically observed in classic dynamic cardiomyoplasty or structural alteration on the cellular level. However, the pattern of spatial connexin 43 distribution at the site of EHT implantation differed from that in uninjured myocardium, as observed by immunochemical staining (Fig. 11.3, See Color Plate 6A). Alterations in connexin 43 immune staining have been previously observed in developing and diseased hearts [Kostin et al., 2002].

1.6. Do EHTs Improve Cardiac Function after Myocardial Infarction?

The most important question is whether implantation of the engineered tissue constructs can improve contractile function after myocardial infarction. We have chosen the model of coronary ligation as being relatively close to human pathology and have set up an experimental model in which the size and the functional consequences of coronary ligation are well defined. Large longitudinal studies using echocardiography, magnetic resonance imaging, and left heart catheterization to analyze in vivo hemodynamics are currently being performed to evaluate the feasibility of our approach.

2. PREPARATION OF REAGENTS

2.1. Complete Culture Medium

Dulbecco's modified minimal essential medium (DMEM) supplemented with 10% horse serum, 2% chick embryo extract, 100 μg/ml streptomycin, and 100 U/ml penicillin G, all sterile, stored at 4 °C for up to 2 weeks.

2.2. Calcium- and Bicarbonate-Free Hanks' Balanced Salt Solution with HEPES (CBFHH)

NaCl 136.9 mM, KCl 5.36 mM, $MgSO_4 \cdot (H_2O)_7$0.81 mM, glucose 5.55 mM, $KH_2PO_4$0.44 mM, $Na_2HPO_4 \cdot (H_2O)_7$0.34 mM, HEPES 20 mM, adjusted to pH 7.5 with NaOH (all chemicals are of analytical grade or best commercially available). Sterilize by filtration.

2.3. Tyrode's Solution (0.2 mM and 0.4 mM Ca^{2+})

NaCl 119.8 mM, KCl 5,4 mM, $MgCl_2$ 1.05 mM, $CaCl_2$ 0.2 mM or 0.4 mM, NaH_2PO_4 0.42 mM, $NaHCO_3$ 22.6 mM, glucose 5.05 mM, Na_2EDTA 0.05 mM, ascorbic acid 0.28 mM. Maintain pH at 7.4 by bubbling with 95% O_2–5% CO_2 (carbogen).

2.4. Trypsin Stock Solution (0.1 g/ml; 50×)

Dissolve 5 g in 50 ml of CBFHH overnight with gentle shaking. Sterilize by filtration. Dilute 1:50 with CBFHH for use.

2.5. DNase Stock Solution (2 mg/ml; 71×)

Dissolve 100 mg in 50 ml of NaCl (0.15 M). Sterilize by filtration and store at −20 °C. Dilute 1 ml with 70 ml of CBFHH for use at 28 μg/ml.

2.6. Concentrated DMEM (2×)

DMEM made up at double strength from powder and filter sterilized, or from 10× concentrate.

2.7. Histologic Fixative

Formaldehyde 4%, methanol 1%, in phosphate-buffered saline (PBS), pH 7.4, containing 1 mM $CaCl_2$ and 30 mM 2,3-butanedione monoxime (BDM).

2.8. Supplemented Tris-Buffered Saline (STBS)

Tris•HCl 0.05 M, NaCl 0.15 M, pH 7.4, containing 10% fetal calf serum, 1% bovine serum albumin, 0.5% Triton X-100, and 0.05% thimerosal.

2.9. Glutaraldehyde Fixative

Glutaraldehyde 2.5% in PBS, pH 7.4 containing 1 mM $CaCl_2$ and 30 mM BDM

3. ISOLATION AND CULTURE METHODOLOGY

The principle of EHT preparation is simple. Freshly isolated neonatal rat heart cardiac myocytes are mixed with freshly neutralized rat tail collagen I, Matrigel® (alternatively Harbor Extracellular Matrix (ECM), which is also extracted from

Engelbreth–Holm–Swarm tumors in mice), and a medium concentrate calculated to yield a final concentration of 1× DMEM, 10% horse serum, and 2% chick embryo extract. This reconstitution mixture is prepared on ice, well mixed by pipetting, and then pipetted into ring-shaped casting molds at room temperature (Fig. 11.4a, See Color Plate 6B). After transfer into the 5% CO_2 incubator at 37 °C the reconstitution mixture quickly (30 min) consolidates, yet remains soft, and thus traps the cells in a geometrically defined 3D environment.

During the first 2–3 days the cells start to spread out, develop cell-cell contacts, and form spontaneously contracting aggregates. Over time aggregates organize into longitudinally oriented strands with diameters of up to 200 μm that become highly interconnected throughout the matrix into a dense muscular 3D network. During the consolidation process EHTs undergo extensive matrix remodeling and change from an initially slack gel that fills the entire circular well (outer/inner diameter: 16/8 mm, height: 5 mm) to a contracted and mechanically stable ring

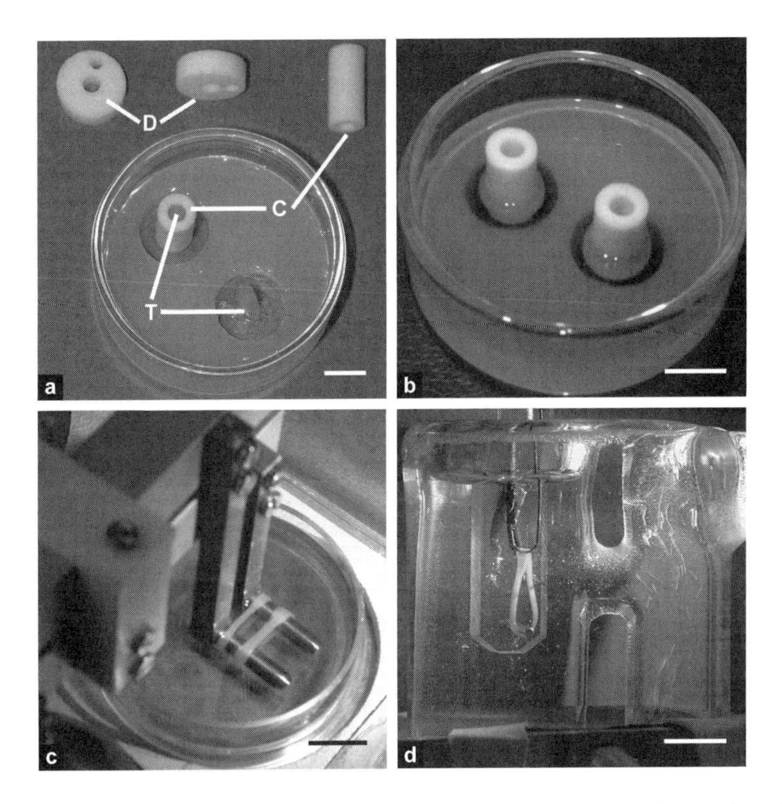

Figure 11.4. Experimental setup for EHT preparation, culture, phasic stretch and analysis of contractile function in the organ bath. a) Casting mold assembly: Silicone tubing (T) was glued to the surface of glass culture dishes. Either Teflon disks (D) or cylinders (C) can be placed over silicone tubing to function as removable spacers during casting mold preparation and EHT culture, respectively. b) EHTs condensed around the central Teflon cylinder in casting molds between culture days 1 and 4. Thereafter, no change of gross morphology was observed. c) EHTs after transfer in a stretch apparatus to continue culture under unidirectional and cyclic stretch (10%, 2 Hz). d) EHT in a thermostatted organ bath. Bars: 10 mm. From Zimmermann et al. [2002a]. (See Color Plate 6B).

(outer/inner diameter: 10/8 mm; thickness: 1 mm) (Fig. 11.4b,c, See Color Plate 6B). During the consolidation process, we transfer EHTs to a stretch device to continue culture under defined phasic strain (10%, 2 Hz) for 5–7 days (Fig. 11.4c, See Color Plate 6B). Under these conditions optimal maturation of EHTs has been observed. Longer periods of stretch often resulted in EHT disruption. Alternatively, EHTs can be cultured under static stretch on simple stainless steel holders, for up to 4 weeks.

3.1. Myocyte Preparation

Cardiomyocytes are isolated from 1- to 3-day-old neonatal rat hearts by a modification of the technique described by Webster et al. [1994].

Protocol 11.1. Isolation of Neonatal Cardiac Myocytes

Reagents and Materials

Sterile

- ❑ Complete medium (See Section 2.1)
- ❑ CBFHH (calcium- and bicarbonate-free HEPES-buffered Hanks' balanced salt solution, pH 7.5, See Section 2.2)
- ❑ Trypsin, 2 mg/ml (See Section 2.4)
- ❑ DNase stock, 2 mg/ml (See Section 2.5)
- ❑ DNase I, working concentration, 28 μg/ml (See Section 2.5)
- ❑ Petri dishes, 15 cm
- ❑ Plastic tube, 50 ml
- ❑ Dissection instruments: scissors, forceps, scalpels
- ❑ Gauze filter: 100-μm mesh

Non-sterile

- ❑ Newborn rat pups

Protocol

(a) Decapitate the pups quickly.
(b) Excise the hearts and remove atria and great vessels.
(c) Keep in 20 ml CBFHH in a 15-cm culture dish at room temperature until all hearts are prepared. For our routine cell isolations, we utilize 30–60 neonatal rat hearts.
(d) Cut the hearts in half.
(e) Wash 3 times with CBFHH.
(f) Mince extensively with scissors in about 5 ml CBFHH to a size of approximately 1 mm^3.
(g) Wash the tissue fragments again in CBFHH.
(h) Transfer to a 50-ml plastic tube for serial trypsin digestion.
(i) Allow tissue fragments to settle and discard the supernate.
(j) Add 10 ml trypsin solution (per total preparation of 30–60 hearts).
(k) Digest at room temperature (~20 °C) for 20 min with gentle agitation.

(l) Allow tissue fragments to settle.
(m) Discard the turbid supernate
(n) Add 9 ml DNase I, 28 μg/ml, to digest traces of DNA that impair sedimentation of tissue fragments and separation from isolated cells. DNase digestion is facilitated by triturating 25 times with a wide-bore pipette.
(o) Allow the tissue fragments to settle, and discard the supernate.
(p) Add 10 ml trypsin for the next cycle of digestion, this time for about 5 min.
(q) Collect the supernates of this trypsin digestion and of the following trituration with DNase I in 50-ml tubes containing 2 ml fetal calf serum (FCS) to inactivate the trypsin, and keep on ice.
(r) Repeat this digestion cycle until nearly complete digestion of the heart tissue is obtained. Over time, gradually reduce the volumes of trypsin and DNase I to 7.5 ml trypsin and 6.5 ml DNase I. After most tissue fragments are disaggregated, the remainder is discarded.
(s) Centrifuge the 50-ml collection tubes (generally 4–6 tubes) at 50 g for 15 min at 4 °C, and discard the supernate.
(t) Resuspend the cell pellet in 2 ml complete medium and pool all the pellets in one of the tubes.
(u) Rinse the other tubes with 2 ml medium each and pool.
(v) Add 250 μl DNase I stock solution per 30-ml cell suspension, triturate the suspension 25×, and centrifuge as above.
(w) Discard the supernate and resuspend the cell pellet in 32 ml medium.
(x) Filter the cell suspension through a 100-μm mesh filter into four 15-cm culture dishes.
(y) Incubate the dishes for 1 h in the CO_2 incubator. During this time, fibroblasts and other noncardiac myocytes adhere, whereas cardiac myocytes remain floating.
(z) Transfer nonadherent cells to a 50-ml tube, sediment by centrifugation, resuspend the pellet in fresh complete medium, and count cells.
(aa) Adjust the cell concentration to 6.4×10^6 cells/ml. Cell yield per animal is generally $2-3 \times 10^6$ viable cells/heart as judged by trypan blue exclusion.

3.2. Engineered Heart Tissue Reconstitution Mixture

A typical reconstitution mixture for 4 EHTs is shown in following Protocol and is prepared on ice in 10-ml tubes.

Protocol 11.2. Preparation of Reconstituted Rat Heart Tissue

Reagents and Materials

Sterile

- ❑ Collagen type I (rat tail)
- ❑ Concentrated DMEM, 2× (See Section 2.6)
- ❑ NaOH

- ❑ Basement membrane proteins: Matrigel® or Harbor ECM
- ❑ Freshly isolated cell suspension
- ❑ Plastic tubes, 10 ml

Non-sterile

- ❑ Ice bath

Protocol

Prepare the following typical reconstitution mixture for 4 EHTs on ice:

	Concentration	Volume	Total
Collagen type I (rat tail)	4.2 mg/ml	847 μl	3.56 mg
Concentrated DMEM	2×	847 μl	
NaOH	0.1 M	184 μl	
Basement membrane proteins (Matrigel®)	As purchased	400 μl	
Freshly isolated cell suspension	6.4×10^6/ml	1722 μl	11×10^6
Total volume		4000 μl	

The volumes are calculated for a total volume of 3.6 ml (900 μl per EHT) plus 10%. Volumes of collagen I and 2× concentrated culture medium must be adapted according to the actual collagen I concentration in the batch. The volume of NaOH is ignored in the calculation of the volume of concentrated medium.

The optimal reconstitution mix has been the result of extensive testing of the effects of cell number, both in the original lattice model [Zimmermann et al., 2000] and in the ring model [Eschenhagen et al., 2002]. As outlined above (See Section 1), an increase in cell number up to 2.5×10^6 cells/EHT in the lattice model improved contractile force, but a further increase sharply deteriorated it. In contrast, cell number could be increased up to 10×10^6 cells/EHT in the ring model without worsening. A cell density $>2.5 \times 10^6$/EHT did not improve function, but led to faster condensation of the cell-matrix mix. Collagen I was found optimal at 0.5–0.8 mg/EHT. A lower collagen content resulted in unstable and only weakly contracting EHTs, whereas a higher collagen content yielded EHTs that were rigid and developed less active force. A collagen content of 0.8 mg/EHT was chosen for mechanical stability, easier handling, and optimal contractile function.

Matrigel was essential for any tissue development in rat EHTs and improved force at concentrations between 5% and 15% (v/v). We chose 10% (v/v) in our standard EHT construction mixture as a compromise between function and costs. Chick embryo extract could be omitted on day 7 without significant effect on force development but was essential during the initial culture period [Zimmermann et al., 2003]. In contrast, even small reductions in horse serum concentration (to 3%) starting on day 7, 8, or 9 of culture prevented condensation of the cell-matrix mix and almost completely abolished contractile function when measured at day 14 [Eschenhagen et al., 2002].

3.3. Cultivation of Engineered Heart Tissue

Circular casting molds (diameter: 16 mm, depth: 5 mm) are prepared in glass culture dishes (diameter: 50 mm).

Protocol 11.3. Preparing Casts and Culturing Engineered Rat Heart Tissue

Reagents and Materials

Sterile

- ❑ Reconstitution mixture, ice-cold (See Protocol 11.2)
- ❑ Complete medium (See Section 2.1)
- ❑ Petri dishes, 3.5 and 5 cm

Non-sterile

- ❑ Silicone tubes (length: 10 mm, diameter: 2 mm)
- ❑ Teflon disks (diameter: 16 mm, depth: 5 mm, central hole diameter: 2 mm)
- ❑ Teflon cylinders (diameter: 8 mm, height: 10 mm)
- ❑ Liquid silicone glue
- ❑ Mechanical stretching device (See Fig. 11.4c, Color Plate 6B)

Protocol

(a) Glue two silicone tubes (length: 10 mm, diameter: 2 mm) to the surface of the 5-cm Petri dish to hold reversible Teflon disks (diameter: 16 mm, depth: 5 mm, central hole diameter: 2 mm) serving as spacers while liquid silicone glue is poured into the culture dish. Teflon spacers can be removed after hardening of the silicone glue to yield circular casting molds (See Fig. 11.4a, Color Plate 6B).

(b) Leave casting molds for at least 2 weeks to ensure that silicone has thoroughly hardened and solvents have evaporated.

(c) Wash molds and boil in distilled water before use. Subsequently, sterilize by autoclaving. Casting molds can be reused infinitely.

(d) Before EHT casting, place removable Teflon cylinders (diameter: 8 mm) over each silicone tubing to yield circular wells (750 mm^3). The size of a casting mold is sufficient to hold the viscous reconstitution mixture (900 μl) in place.

(e) Pipette the ice-cold reconstitution mixture into the mold at room temperature.

(f) Transfer the molds to the 37 °C incubator for 1 hour.

(g) Remove from the incubator, add 6 ml culture medium, and return to the incubator.

(h) Culture EHTs in the casting mold for 7 days at 37 °C, in a 5% CO_2-95% air, humidified atmosphere.

(i) Change the medium after the initial overnight culture and then every other day.

(j) After 7 days, transfer the ring-shaped EHTs to the mechanical stretching device (See Fig. 11.4c, Color Plate 6B) and subject to phasic horizontal stretch at 2 Hz and a lateral strain of 10% from the original EHT length (8- to 8.8-mm internal diameter) in 3.5-cm Petri dishes with 6 ml culture medium for 5–7 days.
(k) Change medium every day during stretching.

Protocol 11.4. Force Measurement in Engineered Heart Tissue

Reagents and Materials

Sterile

- ❑ Precast and cultured EHTs

Non-sterile

- ❑ Modified Tyrode's solution (See Section 2.3) with 0.4 mM Ca^{2+}
- ❑ Organ baths, thermostatically controlled at 37 °C
- ❑ Force transducer

Protocol

(a) Suspend EHTs individually at slack length in 30-ml organ baths at 37 °C fixed between a lower fixed holder and a stainless steel wire connected to the force transducer (See Fig. 11.4d, Color Plate 6B). Tyrode's solution (equilibrated to a pH of 7.4 by continuous bubbling (approximately 30 min) with 95% O_2 – 5% CO_2) with 0.4 mM calcium is used for the initial equilibration and adjustment of EHTs.
(b) After 15 min without pulsing, continue equilibration under electrical stimulation with rectangular pulses (2 Hz, 5 ms, 80–100 mA) until force development and contraction kinetics of EHT reach a steady state (approximately additional 15 min).
   i) Subsequently, preload is adjusted from approximately 0.05–0.1 mN (at slack length) to L_{max}, i.e., the length at which EHTs developed maximal active force.
   ii) Resting tension at L_{max} normally amounts to 0.2–0.3 mN.
   iii) Equilibration and preload adjustment takes about 60 min.
(c) After a change of the Tyrode's solution, measure inotropic (change of force of contraction) and lusitropic (change of contraction kinetics/times) responses to cumulative concentrations of calcium (0.2–2.8 mM) and isoprenaline (0.1–1000 nM) at 0.2 mM baseline calcium concentrations [Zimmermann et al., 2002a]. Twitch tension (TT; systolic force), resting tension (RT; diastolic force), contraction duration (T1: time from 10% to peak force development), and relaxation duration (T2: time from peak contraction to 90% relaxation) are evaluated by BMON software (Ingenieurbüro Jäckel, Hanau, Germany).

---

4. GENE TRANSFER IN ENGINEERED HEART TISSUE

The most efficient way to transfect cardiac myocytes (and other cell types in EHT) is adenoviral gene transfer. We generally purify our virus by cesium chloride density centrifugation and dialysis and store virus stock in 30% glycerol at −20 °C. Virus titration is performed in neonatal rat cardiac myocytes by serial dilutions. The transfection rate of 100% would correspond to 2×10^6 biologically active virus (bav) for 2×10^6 cells. This titer is used to calculate the multiplicity of infection (MOI). A MOI of 1 is, by our definition, sufficient to yield 100% transfection of cultured neonatal rat cardiac myocytes.

In EHT we use an MOI of 50 to get similarly high transfection rates (1.25×10^8 bav/EHT). The necessity to utilize higher virus titers than in 2D cultures of neonatal rat cardiac myocytes could result from the advanced maturation of cardiac myocytes in EHTs. Accordingly, high titers (MOI of 100–10,000) must be utilized to reach comparable transfection rates in adult cardiac myocytes [El-Armouche et al., 2003]. This phenomenon might be attributed to the downregulation of the common coxsackievirus and adenovirus receptor (CAR) during cardiac myocyte differentiation and aging [Communal et al., 2003; Fechner et al., 2003].

Generally, we perform adenovirus infections on EHT culture day 10 under serum-free conditions by directly pipetting adenovirus into the culture medium (5 ml) in which the EHT is (or 2 EHTs are) maintained (See Fig. 11.4c, Color Plate 6B). An aliquot of a virus dilution in serum-free DMEM is pipetted directly into the culture dish containing one or two stretched EHTs (See Fig. 11.4b, Color Plate 6B). After 1 hour, 5 ml of complete medium is added. Functional responses are generally evaluated 48 h after infection. If a virus is used that codes for green fluorescent protein, transfection efficiency can be monitored in the living EHT (Fig. 11.5, See Color Plate 6C) [El-Armouche et al., 2003].

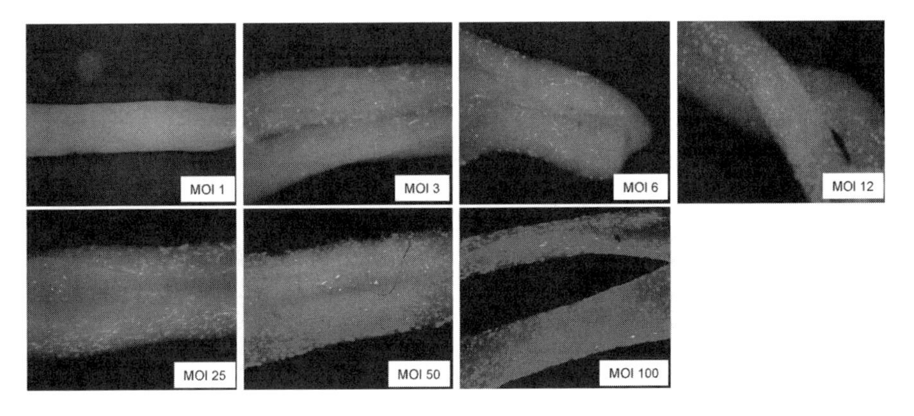

Figure 11.5. Adenoviral gene transfer in EHT. EHTs were infected with increasing virus concentrations/cell of AdPPI-1/GFP. Transfection of cells was visualized by fluorescence of the coexpressed GFP 48 h after infection. Note increasing numbers of cells showing GFP fluorescence throughout EHTs. From El-Armouche et al. [2003]. (See Color Plate 6C.)

5. HISTOLOGIC PROCEDURES IN ENGINEERED HEART TISSUE

Protocol 11.5. Conventional Histology of Engineered Heart Tissue

Reagents and Materials

Non-sterile

- ❑ Fixative (See Section 2.7)
- ❑ PBS
- ❑ Agar, 2%
- ❑ Isopropanol in graded concentrations
- ❑ Paraffin-isopropanol 1:1
- ❑ Melted paraffin wax
- ❑ Embedding oven
- ❑ Microtome
- ❑ Stains: hematoxylin and eosin (H&E), Masson Goldner trichrome, or Sirius red

Protocol

(a) For paraffin embedding fix (See Section 2.7) EHTs overnight at 4 °C.
(b) Wash and soak overnight in PBS.
(c) Embed EHTs in 2% agar blocks.
(d) Dehydrate in graded concentrations of isopropanol, paraffin-isopropanol, and embed in paraffin according to standard procedures.
(e) Stain horizontal and cross sections (4 μm) of EHTs with hematoxylin and eosin (H&E), Masson Goldner trichrome, or Sirius red to evaluate cell-matrix composition of EHTs.

For confocal laser scanning microscopy and immunofluoresence studies EHTs are fixed and embedded in agar as described above or used as whole mount samples after fixation. We found that, in contrast to cultured cells, EHTs tend to create more nonspecific background staining. This is most likely due to a high nonspecific binding of most antibodies to highly charged collagen. Moreover, thick sections or whole mounts that are preferentially used to get a 3D image of the EHT structure add the problem of antibody diffusion. The experimental procedures explained in Protocol 11.6 have been adapted accordingly.

Protocol 11.6. Confocal Immunofluorescence of Engineered Heart Tissue

Reagents and Materials

Non-sterile

- ❑ Supplemented Tris-buffered saline (STBS) (See Section 2.8)
- ❑ Primary and secondary antibodies (See Table 11.1)
- ❑ Vibratome slicer
- ❑ Confocal microscope

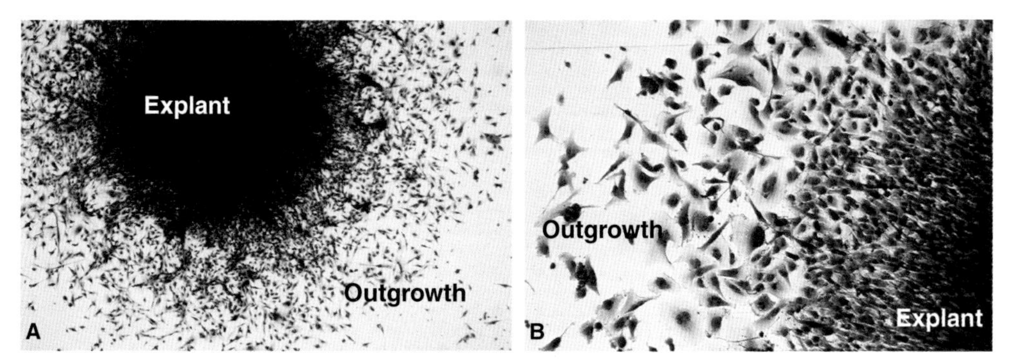

Plate 1. Primary explant and outgrowth. A, 4× objective, B, 10× objective. (See Fig. 1.2 for details.)

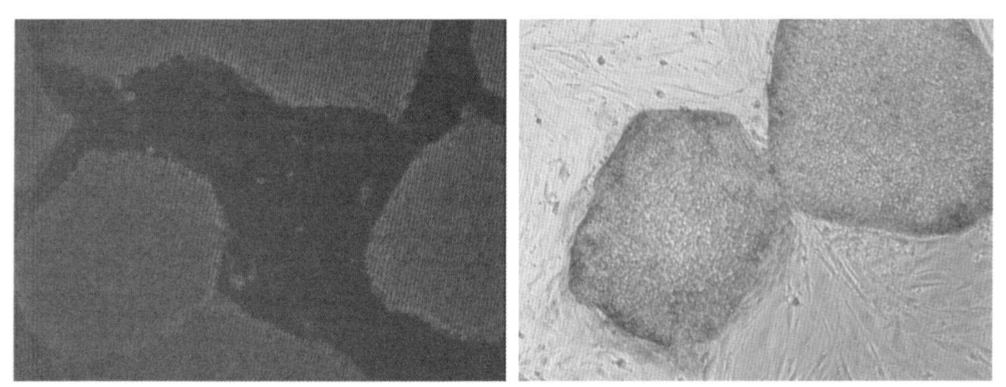

Plate 2. hES cell colonies on mouse embryonic fibroblasts. SSEA-4, red, left and ALP, blue, right. (See Fig. 3.1 for details.)

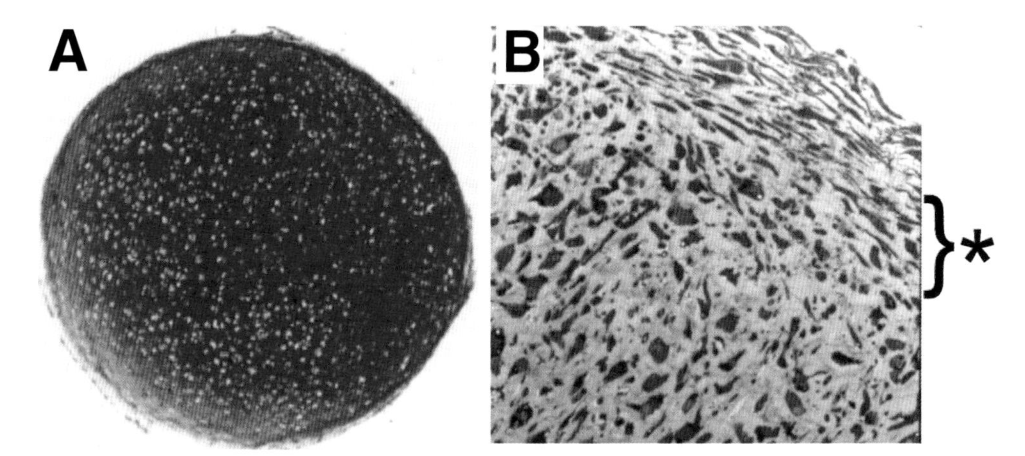

Plate 3. The Toluidine Blue metachromatic matrix of cartilaginous aggregates of human marrow-derived cells after 14 days in chondrogenic medium. A, section of paraffin-embedded whole aggregate; B, higher magnification of edge of a methyl methacrylate-embedded section with the region of flattened cells indicated by asterisk. (See Fig. 4.5 for details.)

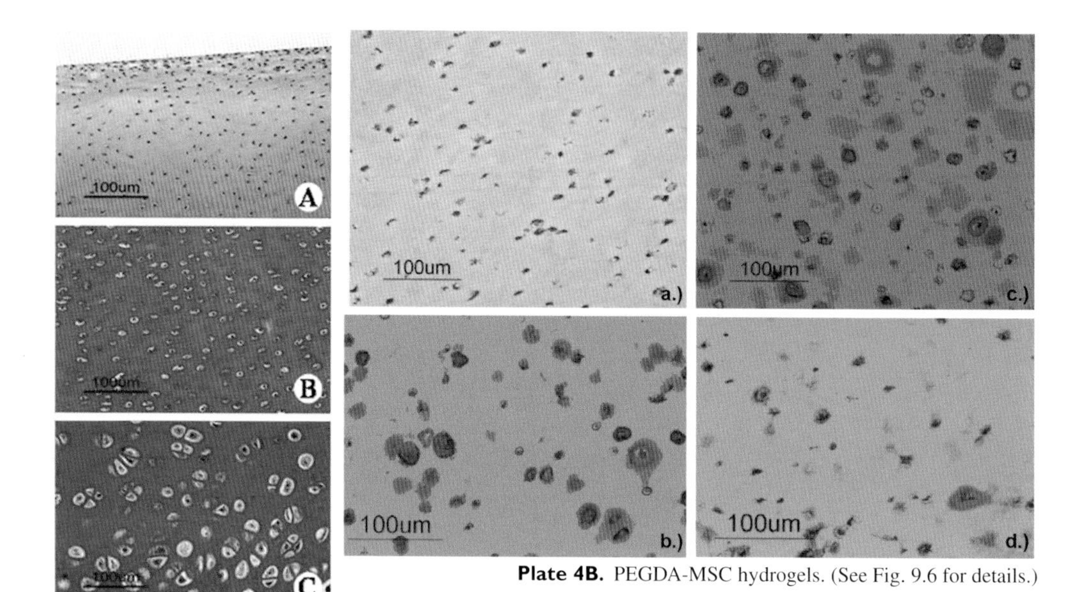

Plate 4B. PEGDA-MSC hydrogels. (See Fig. 9.6 for details.)

Plate 4A. Juvenile bovine cartilage. (See Fig. 9.2 for details.)

Plate 4C. Multilayered PEGDA hydrogel. (See Fig. 9.5 for details.)

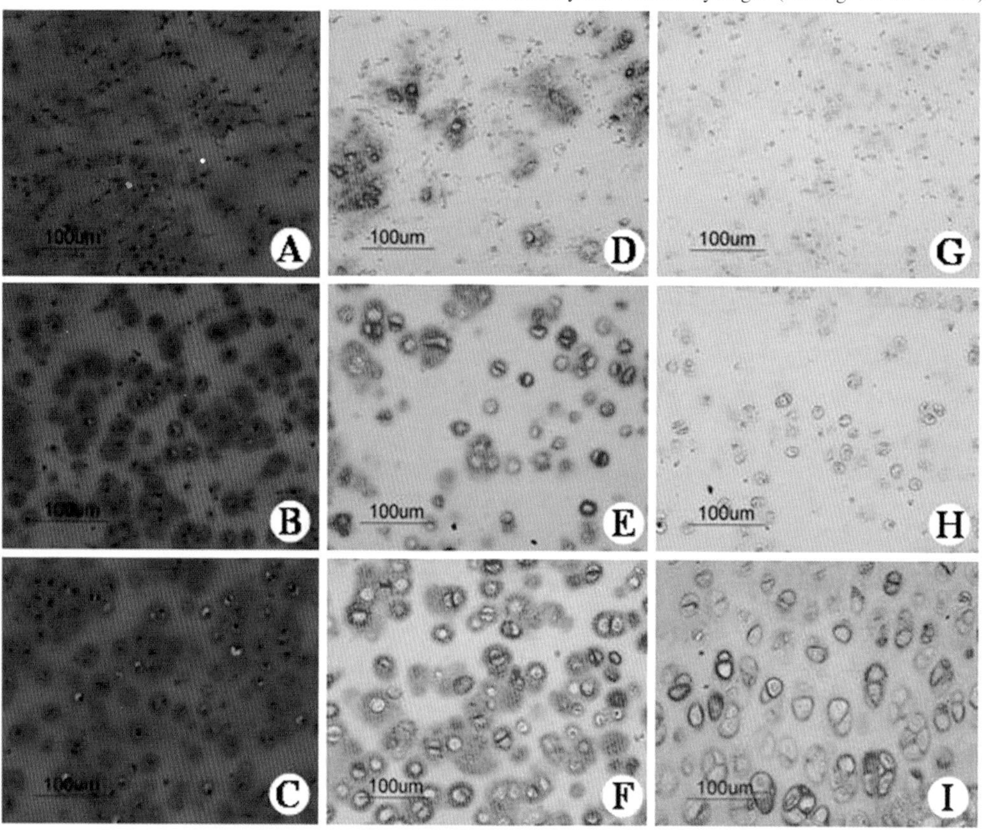

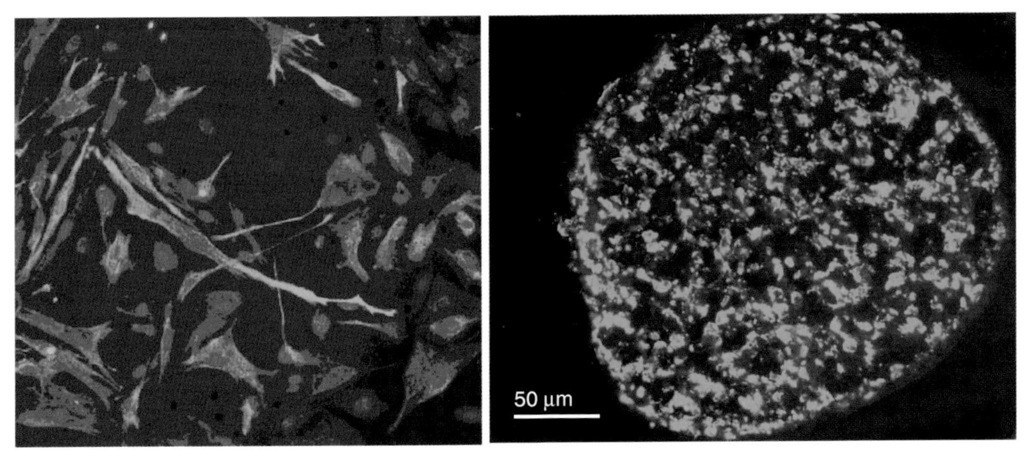

Plate 5A. Human skeletal muscle cells. (See Fig. 10.2.) **Plate 5B.** Cross section of 10-day in vitro HBAM. (See Fig. 10.5.)

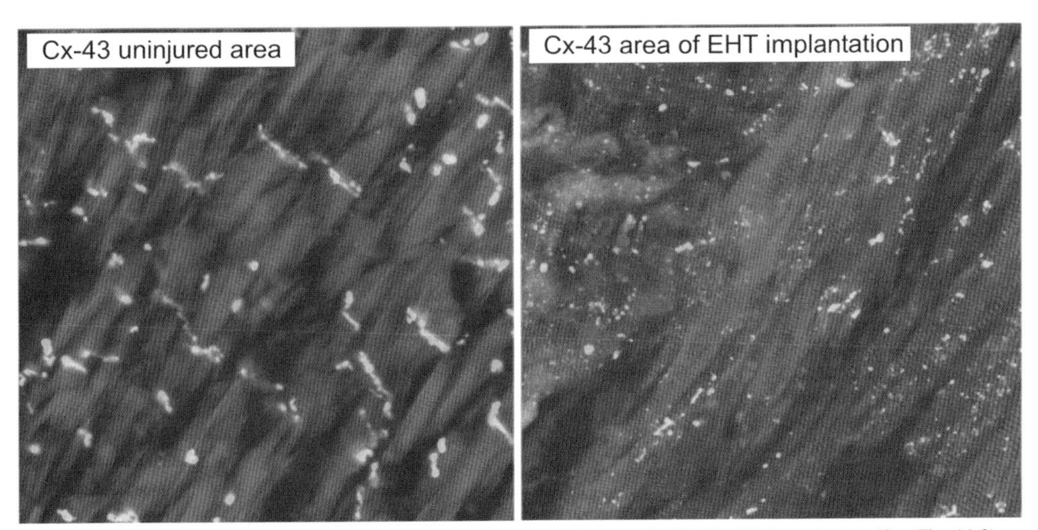

Plate 6A. Effect of EHT implantation of the spatial organization of connexin 43 (Cx-43) in rat hearts. (See Fig. 11.3)

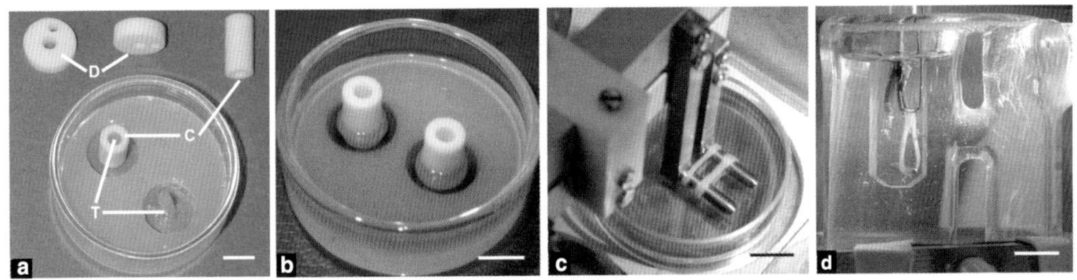

Plate 6B. Experimental setup for EHT preparation, culture, phasic stretch and analysis of contractile function in the organ bath. (See Fig. 11.4 for details.)

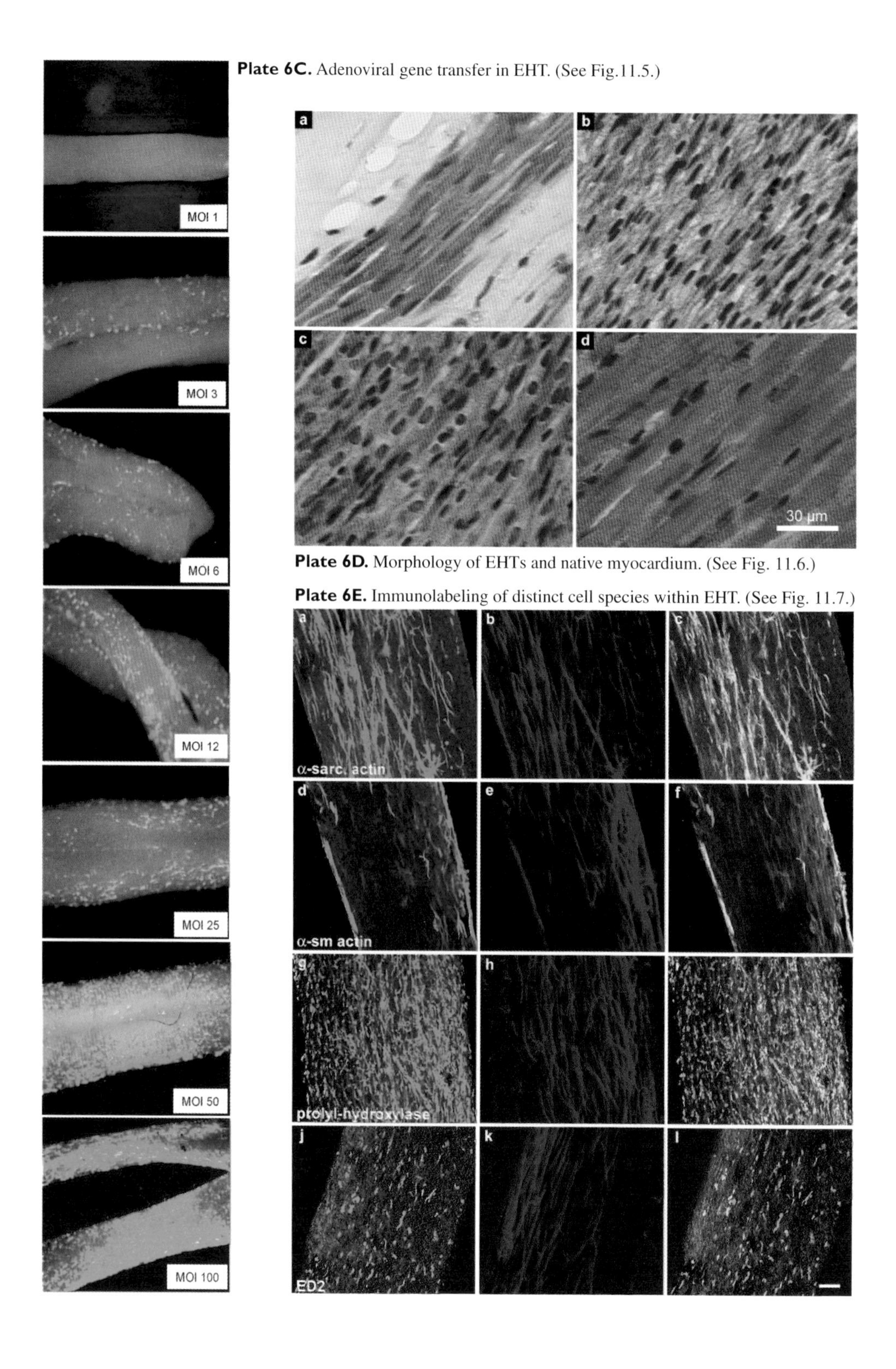

Plate 6C. Adenoviral gene transfer in EHT. (See Fig.11.5.)

Plate 6D. Morphology of EHTs and native myocardium. (See Fig. 11.6.)

Plate 6E. Immunolabeling of distinct cell species within EHT. (See Fig. 11.7.)

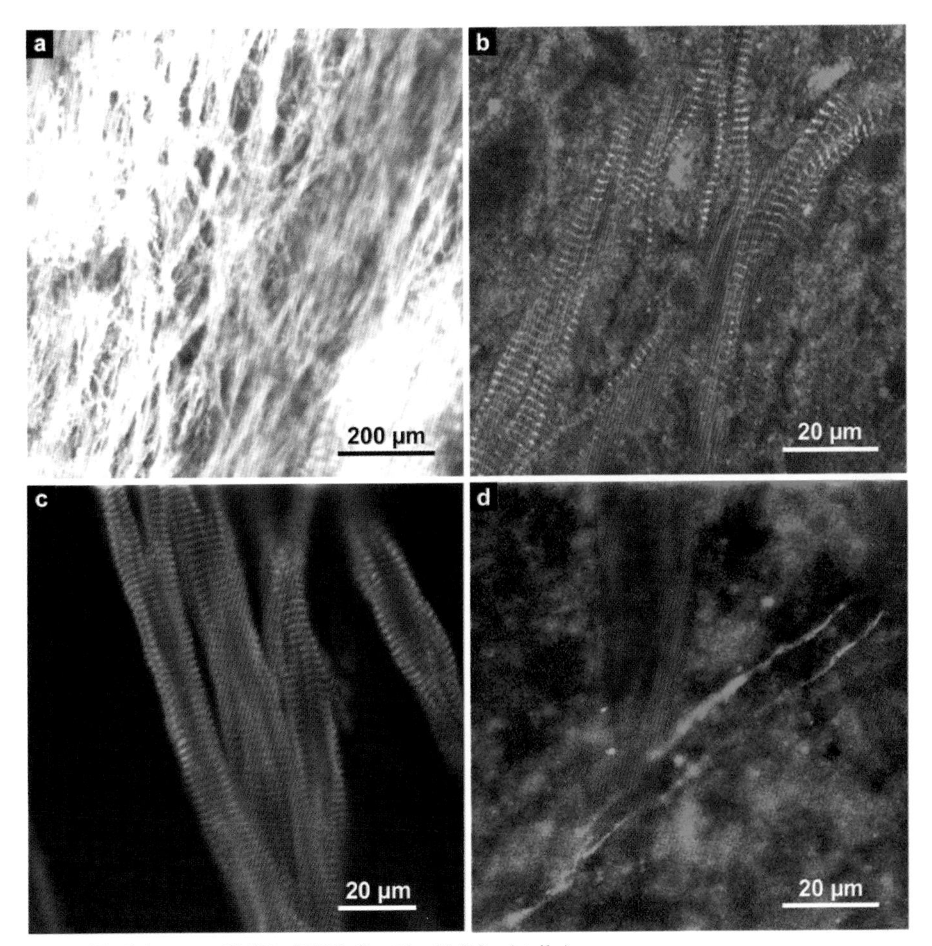

Plate 6F. High-power CLSM of EHT. (See Fig. 11.8 for details.)

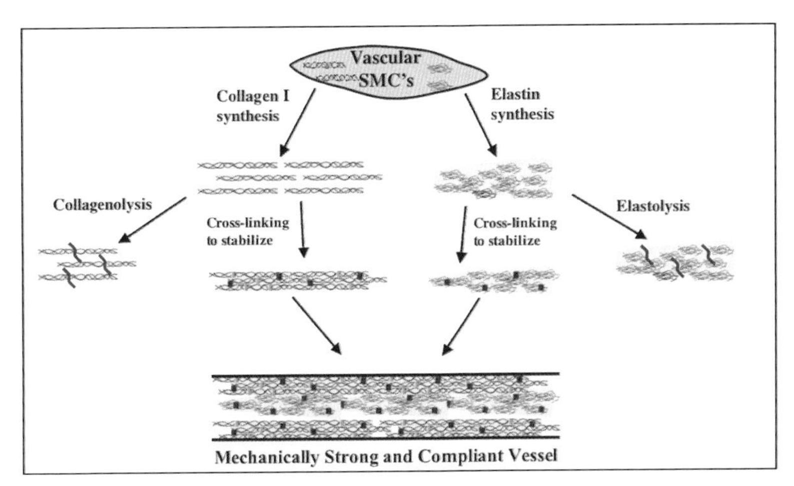

Plate 7A. Secretion of collagen and elastin by smooth muscle cells. (See Figure 12.3.)

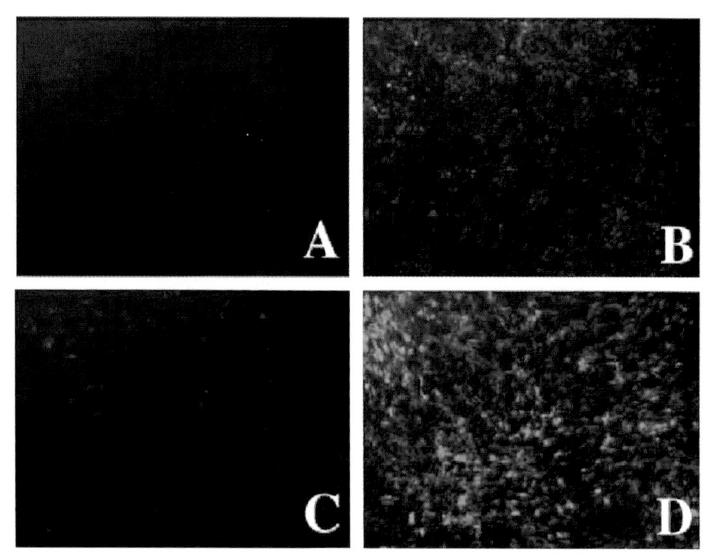

Plate 7B. Enhanced green fluorescent protein (EGFP) expression in cultured ECs. (See Fig. 12.8.)

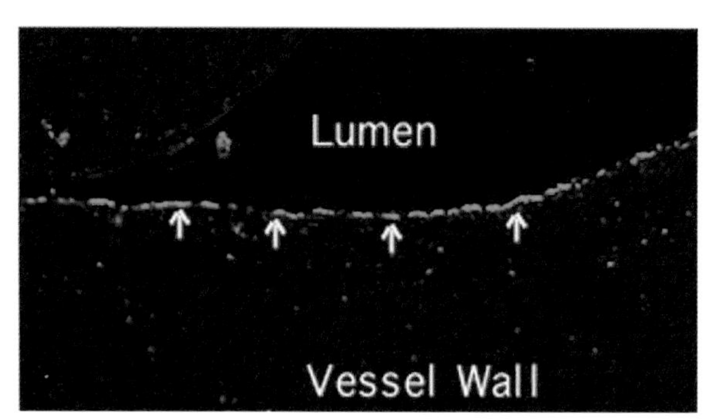

Plate 7C. EGFP expressed on engineered vessel lumen. (See Fig.12.9.)

Plate 8. Characterization of MSCs. C) Chondrocyte differentiation. D) Osteoblast differentiation. (See Fig.13.2.)

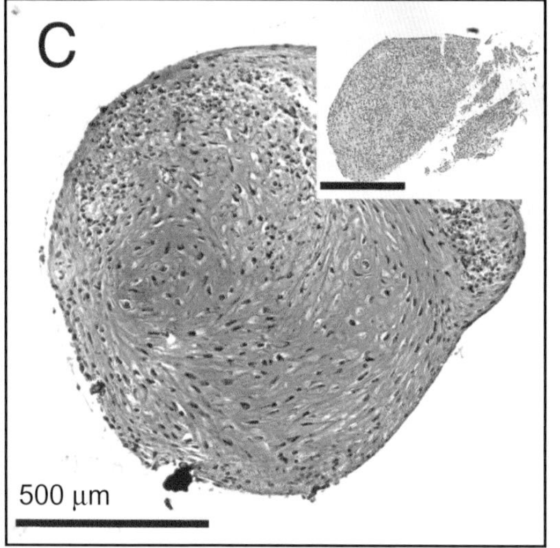

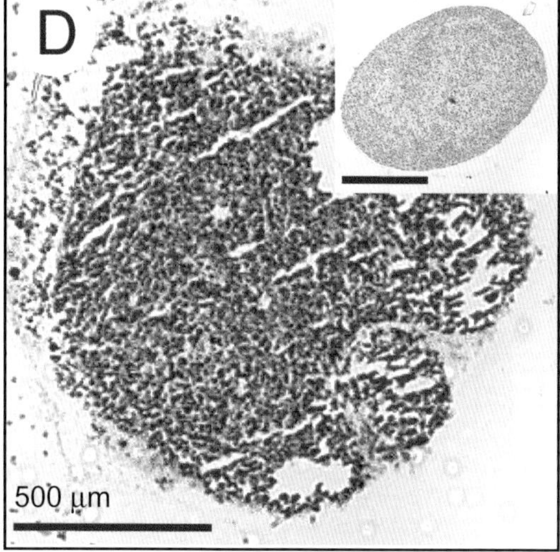

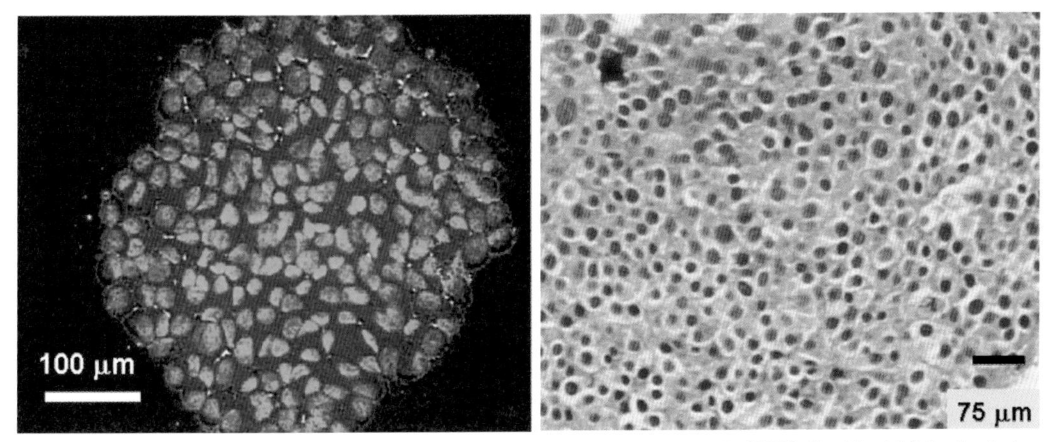

Plate 9A. 3-D assemblies of PC12. Left, static aggregate, right, dynamic aggregate in SLTV. (See Fig. 14.3 for details.)

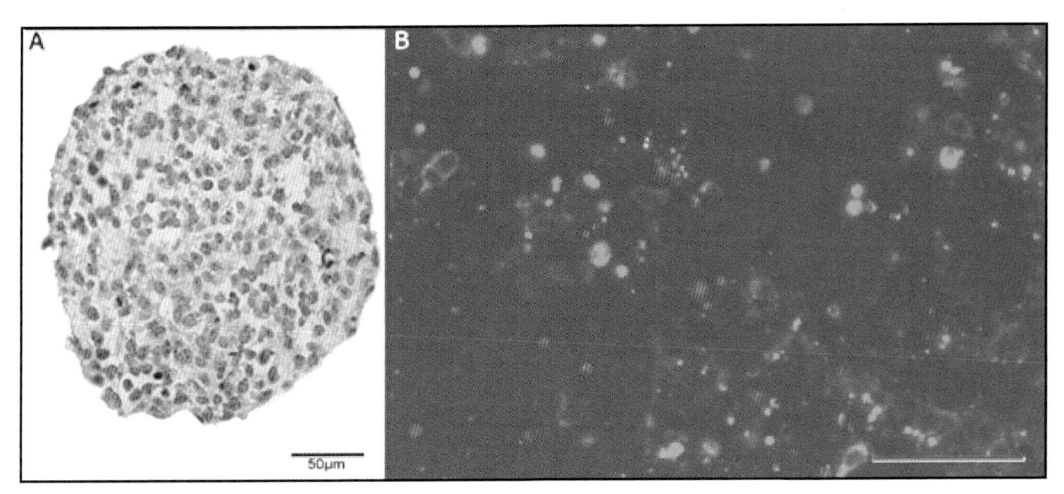

Plate 9B. Morphology and TH content of SNAC tissue constructs. A, SNAC section immumostained for human nuclei, (NT2 cells). B, double immunofluorescence; Sertoli cells, green, TH-positive NT2N neurons red. (See Fig. 14.5 for details.)

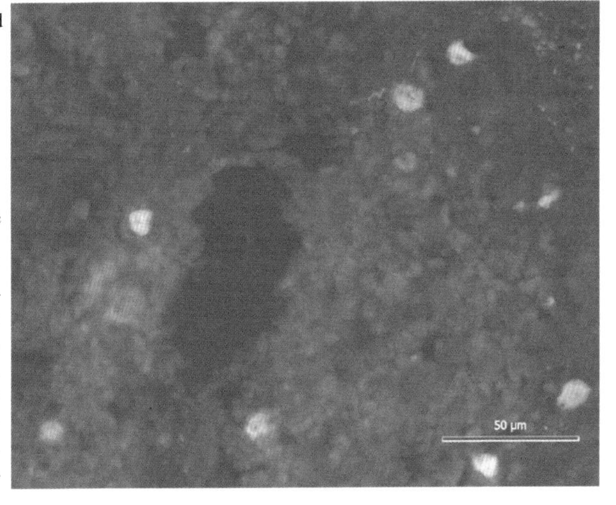

Plate 9C. Photomicrograph through a SNAC tissue construct transplant into the rat striatum 4 weeks postsurgery. Surviving TH-positive NT2N neurons (red) double immunostained with anti-human nuclei antibody (green) can be seen along the course of the penetration. These NT2N neurons contain a green nucleus and lighter green cytoplasm, which now appears yellow because of the double label. Some neurite outgrowth is seen in the TH-positive NT2N neuron near the top right of the photomicrograph.

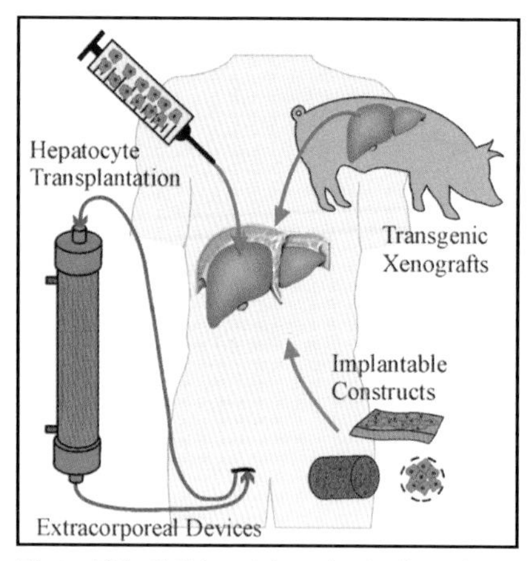

Plate 10A. Cell-based therapies for liver disease. (See Fig. 15.1 for details.)

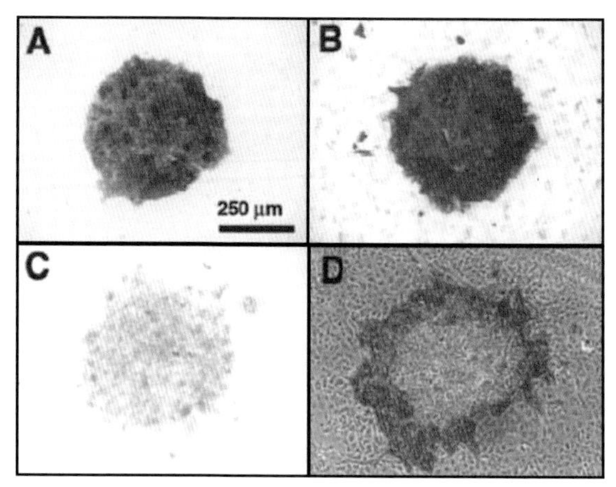

Plate 10B. Intracellular albumin in micropatterned hepatocytes. (See Fig.15.5 for details.)

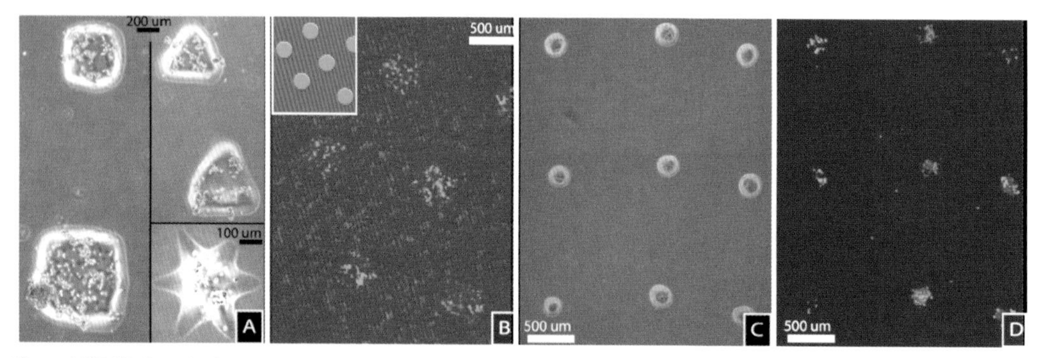

Plate 10C. Hydrogel microstructures containing living cells. (See Fig. 15.10 for details.)

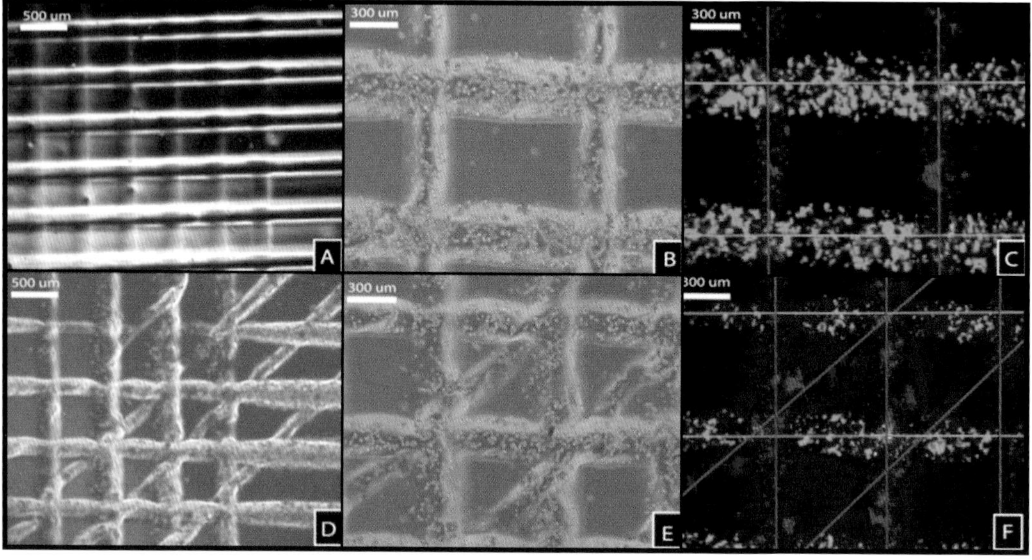

Plate 10D. Multilayer hydrogel microstructures containing living cells. (See Fig. 15.11 for details).

Protocol

(a) Prepare sections (100 μm) of agar-embedded EHTs with a vibratome slicer.
(b) Block nonspecific binding in vibratome and whole mount samples and permeabilize by keeping at 4 °C overnight in STBS.
(c) After two washes in STBS incubate sections/whole mounts with the primary antibodies at 4 °C for 48 h. Table 11.1 gives antibodies, dilutions and sources for α-sarcomeric actinin, α-sarcomeric actin, myomesin, α-smooth muscle actin, prolyl-4-hydroxylase, CD31, and the macrophage marker ED2.
(d) After an overnight wash in STBS incubate sections/whole mounts with secondary antibodies at room temperature for 3 h.
(e) Perform confocal imaging, e.g., with a Zeiss LSM 5 Pascal system using a Zeiss Axiovert microscope.

Protocol 11.7. Transmission Electron Microscopy of Engineered Heart Tissue

Reagents and Materials

Non-sterile

- ❑ Glutaraldehyde, 2.5% (See Section 2.9)
- ❑ Osmium tetroxide-PBS, 1:1
- ❑ PBS
- ❑ Graded ethanol concentrations
- ❑ Acetone
- ❑ Uranyl acetate
- ❑ Lead citrate
- ❑ Epon
- ❑ Ultramicrotome
- ❑ Electron microscope

Protocol

(a) For transmission electron microscopy, fix EHTs in 2.5% glutaraldehyde (See Section 2.9) overnight at 4 °C.
(b) After an overnight wash in PBS, postfix EHTs in osmium tetroxide-PBS, 1:1, for 2 h at room temperature.
(c) After an overnight wash in PBS at 4 °C, dehydrate samples in graded ethanol and acetone, infiltrate with epon/acetone, and embed in epon according to standard protocols.
(d) Stain semithin sections (1 μm) with toluidine blue.
(e) For TEM, cut ultrathin sections (50 nm) and contrast with uranyl acetate and lead citrate.
(f) Examine sections with, e.g., a Zeiss Leo 906 EM system.

---

6. REPRESENTATIVE TISSUE CULTURE STUDY

A recent study characterized EHTs from neonatal rat cardiac myocytes [Zimmermann et al., 2002a]. The main results of this study are presented here. Methods have been presented in detail in Section 5.

6.1. Casting and Culture of Circular EHTs

Circular EHTs can be cast easily in large series. On average we reconstituted 30 EHTs from 30 neonatal rat hearts per week. During culture the EHTs condensed around the removable central Teflon cylinder within the casting molds (See Fig. 11.4b, Color Plate 6B). First contractions of single cells were noted after 24 h, synchronous contractions of cell clusters started at day 2. The size of beating clusters increased until the entire EHT beat synchronously (~1 Hz, day 4–5). Over time, contraction became more regular, more vigorous, and faster (~2 Hz). Physical stability to allow manual handling and mechanical stretch without inflicting damage to the EHT structure was reached after 6–7 days in culture. Vigorous spontaneous contractions of EHTs were noted when the stretch device was turned off and after transfer of EHTs into a culture dish [video sequence at http://circres.ahajournals.org/cgi/content/full/90/2/223/DC1]. At this stage EHTs weighed 29.2 ± 1.4 mg ($n = 12$) and had a diameter of 833 ± 17 μm ($n = 16$).

6.2. Histology

In planar EHT lattices, called cardiomyocyte-populated matrices (CMPMs; [Eschenhagen et al., 1997]), cardiac myocytes were mainly concentrated at the lateral free edges. In contrast, serial sections of paraffin-embedded circular EHTs ($n = 7$) did not reveal a spatial preference of cell distribution. Complexes of multicellular aggregates and longitudinally oriented cell bundles mainly consisting of cardiac myocytes (Fig. 11.6a, See Color Plate 6D) were found throughout circular EHTs. The width of these muscle bundles ranged from 30 to 100 μm. For comparison, paraffin sections of native heart tissue from newborn, neonatal, and adult rats (300 g) were investigated (Fig. 11.6b–d, See Color Plate 6D). In the adult myocardium, compared to the immature tissues, myocytes were larger in width and length, were more intensely stained with eosin, and exhibited clear cross-striation, indicating a higher content of myofilaments. Density of myocyte and nonmyocyte nuclei was approximately threefold lower in the adult tissue, and myocyte nuclei were elongated (length to width 5:1) in contrast to round or oval nuclei in the immature tissue. Surprisingly, histologic features of myocytes forming EHTs resembled those of myocytes within native differentiated myocardium. The intensity of eosin staining was higher than in the immature tissues, cross-striation was visible, albeit to a lesser degree than in the adult tissue, and nuclei had a length-width ratio of 5–6. Other differences from the adult tissue included smaller size of cardiac myocytes and their nuclei and a less compact overall structure.

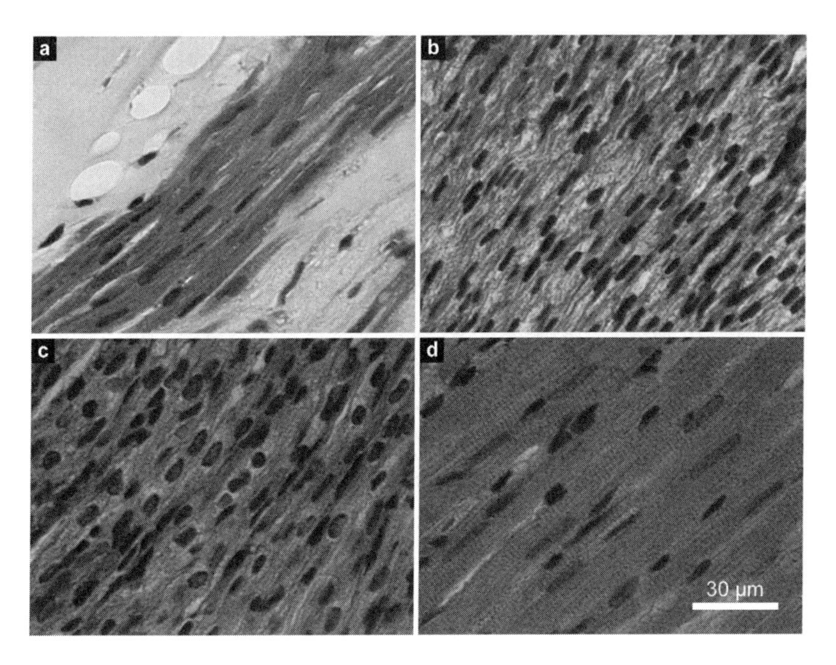

Figure 11.6. Morphology of EHTs and native myocardium. H&E staining of paraffin sections of EHT (a) and native myocardium from newborn (0 dpp; b), 6-day-old (c), and adult (300 g; d) rats. All sections were prepared, fixed, and stained in parallel to allow direct comparison of eosin staining. Note that cardiac muscle bundles in EHTs, even though derived from neonatal rat hearts (0–3 dpp), resemble cardiac morphology in the adult heart more closely than in neonatal myocardium. From Zimmermann et al. [2002a] (See Color Plate 6D).

6.3. Immunoconfocal Characterization of EHTs

To analyze the overall composition and spatial distribution of cell species within EHTs, vibratome sections were immunolabeled to identify cardiac myocytes (α-sarcomeric actin), smooth muscle cells (α-smooth muscle actin), fibroblasts (prolyl-4-hydroxylase), and macrophages (ED2 antigen). Cardiac myocytes (Fig. 11.7a, See Color Plate 6E) constituted the majority of the phalloidin-tetramethylrhodamine isothiocyanate (TRITC)-positive cellular network (Fig. 11.7b,e,h,k, See Color Plate 6E). Smooth muscle cells, positive for α-smooth muscle actin, lined the outer surface of EHTs (Fig. 11.7d, See Color Plate 6E). Some α-smooth muscle actin-positive cells within EHTs may represent smooth muscle cells or immature cardiac myocytes. Fibroblasts and macrophages were scattered throughout EHTs (Fig. 11.7g–l, See Color Plate 6E).

Whole-mount preparations of EHTs were stained with phalloidin-TRITC and examined by confocal laser scanning microscopy (CLSM) (Fig. 11.8, See Color Plate 6F). This technique revealed cell strands forming a network of intensively interconnected cell bundles throughout the entire EHT (Fig. 11.8a, See Color Plate 6F), which condensed to solid muscle bundles at variable positions inside the EHT as depicted in Fig. 11.6a and Color Plate 6D. High-power CLSM demonstrated that the majority of cell bundles were composed of cardiac myocytes with a high

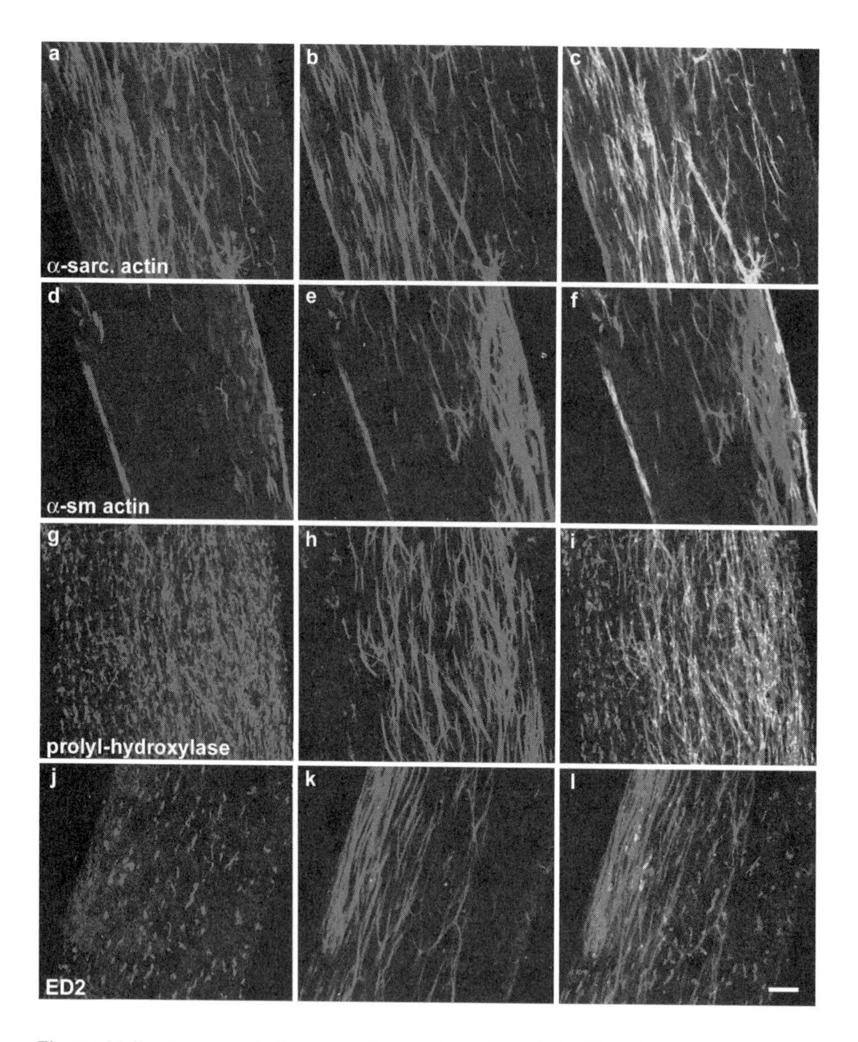

Figure 11.7. Immunolabeling of distinct cell species within EHT. Cellular composition and localization was investigated by immunolabeling (green) of cardiac myocytes (α-sarcomeric actin; a), smooth muscle cells (α-smooth muscle actin; d), fibroblasts (prolyl-4-hydroxylase; g), and macrophages (ED2-antigen; j). Actin filaments were labeled with phalloidin-TRITC (red; b,e,h,k). Superposition of images (c,f,i,l) yields a yellow signal where double fluorescent labeling was achieved. Bar: 100 μm. From Zimmermann et al. [2002a]. (See Color Plate 6E.)

degree of sarcomeric organization (Fig. 11.8b,c, See Color Plate 6F). At high magnification, capillary structures positive for CD31 (platelet endothelial cell adhesion molecule, PECAM) were noted (Fig. 11.8d, See Color Plate 6F).

6.4. Ultrastructural Characterization of EHTs

Ultrastructural hallmarks of cardiac myocyte differentiation are M-band formation, development of T-tubules with dyads/triads, specialized cell-cell junctions, and the reestablishment of an extracellular basement membrane [Anversa et al., 1981; Bishop et al., 1990; Katz, 1992]. Most, but not all, of these features were

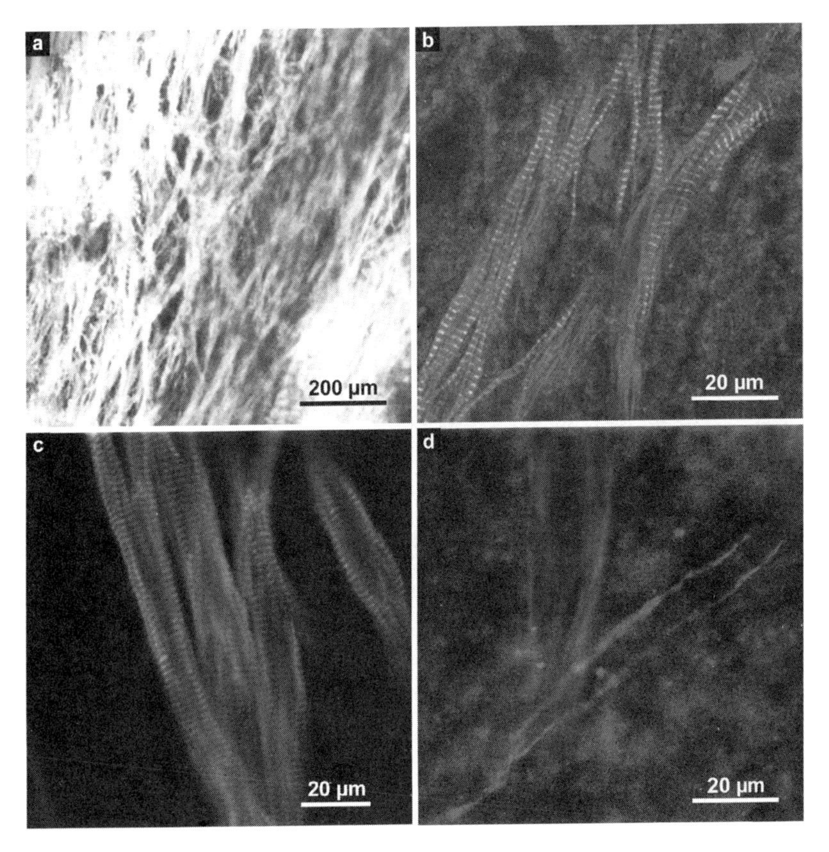

Figure 11.8. High-power CLSM of EHT. a) Extended focus imaging of a whole mount EHT after staining with phalloidin-TRITC. EHTs consist of a dense network of mainly longitudinally oriented cell bundles. b,c) Confocal images (optical slice: 1–2 μm) of vibratome sections (100 μm) of EHTs. Actin (red), myomesin (green; b), actinin (green; c). Note that cell bundles mainly consist of longitudinally oriented cardiac myocytes with sarcomeres in registry. d) Detection of CD31-positive capillary structures in vibratome sections (actin: red; CD31: green). From Zimmermann et al. [2002a] (See Color Plate 6F).

present in the majority of cells (Figs. 11.9, 11.10). Cardiac myocytes within EHTs displayed a predominant orientation of sarcomeres in registry along the longitudinal cell axis (Fig. 11.9a). Cross sections of EHT revealed that most cardiac myocytes were densely packed with myofibrils and mitochondria (Fig. 11.9b).

Morphometric evaluation of 20 longitudinally oriented, mononucleated cardiac myocytes from four EHTs revealed volume fractions as follows: myofibrils (44.7 ± 1.9%), mitochondria (23.9 ± 1.2%), nuclei (8.9 ± 0.9%). The rest (22.5 ± 1.8%) was occupied by sarcoplasmic reticulum (SR), cytoplasm, and undefined structures. Sarcomeres were composed of Z-, I-, A-, and H-bands in most investigated cells. Immature M-bands were noted frequently, but not in all sarcomeres. If present (Fig. 11.10a), they were clearly less developed than in adult myocytes, indicating that cardiac myocytes in EHTs exhibit a high, but not terminal, degree of differentiation. T-tubules were observed at the Z-band level (Fig. 11.10b-d) and often

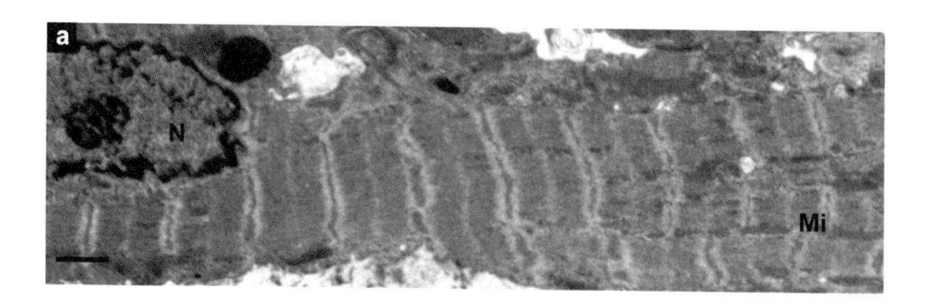

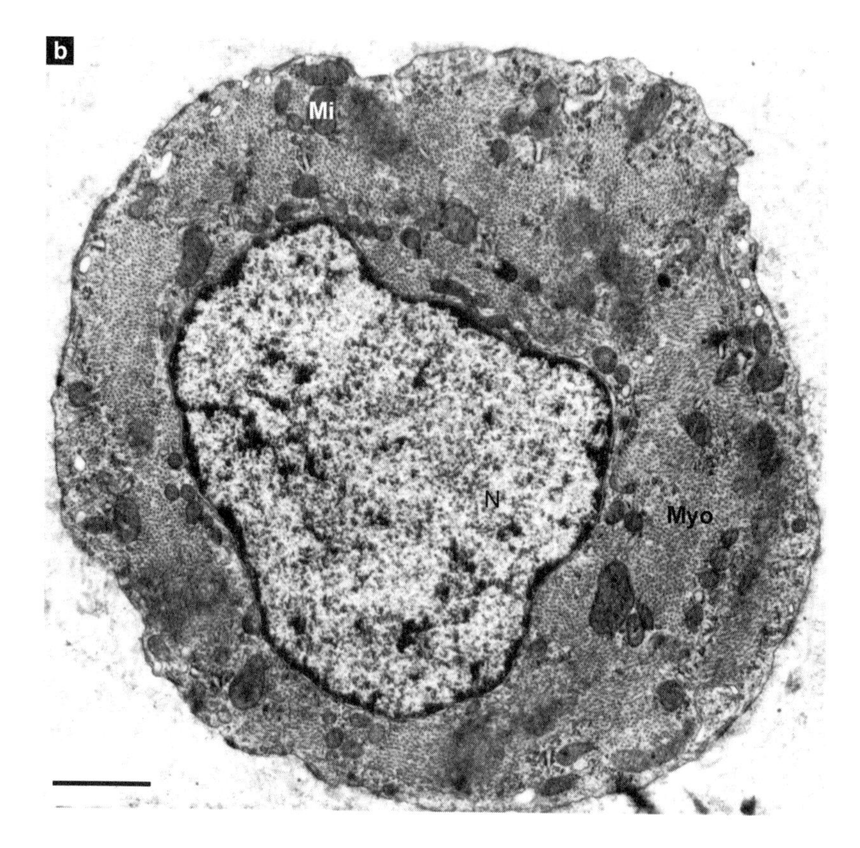

Figure 11.9. TEM of cardiac myocytes within EHT. a) Cardiac myocytes are filled with well-aligned sarcomeres. Note orientation of sarcomeres in registry along the longitudinal axis of the cell. Mitochondria (Mi) are mainly located in the intermyofibrillar space or the perinuclear region (not shown). b) A cross section demonstrates the density of myofibrils (Myo) intermingled with mitochondria in cardiac myocytes reconstituted to EHT. Note the centrally located nucleus (N). Bars: 1 μm. From Zimmermann et al. [2002a].

formed dyads with the SR (Fig. 11.10c, d). Specialized cell-cell junctions responsible for mechanical and electrical coupling of cardiac myocytes (adherens junctions, desmosomes, gap junctions) were found throughout EHTs (Fig. 11.10d, e). Cardiac myocytes often formed a well-developed basement membrane as an additional indication of cardiac myocyte integrity (Fig. 11.10f). Atrial secretory granules characteristic for atrial or undifferentiated ventricular myocytes were absent.

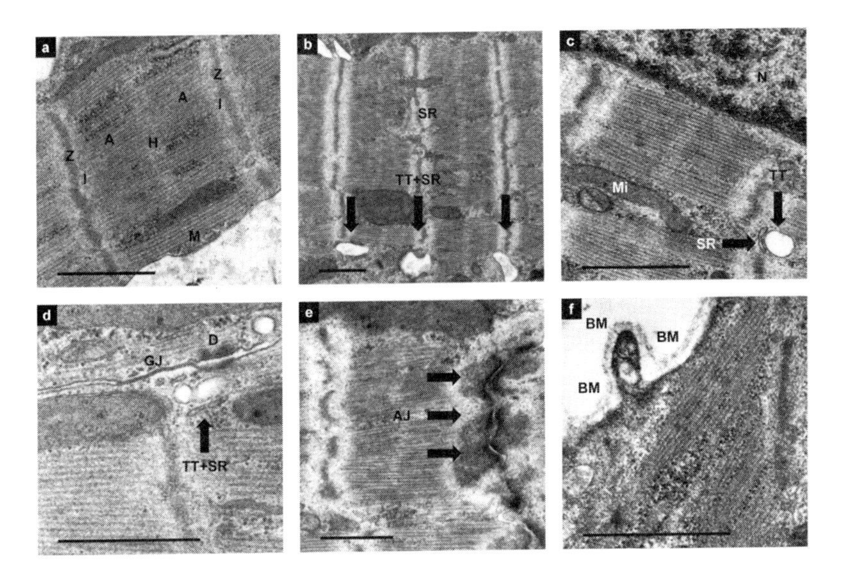

Figure 11.10. TEM of sarcomeric structures, cell-cell junctions, and basal membrane of cardiac myocytes in EHT. a) Formation of (immature) M-bands was noted in some but not all cardiac myocytes, while Z-, I-, A-, and H-bands were clearly distinguishable in most cardiac myocytes in EHT. b–d) T-tubules (TT) of various diameters. Dyad formation with sarcoplasmic reticulum (SR) was found frequently, especially at the Z-band level. d,e) Gap junctions (GJ), desmosomes (D), and adherens junctions (AJ) interconnect 3D reconstituted cardiac myocytes. f) Well-developed basement membrane (BM) around reconstituted cardiac myocytes. Bars: 1 μm. From Zimmermann et al. [2002a].

TEM provided additional evidence that EHTs are reconstituted of various cell species apart from cardiac myocytes resembling an organoid cardiac tissue construct (Fig. 11.11). These cells did not populate EHTs in a random fashion, but formed distinct structures. The outer surface of EHTs was lined with multiple cell layers consisting mainly of nonmyocytes (fibroblasts, smooth muscle cells, endothelial cells, macrophages; Fig. 11.11a). Fibroblasts, sometimes clearly demonstrating secretory activity, were found throughout EHTs (Fig. 11.11b, also Fig. 11.7g–i, See Color Plate 6E). Endothelial cells formed characteristic capillary structures that corresponded to CD31-positive cells observed by CLSM (Fig. 11.11c and Fig. 11.7d, See Color Plate 6E). Cell debris was frequently sequestrated by macrophages (Fig. 11.11d).

6.5. Contractile Properties of Circular EHTs

Contractile force and twitch kinetics of electrically stimulated EHTs were investigated under isometric conditions. At L_{max}, twitch tension (TT) amounted to 0.36 ± 0.06 mN at a resting tension (RT) of 0.27 ± 0.03 mN. Contraction and relaxation time (T1 and T2) were 83 ± 2 ms and 154 ± 9 ms, respectively. An increase in extracellular calcium enhanced TT from 0.34 ± 0.06 to 0.75 ± 0.11 mN, with a maximal inotropic response at 1.6 mM (Fig. 11.12a), RT and twitch kinetics remained unchanged. β-Adrenergic stimulation induced a maximal increase of TT

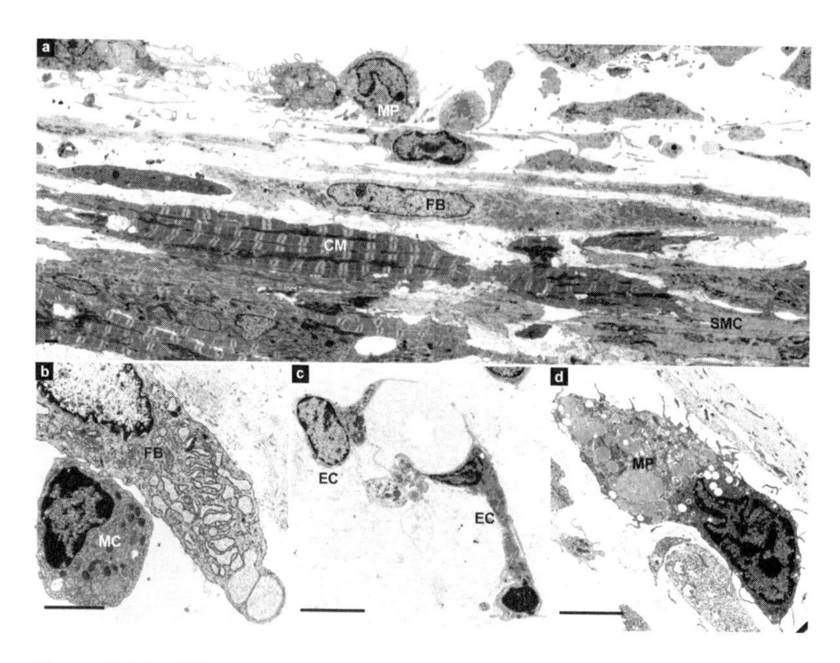

Figure 11.11. TEM of various cell types found to populate EHT. a) Macrophages (MP), fibroblasts (FB), and other cells with a low degree of differentiation lined the outer surface of EHTs. Underneath, a cell layer mainly consisting of cardiac myocytes (CM) and smooth muscle cells (SMC) can be observed. b) Fibroblasts (FB) with and without signs of secretory activities were found throughout EHT. c) Endothelial cells (EC) were noted to form capillary structures within EHT. d) Cells of leukocytotic origin like macrophages and other mononucleated cells (MC; b) were found mainly at the EHT surface but also within EHT. Bars: 1 μm. From Zimmermann et al. [2002a].

from 0.28 ± 0.06 to 0.69 ± 0.09 mN at 1 μmol/l isoprenaline (Fig. 11.12b). Additionally, isoprenaline shortened T1 from 86 ± 4 to 56 ± 2 ms and T2 from 144 ± 8 to 83 ± 3 ms and reduced RT from 0.15 ± 0.02 to 0.05 ± 0.02 mN (Fig. 11.12c,d). The decrease in RT may be mediated by smooth muscle cells that line the surface of EHTs (Fig. 11.12d–f).

6.6. Action Potentials

After equilibration in Tyrode's solution, EHT preparations generated only very infrequent spontaneous action potentials. Electrical stimulation at 1 Hz elicited regular action potentials with fast upstroke velocity (dV/dt_{max} : 66 ± 8 V/s), amplitude of 109 ± 2 mV, and a prominent plateau phase with action potential duration at 20%, 50%, and 90% repolarization of 52 ± 2, 87 ± 4, and 148 ± 3 ms, respectively (Fig. 11.13). In all six experiments resting potential (−73 ± 2 mV) was stable during electrical diastole.

7. DISCUSSION

This chapter describes a method to engineer a 3-dimensional cardiac tissue-like construct in vitro (EHT). Compared to planar lattices, ring-shaped EHTs exhibit

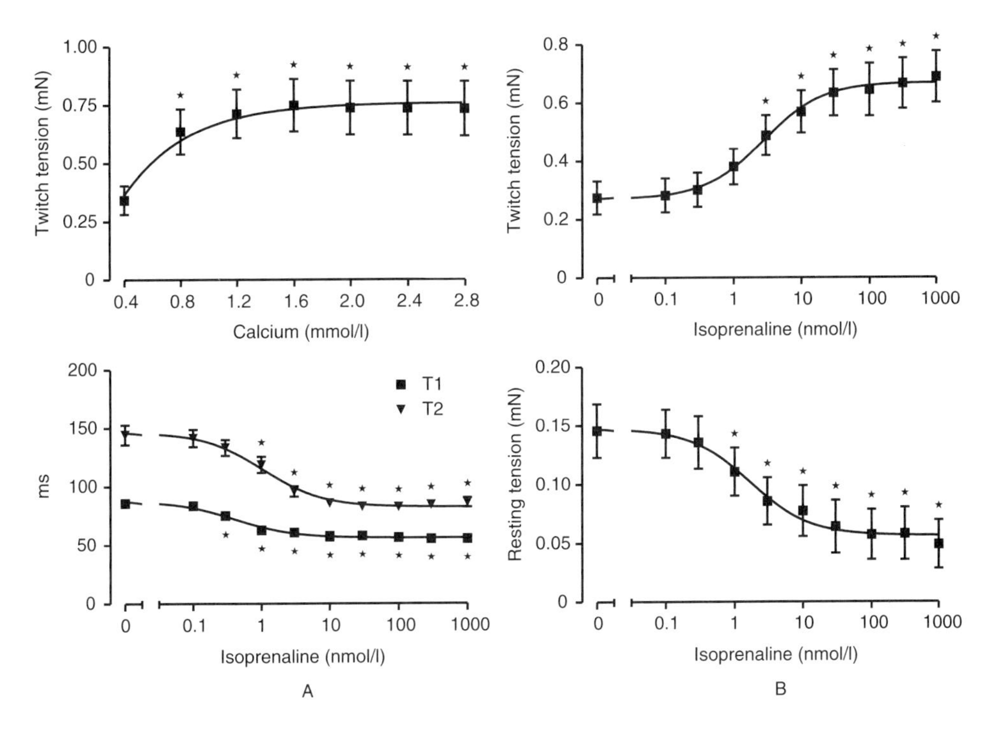

Figure 11.12. Contractile response to calcium and isoprenaline. a) The effect of calcium ($n = 15$, 7 independent cell preparations) was evaluated by cumulatively increasing calcium from 0.4 to 2.8 mM (calculated EC_{50}: 0.46 ± 0.06 mM). b,c) The effect of isoprenaline ($n = 12$, 6 independent cell preparations) was tested at 0.4 mM calcium by cumulatively increasing isoprenaline from 0.1 to 1000 nmol/l (EC_{50}: 2.8 ± 0.6 nmol/l). d) Additionally, isoprenaline lowered RT. *Significant differences ($P < 0.05$; repeated-measures ANOVA with post hoc Bonferroni test) to force at 0.4 mM calcium (a) or predrug value (b–d). From Zimmermann et al. [2002a].

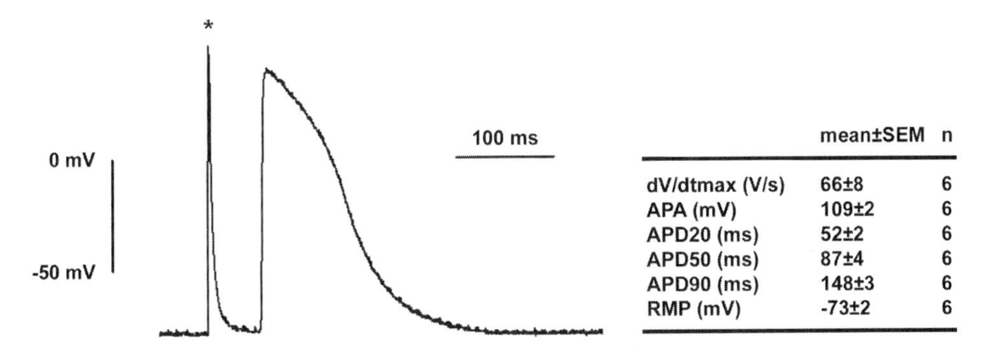

	mean±SEM	n
dV/dtmax (V/s)	66±8	6
APA (mV)	109±2	6
APD20 (ms)	52±2	6
APD50 (ms)	87±4	6
APD90 (ms)	148±3	6
RMP (mV)	-73±2	6

Figure 11.13. Action potential recording from EHT. Original intracellular recording of a representative EHT action potential at 36 °C. *Stimulus artifact. dV/dt_{max}: maximum upstroke velocity; APA: action potential amplitude; APD: action potential duration at 20%, 50%, and 90% repolarization; RMP: resting membrane potential. From Zimmermann et al. [2002a].

a better cardiac tissue-matrix ratio (area fraction occupied by cellular material as compared to pure matrix), improved contractile function, and a high degree of cardiac myocyte differentiation. Action potential recordings revealed electrophysiological properties typical for cardiac tissue. The culture as 3D rings is simple, does not require special equipment, and can therefore be performed in any cell culture laboratory. Importantly, the circular culture format allows for future miniaturization and automation.

7.1. In Vitro Applications

The main advantage of EHTs in our view is that cardiac myocytes in EHTs resemble cardiac myocytes in the intact heart more closely than standard 2D culture systems. This interpretation is supported by the following findings.

1. The cells form a 3D network of intensely interconnected, strictly longitudinally oriented, electrically and mechanically coupled bundles that resemble loose cardiac tissue.
2. The cells are apparently exposed to homogeneous load. The latter feature has not been proven directly (and it would be difficult to do so), but both the fact that the cellular network in EHTs was strictly longitudinally oriented and the geometry of a ring argue for a homogeneous load. As a consequence, tissue formation was much more homogeneous in circular EHTs than in the previously used planar lattices [Eschenhagen et al., 1997; Zimmermann et al., 2000; Fink et al., 2000].
3. In accordance with the organized tissuelike morphology, circular EHTs exhibited a tissuelike ratio of twitch tension to resting tension of 1.33, 3.29, and 14.02 under basal, maximal calcium, and maximal isoprenaline concentrations, respectively. The basal values are in line with similar ratios in intact trabeculae or papillary muscles from humans and rats studied under the same conditions [Weil et al., 1998; Holubarsch et al., 1996; Mende et al., 1992]. This indicates that in circular EHTs the matrix contributes significantly less to mechanical properties than in the planar lattices, where we have described a ratio of twitch tension to resting tension of 0.2–0.3 [Eschenhagen et al., 1997; Zimmermann et al., 2000; Fink et al., 2000].
4. The positive inotropic response to isoprenaline amounted to more than 100% of basal twitch tension in circular EHTs compared to only about 15–30% in the planar lattices. This better resembles the magnitude of the isoprenaline effect in intact rat preparations, where we have reported an isoprenaline-induced increase in twitch tension by 114–145% (EC_{50}: 110 nM) under the same conditions, albeit at a calcium concentration of 1.8 mM [Mende et al., 1992]. The reason for the high sensitivity to isoprenaline in EHTs (EC_{50}: 2.8 nM) remains unclear and parallels the previously observed leftward shift of the calcium response curve in the EHT model [EC_{50}: 0.46 mM vs. 3.1 mM in adult rat papillary muscles; own unpublished data].

5. EHTs are suitable for electrophysiological investigations usually performed on isolated multicellular cardiac preparations, for example, papillary muscles. Six intracellular recordings on EHTs revealed stable resting membrane potentials and action potentials similar to those found in ventricular myocytes from young rats [Kilborn and Fedida, 1990].

6. Cardiac myocytes in EHTs exhibited several morphological features of terminal differentiation [Anversa et al., 1981; Bishop et al., 1990; Katz, 1992]: (a) Densely packed and highly organized sarcomeres; (b) an adult cardiac myocyte-like volume ratio of myofilaments-mitochondria-nucleus of 45:24:9 with the remaining 23% consisting mainly of SR and cytosol, which compares to published data on adult cardiac myocytes of 47:36:2 and 3.5% SR, 11.5% cytosol [Katz, 1992]; (c) all types of normal intercellular connective structures such as adherens junctions, desmosomes, and gap junctions; (d) T-tubules, SR vesicles and T-tubule-SR junctions in form of dyads; and (e) a well-developed basement membrane surrounding cardiac myocytes. It is important to note that some of the features observed in EHTs, especially the T-tubule-SR junctions, were found to be absent in the newborn rat heart [Bishop et al., 1990] and in monolayer cultures of cardiac myocytes [Kostin et al., 1998]. In addition, during the cell isolation procedure, cardiac myocytes lose or disassemble much of their myofilament equipment and appear as rounded cells at the time they are put into the medium-collagen-Matrigel mix. It is remarkable, therefore, that during 14 days of in vitro cultivation they surpass the differentiation state of their source tissue.

7. The electron microscopic investigation also revealed that cardiac cells form not only a myocyte network, but a complex heartlike structure with multiple layers of nonmyocytes at the surface and endothelial cells forming primitive capillaries inside EHTs. Fibroblasts and macrophages were seen throughout the EHTs, suggesting that EHTs represent spontaneously forming cardiac "organoids." The conditions controlling this process or its functional consequences have not been studied here, but the findings may open the possibility of using this system as a model for in vitro cardiac development.

8. Finally, for in vitro applications, technical aspects are also important. The ring system requires only simple casting forms that can be used infinitely and allow the routine production of more precise and highly reproducible EHTs in large series (See contraction experiments, Fig. 11.12). It also opens the way for a multiwell apparatus for drug screening or target validation. Such a device is presently under construction.

7.2. Tissue Engineering

The replacement of defective cardiac tissue by functioning myocardium offers an exciting option in cardiovascular medicine [Li et al., 1998; Reinlib and Field, 2000]. Two principal strategies have been tested so far, mainly in the cryoinjury or in the myocardial infarction model after coronary ligation in mice and rats. One approach uses isolated cells [Li et al., 1996, 1999; Taylor et al., 1998; Etzion et al., 2001; Orlic et al., 2001; Koh et al., 1995; Scorsin et al., 1997; Kobayashi et al., 2000; Menasche et al., 2001; Reinecke et al., 1999], the other in vitro designed tissue equivalents [Leor et al., 2000; Li et al., 1999, 2000]. In most studies, injection of cells into the scar tissue improved global heart function. Surprisingly, the effect appeared to be independent of cell origin, because positive results were reported from fetal or neonatal cardiac myocytes, fibroblasts, endothelial cells, smooth muscle cells, skeletal myoblasts, or pluripotent stem cells [Li et al., 1996, 1999; Taylor et al., 1998; Etzion et al., 2001; Orlic et al., 2001; Koh et al., 1995; Scorsin et al., 1997; Kobayashi et al., 2000]. The concept of expanding autologous skeletal myoblasts ex vivo and injecting them into the postinfarction scar during coronary artery bypass grafting has already been transferred to the human, and the first results appear promising except for an apparent potential to induce arrhythmia [Menasche et al., 2001]. Despite survival and differentiation of implanted cells, mechanical and electrical cell-cell contact between graft and host, a chief requirement for synchronous contractions, was only rarely observed in carefully designed studies [Etzion et al., 2001; Reinecke et al., 1999] and, accordingly, the proof of direct participation of the grafted material in overall cardiac contraction is lacking. Formation of scar tissue inhibiting contact between grafted cells and host tissue appears to account for this problem at least partly [Etzion et al., 2001; Reinecke et al., 1999]. Successful implantation of bone marrow-derived stem cells into the infarction scar was reported in mice [Orlic et al., 2001]. According to this study, stem cells acquired a cardiac phenotype at a surprisingly high frequency. If confirmed by others independently, this study is important in demonstrating the principal potential of an autologous adult stem cell approach.

An alternative approach to cell grafting procedures is tissue replacement with in vitro designed cardiac constructs. For in vitro tissue construction several scaffold proteins and synthetically produced polymers have been tested, including collagen, gelatin, alginate, and polyglycolic acid [Eschenhagen et al., 1997; Zimmermann et al., 2000; Carrier et al., 1999; Leor et al., 2000; Li et al., 2000]. There are some principal problems with this approach:

(i) Scaffold materials often exhibit an intrinsic stiffness that may compromise diastolic function.

(ii) Biodegradation of the scaffold materials remains incomplete, adding to the potential problems with diastolic function.

(iii) Size limitations exist for of all 3D engineered constructs. Li and coworkers [2000] reported that cardiac myocytes seeded on or in gelatin meshes formed a 300-μm-thick cell layer only on the outside. Bursac and coworkers [1999] observed that cardiac myocytes, when seeded on polymer scaffolds, would form cell layers of 50–70 μm. A homogeneous cell distribution within the constructs was not achieved by either group. Core ischemia is well known in papillary muscles with diameters >100 μm [Schouten and ter Keurs, 1986; Gülch and Ebrecht, 1987]. In rat hearts the intercapillary distance is 17–19 μm [Korecky et al., 1982].

EHTs have some principal advantages and share some of these problems. In our view, the advantages are the clearly longitudinally oriented, well-coupled network of muscle bundles, the remarkable degree of differentiation, a cardiac tissue-like contractile function including very low resting tension, and the organoid nature of the construct with a surface lining consisting of nonmyocytes and capillarization. These features should prove advantageous for survival, vascularization, and synchronous beating with the host myocardium. In addition, core ischemia is unlikely because the compact muscle bundles with a diameter of 30–100 μm (See Fig. 6a) were found throughout the EHTs without preferential formation at the outer layers. This indicates that the collagen matrix at the concentration used in this study does not represent a significant diffusion barrier or, alternatively, is rapidly degraded.

However, important limitations remain:

1. The cardiac tissue-like network in EHTs is (with the exception of the compact strands, See Fig. 11.6a, Color Plate 6D) generally much less compact than in native tissues (See Figs. 11.7, Color Plate 6E and 11.8, Color Plate 6F), explaining why contractile force is, in absolute terms, about 10-fold less than in comparable intact cardiac preparations. Very thin cardiac muscle preparations develop maximal twitch tension of >20 mN/mm^2 in ferret, rat, cat, rabbit, and human [Holubarsch et al., 1996]. In contrast, maximal forces in EHTs amounted to 2 mN/EHT, that is, 2 mN/mm^2.

2. The degree of cardiac differentiation, despite being superior to 2D cultures (e.g., T-tubules, SR junctions), is clearly less than in intact adult myocardium (e.g., no mature M-bands).

3. The compact musclelike strands (See Fig. 11.6a, Color Plate 6D) did not exceed 30–100 μm in diameter, which is in line with theoretical considerations and published data. Possibly, optimized culture conditions could allow for thicker and more compact EHTs.

4. Finally, and most importantly, it is unknown at present whether EHTs indeed can serve as a tissue equivalent for replacement therapy and have advantages over cell grafting approaches. These questions are currently under investigation [Zimmermann et al., 2002b].

ACKNOWLEDGMENTS

This study was supported by the German Research Foundation (Deutsche Forschungsgemeinschaft) to T. E. (Es 88/8-2) and the German Ministry for Education and Research to T. E. and W. H. Z. (BMBF FKZ 01GN 0124). We greatly appreciated the support from Prof. W. Neuhuber with electron microscopy and

Drs. G. Wasmeier and U. Nixdorff with echocardiography and the excellent technical assistance of B. Endress, A. Hilpert, and I. Zimmermann, Erlangen. Parts of this study were components of the doctoral theses performed by P. Schubert, K. Schneiderbanger, M. Didié, and I. Melnychenko at the University of Erlangen-Nuremberg.

SOURCES OF MATERIALS

Item	*Supplier*
Antibiotics	Biochrom AG
BMON software	Ingenieurbüro Jäckel
Bovine serum albumin	Sigma
Chick embryo extract	CVSciences
2,3-Butanedione monoxime (BDM)	Sigma
Collagen I from rat tails	CVS GmbH
Confocal microscope	Zeiss
DMEM	Biochrom AG
DNase II, Type V from bovine spleen	Sigma
Epon	Roth
Fetal calf serum	Biochrom AG
Force transducer	Ingenieurbüro Jäckel
Glutamine	Biochrom AG
Glutaraldehyde	Roth
Harbor Extracellular Matrix	Harbor Bio-Products/Tebu
Horse serum	Invitrogen
Inorganic salts and HEPES	Merck
Liquid silicone glue	Dow Corning
Matrigel	BD Biosciences
Osmium tetroxide	Roth
PBS	Invitrogen/GIBCO
Petri dishes, 15 cm, Falcon 3000	BD Biosciences
Thimerosal	Sigma
Tris•HCl	Sigma
Triton X-100	Sigma
Trypsin, certified 1:250 crude	Difco, See Becton Dickinson
Ultramicrotome, Ultracut UCT	Leica
Vibratome slicer	Campden Instruments

REFERENCES

Akins, R.E., Boyce, R.A., Madonna, M.L., Schroedl, N.A., Gonda, S.R., McLaughlin, T.A., Hartzell, C.R. Cardiac organogenesis in vitro: reestablishment of three-dimensional tissue architecture by dissociated neonatal rat ventricular cells. *Tissue Eng.* (1999) 5: 103–118.

Anversa, P., Olivetti, G., Bracchi, P.G., Loud, A.V. (1981) Postnatal development of the M-band in rat cardiac myofibrils. *Circ. Res.* 48: 561–568.

Bishop, S.P., Anderson, P.G., Tucker, D.C. (1990) Morphological development of the rat heart growing in oculo in the absence of hemodynamic work load. *Circ. Res.* 66: 84–102.

Bursac, N., Papadaki, M., Cohen, R.J., Schoen, F.J., Eisenberg, S.R., Carrier, R., Vunjak-Novakovic, G., Freed, L.E. (1999) Cardiac muscle tissue engineering: toward an in vitro model for electrophysiological studies. *Am. J. Physiol. Heart Circ. Physiol.*;277: H433–H444.

Carrier, R.L., Papadaki, M., Rupnick, M., Schoen, F.J., Bursac, N., Langer, R., Freed, L.E., Vunjak-Novakovic, G. (1999) Cardiac tissue engineering: cell seeding, cultivation parameters, and tissue construct characterization. *Biotechnol. Bioeng.*64: 580–589.

Communal, C., Huq, F., Lebeche, D., Mestel, C., Gwathmey, J.K., Hajjar, R.J. (2003) Decreased efficiency of adenovirus-mediated gene transfer in aging cardiomyocytes. *Circulation* 107: 1170–1175.

Condorelli, G., Borello, U., De Angelis, L., Latronico, M., Sirabella, D., Coletta, M., Galli, R., Balconi, G., Follenzi, A., Frati, G., Cusella De Angelis, M.G., Gioglio, L., Amuchastegui, S., Adorini, L., Naldini, L., Vescovi, A., Dejana, E., Cossu, G. (2001) Cardiomyocytes induce endothelial cells to trans-differentiate into cardiac muscle: implications for myocardium regeneration. *Proc. Natl. Acad. Sci. USA* 98: 10,733–10,738.

El-Armouche, A., Rau, T., Zolk, O., Ditz, D., Pamminger, T., Zimmermann, W.H., Jäckel, E., Harding, S.E., Boknik, P., Neumann, J., Eschenhagen, T. (2003) Evidence for protein phosphatase inhibitor-1 playing a positive feedback role in β-adrenergic signaling in cardiac myocytes. *FASEB J.* 17: 437–439

Eschenhagen, T., Fink, C., Remmers, U., Scholz, H., Wattchow, J., Weil, J., Zimmermann, W.H., Dohmen, H.H., Schäfer, H.J., Bischopric, N., Wakatsuki, T., Elson, E.L. (1997) Three dimensional reconstitution of embryonic cardiomyocytes in a collagen matrix: a new heart muscle model system. *FASEB J.* 11: 683–694.

Eschenhagen, T., Didié, M., Münzel, F., Schubert, P., Schneiderbanger, K., Zimmermann, W.H. (2002) 3D engineered heart tissue for replacement therapy. *Basic Res. Cardiol.* 2002(Suppl 1): I146–I152

Etzion, S., Battler, A., Barbash, I.M., Cagnano, E., Zarin, P., Granot, Y., Kedes, L.H., Kloner, R.A., Leor, J. (2001) Influence of embryonic cardiomyocyte transplantation on the progression of heart failure in a rat model of extensive myocardial infarction. *J. Mol. Cell. Cardiol.* 33: 1321–1330.

Fechner, H., Noutsias, M., Tschoepe, C., Hinze, K., Wang, X., Escher, F., Pauschinger, M., Dekkers, D., Vetter, R., Paul, M., Lamers, J., Schultheiss, H.P., Poller, W. (2003) Induction of coxsackievirus-adenovirus-receptor expression during myocardial tissue formation and remodeling: identification of a cell-to-cell contact-dependent regulatory mechanism. *Circulation* 107: 876–882.

Fink, C., Ergün, S., Kralisch, D., Remmers, U., Weil, J., Eschenhagen, T. (2000); Chronic stretch of engineered heart tissue induces hypertrophy and functional improvement. *FASEB J.* 14: 669–679.

Gülch, R.W., and Ebrecht, G. (1987) Mechanics of rat myocardium revisited: investigations of ultra-thin cardiac muscles under high energy demand. *Basic Res. Cardiol.* 82 Suppl 2: 263–274.

Holubarsch, C., Ruf, T., Goldstein, D.J., Ashton, R.C., Nickl, W., Pieske, B., Pioch, K., Ludemann, J., Wiesner, S., Hasenfuss, G., Posival, H., Just, H., Burkhoff, D. (1996) Existence of the Frank-Starling mechanism in the failing human heart. *Circulation* 94: 683–689.

Katz, A.M. (1992) Structure of the heart. In: *Physiology of the Heart*, 2nd ed, New York, NY, Raven Press, pp. 1–36.

Kilborn, M.J., and Fedida, D. (1990) A study of the developmental changes in outward currents of rat ventricular myocytes. *J. Physiol.* 430: 37–60

Kobayashi, T., Hamano, K., Li, T.S., Katoh, T., Kobayashi, S., Matsuzaki, M., Esato, K. (2000) Enhancement of angiogenesis by the implantation of self bone marrow cells in a rat ischemic heart model. *J. Surg. Res.* 89: 189–195.

Koh, G.Y., Soonpaa, M.H., Klug, M.G., Pride, H.P., Cooper, B.J., Zipes, D.P., Field, L.J. (1995) Stable fetal cardiomyocyte grafts in the hearts of dystrophic mice and dogs. *J. Clin. Invest.* 96: 2034–2042.

Koh, G.Y., Soonpaa, M.H., Klug, M.G., Field, L.J. (1993) Long-term survival of AT-1 cardiomyocyte grafts in syngeneic myocardium. *Am. J. Physiol. Heart Circ. Physiol.* 264: H1727–H1733.

Kolodney, M.S., and Elson, E.L. (1993) Correlation of myosin light chain phosphorylation with isometric contraction of fibroblasts. *J. Biol. Chem.* 268: 23,850–23,855.

Korecky, B., Hai, C.M., Rakusan, K. (1982) Functional capillary density in normal and transplanted rat hearts. *Can. J. Physiol. Pharmacol.* 60: 23–32.
Kostin, S., Scholz, D., Shimada, T., Maeno, Y., Mollnau, H., Hein, S., Schaper, J. (1998) The internal and external protein scaffold of the T-tubular system in cardiomyocytes. *Cell Tissue Res.* 294: 449–460.
Kostin, S., Klein, G., Szalay, Z., Hein, S., Bauer, E.P., Schaper, J. (2002) Structural correlate of atrial fibrillation in human patients. *Cardiovasc. Res.* 54: 361–379.
Leor, J., Aboulafia-Etzion, S., Dar, A., Shapiro, L., Barbash, I.M., Battler, A., Granot, Y., Cohen, S. (2000) Bioengineered cardiac grafts: A new approach to repair the infracted myocardium? *Circulation.* 102: III56–III61.
Li, R.K., Mickle, D.A., Weisel, R.D., Zhang, J., Mohabeer, M.K. (1996) In vivo survival and function of transplanted rat cardiomyocytes. *Circ. Res.* 78: 283–288
Li, R.K., Jia, Z.Q., Weisel, R.D., Merante, F., Mickle, D.A. (1999) Smooth muscle cell transplantation into myocardial scar tissue improves heart function. *J. Mol. Cell. Cardiol.* 31: 513–522.
Li, R.K., Yau, T.M., Sakai, T., Mickle, D.A., Weisel, R.D. (1998) Cell therapy to repair broken hearts. *Can. J. Cardiol.* 14: 735–744.
Li, R.K., Yau, T.M., Weisel, R.D., Mickle, D.A., Sakai, T., Choi, A., Jia, Z.Q. (2000) Construction of a bioengineered cardiac graft. *J Thorac. Cardiovasc. Surg.* 119: 368–375.
Menasche, P., Hagege, A.A., Scorsin, M., Pouzet, B., Desnos, M., Duboc, D., Schwartz, K., Vilquin, J.T., Marolleau, J.P. (2001) Myoblast transplantation for heart failure. *Lancet.* 357: 279–280.
Mende, U., Eschenhagen, T., Geertz, B., Schmitz, W., Scholz, H., Schulte am Esch, J., Sempell, R., Steinfath, M. (1992) Isoprenaline-induced increase in the 40/41 kDa pertussis toxin substrates and functional consequences on contractile response in rat heart. *Naunyn Schmiedebergs Arch. Pharmacol.* 345: 44–50.
Most, P., Bernotat, J., Ehlermann, P., Pleger, S.T., Reppel, M., Borries, M., Niroomand, F., Pieske, B., Janssen, P.M., Eschenhagen, T., Karczewski, P., Smith, G.L., Koch, W.J., Katus, H.A., Remppis, A. (2001) S100A1: a regulator of myocardial contractility. *Proc. Natl. Acad. Sci. USA* 98: 13,889–13,894.
Müller-Ehmsen, J., Peterson, K.L., Kedes, L., Whittaker, P., Dow, J.S., Long, T.I., Laird, P.W., Kloner, R.A. (2002a) Rebuilding a damaged heart: long-term survival of transplanted neonatal rat cardiomyocytes after myocardial infarction and effect on cardiac function. *Circulation* 105: 1720–1726.
Müller-Ehmsen, J., Whittaker, P., Kloner, R.A., Dow, J.S., Sakoda, T., Long, T.I., Laird, P.W., Kedes, L. (2002b) Survival and development of neonatal rat cardiomyocytes transplanted into adult myocardium. *J. Mol. Cell. Cardiol.* 34: 107–116.
Orlic, D., Kajstura, J., Chimenti, S., Jakoniuk, I., Anderson, S.M., Li, B., Pickel, J., McKay, R., Nadal-Ginard, B., Bodine, D.M., Leri, A., Anversa, P. (2001) Bone marrow cells regenerate infarcted myocardium. *Nature* 410: 701–705.
Reinecke, H., Zhang, M., Bartosek, T., Murry, C.E. (1999) Survival, integration, and differentiation of cardiomyocyte grafts: a study in normal and injured rat hearts. *Circulation* 100: 193–202.
Reinlib, L., and Field, L. (2000) Cell transplantation as future therapy for cardiovascular disease?: A workshop of the National Heart, Lung, and Blood Institute. *Circulation* 101: E182–E187.
Remppis, A., Pleger, S.T., Most, P., Lindenkamp, J., Ehlermann, P., Schweda, C., Löffler, E., Weichenhan, D., Zimmermann, W.H., Eschenhagen, T., Koch, W.J., Katus, H.A. (2004) S100A1 gene transfer: A strategy to strengthen engineered cardiac grafts. *J. Gene Med.* 6(4): 387–394.
Roell, W., Lu, Z.J., Bloch, W., Siedner, S., Tiemann, K., Xia, Y., Stoecker, E., Fleischmann, M., Bohlen, H., Stehle, R., Kolossov, E., Brem, G., Addicks, K., Pfitzer, G., Welz, A., Hescheler, J., Fleischmann, B.K. (2002) Cellular cardiomyoplasty improves survival after myocardial injury. *Circulation* 105: 2435–2441.
Sakai, T., Li, R.K., Weisel, R.D., Mickle, D.A., Jia, Z.Q., Tomita, S., Kim, E.J., Yau, T.M. (1999) Fetal cell transplantation: a comparison of three cell types. *J. Thorac. Cardiovasc. Surg.* 118: 715–724.

Schouten, V.J., and ter Keurs, H.E. (1986) The force-frequency relationship in rat myocardium. The influence of muscle dimensions. *Pflügers Arch.* 407: 14–17.
Scorsin, M., Hagege, A., Vilquin, J.T., Fiszman, M., Marotte, F., Samuel, J.L., Rappaport, L., Schwartz, K., Menasche, P. (2000) Comparison of the effects of fetal cardiomyocyte and skeletal myoblast transplantation on postinfarction left ventricular function. *J. Thorac. Cardiovasc. Surg.* 119: 1169–1175.
Soonpaa, M.H., Koh, G.Y., Klug, M.G., Field, L.J. (1994) Formation of nascent intercalated disks between grafted fetal cardiomyocytes and host myocardium. *Science* 264: 98–101.
Souren, J.E.M., Schneijdenberg, C., Verkeleij, A.J., Van Wijk, R. (1992) Factors controlling the rhythmic contraction of collagen gels by neonatal heart cells. *In Vitro Cell Dev. Biol.* 28A: 199–204.
Taylor, D.A., Atkins, B.Z., Hungspreugs, P., Jones, T.R., Reedy, M.C., Hutcheson, K.A., Glower, D.D., Kraus, W.E. (1998) Regenerating functional myocardium: improved performance after skeletal myoblast transplantation. *Nat. Med.* 4: 929–933.
Tomita, S., Li, R.K., Weisel, R.D., Mickle, D.A., Kim, E.J., Sakai, T., Jia, Z.Q. (1999) Autologous transplantation of bone marrow cells improves damaged heart function. *Circulation* 100: II247–II256.
Webster, K.A., Discher, D.J., Bishopric, N.H. (1994) Regulation of fos and jun immediate-early genes by redox or metabolic stress in cardiac myocytes. *Circ. Res.* 74: 679–686
Weil, J., Eschenhagen, T., Hirt, S., Magnussen, O., Mittmann, C., Remmers, U., Scholz, H. (1998) Preserved Frank-Starling mechanism in human end stage heart failure. *Cardiovasc. Res.* 37: 541–548.
Zimmermann, W.H., Fink, C., Kralisch, D., Remmers, U., Weil, J., Eschenhagen, T. (2000) Three-dimensional engineered heart tissue from neonatal rat cardiac myocytes. *Biotechnol. Bioeng.* 68: 106–114.
Zimmermann, W.H., Schneiderbanger, K., Schubert, P., Didié, M., Münzel, F., Heubach, J.F., Kostin, S., Neuhuber, W.L., Eschenhagen, T. (2002a) Tissue engineering of a differentiated cardiac muscle construct. *Circ. Res.* 90: 223–230.
Zimmermann, W.H., Didié, M., Wasmeier, G.H., Nixdorff, U., Hess, A., Melnychenko, I., Boy, O., Neuhuber, W.L., Weyand, M., Eschenhagen, T. (2002b) Cardiac grafting of engineered heart tissue. *Circulation* 106: I151–I157.
Zimmermann, W.H., Melnychenko, I., Eschenhagen, T. (2004) Engineered heart tissue for regeneration of diseased hearts. *Biomaterials* 25(9): 1639–1647.
Zimmermann, W.H., Melnychenko, I., Wasmeier, G., Nixdorff, U., Michaelis, B., Dhein, S., Budensky, L., Hess, A., Eschenhagen, T. (2003) Gewebeersatztherapie nach transmuralem Myokardinfarkt. *Z. Kardiol.* 92: I/243 (abstract).
Zolk, O., Marx, M., Jackel, E., El-Armouche, A., Eschenhagen, T. (2003) β-Adrenergic stimulation induces cardiac ankyrin repeat protein expression: involvement of protein kinase A and calmodulin-dependent kinase. *Cardiovasc. Res.* 59: 563–572.

12

Tissue-Engineered Blood Vessels

Rebecca Y. Klinger[1] and Laura E. Niklason[2]

[1] *School of Medicine, and* [2] *Departments of Anesthesiology and Biomedical Engineering, Duke University, Durham, North Carolina 27710*

Corresponding author: nikla001@mc.duke.edu

Culture of Cells for Tissue Engineering, edited by Gordana Vunjak-Novakovic and R. Ian Freshney

1. BACKGROUND

Atherosclerotic vascular disease is the primary cause of morbidity and mortality in the United States [Ross, 1993]. Treatment of cardiovascular disease and its sequelae leads to more than 1.4 million surgical procedures annually that require arterial prostheses, including both coronary and peripheral vascular grafts [American Heart Association, 2002]. Typically, autologous vein or, less frequently, artery is employed in surgical revascularization procedures. In the coronary system, the saphenous vein and internal mammary artery are the most commonly selected conduits. The internal mammary artery offers the highest long-term graft patency rates, with saphenous vein grafts being more prone to progressive intimal hyperplasia and accelerated atherosclerotic change [Eagle et al., 1999]. Individual patients often require multiple coronary artery bypasses, and the more preferable saphenous vein or internal mammary artery grafts may be insufficient or unsuitable because of intrinsic vessel disease or prior use in revascularization [Eagle et al., 1999]. Alternatives for graft material include other autologous venous or arterial sources and cryopreserved nonautologous saphenous vein or umbilical vein; however, both options have inferior patency rates [Eagle et al., 1999]. Although synthetic grafts have been relatively successful in the replacement of large-diameter vessels (6–10 mm), they are rarely used for coronary bypass, as they frequently thrombose early after implantation [Nerem and Seliktar, 2001]. These limitations in the quality and quantity of graftable material have led to a situation in which approximately 100,000 patients requiring revascularization are turned down each year [Williams, 2002]. These individuals then face palliative medical therapy and often suffer myocardial infarctions or endure limb amputations as blood flow becomes progressively constricted. Thus the demand for small-diameter (<6 mm) vascular grafts has prompted many investigators to develop tissue-based vascular replacements that more closely mimic native vascular biology, in the hope of aiding the many thousands of patients with surgically correctable vascular disease.

1.1. Efforts in Vascular Tissue Engineering

Several strategies have been attempted in the construction of autologous small-caliber vascular grafts. The earliest efforts in vascular tissue engineering occurred in the 1980s, when Weinberg and co-workers developed methods to grow arteries by seeding vascular cells onto preformed, tubular collagen gel constructs [Weinberg and Bell, 1986]. Several variations of this collagen gel-based approach have been explored with varying rates of success [Kanda et al., 1993; L'Heureux et al., 1993; Hirai et al., 1994; Kanda and Matsuda, 1994; Tranquillo et al., 1996; Barocas et al., 1998; Girton et al., 1999]. Yet, collagen gel-based constructs can typically only withstand burst pressures of 10–300 mmHg, which is not acceptable for vascular replacement in the coronary system [Girton et al., 2000].

Acellular collagenous grafts have also been investigated as an approach to arterial replacement by many groups [Badylak et al., 1989; Sandusky et al., 1992, 1995; Lantz et al., 1993; Hiles et al., 1995; Wilson et al., 1995; Inoue et al., 1996; Huynh et al., 1999; Roeder et al., 1999]. The group led by Huynh [Huynh et al., 1999] utilized tubes of submucosal collagen derived from porcine small intestine. These tubes were coated with bovine fibrillar collagen and then complexed with heparin-benzalkonium chloride to reduce the propensity for thrombus formation. Although this acellular porcine collagen conduit demonstrated impressive mechanical strength, these types of grafts still possess considerable drawbacks. The potential inflammatory immune response to vascular grafts composed of animal collagens remains largely unknown. Additionally, acellular constructs lack an antithrombogenic endothelium and would have to acquire such an endothelial cell layer after implantation. Endothelial cell recruitment from surrounding tissue is known to be poor in humans [Nerem and Seliktar, 2001].

The first completely biological vascular graft was reported in 1998 by L'Heureux and colleagues. Their approach was to layer continuous sheets of fibroblasts and human umbilical smooth muscle cells around a central mandrel to form tubular vessels [L'Heureux et al., 1998]. After a 6- to 13-week culture period, the inner lumen of the tubular construct was seeded with endothelial cells [L'Heureux et al., 1998]. Although these artificial vessels were able to withstand high pressures, displaying rupture strengths greater than 2000 mmHg, their strength is derived primarily from the adventitial layer rather than the medial layer, which carries the majority of the load in a native artery [Nerem and Seliktar, 2001]. After implantation into dogs, these vessels also suffered a 50% thrombosis rate after 1 week.

We have developed a technique for engineering arteries from explanted autologous vascular cells that are cultured on highly porous, degradable polyglycolic acid (PGA) scaffolds in specially designed bioreactors that subject the tubular scaffolds to physiological pulsatile radial distension that mimics the human cardiovascular system [Niklason et al. 1999; Niklason et al., 1999, 2001]. After 8 weeks of culture time, the polymer scaffold has largely degraded and is replaced by a dense medial layer consisting of smooth muscle cells and collagenous extracellular matrix. Once the lumen has been seeded with endothelial cells, the resulting vascular structure histologically resembles artery and remains patent for up to 4 weeks after implantation in miniature swine. Vessels engineered under such pulsatile conditions also display many of the desired physiological properties of native arteries including burst strengths over 2000 mmHg and contractile response to vasoactive substances.

Although much work is needed to perfect tissue-based vascular constructs, the contributions of the above approaches have brought the field closer to developing biological small-diameter vascular grafts for patients who lack sufficient autologous conduit. Achieving a tissue-engineered vascular graft with the mechanical and physiological properties suitable for implantation into the arterial system could have an enormous impact on surgical interventions for cardiovascular disease.

1.2. Requirements for Tissue-Engineered Vascular Grafts

Our overall goal in developing a biological arterial graft is to closely mimic the structure and properties of native arteries. Muscular arteries are composed of three layers (Fig. 12.1), each conferring specific functional properties [Kelly et al., 1984]. The inner endothelial cell layer functions to prevent spontaneous thrombosis in the vessel and to regulate vascular smooth muscle cell tone. The medial layer is composed of smooth muscle cells and their secreted extracellular matrix

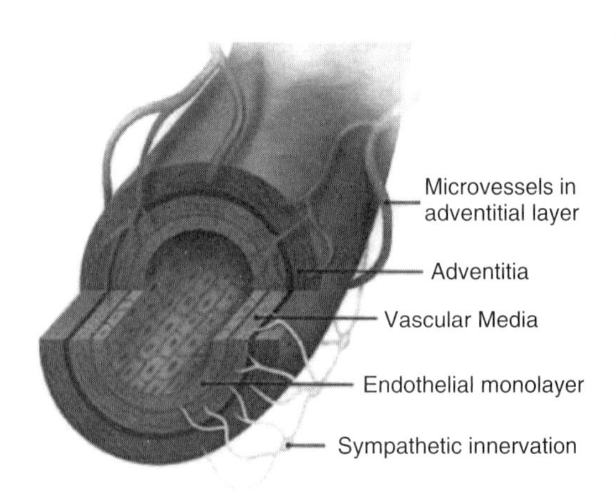

Figure 12.1. Three-layered structure of muscular arteries. [Niklason, 1999].

components including collagen, elastin, and proteoglycans. It is the media that contributes the bulk of the mechanical strength to the vessel as well as its native ability to contract or relax in response to external stimuli. The adventitial layer is composed primarily of fibroblasts and extracellular matrix. Within this outer layer lie the microscopic blood supply to the artery (vasa vasorum) as well as its sympathetic innervation. The ideal tissue-engineered blood vessel will recapitulate this three-layered structure of native arteries.

Specific mechanical requirements for artificial vascular prostheses have been outlined by the American National Standards Institute and the Association for the Advancement of Medical Instrumentation (ANSI/AAMI VP20-1994). According to these guidelines, any implantable arterial construct must be able to withstand normal physiological pressures of 80–120 mmHg, have a burst strength of at least 1680 mmHg, and a suture retention strength of over 273 g [Barron et al., 2003]. Our method of tissue engineering vascular grafts has produced vessels with implantable burst strengths and suture retention strengths up to 91 g [Niklason et al., 1999].

1.3. Endothelium

Engineered blood vessels must possess and maintain a confluent and adherent endothelial cell layer in order to remain nonthrombogenic in vivo. Loosely attached endothelial cells are easily separated from the vessel wall when exposed to shear stresses caused by blood flow in the arterial system. Subendothelial proteins, in particular tissue factor and collagen, are potent stimulators of the coagulation cascade. Thus any denuded areas on the vascular wall are probable sites for thrombus formation. Occlusive thrombus can quickly lead to graft failure and potentially catastrophic downstream consequences including myocardial infarction and limb ischemia. Endothelial cell adherence has been improved by conditioning vascular grafts to shear stress before implantation [Ott and Ballerman, 1995] and by coating vessel lumens with endothelial cell-adherent proteins or sequences [Foxall et al., 1986; Zilla et al.,1989; Thomson et al., 1991]. More recently, fibrin matrices have been conjugated with vascular endothelial growth factor (VEGF) to attract endothelial cells, stimulate their proliferation, and enhance overall attachment of the endothelium [Zisch et al., 2001].

Beyond simply functioning as a physical barrier between the blood and subendothelial tissues, an intact and quiescent endothelium actively inhibits thrombosis (Fig. 12.2) [Maruyama, 1998; Pearson, 1999; van Hinsbergh, 2001]. Thrombomodulin, an integral membrane protein expressed on the surface of endothelial cells, binds thrombin and catalyzes the activation of protein C by thrombin [Wu and Thiagarajan, 1996]. Endothelial cells also synthesize and secrete protein S, a cofactor for protein C. Ultimately, activated protein C exerts its anticoagulant function by proteolytically inactivating factors Va and VIIIa, thereby disrupting the coagulation cascade. The endothelium also produces heparan sulfate proteoglycans that function as a cofactor for antithrombin III, a major serine protease inhibitor found in the plasma that neutralizes the activity of thrombin [Wu and Thiagarajan, 1996].

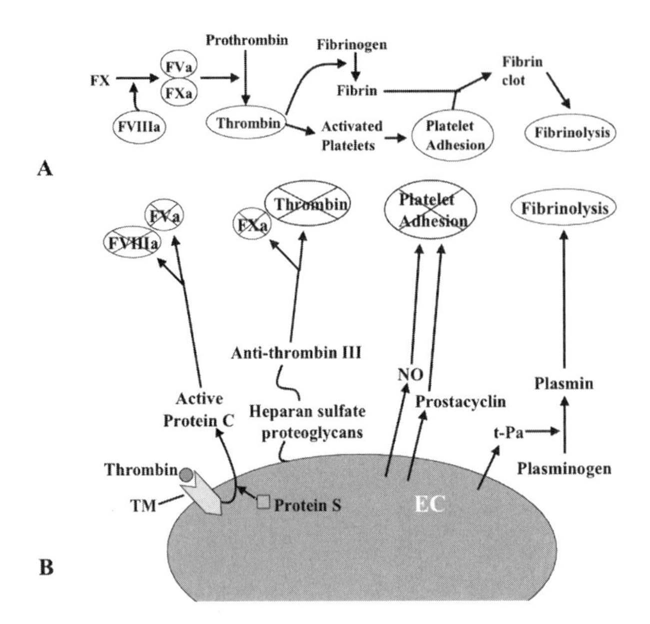

Figure 12.2. Pathway of thrombus formation (A) and endothelial cell antithrombogenic signaling (B). [Mitchell and Niklason, 2002].

Platelet adhesion is suppressed by a number of endothelium-derived factors including nitric oxide, prostacyclin (PGI_2), and negatively charged proteoglycans on the endothelial cell surface. If coagulation occurs, endothelial cells secrete tissue-type plasminogen activator (t-PA), which effects lysis of fibrin clots. The development of a tissue-engineered vascular graft should include the goal of creating a confluent and adherent endothelial cell layer with many of these antithrombogenic properties.

Endothelial cell functionality has been demonstrated in tissue-engineered vessels in vitro. In one case, endothelial cells were shown to cover 92% of the engineered vessel lumen and to express von Willebrand factor (vWF) and prostacyclin after 1 week in culture [Weinberg and Bell, 1986]. In another system, endothelial cells covered 99.2% of the luminal surface and displayed differentiated properties in vitro, including expression of von Willebrand factor, expression of functional thrombin receptors, synthesis of prostacyclin, and active inhibition of platelet aggregation [L'Heureux et al., 1998]. However, these approaches failed to investigate the adhesion of the endothelium in the presences of shear stresses.

Studies of the endothelial cell response to shear stresses have been carried out in vitro by our laboratory. Endothelial cells were seeded onto the lumens of engineered vessels and exposed to $\sim$3 dyn/cm^2 shear flow for 24 to 48 hours. Scanning electron microscopy revealed a slightly rounded endothelial cell morphology and less than complete surface coverage [Niklason et al., 2001].

Several other groups have investigated the in vivo fate of the endothelium in tissue-engineered arteries. In one such experiment, a vascular construct was seeded with canine jugular vein endothelial cells and implanted into the posterior vena cava of the same animal from which the cells were isolated. One week after implantation, endothelial cells were aligned in the direction of blood flow [Hirai and Matsuda, 1996]. Acellular vascular constructs lined with a heparin-benzalkonium chloride complex had luminal endothelial cell coverage within 3 months in rabbits, and the endothelial cells were shown to be oriented in the direction of blood flow [Huynh et al., 1999].

A confluent, thromboresistant endothelial cell layer is an essential component of small-diameter vascular grafts. After implantation into the circulation, mechanically robust vascular grafts are more likely to fail because of thrombotic occlusion than dilation or rupture. Endothelial cells that posses antithrombotic properties—including activation of protein C; expression of heparin sulfate proteoglycans; and production of nitric oxide, prostacyclin, and tissue plasminogen activator—could greatly reduce graft thrombogenic potential. Hence, a successful tissue-engineered vascular graft must contain an endothelial layer that can withstand arterial shear stresses and proactively resist thrombosis.

1.4. Collagenous Extracellular Matrix

In addition to an antithrombogenic endothelium, engineered blood vessels must also possess sufficient mechanical strength to retain anastomotic sutures and resist rupture at arterial pressures. The mechanical strength of native vessels is largely derived from the extracellular matrix components, in particular collagen and elastin [Armentano et al., 1991; Barra et al., 1993; Bank et al., 1996], which are produced by smooth muscle cells in the vascular media (Fig. 12.3, See Color Plate 7A). The stiffness imparted by collagen, with its exceptionally high tensile strength, maintains the structural integrity of blood vessels and prevents their rupture under tension. Without sufficient collagen, vessels cannot remain intact [Dobrin et al.,

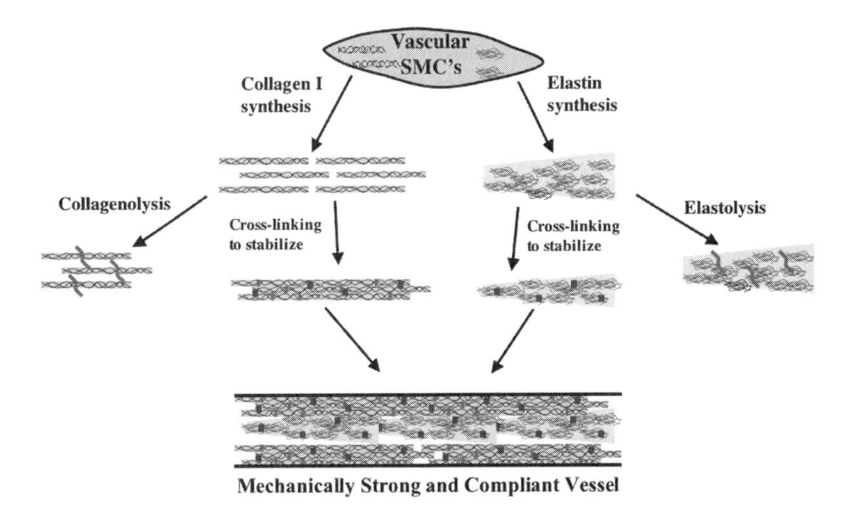

Figure 12.3. Collagen and elastin are secreted by smooth muscle cells. Cross-linking stabilizes collagen and elastin, making them less susceptible to proteolysis. Well-organized layers of insoluble collagen and elastin result in a strong, compliant vessel. [Mitchell and Niklason, 2002]. (See Color Plate 7A)

1984], as the burst strength of both native and engineered vessels is linearly related to the collagen content of the vessel wall [Niklason, 1999].

Cyclic strains [Leung et al., 1976] and various growth factors, including transforming growth factor-β (TGF-β) [Varga et al., 1987], have been shown to increase collagen transcription. Likewise, the addition of ascorbic acid [Geesin et al., 1988] and amino acids to the growth medium enhances collagen synthesis and deposition. The formation of cross-links between and within collagen fibrils stabilizes the proteins [Eyre, 1984] against degradation by matrix metalloproteinases (MMPs) and other enzymes. Cross-link formation may be modulated in vitro by altering the activity of the enzyme lysyl oxidase, which catalyzes collagen cross-linking and is itself activated by copper [Rayton and Harris, 1979] and TGF-β [Shanley et al., 1997]. Highly cross-linked and insoluble collagen fibers are requisite for the development of a strong vascular graft.

As collagen is a prevalent vascular wall protein and is largely responsible for the tensile strength of arteries, several groups have attempted to use collagen substrates to engineer vessels. Most commonly, cells are suspended in a collagen gel mold (Fig. 12.4a). Unfortunately, collagen gels lack tensile strength, because the collagen is neither organized into fibrils nor highly cross-linked. Thus the resulting vascular constructs cannot withstand suturing or physiological blood pressures.

One of the earliest tissue-based vascular grafts, developed by Weinberg and Bell [1986], consisted of several collagen gel layers containing vascular cells. Bovine aortic smooth muscle cells were suspended with medium in a hydrated type I collagen gel that was then cast in an annular mold. Dacron mesh was placed around this smooth muscle cell medial layer for mechanical support, and a suspension of bovine aortic adventitial fibroblasts in collagen gel was cast around the Dacron.

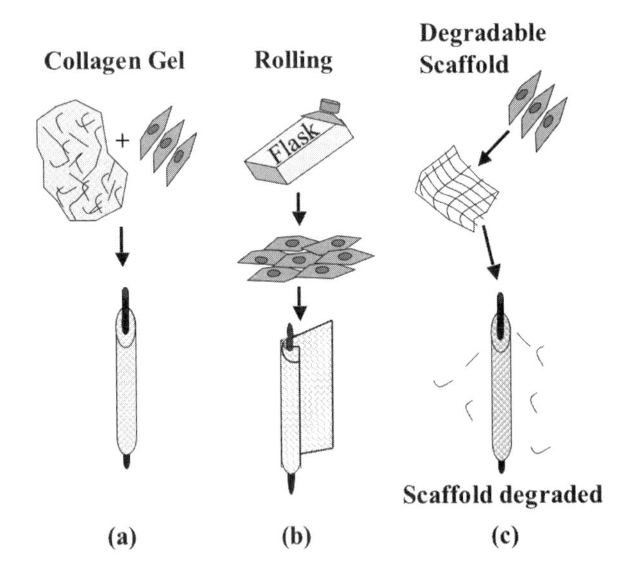

Figure 12.4. Development of collagen gel-based (a), rolled sheet (b), and degradable (c) scaffold vascular grafts. [Mitchell and Niklason, 2002].

The overall strength of these engineered vessels reached only 300 mmHg, which is not within the range required for successful implantation.

Hirai and Matsuda [1996] also employed a collagen gel-based approach to create autologous venous grafts in a canine model. Canine jugular vein smooth muscle cells were mixed with a bovine-derived collagen gel, and the suspension was cast into a cylindrical construct. After 7 days of maturation, the construct was seeded with canine jugular vein endothelial cells and implanted into the posterior vena cava of the same animal from which the cells were derived. Anticipating that the grafts would lack sufficient mechanical integrity on their own, they loosely wrapped a Dacron sleeve around the implant to prevent tearing. Substantial thinning of the vessel occurred over 24 weeks in vivo and resulted in a vessel wall thickness of 96 μm—less than 12% of the original implant thickness. However, no quantitative analyses of the mechanical properties of the grafts were performed, making comparison of the quality of this gel-based construct with other engineered vascular tissues difficult.

Extensive work has been done in an attempt to improve the mechanical properties of collagen gel-based vascular grafts. Organizing the orientation of collagen gel fibrils has been investigated in order to maximize the mechanical strength of these constructs [Tranquillo et al., 1996; Barocas et al., 1998]. Another method to increase mechanical strength is to apply cyclic strain to the collagen gel structure, which promotes synthesis of remodeling enzymes including matrix metalloproteinase-2 [Seliktar et al., 2001].

Methods of creating tissue-engineered arteries have been developed that do not rely on collagen gel suspensions. These constructs are able to withstand much higher burst pressures than their collagen gel-based counterparts. Vessel construction using rolled sheets of cells (Fig. 12.4b), as described by L'Heureux et al. [1993; 1998], achieved burst strengths of greater than 2200 mmHg. As an alternative approach, Huynh and co-workers [Huynh et al., 1999] described the use of decellularized collagen matrices derived from porcine small intestinal submucosa. These constructs demonstrated burst pressures of 931 $\pm$ 284 mmHg, and infiltration of smooth muscle and endothelial cells was observed after implantation into rabbit carotid arteries.

In our laboratory, engineered vessels are constructed with a degradable, biocompatible scaffold (Fig. 12.4c) and a pulsatile culture system [Niklason et al., 1999; Niklason et al., 2001]. During the 6- to 10-week culture periods of this system, smooth muscle cells seeded onto the scaffold secrete collagen (up to 45% of the dry tissue weight) as well as other extracellular matrix proteins. This organized "medial" layer is able to withstand burst pressures as high as 2,300 mmHg, depending on cellular species. Autologous vessels engineered in this manner remained patent for 4 weeks when implanted into swine, although vessel wall dilatation was observed.

Collagen is responsible for the ultimate mechanical strength of engineered vessels, and is thus arguably the most important component of the vessel wall. Critical to the development of successful vascular tissue engineering strategies is the understanding that extracellular matrix deposition and remodeling are dependent on cell sourcing, scaffold material, bioreactor design, nutrient medium composition, growth factor supplementation, mass transfer conditions, and culture time. Our lab has determined that in vitro collagen synthesis can be increased by the application of cyclic strain and culture medium supplementation with growth factors, ascorbic acid, and amino acids. To further increase the mechanical properties of engineered vascular grafts, in particular burst strength and suture retention strength, future investigation must focus on collagen fibril organization and cross-linking.

1.5. Elastin

Elastin functions as a recoil protein that stretches with the artery during each pulse, but then pulls the vessel back to its original diameter. Thus the presence of elastin largely determines the compliance of a vessel in the physiological pressure range. A network of elastin fibers in tissue-engineered arteries will prove to be crucial in preventing vascular dilatation in response to the continuous cyclic strains produced by blood pressure in vivo. Very stiff grafts may prevent adequate transmission of pulsatile wave energy to the downstream vasculature, thereby compromising blood flow. Compliance mismatch has been correlated with graft failure [Davies et al., 1992], showing that larger deviations from arterial compliance make a graft more likely to fail. Hence, the considerable compliance difference between synthetic materials and native arteries may contribute to the high failure rate observed with synthetic grafts [Abbott et al., 1987; Schecter et al., 1997]. Veins exhibit higher compliance than synthetic grafts in the arterial circulation, partially explaining the greater success of venous conduits as vascular grafts.

Although tissue-based vascular grafts are generally more compliant than their synthetic counterparts, there remains much room for improvement. Two groups have been able to demonstrate only scant elastin fibers in their constructs [L'Heureux et al., 1998; Huynh et al., 1999], and this critical component of vascular recoil has been absent in all other engineered grafts examined before implantation. Intriguingly, postimplant in vivo elastin formation has been observed in our engineered constructs [Mitchell and Niklason, 2002]. Reliable deposition of a load-bearing, insoluble elastin network remains a major goal of vascular tissue engineering, to allow for better control of the compliance of engineered grafts. To date, the ability to stimulate the in vitro production of insoluble elastin remains a challenge.

2. PREPARATION OF MEDIA AND REAGENTS

2.1. Endothelial Cell Culture Medium

Dulbecco's modified Eagle's medium (DMEM) supplemented with 10% fetal bovine serum, penicillin 100 U/ml, streptomycin 100 μg/ml, and heparin 125 μg/ml.

2.2. Smooth Muscle Cell Culture Medium

Dulbecco's modified Eagle's medium (DMEM) supplemented with 20% fetal bovine serum, penicillin 100 U/ml, and streptomycin 100 μg/ml.

2.3. Polyglycolic Acid (PGA) Mesh Surface Treatment Solution

Dissolve 40 mg NaOH in 1 L of deionized water to make a 1 N NaOH solution. Make up fresh solution for each PGA treatment.

2.4. Enhanced Culture Medium for Vessel Culture

Dulbecco's modified Eagle's medium (DMEM) supplemented with 20% fetal bovine serum, penicillin G 100 U/ml, HEPES 5 mM, copper sulfate, 12 nM ($CuSO_4 \cdot 5H_2O$, 3 ng/ml), L-proline, 0.43 mM (50 μg/ml), L-glycine, 0.67 mM (50 μg/ml), ascorbic acid, 0.28 mM (50 μg/ml), and L-alanine, 0.22 mM (20 μg/ml).

2.5. Transmission Electron Microscopy Solutions

Fixative: Glutaraldehyde, 0.2 M (2%) in 0.1 M sodium cacodylate buffer.

Postfixation Solution: Osmium tetroxide, 0.079 M (2%) in 0.1 M sodium cacodylate containing 0.1 M (3.37%) sucrose and 1.2 M calcium chloride at pH 7.2 (300 mOsm).

Stain: Uranyl acetate, 0.025 M (1%) in 100% methanol.

2.6. Scanning Electron Microscopy Solutions

Fixative: Glutaraldehyde, 0.25 M (2.5%), in 25 mM sodium cacodylate containing 58 mM (2%) sucrose and 1.2 mM calcium chloride at pH 7.2 (300 mOsm).

Rinse: 100 mM sodium cacodylate containing 0.1 M (3.37%) sucrose and 1.2 mM calcium chloride at pH 7.2 (300 mOsm).

2.7. Krebs-Henseleit Solution

Using deionized water, make a solution consisting of NaCl 118 mM, KCl 4.7 mM, $CaCl_2 \cdot 2H_2O$ 2.5 mM, KH_2PO_4 1.2 mM, $Mg_2SO_4 \cdot 7H_2O$ 1.6 mM, sodium pyruvate 2 mM, $NaHCO_3$ 24.9 mM, glucose 5.6 mM. Maintain pH of 7.4–7.5 as well as adequate oxygen content of the solution by continuously bubbling with 95%O_2-5% CO_2 during use.

3. CULTURE OF TISSUE-ENGINEERED VASCULAR GRAFTS

Our methodology for tissue engineering small-diameter vascular grafts from bovine cells involves: 1) isolation and expansion of smooth muscle cells (SMCs) and endothelial cells (ECs) with high proliferative capacity from vascular tissue, 2) assembly of bioreactor and perfusion system, 3) seeding of SMCs onto a polyglycolic acid (PGA) tubular scaffold within the bioreactor, 4) culture under pulsatile conditions for 8 weeks to produce the vascular graft, and 5) seeding of the luminal surface of the graft with ECs, followed by 3 additional days of culture time. The pulsatile culture conditions are designed to mimic the physical forces present during vasculogenesis and throughout life, which enhance the mechanical properties of the resultant engineered vascular graft.

3.1. Cell Isolation and Culture

Bovine aortic SMCs and ECs are isolated with a technique previously described [D'Amore and Smith, 1993].

Protocol 12.1. Isolation and Culture of Endothelial Cells for Vascular Tissue Engineering

Reagents and Materials

Sterile

- ❑ Endothelial Cell Culture Medium (See Section 2.1)
- ❑ Phosphate-buffered saline (PBSA)
- ❑ Hanks' balanced salt solution (HBSS)
- ❑ Trypsin, 0.25%, 1 mM EDTA 1 × solution
- ❑ Petri dishes: 150 × 25 mm, 60 × 15 mm
- ❑ Plastic conical centrifuge tube, 15 ml
- ❑ Scalpel and No. 10 surgical blade
- ❑ Dissection scissors
- ❑ Tissue forceps

Nonsterile

- ❑ Ice

Protocol

(a) Obtain thoracic aorta from young calves.
(b) Immerse aorta in Hanks' saline.
(c) Place on ice until ready to isolate cells.
(d) In a tissue culture hood, using sterile technique, place aorta into a 15 × 2.5 cm Petri dish for dissection.
(e) Incise the aorta longitudinally with dissection scissors.
(f) To obtain ECs, gently scrape the luminal surface of the aorta with the scalpel blade.
(g) Transfer ECs from the scalpel blade into a 15-ml conical centrifuge tube by pipetting 5 ml PBSA onto the blade, which is held over the opening of the 15-ml conical tube. (Scrape a 3-cm length of vessel and transfer the ECs from that length into a 15-ml conical tube before scraping the next 3 cm length of vessel.)
(h) Intermittently aspirate the ECs up and down with a pipette to break up clumps of cells.
(i) Centrifuge ECs for 5 min at 425 *g*.
(j) Resuspend each EC cell pellet in 5 ml endothelial cell culture medium and plate mixture into a 6 × 1.5 cm Petri dish.

(k) Grow cells in a humidified incubator at 37 °C with 10% CO_2.
(l) Passage cells at subconfluence with 0.25% trypsin-EDTA.

Protocol 12.2. Isolation and Culture of Smooth Muscle Cells for Vascular Tissue Engineering

Materials

Sterile

- ❑ Smooth Muscle Cell Culture Medium (See Section 2.2)
- ❑ 0.25% trypsin-EDTA (See Protocol 12.1)
- ❑ Petri dishes: 15 × 2.5 cm, 6 × 1.5 cm
- ❑ Scalpel and No. 10 surgical blade
- ❑ Dissection scissors
- ❑ Tissue forceps

Protocol

(a) After isolation of ECs, dissect the medial layer of the aorta free from the intimal and adventitial layers.
(b) Cut the medial layer into segments of approximately 1 cm^2.
(c) Place medial segments intimal-side down in 6 × 1.5 cm Petri dishes.
(d) Allow the segments to adhere to the dishes for approximately 10 min.
(e) Add 1 ml Smooth Muscle Cell Culture Medium directly onto the medial segment in each dish.
(f) Place medial segments in humidified incubator at 37 °C and 10% CO_2 overnight.
(g) The following day, add 5 ml fresh Smooth Muscle Cell Culture Medium to each dish, being careful not to detach the medial segment from the dish.
(h) Maintain the medial segments in an incubator for an additional 10 days, after which time the SMCs will have migrated off the segments and become established in two-dimensional culture.
(i) Passage cells at subconfluence, using 0.25% trypsin-EDTA.

3.2. Assembly of Polymer Scaffold

Three-dimensional scaffolds should ensure spatially uniform cell attachment as well as maintain cell phenotype, permit sufficient mass transfer of gases and vital nutrients, and degrade in synchrony with the formation of tissue components. We have chosen a PGA mesh that degrades by passive hydrolysis of ester linkages in the polymer backbone. This FDA-approved material has been shown to be a successful biocompatible and biodegradable polymer for use in tissue engineering. The PGA scaffolds are made as previously described [Freed et al., 1994] and are composed of 13-μm diameter fibers that enable cellular attachment and communication. The mesh is 1 mm thick, with a bulk density of 45 mg/ml and a void volume of 97%, ensuring minimal resistance to nutrient transfer. Before cell seeding, the PGA meshes are base-treated to cleave ester bonds on the surface of the mesh, creating hydroxyl and carboxylic acid groups and thereby increasing surface hydrophilicity for SMC attachment.

Protocol 12.3. Assembly and Treatment of Tubular PGA Scaffold for Vascular Tissue Engineering

Reagents and Materials

Nonsterile

- ❑ PGA Mesh Surface Treatment Solution (See Section 2.3)
- ❑ Deionized water
- ❑ Ethanol (EtOH), 95%
- ❑ Sections of 1-mm-thick PGA mesh, cut to be 7 × 1.5 cm
- ❑ Silicon tubing, measured compliance of 1.5% per 100 mmHg, outer diameter 3.1 mm
- ❑ Dexon® suture, 6-0
- ❑ Dacron® suture, 5-0
- ❑ Dacron® sleeves, 5-mm length × 5-mm internal diameter (2 segments per PGA scaffold)

Protocol

Wear gloves when handling PGA mesh to prevent contamination with skin oils and debris.

(a) Wrap a segment of the PGA mesh along its length around silicon tubing of arbitrary length.
(b) Using 6-0 Dexon suture, sew the PGA along its length around the silicon tubing to form a tubular scaffold that is 7 cm in length and 3.1 mm in internal diameter.
(c) Immerse the PGA scaffold in 95% EtOH.
(d) While the PGA is still wet, transfer the tubular scaffold onto a new piece of silicon tubing that is approximately 25 cm in length. The silicon tubing is replaced in case suturing of the PGA scaffold introduced any nicks into the tubing that could subsequently cause leakage when flow is started.
(e) Immerse the PGA scaffold in 1M NaOH surface treatment solution for 1 min.
(f) Wash the PGA scaffold with copious amounts of deionized water.
(g) Dry the scaffold overnight under vacuum.
(h) To both ends of the PGA tubular scaffold, sew on a sleeve of Dacron with the 5-0 Dacron suture. Dacron sleeves are designed to facilitate attachment of the scaffold to the glass bioreactor (See Protocol 12.4).

3.3. Vessel Culture

Bioreactors for tissue engineering should maintain spatial uniformity of cell seeding, ensure sufficient mass transfer, and supply necessary mechanical stimuli [Freed et al., 1993, 1998]. Our vessels are cultured within hand-blown glass bioreactors that can accommodate two PGA scaffolds each. The PGA scaffolds are attached to glass side arms that are 2 cm in length and 3.5 mm in internal diameter. To maintain sterility, each bioreactor is sealed with a lid that is fitted for gas exchange and medium exchange. Vessels are cultured within the pulsatile bioreactor system for up to 8 weeks, at which time they are luminally seeded with ECs and cultured for an additional 3 days.

Our bioreactor system is specifically designed to expose the vascular cells to pulsatile physical forces during the duration of culture time. This biomimetic system thus mimics the pulsatile stresses that occur in the vasculature throughout life as well as during vasculogenesis [Risau, 1995].

Protocol 12.4. Preparation of Bioreactor for Engineered Blood Vessels

Reagents and Materials

- ❑ Hand-blown glass bioreactor (Fig. 12.5)
- ❑ Magnetic stir bar
- ❑ Dacron® suture, 5-0
- ❑ Preassembled PGA scaffold (from Protocol 12.3)
- ❑ Ethylene oxide or 95% EtOH

Protocol (See Fig. 12.5)

(a) Thread silicone tubing, over which PGA scaffold is sewn, through opposite glass sidearms of the bioreactor and secure.

(b) Trim away excess silicone tubing.

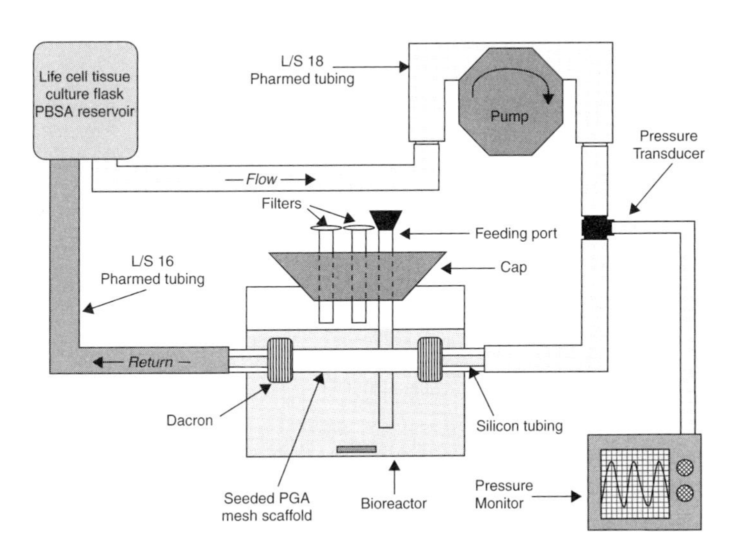

Figure 12.5. Bioreactor and pulsatile perfusion system. Perfusion system consists of a flow side that pumps sterile PBSA from a reservoir in the Lifecell tissue culture flask through silicone tubing that passes through the cell-seeded PGA mesh scaffold. A pressure transducer in the flow side is used to continuously monitor pressures generated within the perfusion system. The sterile PBSA is recycled to the Lifecell tissue culture flask through the return side of the circuit.

(c) Secure each PGA scaffold to the bioreactor sidearms via the Dacron sleeves with 5-0 Dacron suture.
(d) Place a magnetic stir bar into the bioreactor.
(e) Sterilize the entire bioreactor assembly with ethylene oxide or 95% EtOH.
(f) For ethylene oxide sterilization, outgas for 2 or 3 days before cell seeding. For EtOH sterilization, dry for 24 h in a tissue culture hood.

Protocol 12.5. Seeding Smooth Muscle Cells into Vascular Cell Bioreactor

Reagents and Materials

Sterile or aseptically prepared

- ❑ Bovine aortic SMCs, at or below passage 4 to ensure adequate cell proliferation and differentiation
- ❑ 0.25% trypsin-EDTA (See Protocol 12.1)
- ❑ Smooth Muscle Cell Culture Medium (See Section 2.2)
- ❑ Enhanced Culture Medium (See Section 2.4)
- ❑ PBSA (sterile)
- ❑ Bioreactor lid, fitted for gas exchange through 0.2-μm PFTE syringe filters and for medium exchange through Pharmed® tubing capped with an injection port (See Fig. 12.5)
- ❑ Plastic conical centrifuge tube, 50 ml
- ❑ Sterilized bioreactor assembly (from Protocol 12.4)
- ❑ Pharmed® tubing (sizes L/S 16 and L/S 18)
- ❑ Bellows-style pump
- ❑ Lifecell tissue culture flask
- ❑ Medical-grade pressure transducer
- ❑ Clinical pressure monitor
- ❑ Magnetic stirrer
- ❑ Syringes, 10 ml
- ❑ Syringe needles, 21 G
- ❑ HT Tuffryn® syringe filters, 0.2 μm

Nonsterile

- ❑ Ascorbic acid, 25 mg

Protocol (See Fig. 12.5)

The pulsatile flow system is a closed circuit composed of a flow side (yellow tubing in Fig. 12.5)—containing the bellows-style pump, pressure transducer, and pressure monitor—as well as a return side (pink tubing in Fig. 12.5), which recycles PBSA to the Lifecell tissue culture flask reservoir.

(a) Connect the silicone tubing running through one bioreactor sidearm to size L/S 16 Pharmed tubing that is connected to a Lifecell tissue culture flask filled with 300 ml PBSA. This forms the return side of the perfusion circuit.

(b) To assemble the flow side, connect the silicone tubing running through the opposite bioreactor sidearm to size L/S 16 Pharmed tubing.
(c) Introduce the pressure transducer into the flow side between two segments of size L/S 16 tubing.
(d) Connect the segment of L/S 16 tubing beyond the pressure transducer to the larger-diameter L/S 18 tubing, which will be passed through the bellows-style pump.
(e) Use size L/S 16 tubing to connect the L/S 18 tubing passing through the pump back to the PBSA reservoir in the Lifecell tissue culture flask to complete the flow side of the perfusion system.
(f) Trypsinize confluent bovine aortic SMCs with 0.25% trypsin-EDTA.
(g) Centrifuge cells in a 50-ml conical for 5 min at 425 *g*.
(h) Resuspend the cell pellet in Smooth Muscle Cell Culture Medium to a density of 5×10^6 cells/ml
(i) For each engineered vessel, pipette 1.5 ml SMC cell suspension onto the PGA scaffold (See Fig. 12.6a).
(j) Cap the bioreactor and transfer to a tissue culture incubator.
(k) Slowly rotate the bioreactor for 30–45 min to facilitate uniform cell seeding on the scaffold.
(l) Fill the bioreactor with Enhanced Culture Medium.
(m) Remove the bioreactor to a humidified incubator at 37 °C and 10% CO_2 for extended culture.

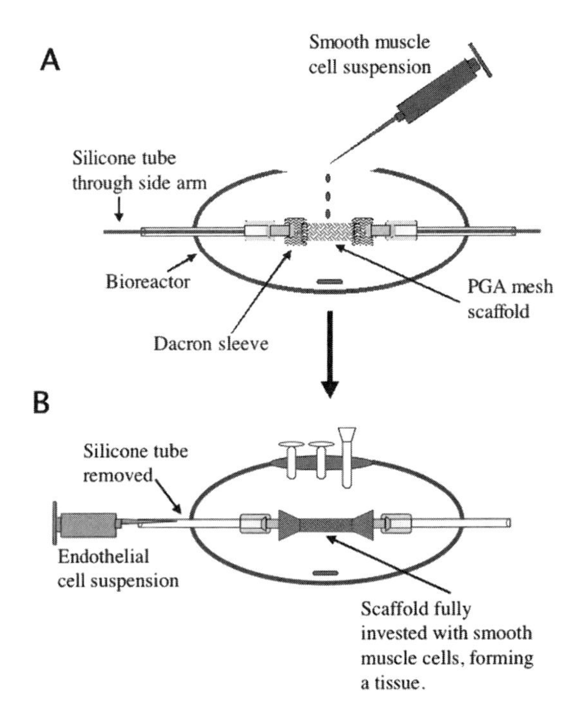

Figure 12.6. Bioreactor and cell seeding. (A) SMCs are seeded onto PGA scaffold by direct pipetting of concentrated cell suspension. Bioreactors are assembled with PGA scaffold and polyester fiber (Dacron) sleeves, which allow formation of fluid-tight connection between vascular tissue and bioreactor. Silicone tubing extends through lumen of vessel and through sidearms of bioreactor to connect with perfusion system. (B) After removal of silicone tubing, EC suspension is injected into engineered vessels via sidearm, and bioreactor is slowly rotated to allow EC seeding in lumen. [Niklason et al., 2001].

(n) Within the incubator, place the bioreactor on a magnetic stirrer to stir the culture medium continuously for the duration of vessel culture.
(o) Turn on the pump to the perfusion system and adjust to the following settings. The bellows-style pump should operate at 165 beats per minute, exerting systolic/diastolic pressures of 270/30 mmHg. The pump circulates sterile PBSA through the silicone tubing that passes through the seeded PGA mesh scaffold in the bioreactor to exert approximately 1.5% radial distension of the tubing with each pulse. PBSA is returned to the Lifecell tissue culture flask, which provides gas exchange to the buffer.
(p) Monitor pressures in the perfusion system continuously with a medical-grade pressure transducer and clinical pressure monitor.
(q) Feed each bioreactor with fresh Enhanced Culture Medium for half of the bioreactor volume twice per week.
(r) Every other day, supplement the culture medium in the bioreactor with 25 mg ascorbic acid. Dissolve the ascorbic acid in 5 ml PBSA and inject through a 0.2-μm HT Tuffryn syringe filter into the medium exchange port, using a 10-ml syringe and a 21 G needle.
(s) After 10–14 days you should observe visible contraction of the vessel constructs under the action of replicating SMCs.
(t) Total culture time is 8 weeks.

Protocol 12.6. Luminal Seeding of Endothelial Cells into Vascular Cell Bioreactor

Cultured vascular grafts must be endothelialized to supply a nonthrombogenic lining to the lumen. After seeding of the endothelial cells, our tissue-engineered vessels possess analogs of the medial and endothelial layers present in native artery.

Reagents and Materials

- ❑ Bovine aortic ECs, at or below passage 4 to ensure adequate cell proliferation and differentiation
- ❑ 0.25% trypsin-EDTA (See Protocol 12.1)
- ❑ Plastic conical centrifuge tube, 50 ml
- ❑ Endothelial Cell Culture Medium (See Section 2.1)
- ❑ Masterflex® modular drive pump
- ❑ Enhanced Culture Medium (See Section 2.4)

Protocol (See Fig. 12.6b)

(a) Trypsinize confluent bovine aortic ECs with 0.25% trypsin-EDTA.
(b) Centrifuge cells in a 50-ml conical centrifuge tube for 5 min at 425 *g*.
(c) Resuspend the cell pellet in Endothelial Cell Culture Medium to a density of 3×10^6 cells/ml.
(d) Inject the EC suspension into the vessel lumen through one sidearm of the bioreactor.

(e) Cap the ends of the bioreactor sidearms and place the bioreactor in an incubator at 37 °C and 10% CO_2.
(f) Rotate the bioreactor around the vessel axis for 90 min to ensure uniform luminal EC seeding.
(g) Connect the bioreactor to a modified perfusion system. The bellows-style pump is replaced with a modular drive pump, and the sterile PBSA is exchanged for Endothelial Cell Culture Medium.
(h) Adjust perfusion rates through the endothelialized vessel lumen gradually from 0.03 to 0.1 ml/s over 3 days of culture to ultimately achieve shear values of 3 dyn/cm^2.

4. CHARACTERIZATION OF ENGINEERED VESSELS

The fabrication of a tissue-engineered vessel that appears to grossly mimic the native arterial structure is not sufficient evidence that a clinically useful conduit has been achieved. Specific analyses must be performed to fully characterize the degree of organization and, importantly, the functionality of these vessel constructs. Analyses should include histology, immunocytochemistry, ultrastructural measurements, pharmacologic responsiveness, and mechanical properties. Vessel architecture, including the presence and organization of extracellular matrix proteins, will be illustrated through histologic and ultrastructural analyses. Immunostaining for cellular markers of differentiation will confirm cellular identity and, in the case of endothelial cells, will help to highlight the degree of endothelial confluence, which can be used to predict long-term patency rates. The response of smooth muscle cells to vasoconstrictive and vasodilatory agents is necessary to further characterize the phenotypic identity of the vascular cells as well as their functionality. Cell and collagen content in the vessel are determined by biochemical analyses (See Chapter 7). In terms of mechanical testing, stress-strain analyses provide the elastic modulus, suture retention strengths provide an indication of the feasibility of clinical implantation, and compliance data demonstrate the graft's ability to resist dilatation and aneurysm formation [Niklason et al., 1999; Niklason et al., 2001]. Despite all these assays, only in vivo implantation studies provide a definitive means of evaluating issues of biocompatibility and determining the practical utility of small-diameter vascular grafts.

Protocol 12.7. Immunostaining for SMC and EC Markers in Vascular Constructs

Reagents and Materials

Nonsterile

- ❑ Mouse monoclonal antibodies: anti-smooth muscle α-actin, anti-calponin, anti-von Willebrand factor
- ❑ Vectastain Elite ABC Kit

- ❑ Biotinylated anti-mouse secondary antibody
- ❑ Normal goat serum, 4% in PBSA
- ❑ Bovine serum albumin (BSA), 3% in PBSA
- ❑ Neutral buffered formalin, 10%
- ❑ EtOH
- ❑ PBSA
- ❑ Paraffin wax
- ❑ Microtome

Protocol

(a) Remove vessels from bioreactor by cutting the silicone tubing at both ends near the glass sidearms so that the vessel is removed with the silicone tubing retained in the lumen. Immediately fix in 10% neutral buffered formalin for 1 h. Fixing the vessel with the silicone tubing still in the lumen prevents shrinkage and subsequent reduction of the size of the vessel lumen.

(b) Dehydrate samples in EtOH and embed in paraffin wax according to standard histologic procedures.

(c) Prepare vessel cross sections of 4-μm thickness with a microtome, and dewax according to standard histologic procedures.

(d) Block with 4% normal goat serum and 3% bovine serum albumin (in PBSA).

(e) Stain for the presence of SMC-specific proteins, using the anti-smooth muscle α-actin (diluted in PBSA 1:500) and anti-calponin (diluted in PBSA 1:10,000) antibodies. EC-specific proteins are stained for with the anti-von Willebrand factor antibody (diluted in PBSA 1:500).

(f) Visualize antibody-antigen complexes with the Vectastain Elite ABC Kit and the included biotinylated anti-mouse secondary antibody (diluted in PBS 1:250) according to the manufacturer's protocol.

Protocol 12.8. Transmission Electron Microscopy of Engineered Vessels

Reagents and Materials

- ❑ Transmission electron microscopy solutions (See Section 2.5)
- ❑ Aqueous EtOH solutions: 50%, 70%, 80%, and 90%
- ❑ EtOH, 100%
- ❑ Propylene oxide (PO) solutions: 50% PO in EtOH and 100% PO
- ❑ Spur resin solutions: 50% resin in PO and 100% resin
- ❑ Reynolds' lead citrate

Protocol

(a) Harvest engineered vessels and fix in 2% glutaraldehyde solution in 0.1 M sodium cacodylate buffer (fixative) for 30–40 min.

(b) Transfer to 2% osmium tetroxide in 0.1 M sodium cacodylate containing 3.37% sucrose and 1.2 mM calcium chloride at pH 7.2 (postfixation solution) for 2 h.

(c) Dehydrate samples with 10-min changes in each of 50%, 70%, 80%, and 90% aqueous EtOH and three times in 100% EtOH.
(d) Change samples to propylene oxide (PO) with 5-min changes in each of 50% PO in EtOH and 100% PO.
(e) Change samples to spur resin with 6-h changes in each of 50% resin in PO and 100% resin.
(f) Stain thin sections (700 Å) with 1% uranyl acetate in 100% methanol and Reynolds' lead citrate.
(g) Examine at 60 kV.

Protocol 12.9. Scanning Electron Microscopy of Engineered Vessels

Reagents and Materials

Nonsterile

- ❑ Scanning Electron Microscopy Solutions (See Section 2.6)
- ❑ Aqueous EtOH solutions: 50%, 70%, 80%, and 90%
- ❑ EtOH, 100%
- ❑ Liquid CO_2
- ❑ Gold, coating quality
- ❑ Sputter coater
- ❑ Critical point dryer

Protocol

(a) Harvest engineered vessels and fix in 2.5% glutaraldehyde in 2.5 mmol/l sodium cacodylate containing 2% sucrose and 1.2 mM calcium chloride at pH 7.2 (fixative) for 2 h.
(b) Rinse in a solution of 100 mM sodium cacodylate containing 3.37% sucrose and 1.2 mM calcium chloride at pH 7.2.
(c) Dehydrate samples with 10-min changes in each of 50%, 70%, 80%, and 90% aqueous EtOH and three times in 100% EtOH.
(d) Change to liquid CO_2 in a pressure chamber and use the critical point dryer to vaporize the CO_2 at 40 °C.
(e) Coat the dried samples with gold, using the sputter coater to an approximate thickness of 20 nm.
(f) Examine with accelerating voltage of 5–15 kV.

The functional responsiveness of engineered vessels to vasoconstrictive and vasodilatory agents can be assessed with a physiological organ bath technique similar to that described by Song et al. [1994; 1995; 2000], as modified from Bateson and Pegg [1994].

Protocol 12.10. Graft Vasoconstrictive and Vasodilatory Function

Reagents and Materials

Nonsterile

- ❑ Two stainless steel wire hooks
- ❑ Custom organ bath
- ❑ Krebs-Henseleit solution (See Section 2.7)
- ❑ Endothelin-1
- ❑ Acetylcholine
- ❑ Isometric force transducer

Protocol

(a) Mount 3-mm segments of engineered vessel between the two stainless steel wire hooks.

(b) Suspend in custom organ bath containing 5 ml Krebs-Henseleit solution, which is gassed continuously with 95% O_2-5% CO_2 at 37 °C.

(c) Fix one hook to the base of the organ chamber and connect the other hook to the isometric force transducer.

(d) After 2 h of equilibration in the Krebs-Henseleit solution, add desired vasoactive agent in increasing concentrations from 1×10^{-9} M to 1×10^{-5} M.

  i) Smooth muscle cell contraction tests (vasoconstriction) may be conducted with endothelin-1 or, alternatively, histamine, bradykinin, angiotensin II, or norepinephrine.

  ii) Smooth muscle cell relaxation (vasodilation) may be elicited by either endothelium-dependent or -independent mechanisms. Endothelium-dependent SMC relaxation (via NO production) is tested with acetylcholine. Sodium nitroprusside induces endothelium-independent SMC relaxation.

(e) Measure isometric forces in response to agent in increasing concentrations.

(f) If testing with more than one vasoactive agent, rinse the vessel segments with Krebs-Henseleit solution and equilibrate for 30 min between tests.

Mechanical properties of engineered vessels—including vessel rupture strengths, compliance, and stress/strain moduli—can be measured with a bench-top system [Dahl et al., 2003], modified from that previously described [Humphrey, 1995].

Protocol 12.11. Mechanical Measurements of Engineered Vessels

Reagents and Materials

Nonsterile

- ❑ PBSA
- ❑ Lifecell tissue culture flask
- ❑ Syringe, 60 ml
- ❑ Pharmed tubing (size L/S 16)
- ❑ Medical-grade pressure transducer

- ❑ Clinical pressure monitor
- ❑ Canon XL1 Digital Video Recorder
- ❑ Power Macintosh with Adobe Photoshop®

Protocol

(a) Attach vessel to flow system equipped with PBSA reservoir (Lifecell tissue culture flask) and pressure transducer as depicted in Fig. 12.7.

(b) Inject PBSA into the flow system with the 60-ml syringe and monitor the pressure increases on the clinical pressure monitor.

(c) Increase the pressure by 50-mmHg increments until the vessel fails, usually by pinhole leak or wall rupture.

(d) At each 50-mmHg pressure increment, record the vessel diameter with the Canon XL1 Digital Video Recorder.

(e) Transfer images to the Power Macintosh and acquire with Adobe Photoshop.

(f) Measure the vessel external diameter at each recorded pressure with Adobe Photoshop.

(g) Obtain measurement of vessel cross-sectional area from histologic preparation of a small vessel segment removed before mechanical testing.

(h) Using the known cross-sectional area of the vessel and the external diameter at each pressure, calculate the internal and external radii of the vessel at each pressure.

(i) Calculate stress (σ) and strain (ε) as follows:

$$\sigma = \frac{8P \times (r_{\text{external}} \times r_{\text{internal}})^2}{(r_{\text{external}}{}^2 - r_{\text{internal}}{}^2) \times (r_{\text{external}} + r_{\text{internal}})^2}$$

$$\varepsilon = \frac{(r_{\text{external}} + r_{\text{internal}})}{(r_{0,\text{external}} + r_{0,\text{internal}})} - 1$$

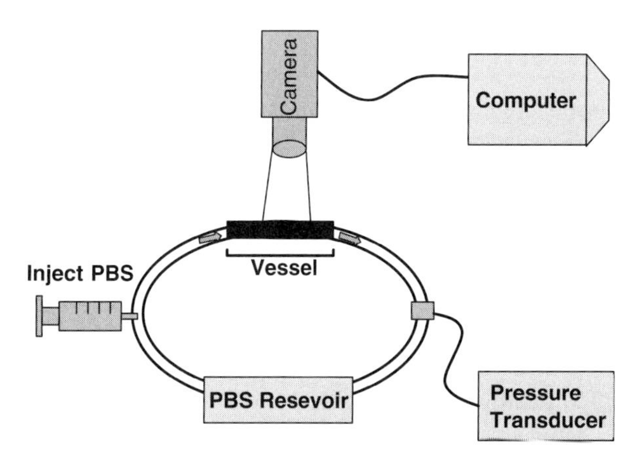

Figure 12.7. Mechanical testing flow system. PBSA is injected into a flow loop with a pressure transducer located downstream of the vessel. A camera records vessel diameters, and images are downloaded onto a computer for analysis. [Dahl et al., 2003].

where P is the pressure inside the vessel, r is the radius at pressure, and r_0 is the radius at zero pressure [Armentano et al., 1991].

(j) Vessel compliance is calculated as follows:

$$\text{compliance} = \frac{\dfrac{(D_2 - D_1)}{D_2}}{(P_2 - P_1)} \times 1000$$

where D_1 and D_2 are the vessel diameters at pressures P_1 and P_2, respectively.

The value obtained when calculating compliance is % per 100 mmHg [Nichols and O'Rourke, 1998].

---

5. DISCUSSION

5.1. Tissue-Engineered Vascular Grafts

In this chapter we have described a biodegradable scaffold-based method of tissue engineering blood vessels with mechanical and functional properties that approach the requisite properties for clinical application. Our constructs, grown under pulsatile conditions, possess many of the physiological and mechanical characteristics of native arteries, including burst strengths of more than 2000 mmHg provided entirely by the medial layer. As our technique also emphasizes the creation of a functional endothelium, we have been the first to implant autologous tissue-engineered arteries that have remained patent and functional for up to 4 weeks. Despite these exciting and promising accomplishments, many hurdles remain on the road to creating a fully functional tissue-based vascular graft. The design of a functional nerve and microvascular supply in vitro to support the vascular tissue remains elusive. Although perhaps not an essential requirement for vascular conduits, achieving their production would mark a significant leap forward in the field of tissue engineering.

5.2. Genetic Manipulation in Tissue-Engineered Blood Vessels

To create functional vascular grafts that mimic the native vasculature as closely as possible, the in vitro loss of optimal cell phenotype will need to be addressed. Endothelial cells in culture, for example, are known to downregulate the expression of many anticoagulant and anti-inflammatory molecules that are expressed on the native, quiescent endothelium [Wu and Thiagarajan, 1996]. Unfortunately, many procoagulant and proinflammatory molecules—including tissue factor, vascular cell adhesion molecule-1 (VCAM-1), and intercellular adhesion molecule-1 (ICAM-1)—are instead upregulated during culture [Carlos and Harlan, 1994; Lin et al., 1997; Allen et al., 1998]. Thus it appears likely that cultured endothelial cells seeded onto the lumen of tissue-engineered blood vessels exhibit a procoagulant rather than antithrombogenic phenotype, ultimately contributing to graft failure due to thrombosis. Not unexpectedly, the critical limitation in the successful implementation of engineered blood vessels to this day remains the high rate of thrombotic occlusion [L'Heureux et al., 1998; Shum-Tim et al., 1999].

One potential method of overcoming these types of phenotypic limitations in tissue engineering may be gene therapy. This technology could allow, for example, life span extension of donor vascular cells [McKee et al., 2003] or the phenotypic manipulation of luminal endothelial cells to a more favorable antithrombogenic character. Work done by our group [Fields et al., 2003] has recently demonstrated the feasibility of stably transfecting endothelial cells with retroviral constructs. Transfection with a marker gene, green fluorescent protein (GFP), revealed infection efficiencies exceeding 60% (Fig. 12.8, See Color Plate 7B) and retained protein expression over time in tissue-engineered vessels. Given the feasibility of this approach, targets such as thrombomodulin may be overexpressed in the endothelial cells used to seed tissue-engineered vessels.(Fig. 12.9, See Color Plate 7C) Loss of thrombomodulin, with its inhibitory effect on tissue factor in the coagulation cascade [Esmon et al., 1983], has been implicated as a cause of decreased resistance to thrombosis in implanted venous grafts [Kim et al., 2002]. Gene therapy is thus a potentially powerful approach in generating ideal cell phenotype for a broad range of tissue engineering purposes.

Although many hurdles still remain in perfecting vascular tissue engineering technologies, functional tissue-based arterial grafts have the potential to revolutionize the surgical treatment of cardiovascular disease. Even beyond this, as the field of tissue engineering rapidly advances, bioengineered vessels will become integral in the development and implantation of other tissues and complex organs,

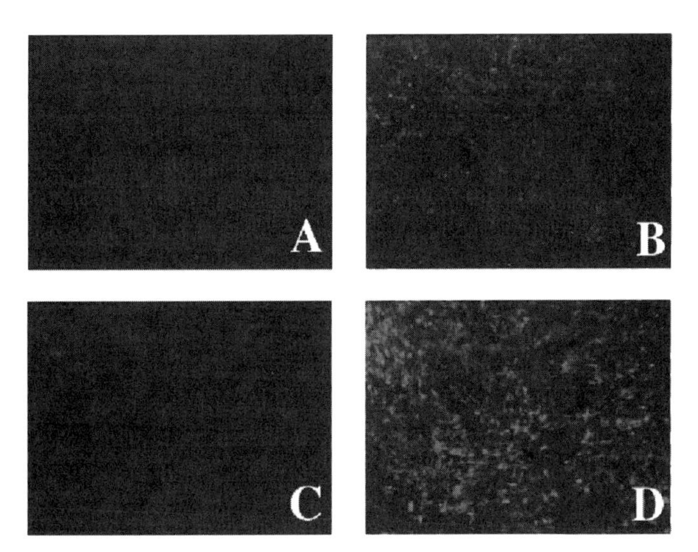

Figure 12.8. Enhanced green fluorescent protein (EGFP) expression in cultured ECs, 7 days after initial infection with PG13-derived retroviral vector. (A) NIH-3T3 cells, not infected with vector, serving as a negative control. (B) Human microvascular ECs (HMECs). (C) Human umbilical vein ECs (HUVECs). (D) Porcine aortic ECs (PAECs). Magnification is 10× for each panel. [Fields et al., 2003] (See Color Plate 7B.)

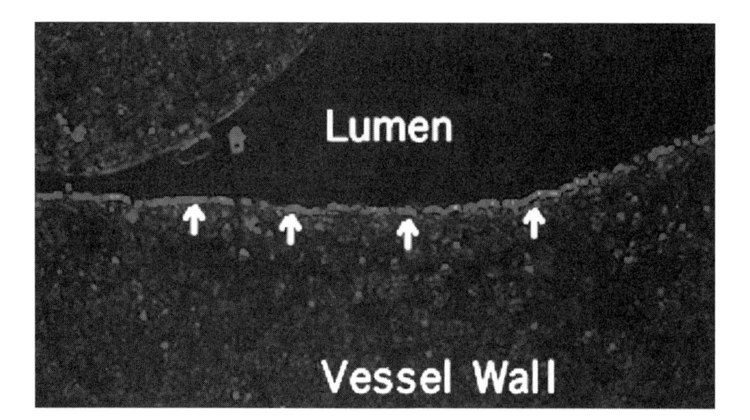

Figure 12.9. EGFP is expressed on engineered vessel lumen. Arrows indicate PAEC monolayer expressing EGFP. Vessel wall and lumen are indicated. [Fields et al., 2003] (See Color Plate 7C.).

all of which ultimately require an intact blood supply for their survival and function in the recipient.

ACKNOWLEDGMENTS

This work was funded by the National Institutes of Health RO1 Grant HL-65766. Rebecca Klinger is a Howard Hughes Medical Institute Medical Student Research Training Fellow.

SOURCES OF MATERIALS

Item	*Supplier*
0.2-μm HT Tuffryn® Membrane syringe filter	Pall
0.2-μm PFTE syringe filters	Cole-Parmer
21G needles	Becton-Dickinson
5-0 Dacron® suture	Davis and Geck
6-0 Dexon® suture	Davis and Geck
Acetylcholine	Sigma
Amino acids	Sigma
Antibiotics	Sigma
Anti-calponin	Sigma
Anti-smooth muscle α-actin	DAKO
Anti-von Willebrand factor	Boehringer Mannheim
Ascorbic acid	Sigma
Bellows-style pump	Gorman-Rupp Industries
Bioreactor	James Glass, Hanover, Mass.
Bioreactor lid	Baxter Healthcare
Biotinylated anti-mouse secondary antibody	Vector
Bovine serum albumin (BSA)	Sigma
Calibrated chart recorder (model 2200S)	Gould
Conical tubes	Corning

Item	*Supplier*
Copper sulfate	Sigma
Critical point dryer	Balzers Union
Custom organ bath	Radnoti
Dacron® vascular graft	Bard
DMEM	GIBCO
Endothelin-1	Sigma
Fetal bovine serum (FBS)	GIBCO
Gold, coating quality	Sigma
Hanks' saline	GIBCO
HEPES	Sigma
Interlink® system injection ports	Becton-Dickinson
Isometric force transducer	Radnoti
Krebs-Henseleit solution components	Sigma
Lifecell tissue culture flask (#420030)	Nexell Therapeutics
Magnetic stirrer	Bellco Glass
Masterflex® modular drive pump (#07553-80)	Cole-Parmer
Normal goat serum	Vector
Petri dishes	Corning
PGA mesh scaffold	Albany International
Pharmed® tubing	Cole-Parmer
Phosphate-buffered saline (PBS)	GIBCO
Pressure monitor (MDE Escort model)	Medical Data Electronics
Pressure transducer (#58-7140-R3-4/03)	Abbott Critical Care Systems
Reynolds' lead citrate (lead (II) citrate tribasic trihydrate)	Sigma
SEM (JOEL 6320 FEGSEM system)	ElectroScan
Silicon tubing	Norton Performance Plastics
Sputter coater	Gatan
Syringes	Becton-Dickinson
Trypsin 0.25%-EDTA	GIBCO
Vectastain Elite ABC Kit	Vector

REFERENCES

Abbott, W.M., Megerman, J., Hasson, J.E., L'Italien, G.J., Warnock, D.J. (1987) Effect of compliance mismatch on vascular graft patency. *J. Vasc. Surg.* 5: 376–382.

Allen, S., et al. (1998). Native low density lipoprotein-induced calcium transients trigger VCAM-1 and E-selectin expression in cultured human vascular cells. *J. Clin. Invest.* 101(5): 1064–1075.

Armentano, R.L., Levenson, J., Barra, J.G., Cabrera Fischer, E.I., Brietbart, G.J., Pichel, R.H., Simon, A. (1991) Assessment of elastin and collagen contribution to aortic elasticity in conscious dogs. *Am. J. Physiol. Heart Circ. Physiol.* 260: H1870–H1877.

American Heart Association (2002) *Heart Disease and Stroke Statistics—2003 Update*. Dallas, TX, American Heart Association.

Badylak, S.F., Lantz, G.C., Coffey, A., Geddes, L.A. (1989) Small intestinal submucosa as a large diameter vascular graft in the dog. *J. Surg. Res.* 47: 74–80.

Bank, A.J., Wang, H., Holte, J.E., Mullen, K., Shammas, R., Kubo, S.H. (1996) Contribution of collagen, elastin, and smooth muscle to in vivo human brachial artery wall stress and elastic modulus. *Circulation* 94: 3263–3270.

Barocas, V.H., Girton, T.S., Tranquillo, R.T. (1998) Engineered alignment in media equivalents: magnetic prealignment and mandrel compaction. *J. Biomech. Eng.* 120: 660–666.

Barra, J.G., Armentano, R.L., Levenson, J., Fischer Cabrera, E.I., Pichel, R.H., Simon, A. (1993) Assessment of smooth muscle contribution to descending thoracic aortic elastic mechanics in conscious dogs. *Circ. Res.* 73: 1040–1050.
Barron, V., Lyons, E., Stenson-Cox, C., McHugh, P.E., Pandit, A. (2003) Bioreactors for cardiovascular cell and tissue growth: a review. *Ann. Biomed. Eng.* 31: 1017–1030.
Bateson, E.A.J., and Pegg, D.E. (1994) Cryopreservation of arteries: selection of a model for human small elastic arteries and preliminary results of preservation of ring-segments with dimethyl sulphoxide. *Cryo-Lett.* 15: 15–26.
Carlos, T., and Harlan, J. (1994) Leukocyte-endothelial adhesion molecules. *Blood* 84(7): 2068–2101.
Dahl, S.L., Koh, J., Prabhakar, V., and Niklason, L.E. (2003) Decellularized native and engineered arterial scaffolds for transplantation. *Cell Transplant.* 12: 659–666.
D'Amore, P.A., and Smith, S.R. (1993) Growth factor effects on cells of the vascular wall: a survey. *Growth Factors* 8: 61–75.
Davies, A.H., Magee, T.R., Baird, R.N., Sheffield, E., Horrocks, M. (1992) Vein compliance: a preoperative indicator of vein morphology and of veins at risk of vascular graft stenosis. *Br. J. Surg.* 79: 1019–1021.
Dobrin, P.B., Baker, W.H., Gley, W.C. (1984) Elastolytic and collagenolytic studies of arteries: implication for the mechanical properties of aneurysms. *Arch. Surg.* 119: 405–409.
Eagle, K., Guyton, R.A., Davidoff, R., Ewy, G.A., Fonger, J., Gardner, T.J., Gott, J.P., Hermann, H.C., Marlow, R.A., Nugent, W.C., O'Connor, G.T., Orszulak, T.A., Riselbach, R.E., Winters, W.L., Yusuf, S. (1999) ACCAHA Guidelines for coronary artery bypass graft surgery. *J. Am. Coll. Cardiol.* 34: 1262–1347.
Esmon, N.L., Carrol, R.C., Esmon, C.T. (1983) Thrombomodulin blocks the ability of thrombin to activate platelets. *J. Biol. Chem.* 258(20): 12238–12242.
Eyre, D. (1984) Cross-linking in collagen and elastin. *Annu. Rev. Biochem.* 53: 717–748.
Fields, R.C., Solan, A., McDonagh, K.T., Niklason, L.E., Lawson, J.H. (2003) Gene therapy of tissue engineered blood vessels. *Tissue Eng.* 9(6): 1281–1287.
Foxall, T.L., Auger, K.R., Callow, A.D., Libby, P. (1986) Adult human endothelial cell coverage of small-caliber dacron and polytetrafluoroethylene vascular prostheses in vitro. *J. Surg. Res.* 41: 158–172.
Freed, L.E., Hollander, A.P., Martin, I., Barry, J.R., Langer, R., Vunjak-Novakovic, G. (1998) Chondrogenesis in a cell-polymer-bioreactor system. *Exp. Cell Res.* 240: 58–65.
Freed, L.E., Vunjak-Novakovic, G., Langer, R. (1993) Cultivation of cell-polymer cartilage implants in bioreactors. *J. Cell. Biochem.* 51: 257–264.
Freed, L.E., Vunjak-Novakovic, G., Biron, R.J., Eagles, D.B., Lesnoy, D.C., Barlow, S.K., Langer, R. (1994) Biodegradable polymer scaffolds for tissue engineering. *Bio/Technology* 12: 689–693.
Geesin, J.C., Darr, D., Kaufman, R., Murad, S., Pinnel, S.R. (1988) Ascorbic acid specifically increases type I and type III procollagen messenger RNA levels in human skin fibroblasts. *J. Invest. Dermatol.* 90: 420–424.
Girton, T.S., Oegema, T.R., Tranquillo, R.T. (1999) Exploiting glycation to stiffen and strengthen tissue equivalents for tissue engineering. *J. Biomed. Mater. Res.* 47: 87–92.
Girton, T.S., Oegema, T.R., Grassl, E.D., Isenberg, B.C., Tranquillo, R.T. (2000) Mechanisms of stiffening and strengthening in media-equivalents fabricated using glycation. *J. Biomech. Eng.* 122(3): 216–223.
Hiles, M.C., Badylak, S.F., Lantz, G.C., Kokini, K., Geddes, L.A., Morff, R.J. (1995) Mechanical properties of xenogeneic small-intestinal submucosa when used as an aortic graft in the dog. *J. Biomed. Mater. Res.* 29: 883–891.
Hirai, J., and Matsuda, T. (1996) Venous reconstruction using hybrid vascular tissue composed of vascular cells and collagen: tissue regeneration process. *Cell Transplant.* 5: 93–105.
Hirai, J., Kanda, K., Oka, T., Matsuda, T. (1994) Highly oriented, tubular hybrid vascular tissue for a low pressure circulatory system. *ASAIO J.* 40: M383–M388.
Humphrey, J.D. (1995) Mechanics of the arterial wall: review and directions. *Crit. Rev. Biomed. Eng.* 23: 1–162.

Huynh, T., Abraham, G., Murray, J., Brockbank, K., Hagen, P.O., Sullivan, S. (1999) Remodeling of an acellular collagen graft into a physiologically responsive neovessel. *Nat. Biotechnol.* 17: 1083–1086.
Inoue, Y., Anthony, J.P., Lleon, P., Young, D.M. (1996) Acellular human dermal matrix as a small vessel substitute. *J. Reconstruct. Microsurg.* 12: 307–311.
Kanda, K., and Matsuda, T. (1994) Mechanical stress-induced orientation and ultrastructural change of smooth muscle cells cultured in three-dimensional collagen lattices. *Cell Transplant.* 3: 481–492.
Kanda, K., Matsuda, T., Oka, T. (1993) Mechanical stress induced cellular orientation and phenotypic modulation of 3-D cultured smooth muscle cells. *ASAIO J.* 39: M686–M690.
Kelly, D.E., Wood, R.L., Enders, A.C. (1984) *Bailey's Textbook of Microscopic Anatomy*. Baltimore, Williams & Wilkins.
Kim, A.Y., et al. (2002) Early loss of thrombomodulin expression impairs vein graft thromboresistance. *Circ. Res.* 90: 205–212.
Lantz, G.C., Badylak, S.F., Hiles, M.C., Coffey, A.C., Geddes, L.A., et al. (1993) Small intestinal submucosa as a vascular graft: a review. *J. Invest. Surg.* 6: 297–310.
Leung, D.Y.M., Glagov, S., Mathews, M.B. (1976) Cyclic stretching stimulates synthesis of matrix components by arterial smooth muscle cells in vitro. *Science* 191: 475–477.
L'Heureux, N., Germain, L., Labbe, R., Auger, F.A. (1993) In vitro construction of a human blood vessel from cultured vascular cells: a morphological study. *J. Vasc. Surg.* 17: 499–509.
L'Heureux, N., Paquet, S., Labbe, R., Germain, L., Auger, F.A. (1998) A completely biological tissue-engineered human blood vessel. *FASEB J.* 12: 47–56.
Lin, M., et al. (1997) Shear stress induction of the tissue factor gene. *J. Clin. Invest.* 99: 737–744.
Maruyama, I. (1998) Biology of endothelium. *Lupus* 7: S41–S43.
McKee, J.A., Banik, S.S., Boyer, M.J., Hamad, N.M., Lawson, J.H., Niklason, L.E., Counter, C.M. (2003) Human arteries engineered in vitro. *EMBO Rep.* 4(6): 633–638.
Mitchell, S.L., and Niklason, L.E. (2002) Requirements for growing tissue-engineered vascular grafts. *Cardiovasc. Pathol.* 12: 59–64.
Nerem, R., and Seliktar, D. (2001) Vascular tissue engineering. *Annu. Rev. Biomed. Eng.* 3: 225–243.
Nichols, W.W., and O'Rourke, M.F. (1998) *McDonald's Blood Flow in Arteries*. New York, Oxford University Press.
Niklason, L.E. (1999) Replacement arteries made to order. *Science* 286: 1493–1494.
Niklason, L.E., Abbott, W., Gao, J., Klagges, B., Hirschi, K.K., Ulubayram, K., Conroy, N., Jones, R., Vasanawala, A., Sanzgiri, S., Langer, R. (2001) Morphological and mechanical characteristics of engineered bovine arteries. *J. Vasc. Surg.* 33(3): 628–638.
Niklason, L.E., Gao, J., Abbott, W.M., Hirschi, K.K., Houser, S., Marini, R., Langer, R. (1999) Functional arteries grown in vitro. *Science* 284: 489–493.
Ott, M., and Ballerman, B.J. (1995) Shear stress-conditioned endothelial cell-seeded vascular grafts: improved cell adherence in response to in vitro shear stress. *Surgery* 117: 334–339.
Pearson, J. (1999) Endothelial cell function and thrombosis. *Balliere's Clin. Haematol.* 12: 329–341.
Rayton, J.K., and Harris, E.D. (1979) Induction of lysyl oxidase with copper. *J. Biol. Chem.* 254: 621–626.
Risau, W. (1995) Differentiation of endothelium. *FASEB J.* 9(10): 926–933.
Roeder, R., Wolfe, J., Lianakis, N., Hinson, T., Geddes, L.A., Obermiller, J. (1999) compliance, elastic modulus, and burst pressure of small-intestine submucosa (SIS), small diameter vascular grafts. *J. Biomed. Mater. Res.* 47: 65–70.
Ross, R. (1993) The pathogenesis of atherosclerosis: a perspective for the 1990s. *Nature* 362: 801–808.
Sandusky, G.E., Lantz, G.C., and Badylak, S.F. (1995) Healing comparison of small intestine submucosa and ePTFE grafts in the canine carotid artery. *J. Surg. Res.* 58: 415–420.
Sandusky, G.E.J., Badylak, S.F., Morff, R.J., Johnson, W.D., Lantz, G. (1992) Histologic findings after in vitro placement of small intestine submucosal vascular grafts and saphenous vein grafts in the carotid artery in dogs. *Am. J. Pathol.* 140: 317–324.

Schecter, A.D., Giesen, P.L.A., Taby, O., Rosenfield, C.L., Rossikhina, M., Fyfe, B.S., Kohtz, D.S., Fallon, J.T., Nemerson, Y., Taubman, M.B. (1997) Tissue factor expression in human arterial smooth muscle cells. TF is present in three cellular pools after growth factor stimulation. *J. Clin. Invest.* 100: 2276–2285.
Seliktar, D., Nerem, R.M., Galis, Z.S. (2001) The role of matrix metalloproteinase-2 in the remodeling of cell-seeded vascular constructs subjected to cyclic strain. *Ann. Biomed. Eng.* 29: 923–934.
Shanley, C.J., Gharee-Kermani, M., Sarkar, R., Welling, T.H., Kriegel, A., Ford, J.W., Stanely, J.C., Phan, S.H. (1997) Transforming growth factor-$\beta 1$ increases lysyl oxidase enzyme activity and mRNA in rat aortic smooth muscle cells. *J. Vasc. Surg.* 25: 446–452.
Shum-Tim, D., et al. (1999) Tissue engineering of autologous aorta using a new biodegradable polymer. *Ann. Thoracic Surg.* 68: 2298–2305.
Song, Y.C., Hunt, C.J., Pegg, D.E. (1994) Cryopreservation of the common carotid artery of the rabbit. *Cryobiology* 31: 317–329.
Song, Y.C., Khirabadi, B.S., Lightfoot, F., Brockbank, K.G.M., Taylor, M.J. (2000) Vitreous cryopreservation maintains the function of vascular grafts. *Nat. Biotechnol.* 18(296–299).
Song, Y.C., Pegg, D.E., Hunt, C.J. (1995) Cryopreservation of the common carotid artery of the rabbit: optimization of dimethyl sulfoxide concentrations and cooling rate. *Cryobiology* 32: 405–421.
Thomson, G.J., Vohra, R.K, Carr, M.H., Walker, M.G. (1991) Adult human endothelial cell seeding using expanded polytetrafluoroethylene vascular grafts: a comparison of four substrates. *Surgery* 1991: 20–27.
Tranquillo, R.T., Girton, T.S., Bromberek, B.A., Triebes, T.G., Mooradian, D.L. (1996) Magnetically orientated tissue-equivalent tubes: application to a circumferentially orientated media-equivalent. *Biomaterials* 17: 349–357.
van Hinsbergh, V. (2001) The endothelium: vascular control of haemostasis. *Eur. J. Obstet. Gynecol. Reprod. Biol.* 95: 198–201.
Varga, J., Rosenbloom, J., Jimenez, S.A. (1987) Transforming growth factor β (TGFβ) causes a persistent increase in steady-state amounts of type I and type III collagen and fibronectin mRNAs in normal human dermal fibroblasts. *Biochem. J.* 247: 597–604.
Weinberg, C.B., and Bell, E. (1986) A blood vessel model constructed from collagen and cultured vascular cells. *Science* 231: 397–400.
Williams, S.K. (2002) Tissue Engineered vascular grafts: From bench to clinical use. *FASEB J.* 14(4): A305.
Wilson, G.J., Courtman, D.W., Klement, P., Lee, J.M., Yeger, H. (1995) Acellular matrix: A biomaterials approach for coronary artery bypass and heart valve replacement. *Ann. Thoracic Surg.* 60: S353–358.
Wu, K.K., and Thiagarajan, P. (1996) Role of endothelium in thrombosis and hemostasis. *Annu. Rev. Med.* 47: 315–331.
Zilla, P., Fasol, R., Preiss, P., Kadletz, M., Deutsch, M., Schima, H., Tsangaris, S., Groscurth, P. (1989) Use of fibrin glue as substrate for in vitro endothelialization of PTFE vascular grafts. *Surgery* 105: 515–522.
Zisch, A.H., Schenk, U., Schense, J.C., Sakiyama-Elbert, S.E., Hubbell, J.A. (2001) Covalently conjugated VEGF-fibrin matrices for endothelialization. *J. Controlled Release* 72: 101–113.

13

Tissue Engineering of Bone

Sandra Hofmann[1,3], David Kaplan[3], Gordana Vunjak-Novakovic[2], and Lorenz Meinel[1–3]

[1]*Institute of Pharmaceutical Sciences, Swiss Federal Institute of Technology, CH-8057 Zürich, Switzerland;* [2]*Department of Biomedical Engineering, Columbia University, 363G Engineering Terrace, Mail Code 8904, 1210 Amsterdam Avenue, New York, NY 10027;* [3]*Department of Bioengineering, Tufts University, Medford, Massachusetts, 02155*

Corresponding author: lorenz.meinel@pharma.ethz.ch

Culture of Cells for Tissue Engineering, edited by Gordana Vunjak-Novakovic and R. Ian Freshney

1. BACKGROUND

With the progressive aging of the population, the need for functional tissue substitutes is increasing. Organ transplantation and mechanical devices have revolutionized medical practice but have limitations. Skeletal tissue loss due to congenital defects, disease, and injury is currently treated by autologous tissue grafting, a method limited by the availability of the host tissue, harvesting difficulties, donor site morbidity, and the clinician's ability to contour delicate 3D shapes [Gross, 2003]. However, autologous grafts are ideal implants as they provide an (i) osteo*con*ductive scaffold (i.e., an environment in which cells can thrive), (ii) osteo*in*ductive growth factors, which are stored in the graft and released during osteoclastic resorption and kick-start the osteogenic process, and (iii) autologous cells, a viable component capable to (re)generate bone tissue and facilitate tissue integration at the implantation site. The generation of autologous bone grafts in vitro, avoiding the harvest of autologous tissue at a second anatomic location, is the ultimate goal in bone tissue engineering. Consequently, scientific strategies utilize and integrate all three components, (i) the scaffold, (ii) the osteoinductive factors, and (iii) the cells, to meet the gold standard for implants, the autologous graft.

Osteoconductive scaffolds facilitate cell attachment and tissue development, and they biodegrade in parallel with the accumulation of tissue components. Therefore, they initially provide a structural and logistical template and degrade matching the rate of bone deposition. The osteoconductive environment is essential for promoting orderly tissue regeneration and in particular benefits from appropriate geometry [Abe et al., 1982; Chu et al., 1995; Caplan et al., 1997; Schaefer et al., 2000; Hunziker, 2002]. Scaffold structure determines the transport of nutrients, metabolites, and regulatory molecules to and from the cells, and pore sizes and interconnectivity influence the geometry of the engineered tissue. These constraints directly impact the uniform distribution of cells throughout the scaffold. Optimized interconnected porosity facilitates cell colonization throughout the construct and is the requirement for fully viable implants. The importance of scaffold geometry has been demonstrated for some porous biodegradable polymers, where new tissue formation has been restricted to a superficial and only 200-μm-thick layer of calcified tissue [Crane et al., 1995; Ishaug et al., 1997].

2. 3D SILK SCAFFOLDS

2.1. Silk as a Biomaterial

The chemical nature of the biomaterial critically affects tissue formation. Physiologically, collagen type I makes up to 90% of the total protein content as present in bone and—following a biomimetic strategy—this biomaterial should be considered the ideal material. A problem in using natural collagen type I is the inadequate biodegradation of this fibrous polymer, resulting in insufficient mechanical properties for bone tissue engineering and a collapse of the engineered tissue resulting in unconnected calcified and unorganized bone clusters [Meinel et al., 2004b]. The biodegradation of collagen type I can be adapted through cross-linkage of the molecules, thereby matching the requirements for bone implants. However, the tissue reaction to cross-linked collagen scaffolds was quite unpredictable and resulted in spontaneous calcification, cytotoxic effects, and scar formation around the implant [van Luyn et al., 1992]. Spontaneous calcification might in particular result in problems for the engineering of osteochondral plugs, when the formation of pure cartilaginous tissue is impaired by the uncontrolled, and not cell-mediated, mineralization. These remarkable differences in cytotoxicity were connected to residual agents as a consequence of processing and the cross-linking agents used [van Luyn et al., 1992, van Wachem et al., 2001].

Among the available natural fibers, silks do have the strongest mechanical properties and even rival the best synthetic high-performance fibers, such as Kevlar, in overall performance (energy absorbed to break) [Cunniff et al., 1994]. The mechanical and thermal properties of dragline silk from the spider *Nephila claviceps* have previously been characterized with reference to silkworm silk [Cunniff et al., 1994]. The best properties of the native fibers collected and tested at quasi-static rates were 60 GPa and 2.9 GPa for initial modulus and ultimate tensile strength, respectively. Based on microscopic evaluations of knotted single fibers, no evidence of kink-band failure on the compressive side of a knot curve was observed [Cunniff et al., 1994]. Synthetic high-performance fibers fail by this mode even at relatively low stress levels; this is a major limitation with synthetic fibers in many applications. Furthermore, silks are mechanically stable up to almost 200 °C in dynamic mechanical evaluations [Cunniff et al., 1994]. In terms of material properties, silks also provide a range of mechanical features that suggest future applications in many different biomaterials needs. Table 13.1 provides some comparisons of the mechanical properties of silks (spider dragline and silkworm), collagen, cross-linked collagen scaffolds, and scaffolds prepared from a synthetic polymer, polylactic acid.

Table 13.1. Comparative mechanical properties of some key fibrous biomaterials for scaffolds.

Materials	Ultimate tensile Str. (mPa)	% Strain at failure	Modulus (MPa)	Reference
Collagen[1]	0.9–7.4	24.1–68.0	1.8–46.0	Pins et al. (1997)
Collagen (cross-linked)[2]	46.8–71.5	11.2–15.6	383–767	Pins et al. (1997)
Collagen[3]	—	—	1820–11,900	Cusack and Miller (1979)
Dragline silk[4]	200–2900	9–39	2000–60,000	Cunniff et al. (1994)
Silkworm silk[5]	600	15–35	5000	Cunniff et al. (1994)
L-PLA[6]	28–50	2.0–6.0	1200–3000	Engelberg and Kohn (1991)

[1] Collagen—tested after stretching from 0 up to 50%.
[2] Collagen—cross-linked and tested after stretching from 0 up to 50%.
[3] Collagen—rat tail: properties determined by light scattering
[4] Dragline silk—from *Nephila claviceps* spider
[5] Silkworm silk—from *Bombyx mori* silkworm cocoon
[6] L-PLA—molecular weights from 50,000 to 300,000, properties for D,L-PLA were in similar ranges

Silks are natural fibers predominantly harvested from the domesticated silkworm, *Bombyx mori*, and have been used traditionally in the form of threads in textiles and sutures for biomedical needs for thousands of years [Halsted, 1892; Lange, 1903, 1907; Ludloff, 1927; Perez-Rigueiro et al., 1998; Sofia et al., 2001]. Recently, silks have also been explored for an extended variety of biomedical applications including osteoblast and fibroblast cell support matrices and for ligament tissue engineering [Sofia et al., 2001; Altman et al., 2002; Meinel et al., 2004b,c].

Aside from existing natural sources of silk (*B. mori* silkworm silk from sericulture), future options from the availability of reasonable quantities of genetically engineered silk variants (such as from Polymer Technologies, Nexia Biotechnologies, and others) would expand the set of structures available for use in vivo.

Silkworm silk contains a fibrous protein termed fibroin (both heavy and light chains) that forms the thread core, which is encased in a sericin coat. Sericins are a family of gluelike proteins that hold the individual fibroin fibers together to form the composite fibers of the cocoon case to protect the metamorphosing worm. Fibroin is a protein up to 90% of which consists of the amino acids glycine, alanine, and serine that form water insoluble crystalline β-sheets on shearing, drawing, heating, spinning, or exposure in an electric field or to polar solvents such as methanol [Guhrs et al., 2000].

Silks from silkworm have been extensively characterized for biocompatibility, because they have been used for decades as sutures in vivo. Initially, adverse immunological reactions were found, and this prompted the replacement of silkworm silk sutures with nylons approximately 20 years ago [Soong and Kenyon, 1984]. However, it was clearly shown that adverse reactions to silks were due to the presence of residual sericin and not the fibroin itself [Soong and Kenyon, 1984]. The biocompatibility was also evaluated for mesenchymal stem cells. The inflammatory reaction was significantly lower in cells seeded on silk films as compared to collagen and polylactide (PLA) films. Furthermore, the accumulation of inflammatory cells around films implanted intramuscularly was again lower for silk as compared to collagen and PLA [Meinel et al., 2005; Meinel et al., 2003]. These data, both in vitro [Santin et al., 1999], and in vivo [Soong and Kenyon, 1984], illustrate that biocompatible silkworm silk rivals other biomaterials in use today.

A misconception with silk revolves around its degradation in vivo. The U.S. Pharmacopoeia defines an absorbable material as one that loses "most of its tensile strength within 60" days in vivo. By this definition, silk is correctly classified as nondegradable. However, according to the literature, silk is degradable over longer time frames as a function of proteolytic degradation and matrix mechanical fatigue [Soong and Kenyon, 1984; Rossitch, 1987; Bagi et al., 1995; Altman et al., 2003]. Several studies detail silk degradation in vivo with variable rates dependent on the animal model and implantation site [Postlethwait, 1970; Salthouse et al., 1977; Greenwald et al., 1994; Lam et al., 1995; Bucknall et al., 1983]. In general, silks lost the majority of their tensile strength within 1 year in vivo, and failed to be recognized in the implantation site within 2 years.

Taken together, silk protein-based polymers combine several advantages in particular as they potentially address the needs for bone tissues regrown in vitro for subsequent implantation.

- The *natural role* of structural/fibrous proteins in tissue remodeling, including collagens in the extracellular matrix (ECM).
- *Biocompatibility* with *degradability* properties for ingrowth and reintegration into native tissues, the need for materials that can be degraded or resorbed that will have minimal negative impact on surrounding tissues. Silks have been used as sutures for decades and been shown to be biocompatible and degradable [Meinel et al., 2003].
- The need for materials with robust *mechanical integrity* until new tissue is regenerated. Silks are unique in this feature, exhibiting strength, flexibility, and compression properties that exceed those of all other natural fibers. This is particularly important for in situ pairs where a matrix will need to be formed and retain mechanical integrity during osseoreintegration.
- The need for *matrices that can be functionalized* with cell growth factors, with control over placement and density of these factors using chemistries that are nontoxic and biocompatible. This approach has already been used successfully with silks in our prior studies [Sofia et al., 2001].
- The ability to *self-assemble to establish conformal fill-ins* in vivo *during tissue regeneration*, thus avoiding gaps leading to fibrous encapsulations and scar tissue. This feature would be required if the scaffolds were to be used in vivo as temporary matrices to fill in defects during osseointegration, and this is a characteristic of silks.

2.2. Silk Purification

It is apparent from the above that silk protein-based polymers are logical choices for biocompatible scaffolding for the formation of advanced matrices to induce bone tissue repair because of their mechanical properties as well as a wide range of other advantages.

Protocol 13.1. Purification of Silk for Bone Tissue Scaffolds

Reagents and Materials

- ❑ Silk cocoons from *Bombyx mori*
- ❑ Sodium carbonate
- ❑ Lithium bromide
- ❑ UPW
- ❑ Glass beakers, 2 L
- ❑ Falcon tubes, 50 ml
- ❑ Glass bottle, 100 ml

- ❑ Syringe 18G, 20 ml, and needle
- ❑ Millex-SV Syringe-driven filter unit, pore size 5 μm
- ❑ Slide-a-Lyzer® Dialysis Cassette 3–12 ml
- ❑ Oven (set at 55 °C)
- ❑ Magnetic heating/stirring plate with magnet
- ❑ Freezer at −70 °C to −80 °C
- ❑ Lyophilizer

Protocol

(a) Heat 0.75 L UPW in a glass beaker until it boils.
(b) Dissolve 1.59 g Na_2CO_3 in the 0.75 L UPW, up to a final molarity of 0.02 M.
(c) Cut 3 cocoons in 8 parts each, clean from larvae and debris.
(d) Transfer the cocoons into the boiling solution and boil under stirring for 1 h. Replace evaporated water.
(e) Rinse the silk with 1 L hot UPW.
(f) Rinse 10 times with 1 L cold UPW.
(g) Dry the silk overnight in a fume hood and note the dry weight.
(h) Prepare a 9 M solution (781.6 mg/ml) of LiBr in UPW.
(i) Prepare a 10% (w/v) solution of dried silk in 9 M LiBr in a glass bottle and leave silk at 55 °C for 4–5 h, until it is completely dissolved (some debris may be left).
(j) Filter the solution through a 5-μm pore size syringe filter.
(k) With a syringe, insert 6–8 ml of the filtered solution into a dialysis cassette and dialyze against 1 L of UPW (per cassette) on a magnetic stirrer plate.
(l) Replace UPW after 1 h, after 3 h, after 12 h, after 24 h and after 36 h (a total of 5 changes).
(m) Pipette the dialyzed solution into 25-ml aliquots in 50-ml Falcon tubes.
(n) Freeze the solution for 2 h at −70 °C to −80 °C.
(o) Lyophilize until silk has completely dried. This takes up to 3 days.
(p) Lyophilized silk can be stored at room temperature.

2.3. Preparation of Silk-RGD

Protocol 13.2 is for an efficient two-step coupling of proteins in solution using EDC and Sulfo-NHS. The procedure allows for sequential coupling of two proteins without exposing the second protein to EDC and thus affecting carboxyls on the second protein. This procedure quenches the first reaction with a thiol compound.

Protocol 13.2. Coupling of RGD Motif to Silk

Reagents and Materials

- ❑ Silk from Protocol 13.1 Step (k)
- ❑ Modified MES buffer

For a volume of 500 mL, pour 1 package of BupH™ MES buffered saline in a beaker and rinse empty package once with UPW. Add 10.11 g sodium chloride. Fill up with UPW to 450 ml. Titrate pH with a highly concentrated sodium hydroxide solution to pH 6. Add UPW to a final volume of 500 ml.

- ❑ GRGDS
- ❑ Sulfo-NHS
- ❑ EDC
- ❑ 2-Mercaptoethanol
- ❑ Hydroxylamine HCl
- ❑ UPW
- ❑ Glass beakers, 2 L
- ❑ Falcon tubes, 50 ml
- ❑ Syringe, 20 ml, and needle 18G
- ❑ Slide-a-Lyzer® Dialysis Cassette, 3–12 ml
- ❑ Magnetic heating/stirring plate with magnet
- ❑ Freezer at −70 °C to −80 °C
- ❑ Lyophilizer
- ❑ Glass bottle, 100 ml

Protocol

(a) Follow Protocol 13.1 until Step (k).

(b) Change UPW after 1 h, after 3 h, and the next morning.

(c) The next evening, discard the water and add 1 L modified MES buffer per dialysis cassette.

(d) Change modified MES buffer after 12 h.

(e) The next evening, carefully transfer the solution from the dialysis cassette into a glass bottle with a syringe.

(f) Per mg of dry silk protein—as measured in Protocol 13.1, step (g)—add 0.4 mg EDC and 1.1 mg Sulfo-NHS. Allow reaction to take place at room temperature for 15 min.

(g) In a fume hood: Per ml of solution, add 1.4 μl 2-mercaptoethanol per ml solution to quench the EDC.

(h) For 1.1 g of dry silk—as measured in Protocol 13.1, step (g)—add 7.5 mg GRGDS protein to the reaction mixture.

(i) Allow reaction at room temperature for 2 h.

(j) Add hydroxylamine HCl powder to a final concentration of 10 mM. This method of quenching hydrolyzes any unreacted NHS present on the surface of the silk and results in regeneration of the original carboxyls.

(k) Transfer 6–8 ml of the silk-RGD solution into a dialysis cassette and dialyze against 1 L UPW (per cassette) on a magnetic stirrer plate.

(l) Change UPW after 1 h, after 3 h, after 12 h, after 24 h and after 36 h (total of 5 changes).

(m) Transfer the dialyzed solution into 25-ml aliquots in 50-ml Falcon tubes.

(n) Freeze the solution for 2 h at −70 °C to −80 °C.

(o) Lyophilize until silk-RGD has completely dried. This may take up to 3 days.
(p) Lyophilized silk-RGD can be stored at room temperature.

Protocol 13.3. Preparation of Silk Scaffolds

Reagents and Materials

- ❑ Lyophilized silk (or silk-RGD) from Protocol 13.1 (or 13.2)
- ❑ HFIP

Note: HFIP is volatile, store bottle at 2–8 °C for 30 min prior to use. Precool pipette tips to 2–8 °C. Work under fume hood with HFIP. Close vessels immediately after use and wrap with Parafilm. Avoid skin contact or inhalation of HFIP. Read security sheets before working with HFIP.

- ❑ Sodium chloride USP, granular
- ❑ Methanol
- ❑ UPW
- ❑ Glass container, 20 ml, with snap cap
- ❑ Teflon container with snap cap
- ❑ Glass beaker, 1 L
- ❑ Parafilm
- ❑ Razor blade

Protocol

(a) Prepare a 17% (w/v) solution of lyophilized silk in HFIP in a glass container. Close the container firmly and seal with Parafilm. The dissolution procedure may take up to 1 day at room temperature.
(b) Add 3.4 g porogen (e.g., NaCl) into a Teflon container.
(c) Add 1 ml dissolved silk on the porogen and immediately cover the Teflon container (minimize evaporation of HFIP). The general ratio of NaCl to silk is 20:1 (w/w).
(d) Leave container closed for 6 h.
(e) Then open the Teflon container and allow evaporation of the solvent only in a fume hood. This may take up to 4 days.
(f) Carefully remove the salt-silk composite from the container by tapping it upside down on the bench. If the scaffold doesn't come out easily, let it dry for another 1 or 2 days.
(g) Immerse the salt-silk composite in 90% methanol in H_2O (v/v) for 30 min. Polar solvents such as methanol induce a conformational change of the water-soluble silk I into the water-insoluble silk II conformation.
(h) Remove the scaffold from the methanol solution and dry overnight in a fume hood.
(i) Transfer the dry scaffold into a 1-L beaker with UPW for at least 24 h for salt leaching. Change the water at least 5 times.

(j) Air dry scaffold. This may take up to 3 days.
(k) To cut the scaffold into the desired shape, completely soak scaffolds in UPW for 10 min. Cut scaffold with a sharp razor blade.

---

3. PREPARATION OF REAGENTS AND MEDIA

For the engineering of bone from mesenchymal stem cells, medium composition guides differentiation along the osteogenic lineage. This differentiation process is accompanied by the expression/activity of transcription factors, regulating lineage restriction, commitment, and/or differentiation within some of the mesenchymal lineages including osteoblasts [Ducy et al., 1997; Ducy, 2000; Liu et al., 2001]. A master regulatory transcription factor for osteogenic differentiation is cbfa1. Cbfa1 gene deletion results in complete absence of bone formation and mature osteoblasts [Komori et al., 1997; Otto et al., 1997]. This demonstrates the importance of this master transcription factor for osteoblast differentiation and bone formation. Notably, cbfa1 overexpression blocks osteoblast maturation [Liu et al., 2001]. In analogy to the consequences of overexpression or deletion of cbfa1, many reports document a stimulatory effect of BMPs on osteoblasts, and a few show inhibitory effects. A possible explanation may be the dependence of the actions of growth factors on the relative stage of differentiation of the target cells.

BMP-2 is generally used at high concentrations of 1 μg/ml, but significant effects on total mineralization were observed at concentrations as low as 40 ng/ml compared to medium without supplementation of BMP-2 [Meinel et al., 2004a]. Another approach, avoiding cost-intensive use of recombinant growth factors, is the use of adenoviral transfection. However, the transduction of MSCs with the adenoviral vector carrying the BMP-2 gene resulted in significantly lower mineralization compared to untransfected MSCs cultured in the presence of BMP-2 concentrations as measured for the transduced cells. A possible explanation is that the dual role of the transduced cells—BMP-2 production and mineralization—prevents them from performing equally well as untransduced cells exposed to BMP-2-supplemented medium.

Glucocorticoids such as dexamethasone, added routinely in assays of osteoprogenitors (CFU-O), also have both inhibitory and stimulatory effects on skeletal cells, and an emerging view is that this reflects opposite effects on precursors versus more mature cells in the lineages [Gronthos et al., 1994; Aubin and Liu, 1996; Aubin and Triffitt, 2002]. It has also been suggested that even a transient exposure of stem cells to dexamethasone may be effective in inducing and maintaining the osteoblastic phenotype [Jaiswal et al., 1997]. Glycerophosphate has been found to have a significant effect to induce osteogenic differentiation and increases alkaline phosphatase activity and osteocalcin production. L-Ascorbic acid (AA, vitamin C) increases cell viability and is a cofactor in the hydroxylation of proline and lysine residues and is therefore necessary for the production of collagen. AA has also been demonstrated to increase alkaline phosphatase activity [Choong et al., 1993]. Together with β-glycerophosphate, AA was found to be a prerequisite for the formation and mineralization of the extracellular matrix [Maniatopoulos et al., 1988]. The composition of the medium directly affects the osteogenic differentiation process. The protocols suggest the use of a metabolically stable variant of ascorbic acid, ascorbic acid 2-phosphate.

Prepare culture media aseptically. Store at 4 °C; preheat to 37 °C prior to use.

3.1. Control Medium

Dulbecco's modified Eagle's medium (DMEM)
Fetal bovine serum (FBS), 10%
Penicillin, 100 U/ml, streptomycin, 100 μg/ml
Fungizone, 0.5 μg/ml

Can be stored at 4 °C for up to 1 week.

3.2. Expansion Medium

Dulbecco's modified Eagle's medium (DMEM)
Fetal bovine serum (FBS), 10%
Nonessential amino acids solution, 1%
Penicillin, 100 U/ml, streptomycin, 100 μg/ml
Fungizone, 0.5 μg/ml
bFGF (human, recombinant), 1 ng/ml

Can be stored at 4 °C for up to 1 week.

3.3. Osteogenic Medium

Dulbecco's modified Eagle's medium (DMEM)
Fetal bovine serum (FBS), 10%
Penicillin, 100 U/ml, streptomycin, 100 μg/ml
Fungizone, 0.5 μg/ml
Ascorbic acid 2-phosphate, 50 μg/ml
Dexamethasone, 10 nM
β-Glycerophosphate, 10 mM
BMP-2, 1 μg/ml

Use medium immediately; do not store or reuse.

3.4. Osteogenic Medium Double Concentration

Dulbecco's modified Eagle's medium (DMEM)
fetal bovine serum (FBS), 10%
Penicillin, 100 U/ml, streptomycin, 100 μg/ml
Fungizone, 0.5 μg/ml
Ascorbic acid 2-phosphate, 100 μg/ml
Dexamethasone, 20 nM

β-Glycerophosphate, 20 mM
BMP-2, 2 μg/ml

Use medium immediately; do not store or reuse.

3.5. Cartilage Medium (for Cell Characterization)

Dulbecco's modified Eagle's medium (DMEM)
Fetal bovine serum (FBS), 10%
Penicillin, 100 U/ml, streptomycin, 100 μg/ml
Fungizone, 0.5 μg/ml
Ascorbic acid 2-phosphate, 50 μg/mL
Dexamethasone, 10 nM
Nonessential amino acids solution, 1%
Insulin, 5 μg/ml
TGF-β_1, 5 ng/ml

Use medium immediately; do not store or reuse.

3.6. Cartilage Medium Double Concentration

Dulbecco's modified Eagle's medium (DMEM)
Fetal bovine serum (FBS), 10%
Penicillin, 100 U/ml, streptomycin, 100 μg/ml
Fungizone, 0.5 μg/ml
Ascorbic acid 2-phosphate, 100 μg/ml
Dexamethasone, 20 nM
Nonessential amino acids solution, 2%
Insulin, 10 μg/ml
TGF-β_1, 10 ng/ml

Use medium immediately; do not store or reuse.

3.7. Preparation of bFGF

Lyophilized samples should be reconstituted with sterile 10 mM Tris pH 7.6 containing 1% BSA to a final bFGF concentration of 0.1 mg/ml. For longer term storage, aliquot and store in polypropylene vials at $-20\,^{\circ}$C. Avoid repeated freeze-thaw cycles. In applications requiring long-term use of this growth factor in cell cultures, refilter material after dilution in protein (BSA or FBS)-containing buffer, through a 0.22-μm low-protein-binding filter.

3.8. Preparation of BMP-2

Prepare a 1 mg/ml BMP-2 stock solution in a buffer containing 0.5% sucrose, 2.5% glycine, 5.0 mM glutamic acid, 5.0 mM sodium chloride, 0.01% Tween 80, pH 4.5. Store in single-use aliquots at $-80\,^{\circ}$C or $-20\,^{\circ}$C. Avoid repeated freeze-thawing. To avoid loss of protein due to adherence to surfaces, BMP-2 should always be added to culture medium after the addition of carrier protein (0.1% BSA or 1–10% appropriate serum).

3.9. Preparation of TGF-β_1

Purified recombinant human TGF-β_1 is an extremely hydrophobic protein that adheres strongly to surfaces. To ensure recovery, lyophilized samples should be reconstituted with sterile 4 mM HCl containing 1 mg/ml BSA to a final TGF-β_1 concentration of no less than 1 μg/ml. Upon reconstitution, this cytokine can be stored under sterile conditions at 2–8 °C for 1 month or at −20 °C to −80 °C in a manual defrost freezer for 3 months without detectable loss of activity. Avoid repeated freeze-thaw cycles.

4. ISOLATION AND CULTURE METHODOLOGY OF HMSC

Bone marrow contains at least two distinct populations of stem cells, one hematopoietic and the other nonhematopoietic mesenchymal. Hematopoietic stem cells in the adult give rise to all components of the immune and blood system [Lagasse et al., 2000], whereas mesenchymal stem cells (MSC) can differentiate into bone, cartilage, or adipose tissue [Pittenger et al., 1999]. However, MSCs are present within adult bone marrow at an exceedingly low frequency of <1 MSC per 10^6 bone marrow cells [Bruder et al., 1994]. The low frequency of mesenchymal cells within the marrow compartment has led investigators to develop a variety of techniques for stem cell isolation and culture. The described isolation procedure is based on Histopaque gradient centrifugation isolating mononuclear cells with a density of 1.077 g/cm^3, the cells' ability to adhere to the tissue culture plate surface, and the ability of the cells to undergo chondrogenic and osteogenic differentiation. A thorough characterization of the isolated cells is essential for comparison with other experiments performed with cells harvested from other volunteers. A routine assessment can involve the analysis of the cell morphology, as mesenchymal cells are typically adherent marrow cells, generally considered to be spindle shaped. Further analysis involves the surface antigen expression by flow cytometry and their ability to differentiate along skeletal lineages [Pittenger et al., 1999; Meinel et al., 2004b]. The analysis of cells in different passages can provide information about the number of expansion cycles possible without a significant reduction in differentiation capacity. We generally observe stable differentiation capacity up to 4 passages, using the techniques as described in Protocols 13.5, 13.6, 13.8 and Fig 13.1.

To date, an antigen surface determinant considered specific for mesenchymal stem cells has not been found. The so-called SH2 antibodies originally developed by Caplan and colleagues recognize CD105, which is also known as endoglin and is a regulatory component of the TGF-β receptor complex. Many investigators have explored CD105 as an important antigenic determinant in the identification of mesenchymal stem cells. However, it is a matter of debate whether mesenchymal stem cells can be identified solely by flow cytometry and, thereby, clearly distinguished

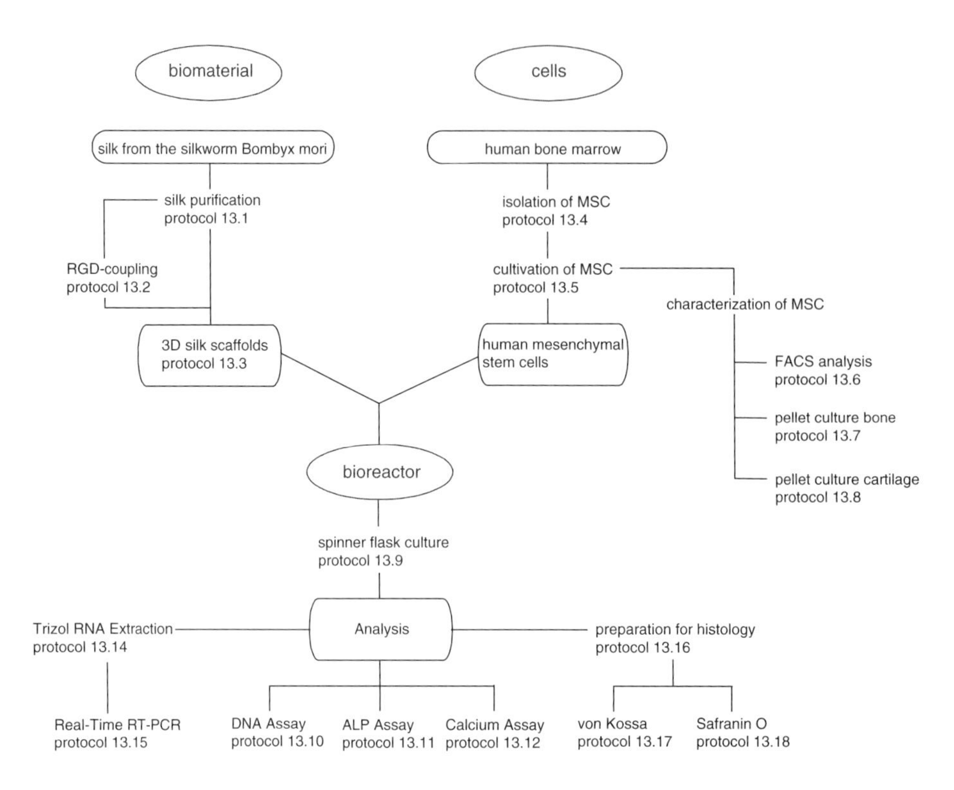

Figure 13.1. Schematic overview of experimental approach.

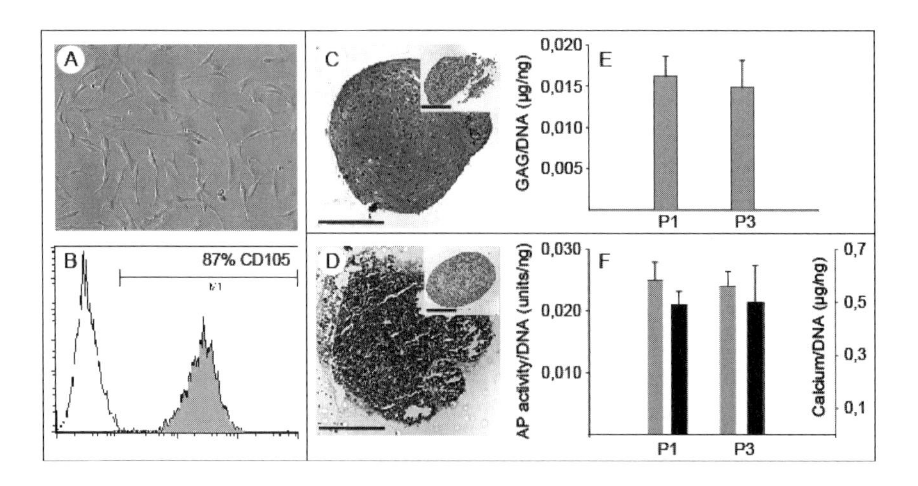

Figure 13.2. Characterization of MSCs. A) phase-contrast photomicrographs of putative passage 2 MSCs at approximately 40% confluence at an original magnification of ×20. B) Surface CD105 expression of passage 2 MSCs. C) Characterization of chondrocyte differentiation of MSCs in pellet culture, treated either with chondrogenic medium or with control medium (insert). Pellet diameter is approximately 2 mm, stained with safranin O. D) Characterization of osteoblast differentiation of MSCs in pellet culture, treated either with osteogenic medium or with control medium (insert). Pellet diameter is approximately 2 mm, stained according to von Kossa. E) Sulfated GAG/DNA (µg/ng) deposition of passage 1 and 3 MSC pellet culture after 4 weeks. F) Calcium deposition/DNA (µg/ng) and AP activity/DNA (units/ng) of passage 1 and 3 MSC pellet culture. Reprinted with the permission of the Biomedical Research Society. (See Color Plate 8.)

from other adherent mesenchymal cells solely by antigen expression [Prockop et al., 2001, 2003]. Therefore, we suggest evaluating the absence of adherent and nonadherent cells, in particular of hematopoietic (CD34) or endothelial origin (CD31) [Spangrude et al., 1988; DeLisser et al., 1993; Negrin et al., 2000]. The protocols also recommend the analysis of CD71, the transferrin receptor, to assess the proliferative state of the isolated cells.

The ability of hMSCs to undergo selective differentiation in response to environmental factors is documented by chondrogenic or osteogenic differentiation in chondrogenic or osteogenic culture medium, respectively, and the lack of differentiation in control medium. Similar per-cell amounts of glycosaminoglycan (GAG) and calcium measured in cells of higher passages (e.g., P_1, P_3, and P_5) are also important as they enable the use of small initial bone marrow aspirates to obtain sufficient amounts of cells for seeding clinically sized scaffolds. Together, the capacity for expansion in the undifferentiated state and the maintained ability for subsequent chondrogenic or osteogenic differentiation establish the feasibility of using hMSCs for bone tissue engineering (Fig. 13.2, See Color Plate 8) [Meinel et al., 2004a].

4.1. Isolation of Mesenchymal Stem Cells

Protocol 13.4. Isolation of hMSCs from Bone Marrow

Reagents and Materials

Sterile

- ❑ Fresh bone marrow, EDTA- or heparin treated; keep on ice
- ❑ Expansion medium (See Section 3.2)
- ❑ Control medium (See Section 3.1)
- ❑ RPMI 1640 medium supplemented with 5% FBS
- ❑ Histopaque®-1077
- ❑ PBSA
- ❑ Trypan Blue
- ❑ Trypsin-EDTA
- ❑ FBS supplemented with 10% DMSO
- ❑ Culture flasks, 150 cm^2
- ❑ Accuspin™ tube
- ❑ Polypropylene Falcon tubes, 50 ml
- ❑ Cryovials, 2 ml

Nonsterile

- ❑ Hemocytometer
- ❑ Nalgene™ Cryo 1 °C Freezing Container
- ❑ Centrifuge
- ❑ Incubator at 37 °C, 5% CO_2

Protocol

(a) Add 15 ml Histopaque®-1077 (room temperature) to the upper chamber of the Accuspin™ tube.

(b) Centrifuge at 800 *g* for 30 s at room temperature. Histopaque®-1077 is now in the lower chamber of the tube.
(c) Add 10 ml RPMI-FBS solution to 5-ml bone marrow aliquots and gently mix by pipetting.
(d) Pipette suspension carefully to the upper chamber of the prepared Accuspin™ tube.
(e) Centrifuge at 800 *g* for 15 min at room temperature.
(f) Carefully transfer the opaque interface, which contains the mononuclear cells, into a clean Falcon tube with a pipette.
(g) Pellet cells by centrifugation at 300 *g* for 10 min and 4 °C, aspirate supernate, and resuspend cells in RPMI-FBS medium.
(h) Count an aliquot of the cell suspension using a hemocytometer and determine viability by dye exclusion with Trypan Blue.
(i) Centrifuge cells at 300 *g* for 10 min and 4 °C, aspirate supernate, and resuspend cells in expansion medium.
(j) Plate suspension at a density of 5×10^3 cells/cm^2 in expansion medium in a cell culture flask (40 ml per flask).
(k) Incubate the cells in a humidified 37 °C, 5% CO_2 incubator.
(l) At days 3 and 5: add 20 ml expansion medium to each flask.
(m) When cells reach 60% confluence: wash the cells 2 times with PBSA and add fresh expansion medium.
(n) Change medium twice a week until cells reach 80% confluence (days 12–22), then aspirate expansion medium and wash cells twice with PBSA at 37 °C.
(o) Add 8 ml trypsin-EDTA to each flask and distribute the solution evenly.
(p) Incubate for 5 min or until cells start to detach at 37 °C.
(q) Inactivate the trypsin by adding 8 ml control medium cooled to 4 °C, mix, and transfer into a clean 50-ml Falcon tube. Use cold serum-containing medium at this step to block trypsin activity.
(r) Wash the flask once again by adding 8 ml control medium cooled to 4 °C and combine with the mix from Step (q) in the Falcon tube.
(s) Centrifuge at 300 *g* for 10 min at 4 °C.
(t) Aspirate the supernate and resuspend the cells in 10 ml control medium.
(u) Combine all cell suspensions in one Falcon tube, mix well by pipetting.
(v) Count an aliquot of the cell suspension using a hemocytometer and determine cell viability by dye exclusion with Trypan Blue. Count unstained cells.
(w) Centrifuge cells at 300 *g* for 10 min at 4 °C, aspirate the supernate, and resuspend cells at a concentration of 5×10^6 in FBS-DMSO solution. Work rapidly as unfrozen cells should be exposed to DMSO for the shortest time possible.
(x) Freeze 1 ml per cryovial at a rate of 1 °C/min in a freezing container for 6 h at −80 °C.
(y) Transfer cells into liquid nitrogen freezer.

4.2. Thawing and Maintenance of hMSC Cell Culture

Bone marrow mononuclear cells are obtained by Histopaque density gradient centrifugation. hMSCs, or possibly a specific subpopulation, can be expanded in vitro as undifferentiated cells responsive to environmental cues inducing differentiation toward mesenchymal lineages as well as other lineages [Cancedda et al., 2003a,b]. bFGF supplementation of the culture medium promotes hMSC proliferation and maintains their multilineage potential during expansion [Bruder et al., 1997; Freed et al., 1997; Meinel et al., 2004b,c].

Protocol 13.5. Expansion of Human Mesenchymal Stem Cells

Reagents and Materials

Sterile

- ❑ Expansion medium (See Section 3.2)
- ❑ Control medium (See Section 3.1)
- ❑ Trypan blue
- ❑ PBSA
- ❑ Trypsin-EDTA
- ❑ Polypropylene Falcon tubes, 50 ml
- ❑ Triple flasks (Nalgene Nunc)

Unless noted otherwise, reagent temperatures are 37 °C.

Nonsterile

- ❑ Hemocytometer
- ❑ Centrifuge
- ❑ Water bath, 37 °C
- ❑ Incubator at 37 °C, 5% CO_2

Protocol

(a) To recover the frozen cells, thaw rapidly at 37 °C (water bath). Unfrozen cells should be exposed to DMSO for the shortest time possible.
(b) Transfer the cells to a 50-ml Falcon tube containing 30 ml DMEM at 4 °C.
(c) Centrifuge at 300 g for 10 min at 4 °C, aspirate the supernate, and resuspend the cells in expansion medium.
(d) Count an aliquot of the cell suspension using a hemacytometer and determine cell viability through dye exclusion with Trypan Blue. Count unstained cells only.
(e) Plate cells at a density of 5×10^3 cells/cm^2 in expansion medium in a Triple flask (100 ml).
(f) At day 3, add 50 ml expansion medium per Triple flask.
(g) At day 5, replace medium with 100 ml fresh expansion medium.
(h) When cells reach confluence (around day 7), aspirate expansion medium and wash cells twice with PBSA at 37 °C.

(i) Add 15 ml of Trypsin-EDTA to each flask and distribute the solution evenly.
(j) Incubate for 5 min or until cells start to detach at 37 °C.
(k) Inactivate the trypsin by adding 15 ml control medium at 4 °C, mix, and pour into a clean 50-ml Falcon tube. As the antitryptic activity of serum-free medium may be insufficient to quench the proteolytic activity of the trypsin, it is important to use serum-containing medium at this step.
(l) Wash the flask once again by adding 15 ml of control medium at 4 °C and combine with the mix from Step (j) in the Falcon tube.
(m) Centrifuge at 300 *g* for 10 min at 4 °C.
(n) Aspirate the supernate and resuspend the cells in 10 ml control medium.
(o) Combine all cell suspensions in one Falcon tube; mix well by pipetting.
(p) Count an aliquot of the cell suspension using a hemocytometer and determine viability through dye exclusion with Trypan Blue. Count unstained cells.
(q) Centrifuge cells at 300 g for 10 min at 4 °C, aspirate the supernate, and resuspend cells at the desired concentration.

5. CHARACTERIZATION OF HMSCS

Bone marrow is a complex tissue comprised of hematopoietic precursors, their differentiated progeny, and a connective tissue network referred to as stroma. The stroma itself is a heterogenous mixture of cells including adipocytes, reticulocytes, endothelial cells, and fibroblastic cells that are in direct contact with the hematopoietic elements. Since it was well established that the stroma contains cells that differentiate into bone, cartilage, fat, and a connective tissue that supports the differentiation of hematopoietic stem cells, identification of the progenitor cells for these mesenchymal tissues has been an area of investigation [Bruder et al., 1997].

hMSCs identified with established isolation techniques (plastic adhesion, flow cytometry) are highly heterogeneous and have a variable potential for mesenchymal tissue development. Recently identified markers of mesenchymal and bone [Gronthos et al., 2003] progenitors have been used to address this problem. Consequently, we characterize our isolated and expanded hMSCs by their surface antigen pattern with fluorescence-activated cell sorting (FACS) and their capacity to differentiate into bone or cartilage depending on the number of culture passages before their use.

Purified hMSCs have been extensively characterized with respect to their complement of cell surface and extracellular matrix molecules, as well as their secretory cytokine profile in control and experimental conditions [Haynesworth et al., 1992].

The low frequency of mesenchymal stem cells within the marrow compartment (less than 1 MSC per 10^6 bone marrow mononuclear cells [Caplan, 1994]) has led investigators to the development of a variety of methods for stem cell isolation and culture. It is important to show that the cells are indeed hMSCs.

5.1. Authentication of Surface Antigen Analysis

Protocol 13.6. FACS Analysis of Human Mesenchymal Stem Cells

Reagents and Materials

Sterile

- ❑ RPMI 1640 medium supplemented with 5% FBS
- ❑ Centrifuge tubes, 50 ml (Falcon)
- ❑ Polystyrene tubes, 5 ml (Falcon)
- ❑ Antibodies: CD14-FITC, CD31-PE, CD34-APC, CD44-FITC, CD45-APC, CD71-APC, mouse anti-human CD105, goat anti-mouse IgG-FITC

Nonsterile

- ❑ Formalin 2% in PBSA (prepare fresh, keep at 4 °C)
- ❑ Incubator at 37 °C, 5% CO_2
- ❑ Hemocytometer
- ❑ Vortex mixer

Protocol

(a) Prepare cells according to Protocol 13.5 until step (m).
(b) Wash cells twice in RPMI-FBS medium.
(c) Count an aliquot of the cell suspension using a hemocytometer and determine cell viability through dye exclusion with Trypan Blue. Count unstained cells.
(d) Centrifuge cells at 300 *g* for 10 min at 4 °C, aspirate supernate, and resuspend cells in RPMI-FBS at a concentration of 1×10^7 cells/ml in a 50-ml Falcon tube.
(e) Add 50 µl of this cell suspension (500,000 cells) in each 5-ml Falcon tube.
(f) Add 2 µl of each desired CD stock solution (CD45-APC, CD44-FITC, CD31-PE, CD34-APC, CD14-FITC, CD71-APC, CD105; concentration of each: 0.5 µg/µl) into each corresponding 5-ml Falcon tube.
(g) Incubate for 30 min on ice in the dark.
(h) Centrifuge at 300 *g* for 5 min at 4 °C and aspirate supernate.
(i) Wash in 1 ml RPMI-FBS medium.
(j) Centrifuge at 300 *g* for 5 min at 4 °C.
(k) Aspirate medium.
(l) Add 1 ml RPMI-FBS medium and resuspend cells gently.
(m) Centrifuge at 300 *g* for 5 min at 4 °C.
(n) Carefully aspirate supernate.
(o) Resuspend cells in 1 ml RPMI-FBS medium; place on ice, except for tubes designated CD105 (go to Step (p) for non-labelled antibodies; here for CD105; and for others continue at (u)).
(p) All tubes designated CD105: add 50 µl RPMI-FBS and 2 µl secondary antibody (polyclonal goat anti-mouse IgG-FITC); incubate on ice for 30 min in the dark.

(q) All tubes designated CD105: Centrifuge at 300 *g* for 5 min at 4 °C, aspirate supernate.
(r) All tubes designated CD105: Wash in 1 ml RPMI-FBS medium.
(s) All tubes designated CD105: Centrifuge at 300 *g* for 5 min at 4 °C, aspirate supernate.
(t) All tubes designated CD105: Resuspend cells in 1 ml RPMI-FBS medium, place on ice.
(u) Centrifuge all tubes at 300 *g* for 5 min at 4 °C.
(v) Aspirate supernate and resuspend in 1 ml PBSA.
(w) Centrifuge all tubes at 300 *g* for 5 min at 4 °C.
(x) Carefully aspirate supernate.
(y) All tubes: Add 100 μl Formalin 2% in PBSA.
(z) All tubes: Store at 4 °C in the dark (aluminum foil) and measure within a day.

5.2. Pellet Culture of Bone

Protocol 13.7. Osteogenic Differentiation in Pellet Culture

Reagents and Materials

Sterile or aseptically prepared

- ❑ hMSCs, prepared according to Protocol 13.5
- ❑ Osteogenic medium
- ❑ Screw cap microcentrifuge tubes, 2 ml

Nonsterile

- ❑ Incubator at 37 °C, 5% CO_2

Protocol

(a) Prepare cells according to Protocol 13.5.
(b) Dilute cells to 2×10^5 cells/ml in osteogenic medium.
(c) Pipette 1 ml of the cell suspension (200,000 cells) into each screw cap tube. Make sure cells are well resuspended, to get the same number of cells per tube.
(d) Centrifuge at 300 *g* for 10 min at 4 °C.
(e) Unscrew the cap to allow ventilation and prevent contamination.
(f) Incubate cells at 37 °C, 5% CO_2.
(g) Change medium 3 times per week by carefully aspirating 0.5 ml medium with a pipette and adding 0.5 ml osteogenic medium double concentration (See Section 3.4).

Note: For analysis of the osteogenic differentiation of the cell pellet use Protocols 13.10, 13.11, 13.12, 13.14, 13.15, 13.16 and 13.17.

5.3. Pellet Culture of Cartilage

Protocol 13.8. Chondrogenic Differentiation in Pellet Culture

Reagents and Materials

Sterile or aseptically prepared

- ❑ hMSCs, prepared according to Protocol 13.5
- ❑ Cartilage medium
- ❑ Screw cap microcentrifuge tubes, 2 ml

Nonsterile

- ❑ Incubator at 37 °C, 5.0% CO_2

Protocol

(a) Prepare cells according to Protocol 13.5.
(b) Dilute cells to 2×10^5 cells/ml in cartilage medium.
(c) Pipette 1 ml of the cell suspension (200,000 cells) into each screw cap tube. Make sure cells are well resuspended, to get the same number of cells per tube.
(d) Centrifuge at 300 g for 10 min at 4 °C.
(e) Unscrew the cap to allow ventilation and prevent contamination.
(f) Incubate cells at 37 °C, 5% CO_2.
(g) Change medium 3 times per week by carefully aspirating 0.5 ml medium with a pipette and adding 0.5 ml Cartilage medium double concentration (See Section 3.6).

Note: For analysis of the differentiation of the cell pellet into cartilage use Protocols 13.10, 13.13, 13.14, 13.15, 13.16 and 13.18.

6. TISSUE ENGINEERING OF BONE

Mechanically active environments generally result in better engineered tissue composition, morphology, and mechanical properties than static environments. Most likely, this is due to enhanced mass transport at tissue surfaces [Freed and Vunjak-Novakovic, 2000]. Transport limitations are a significant problem in the engineering of bone, a highly vascularized tissue, which could be cultured only to thicknesses of 250–500 μm in the best case in static culture [Ishaug et al., 1997; Martin et al., 2001].

The mechanical environment is modulated through the use of bioreactors, including spinner flasks, with tissue constructs fixed in place and cultured either statically or in well-mixed medium [Vunjak-Novakovic et al., 1996]. Other bioreactors are rotating vessels, in which the tissue constructs are suspended in dynamic laminar flow [Freed et al., 1997], and perfused chambers with flow of medium around [Glowacki et al., 1998] or through the constructs [Pazzano et al., 2000; Meinel et al., 2004b]. The presented protocols describe the use of spinner flasks, which are easy to set up and lead to higher mineralization rates than static cultures

[Pazzano et al., 2000; Meinel et al., 2004b,c]. The protocols suggest the use of scaffolds with a diameter of 12 mm, and diameter can be chosen between 3 and 20 mm with the same protocol. The thickness of the construct is the limiting factor. Generally, silk scaffolds prepared according to Protocol 13.3 and cut to a final thickness between 0.5 and 3 mm result in a homogenous mineralization throughout the scaffold, compared to a less dense mineralized zone in the center for scaffolds exceeding a thickness of 5 mm.

In addition to biological stimuli, bioreactors provide physiologically relevant physical signals (e.g., interstitial fluid flow, shear, pressure, mechanical compression). A recent study details the geometry and extent of mineralized tissue under laminar and dynamic flow conditions [Meinel et al., 2004b].

6.1. Preparation of Spinner Flask

Protocol 13.9. Culture of hMSCs on Silk-RGD Scaffolds in Spinner Flasks

Reagents and Materials

Sterile

- ❑ Silk or Silk-RGD scaffolds prepared according to Protocol 13.3
- ❑ Cells prepared according to Protocol 13.5
- ❑ PBSA
- ❑ Matrigel
- ❑ Ethanol 70%
- ❑ Ethanol pads
- ❑ Control medium (See Section 3.1)
- ❑ Osteogenic medium (See Section 3.3)
- ❑ Osteogenic medium double concentration (See Section 3.4)
- ❑ Autoclaved dermal puncher (diameter 3 mm–1 cm)
- ❑ Razor blades
- ❑ Tube for stoppers
- ❑ Spinner flasks
- ❑ Tweezers
- ❑ Petri dish
- ❑ Plate, 6 well
- ❑ Snap cap centrifuge tubes, 1.5 or 2 ml
- ❑ Syringe, 5 ml
- ❑ Gloves
- ❑ Tubing (for stoppers)
- ❑ Autoclaved towel (as an underlay)

Nonsterile

- ❑ Centrifuge
- ❑ Magnetic stirrer, placed in the incubator
- ❑ Incubator 37 °C, 5% CO_2

Protocol

A. Preparation of silk scaffolds

(a) Soak the scaffolds in PBSA for 10 min (e.g., in a Petri dish), until they are completely soaked.
(b) Cut scaffolds into 1- to 2-mm-thick disks with razor blades.
(c) Punch out scaffolds with diameters of 3–10 mm.
(d) Transfer scaffolds into a Petri dish and air dry overnight.
(e) Autoclave the scaffolds (121 °C, 15 min).

Note: From now on work under aseptic conditions.

(f) Transfer the scaffolds with sterile tweezers into a 6-well plate; add DMEM and transfer to incubator.

B. Preparation of spinner flasks

(a) Clean and assemble the spinner flask.
(b) Check if 20G needle can be slipped over the wire of the spinner flask.
(c) Mark 75 and 150 ml levels.
(d) Close spinner flask, wrap in aluminum foil, and autoclave.

C. Preparation of stoppers

(a) Cut tubing into 0.5-mm-thick disks.
(b) Cut disks into 4 pieces (2 mm long and 2 mm wide).
(c) Autoclave stoppers.

D. Preparation of cells

Note: Matrigel must be kept on ice to prevent from gelling. Defrost the Matrigel on ice in a 2–8 °C refrigerator (takes about 2 h)

(a) Resuspend cells from Protocol 13.5 in control medium at a concentration of 5×10^6 cells/ml.
(b) Pipette 1 ml of the cell suspension (5×10^6 cells) into each snap cap tube. Make sure you always resuspend the cells well to get the same number of cells per well.
(c) Centrifuge tubes at 300 g for 10 min at 4 °C.
(d) Aspirate supernate.
(e) Add 20 μl Matrigel to each cell pellet; keep suspension on ice.
(f) Take 6-well plate with scaffolds out of incubator.
(g) Aspirate the medium completely.
(h) Carefully resuspend the cells in the Matrigel with the pipette tip.
(i) Transfer cells from one tube onto one scaffold.

(j) After completing one 6-well plate, put it into the incubator for 15 min without medium to allow gel formation.
(k) Add 5 ml control medium per well.

E. Preparation of spinner flask culture

Note: Work with sterile gloves and under aseptic conditions. Work with an assistant.

(a) Let the assistant unwrap the autoclaved towel without touching it; take the towel, unwrap and use as an underlay.
(b) Same procedure with the wrapped spinner flask, tweezers, the 5-ml syringe, and needles. Place everything on the towel.
(c) Open the spinner flask; assemble syringe and needle.
(d) The assistant gets the cells seeded on the scaffolds from the incubator and opens the 6-well plate in the laminar flow hood. Everything that has been touched by the assistant is now under the hood, just next to the sterile towel ("your" aseptic area).
(e) With syringe and needle, thread one stopper; fix it with the tweezers in the middle of the needle.
(f) Use tweezers for the first scaffold to get it from the 6-well plate. Thread the cell-seeded scaffold onto the needle. Make sure the cells are on the top side!
(g) Thread a second and a third stopper on the same needle, then the second scaffold, then the fourth stopper.
(h) Slip the needle over the needle of the spinner flask and position the tip of it at the position where the upper scaffold should be. Use tweezers to slide over the first scaffold plus upper and lower stopper on the flask's needle (scaffold must be below the 150 ml mark).
(i) Repeat for the second cell-seeded scaffold.
(j) Repeat Steps (f)–(i) 3 times, so you have a total of 8 scaffolds per spinner flask thread on 4 wires.
(k) Carefully close the spinner flask.
(l) Add 150 ml medium through one sidearm.
(m) Transfer the spinner flask on the magnetic stirrer plate in the incubator.
(n) Unscrew the screw cap on both sides a little for ventilation. Do not open completely, to avoid contamination.
(o) Stir at 60 rpm.
(p) Change medium 3 times a week.

F. Medium change
(a) Close screw caps and turn off the magnetic stirrer.
(b) Take the spinner flask from the incubator into the hood; spray well with ethanol 70%.
(c) Clean sidearms with an ethanol pad and open one sidearm.

(d) With a 2-ml sterile pipette, take out a medium sample, e.g., for determination of secreted proteins.
(e) With a Pasteur pipette, aspirate medium until the liquid level equals the 75 ml mark. Avoid touching the scaffolds with the pipette.
(f) Add 75 ml fresh medium—either control medium or osteogenic medium with a double concentration—using a 25-ml pipette.
(g) Close the screw cap.
(h) Put the spinner flask back into the incubator.
(i) Unscrew the screw cap for aspiration; do not open completely to avoid contamination.
(j) Stir at 60 rpm.

---

7. ANALYTICAL ASSAYS

A thorough analysis of mineralization and the progression of differentiation relies on biochemical assays and quantification of gene expression, respectively.

Routine assessments involve the quantification of total calcium content and the activity of alkaline phosphatase (ALP), a cell surface protein bound to the plasma membrane through phosphatidylinositol phospholipid complexes. High ALP activities are associated with active formation of mineralized matrix, and highest levels are found in the mineralization front in bone healing [Bruder and Caplan, 1990].

Particularly important in defining the phenotype of the differentiating stem cells is an understanding of bone tissue development in relation to gene expression of the cells.

The mRNA expression of the genes encoding important proteins during osteogenic differentiation of hMSC in vitro has not yet been systematically investigated, and results from different studies are often contradictory. Identification of mRNA markers characterizing the progression of hMSC toward the osteogenic lineage is further complicated by the known variability of cells from different individuals or due to differences in the isolation protocols [Phinney et al., 1999] or limitations inherent to the techniques used (i.e., semiquantitative techniques like Northern blots or conventional RT-PCR) [Frank et al., 2002].

Recently established real-time quantitative RT-PCR technology has made mRNA analysis more reproducible, precise, and sensitive than conventional RT-PCR, because it allows (i) measurement of the amount of amplified product with a quantitative laser-based method and (ii) data collection in the early exponential phase of the PCR reaction, when none of the reagents is rate-limiting [Gibson et al., 1996].

We developed real-time quantitative RT-PCR assays for genes encoding for (i) osteoblast-related membrane and extracellular matrix molecules (i.e., alkaline phosphatase (ALP), aggrecan (Agg), bone sialoprotein (BSP), osteopontin (OP), type 1 collagen (Col1)); (ii) bone morphogenetic protein 2 (BMP-2), an important growth factor determining cell mesenchymal precursors; and (iii) cbfa1 (a genetic regulator of osteoblast function and also known as Osf2 or RUNX2), a transcription factor related to osteogenesis.

BMP-2 is a molecule attracting MSCs, and it induces proliferation and differentiation of mesenchymal progenitor cells, producing bone tissue even at ectopic sites [Yamagiwa et al., 2001].

BSP has been proposed as the main nucleator of hydroxyapatite crystal formation and correlates with the initial phase of matrix mineralization [Ganss et al., 1999]. BMP-2 overexpression, as found in newly formed osteoblasts, is regulated through hormones and cytokines that promote bone formation [Wozney, 1989]. Osteocalcin comprises 10–20% of the noncollagenous proteins in bone, depending on age and species. Levels of ostecalcin are low at early stages and increase with increasing age. The function of ostecalcin may be to inhibit calcification until the appropriate temporal and spatial conditions are met. This was supported in vitro because ostecalcin inhibits hydroxyapatite crystal growth in solution [Luo et al., 1995].

Osteopontin is one of the most abundant noncollagenous proteins in bone; it binds to various extracellular molecules, including type I collagen, fibronectin, and osteocalcin, and may add physical strength to extracellular matrices [Denhardt and Noda, 1998].

7.1. DNA Assay

Protocol 13.10. DNA Assay of Cells Cultured on Scaffolds or as Pellets

Reagents and Materials

Nonsterile

- ❑ Cell lysis solution: Triton X-100, 0.1% (v/v) in UPW (if samples are also used for ALP Assay, use Triton X-100, 0.2% (v/v) + 5 mM Magnesium chloride in UPW)
- ❑ PicoGreen® dsDNA Quantitation Kit
- ❑ Paper towel
- ❑ Razor blade
- ❑ Microcentrifuge tubes with screw cap, 2 ml
- ❑ Steel balls
- ❑ Parafilm
- ❑ Analytical balance
- ❑ MinibeadBeater™
- ❑ Microcentrifuge
- ❑ Microtiter plate for fluorescence
- ❑ Fluorescence microplate reader

Protocol

A. Preparation of the scaffolds

(a) Grow MSCs on scaffolds in spinner flasks as described in Protocol 13.9.

(b) Dry scaffolds on a clean paper towel for 3 min on both sides.
(c) Measure wet weight of the scaffolds. Eventually cut the scaffolds into halves and measure the wet weight of each halves.
(d) Put each scaffold into a labeled microcentrifuge tube.
(e) Add 2 steel balls and 1 ml cell lysis solution per tube.
(f) Close tube firmly and wrap with Parafilm.
(g) Disintegrate scaffold by using a MinibeadBeater™ 3 times at 25,000 rpm for 10 s each time. Place on ice between cycles for cooling.
(h) Incubate at room temperature for 48 h.
(i) Transfer the content of the tube into a clean labeled tube without transferring the steel balls.
(j) Centrifuge at 3000 *g* for 10 min.
(k) Pipette the supernate into a clean, labeled tube without destroying the pellet.
(l) Proceed with the DNA Assay or store samples at $-20\,^{\circ}$C until measurement.

B. Preparation of cell pellets

(a) Grow MSCs in pellet culture according to Protocol 13.7 or 13.8.
(b) Centrifuge tube with cell pellet at 3000 *g* for 10 min at $4\,^{\circ}$C.
(c) Aspirate supernate carefully without destroying the cell pellet.
(d) Add 300 μl cell lysis solution to each tube.
(e) Disintegrate the pellet with a pipette tip by pipetting 20 times up and down.
(f) Incubate at room temperature for 48 h.
(g) Centrifuge tube at 3000 *g* for 10 min.
(h) Transfer the supernate with a pipette into a clear microcentrifuge tube without destroying the cell pellet.
(i) Proceed with DNA Assay or store samples at $-20\,^{\circ}$C until measurement.

C. DNA Assay

(a) Thaw your samples if frozen and centrifuge at 3000 *g* for 10 min.
(b) Prepare a sample/standard setup table in your lab-book.
(c) Calculate the amount of reagents needed as follows:
provided in the kit:

Reagent A: PicoGreen dsDNA quantitation reagent (in DMSO—toxic)
Reagent B: 20× TE buffer
Reagent C: λ-DNA standard

To prepare:

Working A: Bring Reagent A to room temperature in the dark.
Amount (in ml) of Working A needed = (number of samples)/2 + 6.

Prepare a 200-fold dilution of Reagent A in Working B in a plastic container wrapped in aluminum foil.

Working B: Make a 1:20 dilution of Reagent B in DNase-free UPW. Working B is required to prepare solutions A and C and for the dilution of standards and samples. Prepare sufficient amounts according to the following equation: Amount (in ml) of Working B needed : $0.875 \times$ (number of samples) $+ 10$

Working C: Add 30 μl Reagent C to 1.47 ml Working B.

(d) Prepare the needed Working solutions.
(e) Prepare the standards according to the following table in Eppendorf tubes. Do not add Working A at this step:

DNA Concentration (ng/ml)	Working C (μl)	Working B (μl)	Working A (μl)
0 (zero standard)	0	500	500
10	5	495	500
50	25	475	500
250	125	375	500
1000	500	13050	500

(f) Add 125 μl of each sample to a new and labelled Eppendorf tube.
(g) Add 375 μl of Working B to each tube.
(h) Add 500 μl Working A into each standard or sample tube. From now on, minimize light exposure of your samples.
(i) Vortex tubes.
(j) Incubate tubes in the dark at room temperature for 3 min.
(k) Pipette 100 μl of each tube into the corresponding well of a microtiter plate.
(l) Read fluorescence with a fluorescence microplate reader set at excitation 480 nm, emission 520 nm.

△ *Safety note.* PicoGreen dsDNA quantitation reagent is toxic. Carefully read manufacturer's instructions.

D. Calculation of the results

(a) Average multiple readings for each standard or sample (if performed) and subtract the average zero standard fluorescence.
(b) Create a standard curve by reducing the data with computer software capable of generating a linear curve fit. Alternatively: plot mean fluorescence for each standard on the *y*-axis against the DNA concentration on the *x*-axis.
(c) As samples have been diluted, the concentration read from the standard curve must be multiplied by the chosen dilution factor.

7.2. ALP Assay

Protocol 13.11. Assay of Alkaline Phosphatase Activity in Bone Constructs

Reagents and Materials

Nonsterile

- ❑ Cell lysis solution: Triton X-100, 0.2% (v/v) + 5 mM magnesium chloride in UPW
- ❑ Alkaline phosphatase (ALP) Kit
- ❑ Sodium hydroxide solution, 0.2 M
- ❑ Paper towel
- ❑ Razor blade
- ❑ Eppendorf tubes
- ❑ Microcentrifuge tubes with screw cap, 2 ml
- ❑ Steel balls
- ❑ Parafilm
- ❑ Analytical balance
- ❑ MiniBeadBeater™
- ❑ Microcentrifuge
- ❑ Microtitration plate
- ❑ Microplate reader
- ❑ Waterbath

Protocol

Note: Work on ice all times unless otherwise stated.

A. Scaffolds

(a) Grow MSCs on scaffolds in spinner flasks as described in Protocol 13.9.
(b) Dry scaffolds on a clean paper towel for 3 min on both sides.
(c) Measure wet weight of the scaffolds. Eventually cut the scaffolds into halves and measure the wet weight of halves.
(d) Put each scaffold into a labeled microcentrifuge tube.
(e) Add 2 steel balls and 1 ml cell lysis solution per tube.
(f) Close tube firmly and wrap with Parafilm.
(g) Disintegrate scaffold by using a MinibeadBeater™ 3 times at 25,000 rpm for 10 s. Place on ice between cycles for cooling.
(h) Transfer the content of the tube into a clean labeled tube without transferring the steel balls.
(i) Centrifuge at 300 *g* for 10 min at 4 °C.
(j) Transfer the supernate into a clean, labeled tube. Avoid destruction of the pellet.
(k) Run ALP assay immediately.

B. Cell pellets

(a) Grow MSCs in pellet culture according to Protocol 13.7 or 13.8.

(b) Centrifuge tube with cell pellet at 300 *g* for 10 min at 4 °C.
(c) Aspirate supernate carefully without destroying the cell pellet.
(d) Add 300 μl cell lysis solution to each tube.
(e) Disintegrate the pellet with a pipette tip by pipetting 20 times up and down.
(f) Incubate at room temperature for 48 h.
(g) Centrifuge tube at 300 *g* for 10 min at 4 °C.
(h) Transfer the supernate into a clean, labeled tube. Avoid disruption of the pellet.
(i) Run ALP assay immediately.

Note: If you intend to perform the DNA Assay on the same samples, take the amount of sample needed for the ALP Assay (80 μl) out and incubate the rest of the sample for 48 h at room temperature. Proceed with Protocol 13.10A, step (i) for scaffolds and with Protocol 13.10B, step (g) for pellets.

C. ALP Assay

(a) Prepare a sample/standard setup table in your lab-book and dilute *p*-nitrophenol standards as follows.

1 mM p-Nitrophenol (ml) in Triton X-100 + $MgCl_2$	Triton X-100 0.2% (v/v) + 5 mM $MgCl_2$ (ml)	Final P-Nitrophenol Concentration (μg/ml)
0	1	0
0.05	0.95	6.955
0.2	0.8	27.82
0.6	0.4	83.46
0.9	0.1	125.19

(b) Standards can be stored at 2–8 °C for 2 months.
(c) Prepare reagents:

Working A: 0.75 M 2-Amino-2-methylpropanol (AMP):
Amount of Working A needed (in μl) :
(number of samples + number of standards + 2) × 40
Working B: 10 mM p-nitrophenylphosphate in 0.15 M AMP.
Amount of Working B needed (in μl): (number of samples + number of standards + 2) × 100

(d) Pipette 80 μl of samples or standards in labeled Eppendorf tubes at room temperature.
(e) Add 20 μl Working A to each tube.
(f) Add 100 μl of Working B to each tube.
(g) Incubate samples in a water bath at 37 °C until color develops and record time.
(h) As soon as a yellow color develops, add 100 μl of NaOH 0.2 M to stop reaction.
(i) According to the plate setup, pipette 100 μl from each tube into the corresponding well of the microtitration plate.

(j) Determine the optical density of each well with a microplate reader set to 405 nm.

D. Calculation of the results

(a) Average multiple readings for each standard and sample (if performed) and subtract the average zero standard optical density.
(b) Create a standard curve by reducing the data with computer software capable of generating a linear curve fit. Alternatively: plot mean fluorescence for each standard on the y-axis against the DNA concentration on the x-axis.
(c) As samples have been diluted, the concentration read from the standard curve must be multiplied by the dilution factor.
(d) Calculate the amount of p-nitrophenol produced per minute using the standard curve and the time required until color development. The incubation time chosen should fall within the linear portion of the OD versus time curve (See step (h) of Protocol C).

7.3. Calcium Assay

Protocol 13.12. Calcium Assay in Bone Constructs

Reagents and Materials

Nonsterile

- ❑ TCA 5%: trichloroacetic acid 5% (v/v) in UPW
- ❑ Calcium-phosphorus combined standard, 10 mg/dl calcium
- ❑ Calcium binding reagent
- ❑ Calcium buffer reagent
- ❑ Paper towel
- ❑ Razor blade
- ❑ Microcentrifuge tubes with screw cap, 2 ml
- ❑ Steel balls
- ❑ Parafilm
- ❑ Analytical balance
- ❑ MiniBeadBeater™
- ❑ Microcentrifuge
- ❑ Microtiter plate
- ❑ Microplate reader

Protocol

A. Scaffolds

(a) Grow MSCs on scaffolds in spinner flasks as described in Protocol 13.9.
(b) Dry scaffolds on a clean paper towel for 3 min on both sides.
(c) Measure wet weight of the scaffolds. Eventually cut the scaffolds into halves and measure the wet weight of halves.
(d) Transfer each scaffold into a labeled microcentrifuge tube.

(e) Add 2 steel balls and 1 ml TCA 5% per tube.
(f) Close tube firmly and wrap with Parafilm.
(g) Disintegrate scaffold by using a MinibeadBeater™ 3 times at 25,000 rpm for 10 s each time. Place on ice between cycles for cooling.
(h) Incubate the tubes at room temperature for 30 min.
(i) Centrifuge at 3000 *g* for 10 min.
(j) Pipette the supernate into a clean, labeled tube without disrupting the pellet.
(k) Add 1 ml of TCA 5% into the tube with the steel balls.
(l) Incubate the tubes at room temperature for 30 min.
(m) Combine the 2 corresponding TCA solutions.
(n) Centrifuge at 3000 *g* for 10 min.
(o) Transfer supernate into a clean, labeled tube without disrupting the pellet.

B. Cell pellets

(a) Grow MSCs in pellets according to Protocols 13.7 or 13.8.
(b) Centrifuge tube with cell pellet at 3000 *g* for 10 min at 4 °C.
(c) Aspirate supernate carefully without destroying the cell pellet.
(d) Add 250 μl of TCA 5% to each tube.
(e) Disintegrate the pellet with a pipette tip by pipetting 20 times up and down.
(f) Incubate the tubes at room temperature for 30 min.
(g) Centrifuge tube at 3000 *g* for 10 min at 4 °C.
(h) Transfer supernates in clear microcentrifuge tubes without destroying the cell pellet.
(i) Add 250 μl of TCA 5% to each tube containing the pelleted cell debris.
(j) Disintegrate the pellet with a pipette tip by pipetting 20 times up and down.
(k) Let the tube stand at room temperature for 30 min.
(l) Centrifuge tube at 3000 *g* for 10 min at 4 °C.
(m) Combine the supernate with the one previously collected at (h).

C. Calcium Assay

(a) Prepare a sample/standard setup table in your lab-book and dilute calcium standards as follows

Calcium Standard Sigma Ca^{2+} 10 mg/dl	100% TCA	UPW	Final Ca^{2+} concentration
1.6 ml	0.1 ml	0.3 ml	80 μg/ml
1.2 ml	0.1 ml	0.7 ml	60 μg/ml
0.8 ml	0.1 ml	1.1 ml	40 μg/ml
0.4 ml	0.1 ml	1.5 ml	20 μg/ml
0.2 ml	0.1 ml	1.7 ml	10 μg/ml
0 ml	0.1 ml	2.0 ml	0 μg/ml

(b) Standards can be stored at 2–8 °C.
(c) Pipet 10 μl of each standard or sample in duplicate into microtiter plate wells.

(d) Count total well number.
(e) Freshly prepare a 1:1 dilution of calcium binding reagent and calcium buffer reagent; 100 μl solution is required per well.
(f) Pipette 100 μl of this solution into each well.
(g) Incubate plate at room temperature in the dark for 5 min.
(h) Determine the optical density of each well with a microplate reader set to 575 nm.

Note: If absorption of sample exceeds the highest standard, dilute the sample in TCA 5% and read again.

D. Calculation of the results
(a) Average the duplicate readings for each standard and sample and subtract the average zero standard optical density.
(b) Create a standard curve by reducing the data with computer software capable of generating a linear curve fit. Alternatively: plot mean absorbance for each standard on the *y*-axis against total Ca^{2+} concentration on the *x*-axis.
(c) As samples have been diluted, the concentration read from the standard curve must be multiplied by the dilution factor.

7.4. Glycosaminoglycan (GAG) Assay

Protocol 13.13. Assay of GAGs in Cartilage and Bone Constructs

Reagents and Materials
Nonsterile
- ❑ PBE buffer in UPW:
  0.1 M Disodium hydrogen phosphate ($M_r = 141.95$)
  0.01 M EDTA disodium salt ($M_r = 372.24$)
  Adjust pH to 6.5, filter sterilize, store at 4 °C; shelf life 6 months
- ❑ Chondroitin sulfate standard stock solution 50 mg/ml (CS standard) in PBE buffer
  50 mg/ml chondroitin sulfate
  Cysteine 14.4 mM ($M_r = 121.2$)
  Can be stored at −20 °C for 1 year. On the experiment day, dilute with PBE to a final concentration of 100 μg/ml
- ❑ Dimethylmethylene blue solution (DMMB) in UPW:
  40.5 mM Glycine ($M_r = 75.07$)
  Sodium chloride ($M_r = 58.45$)
  Dissolve 2.37 g sodium chloride and 3.04 g glycine in 905 ml UPW, then add 95 ml 0.1 M hydrochloric acid. While stirring, add 16 mg DMMB to the solution. Check pH = 3 and optical density at 525 nm = 0.31–0.34. Store at room temperature in the dark (aluminum foil).
- ❑ Papain solution (freshly prepared): 14.4 mM Cysteine ($M_r = 121.15$) in 20 ml PBE buffer and adjust to pH 6.3; filter through 0.22-μm filter. Add 0.1 ml sterile papain stock.

- ❑ Paper towel
- ❑ Razor blade
- ❑ Microcentrifuge tubes, 2 ml, with screw cap
- ❑ Needle, 20 G
- ❑ Water bath at 60 °C
- ❑ Parafilm
- ❑ Vortex mixer
- ❑ Photometer and cuvettes

Protocol

A. Scaffold

(a) Grow MSCs on scaffolds in spinner flasks as described in Protocol 13.9.
(b) Dry the scaffold on a clean paper towel for 3 min on both sides.
(c) Measure wet weight of the scaffolds. Eventually cut scaffolds into halves and note the wet weight of halves as well.
(d) Put each scaffold into a labeled microcentrifuge tube.
(e) Punch 10 little holes into the cap of a microcentrifuge tube.
(f) Freeze constructs at −80 °C overnight.
(g) Lyophilize for 2–3 days and measure the dry weight of each scaffold.
(h) Add 1 ml papain solution per tube.
(i) Close tube with a new cap; wrap with Parafilm.
(j) Incubate at 60 °C for 16 h, vortexing 5 times.
(k) Centrifuge at 3000 *g* for 10 min at 4 °C.
(l) Perform GAG assay with supernate.

B. Pellets

(a) Grow MSCs in pellets according to Protocols 13.7 or 13.8.
(b) Centrifuge tube with cell pellet at 3000 *g* for 10 min at 4 °C.
(c) Aspirate supernate.
(d) Punch 10 little holes into the cap of a microcentrifuge tube.
(e) Freeze constructs at −80 °C overnight.
(f) Lyophilize for 2–3 days.
(g) Add 0.5 ml of papain solution per tube.
(h) Close tube with a new cap; wrap with Parafilm.
(i) Incubate at 60 °C for 16 h, vortexing 5 times.
(j) Centrifuge at 3000 *g* for 10 min at 4 °C.
(k) Perform GAG assay with supernate.

C. GAG Assay

Note: DMMB solution must be checked before use. Keep indicated times constant.

(a) Set photometer to 525 nm.

(b) Pipet 0.1 ml standard solution (See table below) or sample into cuvette.
(c) Add 2.4 ml DMMB solution.
(d) Cover cuvette with Parafilm and vortex.
(e) Read absorbance against blank
(f) Standards:

STD No.	CS (100 μg/ml) (ml)	PBE (ml)	DMMB solution (ml)	CS (μg)
Blank	0	0.1	2.4	0
1	0.01	0.09	2.4	1
2	0.02	0.08	2.4	2
3	0.03	0.07	2.4	3
4	0.04	0.06	2.4	4
5	0.05	0.05	2.4	5
6	0.07	0.03	2.4	7
7	0.1	0	2.4	10

(g) Prepare all standards and samples in triplicates.

D. Calculation of the results

(a) Average the triplicate readings for each standard and sample and subtract the average blank optical density.
(b) Create a standard curve by reducing the data with computer software capable of generating a linear curve fit. Alternatively, plot mean absorbance for each standard on the *y*-axis against the CS concentration on the *x*-axis.
(c) As samples have been diluted, the concentration read from the standard curve must be multiplied by the dilution factor.

7.5. TRIzol RNA Extraction

Protocol 13.14. RNA Extraction from Bone Constructs

Reagents and Materials

Nonsterile

- ❑ TRIzol
- ❑ Chloroform
- ❑ Ethanol, 70%
- ❑ RNAseZap®
- ❑ Qiagen RNeasy Mini Kit

△ *Safety notes.* TRIzol reagent is carcinogenic. Use gloves and eye protection. Avoid contact with skin or inhalation. Always work under a fume hood. Read manufacturer's safety instructions.

Chloroform is carcinogenic; wear safety glasses and gloves and always work under a fume hood. Read safety data instructions.

- ❑ Paper towel
- ❑ Microcentrifuge tubes, 2 ml, with screw cap
- ❑ Microcentrifuge

Protocol

Note: Contamination with RNA must be minimized. Please note: Human skin is the major source of contamination!

Precautions to prevent RNase contamination

- ♦ Always wear disposable gloves. Change gloves frequently.
- ♦ Use sterile RNAse-free, disposable plasticware and automatic pipettes reserved for RNA work.
- ♦ RNA is protected from RNase contamination in TRIzol reagent. Downstream sample handling requires RNase-free glassware or plasticware. Glass and metal items (tweezers, steel balls, beakers) can be exposed to 300 °C for 4 h, and plastic items (tubes) can be soaked for 10 min in 0.1 M NaOH + 1 mM EDTA, rinsed thoroughly with RNAse-free water, and autoclaved. RNAse-free containers are also commercially available.
- ♦ To prepare RNase-free water, 0.01% (v/v) diethylpyrocarbonate (DEPC) into UPW in RNase-free glass bottles. Incubate for 12 h at room temperature, then autoclave bottle (read instruction for autoclaving liquids).
- ♦ Clean all surfaces (for example, bench top) thoroughly with RNAseZap®, wipe with paper towel, rinse with water, and dry with clean paper towel.

A. Scaffolds

(a) Grow MSCs on scaffolds in spinner flasks according to Protocol 13.9.
(b) Dry scaffolds on a clean paper towel for 3 min on both sides.
(c) Measure wet weight of the scaffolds. Eventually cut the scaffolds into halves and measure the wet weight of halves.
(d) Transfer each scaffold into a labeled microcentrifuge tube.
(e) Add 2 steel balls and 1 ml TRIzol per tube.
(f) Close tube firmly and wrap with Parafilm.
(g) Disintegrate scaffolds by using a MinibeadBeater™ 6 times at 25,000 rpm for 10 s. Place on ice between cycles for cooling.
(h) Transfer the content of the tube into a clean labeled tube without transferring the steel balls.
(i) Centrifuge at 12,000 *g* for 10 min at 4 °C.
(j) Transfer supernate into a clean, labeled tube without destroying the pellet, and immediately perform RNA Extraction or freeze samples at −80 °C for later analysis.

B. Cell pellets

(a) Grow MSCs in pellets according to Protocols 13.7 or 13.8.
(b) Centrifuge tube with cell pellet at 3000 *g* for 10 min at 4 °C.
(c) Aspirate supernate carefully without destroying the cell pellet.
(d) Add 1 ml TRIzol to each tube.
(e) Disintegrate the pellet with a pipette tip by pipetting 20 times up and down.
(f) Centrifuge tube at 12,000 *g* for 10 min at 4 °C.
(g) Transfer supernate into a clear microcentrifuge tube without destroying the cell pellet, and immediately perform RNA Extraction or freeze samples at −80 °C for later analysis.

C. RNA Extraction

Before starting to work, clean all working surfaces with RNAse Zap® as above. Change gloves frequently. If not mentioned otherwise, work on ice.

(a) Incubate sample at room temperature for 5 min.
(b) Add 190 μl of chloroform to each tube. A clear layer should occur.
(c) Vortex tube vigorously for 15 s.
(d) Incubate sample at room temperature for 3 min.
(e) Centrifuge tube at 12,000 *g* for 15 min at 4 °C. An upper clear aqueous phase containing the RNA appears.
(f) Carefully transfer upper aqueous phase to a fresh tube (do not transfer the interface); estimate the transferred volume you can take out. It is important not to disturb the lower layers during extraction—this will contaminate your sample and may influence enzyme activity of later experiments.
(g) Add the same volume of 70% ethanol to the homogenized lysate (estimated in the previous step), and mix well by pipetting. Do not centrifuge.
(h) Apply up to 700 μl sample, including precipitate that may have formed, to an RNeasy mini spin column sitting in a 2-ml collection tube and centrifuge at 8000 *g* for 15 s. If the volume of your sample exceeds 700 μl, successively load aliquots onto the RNeasy column and centrifuge as above. Discard flow-through and reuse. Reuse the collection tube for the following step.
(i) Pipette 700 μl Buffer RW1 onto the RNeasy column, and centrifuge at 8000 *g* for 15 s.
(j) Transfer RNeasy column into a new 2-ml collection tube.
(k) Discard flow-through and old collection tube.
(l) Pipet 500 μl Buffer RPE onto the RNeasy column.
(m) Centrifuge at 8000 *g* for 15 s.
(n) Discard flow-through and reuse collection tube.
(o) Pipet 500 μl Buffer RPE onto the RNeasy column.
(p) Centrifuge at maximum speed for 2 min to dry the membrane.
(q) Transfer the RNeasy column into a new 1.5-ml Eppendorf tube and pipette 30 μl RNAse-free water directly onto the RNeasy membrane.

(r) Centrifuge at 8000 *g* for 1 min. The RNA is now in the collection tube.
(s) Close tube and immediately freeze at −80 °C.

7.6. Real-Time RT-PCR

Protocol 13.15. Gene Expression Analysis in Bone and Cartilage Constructs

Reagents and Materials

Nonsterile

- ❑ RNA from Protocol 13.14
- ❑ SuperScript™ First-Strand Synthesis System for RT-PCR
- ❑ Custom designed primers and probes or Assay on Demand (Applied Biosystems)
- ❑ TaqMan Universal PCR Mix
- ❑ Microcentrifuge tubes, 0.2 ml, or thin-walled PCR tubes, autoclaved
- ❑ PCR tubes or plates with caps
- ❑ Programmable thermal cycler
- ❑ Microcentrifuge
- ❑ ABI PRISM® 7000 Sequence Detection System
- ❑ Vortex mixer

Protocol

A. cDNA synthesis with Oligo(dT)

(a) Mix and briefly centrifuge each component prior to use.
(b) Prepare RNA-primer mixtures in a 0.2-ml PCR tube as follows:

RNA-primer mixture:	μl
RNA	8
dNTP mix	1
Oligo(dT)$_{12-18}$ (0.5 μg/μl)	1

(c) Incubate each sample at 65 °C for 5 min.
(d) Place samples on ice for at least 1 min.
(e) Prepare the following reaction mixture, adding each component in the indicated order. Prepare reaction mixture for total number of samples +3 reactions.

Reaction mixture	Per sample (μl)
10× RT buffer	2
25 mM $MgCl_2$	4
0.1 M DTT	2
RNaseOUT™	1

(f) Add 9 μl reaction mixture to each RNA-primer mixture, mix gently, and collect by brief centrifugation.

(g) Incubate at 42 °C for 2 min.
(h) Add 1 μl (50 units) of SuperScript™ II RT to each tube and mix by pipetting.
(i) Incubate at 42 °C for 50 min.
(j) Terminate the reactions at 70 °C for 15 min.
(k) Chill on ice.
(l) Add 1 μl RNase H to each tube and incubate at 37 °C for 20 min.
(m) Freeze samples at −80 °C or proceed directly with the PCR reaction protocol.

B. Real-time RT-PCR Reaction Protocol

Note: Keep all reagents on ice at all times; probes are light sensitive. Darken the room.

(a) Prepare Master Mix for each marker of interest in the order as outlined in the tables below. Prepare all mixes for your number of samples +3 reactions.
Custom designed Primers and Probes Master Mix:

Reagents	Volume (μl) per sample
H_2O	13
Forward primer	4.5
Reverse primer	4.5
Probe	2
2× TaqMan Universal Master Mix	25

Assay on Demand Master Mix:

Reagents	Volume (μl) per sample
H_2O	21.5
AoD Kit	2.5
2× TaqMan Universal Master Mix	25

(b) Add 1 μl cDNA to each reaction tube.
(c) Add 49 μl Master Mix to each tube and mix well by pipetting.
(d) Close wells tightly with cover strips.
(e) Perform RT-PCR according to the manufacturer's protocol.

Calculations:
(a) ct Gene of interest—ct GAPDH $= \Delta$ ct Gene of interest
(b) Δ ct zero point (control)—Δ ct Gene of interest $= \Delta\Delta$ ct
$2^{\Delta\Delta ct}$ = expression

Custom designed primers and probes (Applied Biosystems):

Forward GAPDH primer:	ATG GGG AAG GTG AAG GTC G
Reverse GAPDH primer:	TAA AAG CCC TGG TGA CC
GAPDH probe:	VIC CGC CCA ATA CGA CCA AAT CCG TTG AC TAMRA
Forward Collagen 1 primer:	CAG CCG CTT CAC CTA CAG C
Reverse Collagen 1 primer:	TTT TGT ATT CAA TCA CTG TCT TGC C
Collagen 1 probe:	6FAM CCG GTG TGA CTC GTG CAG CCA TC TAMRA
Forward Collagen 2 primer:	GGC AAT AGC AGG TTC ACG TAC A
Reverse Collagen 2 primer:	CGA TAA CAG TCT TGC CCC ACT T
Collagen 2 probe:	6FAM ATG GAA CAC GAT GCC TTT CAC CAC GA TAMRA

Dilute all of these in 1× TE buffer to the appropriate concentration (according to the table below).

Reference for all sequences: Eckstein et al. (2001).

Marker	Aliquoted Primers (μM)	Aliquoted Probe (μM)
GAPDH	10	6.25
Collagen 1	10	6.25
Collagen 2	10	6.25

Assay on Demand (Applied Biosystems):

Human aggrecan 1:	Hs00153936_m1
Human ALP:	Hs00240993_m1
Human BMP-2:	Hs00154192_m1
Human BSP:	Hs0000173720_m1
Human cbfa1 (RUNX2):	Hs00231692_m1
Human COX-2:	Hs00153133_m1
Human IGF-I:	Hs00153126_m1
Human IL-1β:	Hs00174097_m1
Human iNOS:	Hs00167248_m1
Human OP:	Hs00167093_m1
Human SRY-box-9:	Hs00165814_m1

Note: Always check updates on the homepage of Applied Biosystems: http://www.appliedbiosystems.com

8. HISTOLOGY

Protocol 13.16. Fixation and Paraffin Embedding of Bone and Cartilage Constructs

Reagents and Materials

Non-sterile

❑ 10% Neutral buffered formalin
❑ PBSA
❑ HistoGel™
❑ Ethanol
❑ Xylene (mixed xylene isomers)

△ *Safety note.* Xylenes and formalin is harmful, always work under fume hood. Read safety sheets carefully.

❑ Histology embedding cassettes
❑ Paraffin embedding station
❑ Lens paper
❑ Oven

Protocol

A. Scaffolds

(a) Grow MSCs on scaffolds in spinner flasks according to Protocol 13.9.
(b) Transfer each scaffold into a labeled histocassette.
(c) Incubate in 10% neutral buffered formalin at 4 °C for 24 h.
(d) Place in PBS at 4 °C until further processing.

B. Preparation of cell pellets

(a) Grow MSCs in pellets according to Protocol 13.7 or 13.8.
(b) Wrap each pellet into lens paper
(c) Transfer each package into a labeled histocassette
(d) Incubate in 10% neutral buffered formalin at 4 °C for 24 h.
(e) Place in PBS at 4 °C for at least 2 h.
(f) Encapsulate each pellet into HistoGel™ according to the manufacturer's instruction.
(g) Transfer to labeled histocassette and store in PBS at 4 °C until further processing.

C. Preparation for histology

(a) Dehydrate specimens in histocassettes with a series of alcohols: 70% ethanol for 24 h.
(b) 90% Ethanol for 1 h.

(c) 2 times 95% ethanol for 1 h each.
(d) 3 times 100% ethanol for 1 h each.
(e) 3 times xylene for 30 min each.
(f) Paraffin at 60 °C for 1 h.
(g) Paraffin at 60 °C overnight.
(h) Paraffin at 60 °C for 1 h.
(i) Embed specimens in paraffin blocks at an embedding station. If desired, cut scaffolds and show cross-sections by embedding accordingly.
(j) Cut 5-μm thin sections from the block with a microtome.
(k) Lift the cut sections onto a glass slide and air dry

Protocol 13.17. Von Kossa Staining of Sections of Bone Constructs

Note: Long exposure to strong light can lead to false positive stain.

Reagents and Materials

- ❑ Paraffin slides of specimens from Protocol 13.16
- ❑ Ethanol
- ❑ Xylene
- ❑ Freshly prepared 5% silver nitrate solution in UPW
- ❑ 5% Hypo solution (sodium thiosulfate) in UPW
- ❑ Nuclear-fast Red
- ❑ Cytoseal™ 60
- ❑ Oven
- ❑ Lamp, 60 W
- ❑ Aluminum foil
- ❑ Holder for histoslides
- ❑ Coverslips

Protocol

A. Deparaffinization and Hydration

(a) Dewax slides in oven at 55 °C for 30 min.
(b) 2 times xylene for 5 min each.
(c) 2 times 100% ethanol for 5 min each.
(d) 2 times 95% ethanol for 3 min each.
(e) 70% Ethanol for 3 min.
(f) UPW for 1 min.
(g) Proceed directly to staining.

B. Staining

(a) Incubate in 5% silver solution. Place in front of a 60 W lamp, place aluminum foil behind jar to reflect the light for 1 h or until calcium turns black.

(b) Rinse 3 times in UPW.
(c) 5% Hypo solution for 5 min.
(d) Wash in tap water.
(e) Rinse in UPW.
(f) Counterstain in Nuclear Fast Red for 5 min.
(g) Wash in UPW.
(h) Continue with dehydration and mounting step.

C. Dehydration and mounting
(a) Dehydrate slides with a series of alcohols: 2 times 5 dips in ethanol 95%.
(b) 2 times 5 dips in ethanol 100%.
(c) 2 times 3 min in xylene.
(d) Mount with Cytoseal™ 60 and cover with coverslip.

Protocol 13.18. Safranin-O Staining of Sections of Cartilage Constructs

Reagents and Materials

Non-sterile

- ❑ Paraffin slides of specimens from Protocol 13.16
- ❑ Ethanol
- ❑ Xylene (mixed xylene isomers)
- ❑ Harris hematoxylin
- ❑ 0.5% acetic acid in 70% ethanol
- ❑ 0.02% Fast Green in UPW
- ❑ 1% acetic acid in UPW
- ❑ 0.1% Safranin O in UPW
- ❑ Cytoseal™ 60
- ❑ Oven
- ❑ Holder for histoslides
- ❑ Coverslips

Protocol

(a) Deparaffinize and hydrate according to Protocol 13.17 A Steps (a) to (g).
(b) Incubate in Harris hematoxylin for 8 min.
(c) Rinse in UPW.
(d) Dip 2 times in 0.5% ethanolic acetic acid.
(e) Rinse with running water to enhance blue staining of nucleus for at least 5 min.
(f) 0.02% Aqueous Fast Green for 4 min.
(g) 3 Dips in 1% acetic acid.
(h) 0.1% Safranin O for 6 min.
(i) Continue with dehydration and mounting step from Protocol 13.17C, Steps (a) through (d).

---

9. IN VITRO APPLICATIONS

A number of methods and techniques have been established to treat large bone defects after, for example, trauma or in the context of tumors. These methods do not restore major damage to a tissue or organ in a truly satisfactory way. At the same time, the need for bone substitutes increases with the progressive aging of the population. Routine techniques rely on the harvest of autologous bone, a method limited by the availability of transplant material, harvesting difficulties, donor site morbidity, additional pain, longer hospital stays, and higher treatment costs, with a direct impact on function, availability, and psychological and social well-being of patients. Novel methods currently being studied include conduction (by a scaffold) and induction (by bioactive molecules) of cell migration to repair relatively small defects and cell transplantation into the defect site (with or without biomaterial) to repair larger defects [Alsberg et al., 2001]. The tissue engineering approach as described in this chapter involves the use of mechanically robust and osteoconductive biomaterials seeded with mesenchymal stem cells, exposed to osteogenic stimuli [Meinel et al. 2004b,c]. Compared to the transplantation of cells alone, in vitro-grown tissue constructs offer the potential advantage of immediate functionality. The in vitro environment is indeed heavily simplified as compared to the complex physiological situation, and therefore tissue engineers search for simplifying principles that allow at least the recapitulation of some aspects of tissue morphogenesis and cellular assembly into tissue structures [Lauffenburger and Griffith, 2001]. The proposed approach uses silk as a biomaterial to provide an osteoconductive environment. The advantages are good biocompatibility, cell attachment, and cell differentiation along with unique mechanical properties. The sustained degradation of silk provides a robust template for the cells to deposit bone on the lattice surface, thereby resulting in a biomimetic and three-dimensional trabecular network.

We did not observe this orientation of the deposited bone in faster degrading collagen protein or on PLGA composite scaffolds. Therefore, silks can provide a blueprint for the desired bone geometry through the design of scaffold structures. In principle, this would allow the engineering of any trabecular geometry, but in reality it does not. We have shown that a network with pore sizes ranging between 200 and 700 μm can be engineered, and probably this can be extended to bigger pore sizes. Geometries less than 200 μm result in the formation of a dense and continuous bone plate, originating from a coalescing trabecular network (Fig. 13.3). This feature may be particularly interesting for the subchondral bone plate, present at the interface between the epiphysis and the overlying hyaline cartilage. Tissue engineering of cartilage with silk scaffolds has been demonstrated before [Meinel et al. 2004a], and therefore the presented results demonstrate all elements necessary to engineer ostechondral grafts. However, the engineering of complex composite tissues within a single scaffold requires the exposure of one cell source—mesenchymal stem cells—to different growth factors to drive the differentiation along the desired lineages. This requires a spatially restricted functionalization of the scaffold with growth factors and, consequently, functional

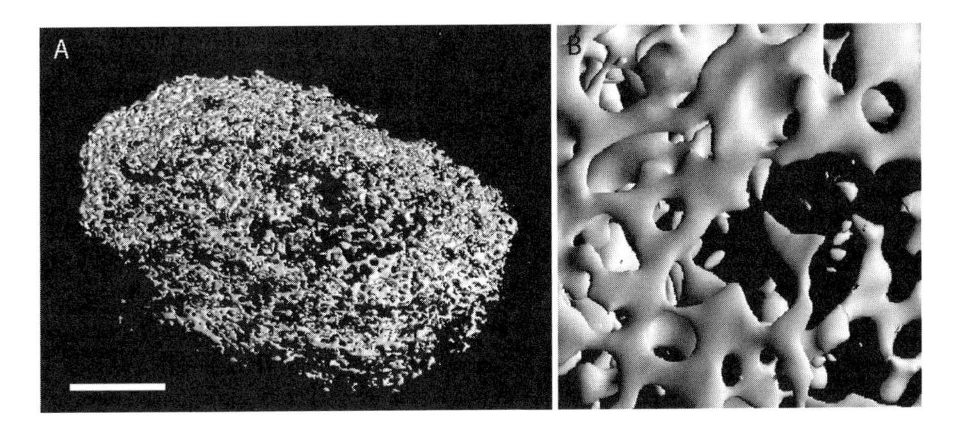

Figure 13.3. Tissue engineering of bone-like tissue. Bar = 5 mm (A). Magnification shows trabecula-like structure (B).

groups that allow an easy manipulation of the biomaterial. Silks, as do other proteins, provide all kinds of functional groups, allowing covalent decoration through easy chemical reaction. Therefore, silks offer an advantage because of their easy manipulation to serve as functionalized biomaterials for the engineering of complex composite tissues.

Diffusional limitations during culture, although in part overcome by the use of bioreactors, still remain the major challenge in tissue engineering and generally result in bone tissues with a thickness of less than 0.5 mm [Ishaug et al., 1997]. Although these protocols serve as a basis to engineer trabecular networks of more than 3 mm, the connectivity, trabecular surface, and volume are higher at the outer areas of the scaffold as compared to the center section. We have tried to overcome these limitations in earlier perfusion studies by forcing the medium flow through collagen [Meinel et al. 2004b] and silk scaffolds (unpublished data). However, mineralization was significantly less compared to the use of spinner flasks, and bone formation was restricted to mineralized rods with the absence of a trabecular network. A possible avenue to address the problem of restricted diffusion mass transport may be mechanical stimulation. Silks may again be particularly well suited for mechanical stimulation, which would allow a series of mechanical simulations without altering the mechanical integrity. In addition to the restricted mass transport, inhomogeneous seeding of the scaffold with cells can contribute to the variances in mineralization [Vunjak-Novakovic et al., 1996, 1998, 1999; Vunjak-Novakovic, 2003].

Despite all the advantages, silks as natural products have an intrinsic batch-to-batch variability, depending on the sericulture of the silkworms, nutrition, and temperature of silkworm culture. Aside from existing natural sources of silk, future options provided by the availability of reasonable quantities of genetically engineered silk variants would expand the set of structures available for use in vivo, including spider silk. These techniques result in homogenous and reproducible fabrication of silks, overcoming natural variability.

In summary, silks provide interesting and unique properties for bone tissue engineering promoting orderly tissue regeneration. Bone formation results in complex trabecular networks, predetermined by the lattice structure of the silk scaffolds. This allows the custom-made fabrication of cancellous bone implants, highly resembling the geometry at the implantation site. Furthermore, the unique properties of silks offer the advantage of prolonged mechanical competence together with the possibility of exposing the biomaterial to repetitive cycles of mechanical stimulation without compromising mechanical integrity.

SOURCES OF MATERIALS

Materials	*Suppliers*
Acetic acid	Sigma
Accuspin™ tubes	Sigma Diagnostics
ALP Kit: Alkaline phosphatase (ALP)	Sigma Diagnostics
AMP (2-Amino-2-methylpropanol), alkaline buffer solution A9226 or as provided in the ALP Kit	Sigma
Antibodies for FACS analysis:	
CD31-PE (555446)	Pharmingen (BD Biosciences)
CD34-APC (555824)	Pharmingen (BD Biosciences)
CD14-FITC (555397)	Pharmingen (BD Biosciences)
CD71-APC (551374)	Pharmingen (BD Biosciences)
Mouse anti-human CD105 (555690)	Pharmingen (BD Biosciences)
CD45-APC (555485)	Pharmingen (BD Biosciences)
CD44-FITC (555478)	Pharmingen (BD Biosciences)
Ascorbic acid phosphate	Sigma
bFGF (human, recombinant) (basic fibroblast growth factor)	Invitrogen
BMP-2, bone morphogenetic protein	Wyeth
Bone marrow	Cambrex
BupH™ MES buffered saline	Pierce
Calcium binding reagent	Sigma Diagnostics
Calcium buffer reagent	Sigma Diagnostics
Calcium-phosphorus combined standard, 10 mg/dl calcium	Sigma Diagnostics
Chloroform	Sigma
Chondroitin sulfate A, sodium salt from bovine trachea, 70%	Sigma
Cocoons from *Bombyx mori* (Linne, 1758)	Individual sources
Collagen: Avitene Ultrafoam collagen sponge	Bard
Cysteine	Sigma
Cytoseal™ 60	Microm
DEPC, diethylpyrocarbonate	Sigma
Dermal puncher	Miltey
Dexamethasone	Sigma
1,9-Dimethylmethylene blue dye ("Eschenmoser's salt")	Sigma-Aldrich
Disodium hydrogen phosphate	Sigma

Materials	Suppliers
DMSO, dimethyl sulfoxide	Sigma-Aldrich
Dulbecco's modified Eagle's medium (DMEM)	Invitrogen
EDC, 1-Ethyl-3-(3-dimethylaminopropyl)carbodiimide	Pierce
EDTA disodium salt	Sigma
Ethanol	Sigma
Fast Green FCF	Sigma
Fetal bovine serum (FBS)	Invitrogen
Formalin, 10% neutral buffered	Sigma
Fungizone	Invitrogen
Glutamic acid	Sigma
β-Glycerophosphate	Invitrogen
Glycine	Sigma
Hematoxylin solution, Harris modified	Sigma
HFIP, 1,1,1,3,3,3-hexafluoro-2-propanol	Sigma-Aldrich
HistoGel™	Lab Storage Systems Inc.
Histopaque®-1077	Sigma
Hydroxylamine hydrochloride	Pierce
Lithium bromide	Sigma
Matrigel, basement membrane matrix, from BD Biosciences	
Magnesium chloride	Sigma
MEM nonessential amino acids solution	Invitrogen
2-Mercaptoethanol	Sigma
Methanol	Sigma
Millex-SV syringe-driven filter unit, pore size 5 μm	Millipore
MiniBeadBeater™	Biospec
Nalgene™ Cryo 1 °C Freezing Container	Nalge Nunc
Nuclear Fast Red	Sigma
Papain suspension, activates to ≥20 units per mg protein	Worthington
PBSA: phosphate-buffered saline, w/o calcium or magnesium	Invitrogen
Penicillin-streptomycin, liquid	Invitrogen
PicoGreen® dsDNA Quantitation Kit	Molecular Probes
RGD: GRGDS: H-Gly-Arg-Gly-Asp-Ser-OH	Calbiochem
RNAseZap®	Ambion
Rneasy Mini Kit	Qiagen
RPMI 1640	Invitrogen
Safranin O	Sigma
Silver nitrate	Sigma
Slide-a-Lyzer® Dialysis Cassette, cut-off 2000 g/mol	Pierce
Sodium carbonate	Sigma
Sodium chloride USP granular	Fisher
Sodium chloride	Sigma
Sodium hydroxide	Sigma
Sodium thiosulfate	Sigma
Spinner flask	Bellco
Sucrose	Fluka
Sulfo-NHS, *N*-Hydroxysulfosuccinimide	Pierce

Materials	*Suppliers*
SuperScript™ First-Strand Synthesis System for RT-PCR	Invitrogen
TaqMan 2× PCR Master Mix	Applied Biosystems
TCA, trichloroacetic acid	Sigma
Teflon PFA container, 5-ml round vials with snap cap	Cole Parmer
TGF-β_1, transforming growth factor-β_1, human recombinant	R&D systems
Triton X-100	Sigma
TRIzol	Invitrogen
Trypan Blue	Invitrogen
Trypsin-EDTA, liquid, 0.25% trypsin, 1 mM EDTA	Invitrogen
Tween 80	Sigma
Xylene (mixed xylene isomers, X2377)	Sigma

REFERENCES

Abe, K., Shimada, Y., et al. (1982) Long-term administration of prostaglandin E1: report of two cases with tetralogy of Fallot and esophageal atresia. *Crit. Care. Med.* 10(3): 155–158.

Alsberg, E., Hill, E.E., et al. (2001) Craniofacial tissue engineering. *Crit. Rev. Oral Biol. Med.* 12(1): 64–75.

Altman, G.H., Diaz, F., et al. (2003) Silk-based biomaterials. *Biomaterials* 24(3): 401–416.

Altman, G.H., Horan, R.L., et al. (2002) Silk matrix for tissue engineered anterior cruciate ligaments. *Biomaterials* 23(20): 4131–4141.

Aubin, J., and Liu, F. (1996) The osteoblast lineage. In Bilezikian, J., Raisz, L. Rodan, G., eds., *Principles of Bone Biology*. New York, Academic Press, pp. 51–67.

Aubin, J., and Triffitt, J.T. (2002) Mesenchymal stem cells and the osteoblast lineage. In Bilezikian, J., Raisz, L., Rodan, G., eds., *Principles of Bone Biology*. New York, Academic Press, p. 68–81.

Bagi, C., van der Meulen, M., et al. (1995) The effect of systemically administered rhIGF-I/IGFBP-3 complex on cortical bone strength and structure in ovariectomized rats. *Bone* 16(5): 559–565.

Bruder, S.P., and Caplan, A.I. (1990) Terminal differentiation of osteogenic cells in the embryonic chick tibia is revealed by a monoclonal antibody against osteocytes. *Bone* 11(3): 189–198.

Bruder, S.P., Fink, D.J., et al. (1994) Mesenchymal stem cells in bone development, bone repair, and skeletal regeneration therapy. *J. Cell Biochem.* 56(3): 283–294.

Bruder, S.P., Jaiswal, N., et al. (1997) Growth kinetics, self-renewal, and the osteogenic potential of purified human mesenchymal stem cells during extensive subcultivation and following cryopreservation. *J. Cell Biochem.* 64(2): 278–294.

Bucknall, T.E., Teare, L., Ellis, H. (1983) The choice of a suture to close abdominal incisions. *Eur. Surg. Res.* 15: 59–66.

Cancedda, R., Dozin, B., et al. (2003a) Tissue engineering and cell therapy of cartilage and bone. *Matrix Biol.* 22(1): 81–91.

Cancedda, R., Mastrogiacomo, M., et al. (2003b) Bone marrow stromal cells and their use in regenerating bone. *Novartis Found. Symp.* 249: 133–143; discussion 143–147, 170–174, 239–241.

Caplan, A.I. (1994) The mesengenic process. *Clin. Plast. Surg.* 21(3): 429–435.

Caplan, A.I., Elyaderani, M., et al. (1997) Principles of cartilage repair and regeneration. *Clin. Orthop.* (342): 254–269.

Choong, P.F., Martin, T.J., et al. (1993) Effects of ascorbic acid, calcitriol, and retinoic acid on the differentiation of preosteoblasts. *J. Orthop. Res.* 11(5): 638–647.

Chu, C.R., Coutts, R.D., et al. (1995) Articular cartilage repair using allogeneic perichondrocyte-seeded biodegradable porous polylactic acid (PLA): a tissue-engineering study. *J. Biomed. Mater. Res.* 29(9): 1147–1154.
Crane, G.M., Ishaug, S.L., et al. (1995) Bone tissue engineering. *Nat. Med.* 1(12): 1322–1324.
Cunniff, P.M., Fossey, S.A., et al. (1994) Mechanical and thermal properties of the dragline silk from the spider *Nephila claviceps*. *Poly. Adv. Technol.* 5: 401–410.
Cusack, S., and Miller, A. (1979) Determination of the elastic constants of collagen by Brillouin light scattering. *J. Mol. Biol.* 135(1): 39–51.
DeLisser, H.M., Newman, P.J., et al. (1993) Platelet endothelial cell adhesion molecule (CD31). *Curr. Top. Microbiol. Immunol.* 184: 37–45.
Denhardt, D.T., and Noda, M. (1998) Osteopontin expression and function: role in bone remodeling. *J. Cell Biochem. Suppl.* 30–31: 92–102.
Ducy, P. (2000) Cbfa1: a molecular switch in osteoblast biology. *Dev. Dyn.* 219(4): 461–471.
Ducy, P., Zhang, R., et al. (1997) Osf2/Cbfa1: a transcriptional activator of osteoblast differentiation. *Cell* 89(5): 747–754.
Eckstein, F., Winzheimer, M., et al. (2001) Interindividual variability and correlation among morphological parameters of knee joint cartilage plates: analysis with three-dimensional MR imaging. *Osteoarthritis Cartilage* 9: 101–111.
Engelberg, I., and Kohn, J. (1991) Physico-mechanical properties of degradable polymers used in medical applications: a comparative study. *Biomaterials* 12(3): 292–304.
Frank, O., Heim, M., et al. (2002) Real-time quantitative RT-PCR analysis of human bone marrow stromal cells during osteogenic differentiation in vitro. *J. Cell Biochem.* 85(4): 737–746.
Freed, L.E., Langer, R., et al. (1997) Tissue engineering of cartilage in space. *Proc Natl Acad Sci USA* 94(25): 13,885–13,890.
Freed, L.E., and Vunjak-Novakovic, G. (2000) Tissue engineering bioreactors. In Lanza, R.P., Langer, R., Vacanti, J., eds., *Principles of Tissue Engineering.* San Diego, Academic Press, pp. 143–156.
Ganss, B., Kim, R.H., et al. (1999) Bone sialoprotein. *Crit. Rev. Oral Biol. Med.* 10(1): 79–98.
Gibson, U.E., Heid, C.A., et al. (1996) A novel method for real time quantitative RT-PCR. *Genome Res.* 6(10): 995–1001.
Glowacki, J., Mizuno, S., et al. (1998) Perfusion enhances functions of bone marrow stromal cells in three-dimensional culture. *Cell Transplant.* 7(3): 319–326.
Greenwald, D., Shumway, S., Albear, P., Gottlieb, L. (1994) Mechanical comparison of 10 suture materials before and after in vivo incubation. *J. Surg. Res.* 56: 372–377.
Gronthos, S., Graves, S.E., et al. (1994) The STRO-1+ fraction of adult human bone marrow contains the osteogenic precursors. *Blood* 84(12): 4164–4173.
Gronthos, S., Zannettino, A.C., et al. (2003) Molecular and cellular characterisation of highly purified stromal stem cells derived from human bone marrow. *J. Cell Sci.* 116(9): 1827–1835.
Gross, A.E. (2003) Cartilage resurfacing: filling defects. *J. Arthroplasty* 18(3 Suppl 1): 14–17.
Guhrs, K.H., Weisshart, K., et al. (2000) Lessons from nature—protein fibers. *J. Biotechnol.* 74(2): 121–134.
Halsted, W. (1892) The employment of fine silk in preference to catgut and the advantage of transfixing tissues and vessels in controlling hemorrhage. *Ann. Surg.* 16: 505.
Haynesworth, S.E., Goshima, J., et al. (1992) Characterization of cells with osteogenic potential from human marrow. *Bone* 13(1): 81–88.
Hunziker, E.B. (2002) Articular cartilage repair: basic science and clinical progress. A review of the current status and prospects. *Osteoarthritis Cartilage* 10(6): 432–463.
Ishaug, S.L., Crane, G.M., et al. (1997) Bone formation by three-dimensional stromal osteoblast culture in biodegradable polymer scaffolds. *J. Biomed. Mater. Res.* 36(1): 17–28.
Jaiswal, N., Haynesworth, S.E., et al. (1997) Osteogenic differentiation of purified, culture-expanded human mesenchymal stem cells in vitro. *J. Cell Biochem.* 64(2): 295–312.
Komori, T., Yagi, H. et al. (1997) Targeted disruption of Cbfa1 results in a complete lack of bone formation owing to maturational arrest of osteoblasts. *Cell* 89(5): 755–764.
Lagasse, E., Connors, H., et al. (2000) Purified hematopoietic stem cells can differentiate into hepatocytes in vivo. *Nat. Med.* 6(11): 1229–1234.
Lam, K.H., Nijenhuis, A.J., et al. (1995) Reinforced poly(L-latic acid) fibres as suture material. *J. Appl Biomater* 6(3): 191–197.

Lange, F. (1903) Über die Sehenplastik. *Verh. Dtsch. Orthop. Ges.* 2: 10–12.
Lange, F. (1907) Künstliche Bänder aus Seide. *Münch. Med. Wochenschr.* 17: 834–836.
Lauffenburger, D.A., and Griffith, L.G. (2001) Who's got pull around here? Cell organization in development and tissue engineering. *Proc. Natl. Acad. Sci. USA* 98(8): 4282–4284.
Liu, W., Toyosawa, S., et al. (2001) Overexpression of Cbfa1 in osteoblasts inhibits osteoblast maturation and causes osteopenia with multiple fractures. *J. Cell Biol.* 155(1): 157–166.
Ludloff, K. (1927) Der operative Ersatz des vorderen Kreuzbandes am Knie. *Zentralbl. Chir.* 54: 3162–3166.
Luo, G., D'Souza, R., et al. (1995) The matrix Gla protein gene is a marker of the chondrogenesis cell lineage during mouse development. *J. Bone Miner. Res.* 10(2): 325–334.
Maniatopoulos, C., Sodek, J., et al. (1988) Bone formation in vitro by stromal cells obtained from bone marrow of young adult rats. *Cell Tissue Res.* 254(2): 317–330.
Martin, I., Shastri, V.P., et al. (2001) Selective differentiation of mammalian bone marrow stromal cells cultured on three-dimensional polymer foams. *J. Biomed. Mater. Res.* 55(2): 229–235.
Meinel, L., Zoidis, E., et al. (2003) Localized insulin-like growth factor I delivery to enhance new bone formation. *Bone* 33(4): 660–672.
Meinel, L., Hofmann, S., et al. (2004a) Engineering cartilage-like tissue using human mesenchymal stem cells and silk protein scaffolds. *Biotechnol. Bioeng.* 88(3): 379–391
Meinel, L., Karageorgiou, V., et al. (2004b) Bone tissue engineering using human mesenchymal stem cells: effects of scaffold material and medium flow. *Ann. Biomed. Eng.* 32(1): 112–122.
Meinel, L., Karageorgiou, V., et al. (2004c) Engineering bone-like tissue using human mesenchymal stem cells and silk scaffolds. *J. Biomed. Mater. Res.* 71A: 25–34
Meinel, L., Hofmann, S., et al. (2005) Inflammatory responses in vivo and in vitro to silk and collagen films. *Biomaterials* 26 (2): 147–155
Negrin, R.S., Atkinson, K., et al. (2000) Transplantation of highly purified CD34+Thy-1+ hematopoietic stem cells in patients with metastatic breast cancer. *Biol. Blood Marrow Transplant.* 6(3): 262–271.
Otto, F., Thornell, A.P., et al. (1997) Cbfa1, a candidate gene for cleidocranial dysplasia syndrome, is essential for osteoblast differentiation and bone development. *Cell* 89(5): 765–771.
Pazzano, D., Mercier, K.A., et al. (2000) Comparison of chondrogensis in static and perfused bioreactor culture. *Biotechnol. Prog.* 16(5): 893–896.
Perez-Rigueiro, J., Viney, C., et al. (1998) Silkworm silk as an engineering material. *J. Appl. Polym. Sci.* 70: 2439–2447.
Phinney, D.G., Kopen, G., et al. (1999) Plastic adherent stromal cells from the bone marrow of commonly used strains of inbred mice: variations in yield, growth, and differentiation. *J. Cell Biochem.* 72(4): 570–585.
Pins, G.D., Huang, E.K., et al. (1997) Effect of static axial strain on the tensile properties and failure mechanism of self assembled collagen fibres. *J. Appl. Polym. Sci.* 63: 1429–1440.
Pittenger, M.F., Mackay, A.M., et al. (1999) Multilineage potential of adult human mesenchymal stem cells. *Science* 284(5411): 143–147.
Postlethwait, R.W. (1970) Long-term comparative study of nonabsorbable sutures. *Ann. Surg.* 171: 892–898.
Prockop, D.J., Gregory, C.A., et al. (2003) One strategy for cell and gene therapy: harnessing the power of adult stem cells to repair tissues. *Proc. Natl. Acad. Sci. USA* 100 Suppl 1: 11,917–11,923.
Prockop, D.J., Sekiya, I., et al. (2001) Isolation and characterization of rapidly self-renewing stem cells from cultures of human marrow stromal cells. *Cytotherapy* 3(5): 393–396.
Rossitch, E., Jr. (1987) On delayed reactions to buried silk sutures. *NC Med. J*/ 48(12): 669–670.
Salthouse, T.N., Matlaga, B.F., Wykoff, M.H. (1977) Comparative tissue response to six suture materials in rabbit cornea, sclera, and ocular muscle. *Am. J. Ophthalmol.* 84: 224–233.
Santin, A.D., Hermonat, P.L., et al. (1999) Expression of surface antigens during the differentiation of human dendritic cells vs. macrophages from blood monocytes in vitro. *Immunobiology* 200(2): 187–204.
Schaefer, D., Martin, I., et al. (2000) In vitro generation of osteochondral composites. *Biomaterials* 21(24): 2599–2606.

Sofia, S., McCarthy, M.B., et al. (2001) Functionalized silk-based biomaterials for bone formation. *J. Biomed. Mater. Res.* 54(1): 139–148.
Soong, H.K., and Kenyon, K.R. (1984) Adverse reactions to virgin silk sutures in cataract surgery. *Ophthalmology* 91(5): 479–483.
Spangrude, G.J., Heimfeld, S., et al. (1988) Purification and characterization of mouse hematopoietic stem cells. *Science* 241(4861): 58–62.
van Luyn, M.J., van Wachem, P.B., et al. (1992) Relations between in vitro cytotoxicity and crosslinked dermal sheep collagens. *J. Biomed. Mater. Res.* 26(8): 1091–1110.
van Wachem, P.B., Plantinga, J.A., et al. (2001) In vivo biocompatibility of carbodiimide-crosslinked collagen matrices: Effects of crosslink density, heparin immobilization, and bFGF loading. *J. Biomed. Mater. Res.* 55(3): 368–378.
Vunjak-Novakovic, G. (2003) The fundamentals of tissue engineering: scaffolds and bioreactors. *Novartis Found. Symp.* 249: 34–46; discussion 46–51, 170–174, 239–241.
Vunjak-Novakovic, G., Freed, L.E., et al. (1996) Effects of mixing on the composition and morphology of tissue-engineered cartilage. *AIChE J.* 42: 850–860.
Vunjak-Novakovic, G., Martin, I., et al. (1999) Bioreactor cultivation conditions modulate the composition and mechanical properties of tissue-engineered cartilage. *J. Orthop. Res.* 17(1): 130–138.
Vunjak-Novakovic, G., Obradovic, B., et al. (1998) Dynamic cell seeding of polymer scaffolds for cartilage tissue engineering. *Biotechnol. Prog.* 14(2): 193–202.
Wozney, J.M. (1989) Bone morphogenetic proteins. *Prog. Growth Factor Res.* 1(4): 267–280.
Yamagiwa, H., Endo, N., et al. (2001) In vivo bone-forming capacity of human bone marrow-derived stromal cells is stimulated by recombinant human bone morphogenetic protein-2. *J. Bone Miner. Metab.* 19(1): 20–28.

14

Culture of Neuroendocrine and Neuronal Cells for Tissue Engineering

Peter I. Lelkes[1], Brian R. Unsworth[2], Samuel Saporta[3], Don F. Cameron[3], and Gianluca Gallo[4]

[1]School of Biomedical Engineering, Science and Health Systems, Drexel University, Philadelphia, Pennsylvania; [2]Department of Biology, Marquette University, Milwaukee, Wisconsin; [3]Department of Anatomy and Center for Excellence for Aging and Brain Repair, University of South Florida Health Sciences Center, College of Medicine, Tampa, Florida; [4]Department of Neurobiology and Anatomy, Drexel College of Medicine, Philadelphia, Pennsylvania

Corresponding authors: pilelkes@drexel.edu, brian.unsworth@marquette.edu, ssaporta@hsc.usf.edu, gg48@drexel.edu

Culture of Cells for Tissue Engineering, edited by Gordana Vunjak-Novakovic and R. Ian Freshney

This chapter represents the joint effort of three distinct laboratories to describe their different approaches and methodologies for engineering neuroendocrine/neuronal tissues for the possible treatment of neurodegenerative diseases, such as Parkinson disease, which are related to the loss of catecholaminergic neurons. In the first part, Lelkes and Unsworth discuss the generation of three-dimensional catecholaminergic, adrenal medullary organoids with PC12 pheochromocytoma cells. In the second part, Saporta and Cameron focus on 3-D tissue constructs comprised of Sertoli cells and a neuronal cell line, NT2N. Finally, in the third part, Gallo discusses methods for isolating primary neuronal cells, using the chick dorsal root ganglion as a model system. Although each part was written by the authors mentioned above the overall responsibility for this enterprise rests with Peter I. Lelkes. The reader is advised to refer to each of the individual authors for specific questions regarding details of their particular methods.

1. THREE-DIMENSIONAL NEUROENDOCRINE/NEURONAL DIFFERENTIATION OF PC12 PHEOCHROMOCYTOMA CELLS

This section was contributed by Peter I. Lelkes and Brian R, Unsworth, to whom all related correspondence should be addressed.

1.1. Introduction

PC12 pheochromocytoma cells are cloned catecholaminergic cells derived from a spontaneous tumor (pheochromocytoma) of adrenal medullary chromaffin cells isolated from New England Deaconess rats [Greene and Tischler, 1976; Tischler and Greene, 1978]. Embryologically, chromaffin cells are derived from the sympathoadrenal (SA) lineage in the neural crest. In the rat, during migration out of the neural crest, some SA cells arrive in the adrenal anlagen around embryonic day 10.5 and are arrested there both developmentally and functionally (a process termed neoteny), while other SA cells continue to migrate and differentiate eventually into sympathetic neurons [Anderson and Axel, 1986; Anderson, 1989, 1993]. In the adrenal anlagen, these SA cells differentiate into two distinct catecholaminergic chromaffin cell types, which are characterized as either noradrenergic or adrenergic, that is, each of these related, yet distinct, cell types synthesizes and secretes either noradrenaline (norepinephrine, NE) or adrenaline (epinephrine, E) [Moro et al., 1990; Weiss et al., 1996]. Recent studies indicate that the expression of noradrenaline in the NE cells during differentiation is independent of the presence of glucocorticoids (synthesized in the neighboring adrenal cortex). By contrast, the conversion of NE to E, which requires the functional expression of phenylethanolamine-*N*-methyltransferase (PNMT), the final enzyme in the catecholamine synthesizing cascade, might require induction by glucocorticoids, such as dexamethasone [Doupe et al., 1985; Anderson and Michelson, 1989; Finotto et al., 1999; Wong, 2003]. The details of the chromaffinergic differentiation process, as fascinating as they may be, are beyond the scope of this article and can be found in a recent review [Huber et al., 2002]. Suffice it to say that, regardless of the effect of glucocorticoids, homotypic and heterotypic 3-D cell-cell interactions during organogenesis appear to be pivotal for the organotypic differentiation of the parenchymal cells in the adrenal medulla [Lelkes and Unsworth, 1992]. Chromaffin cells have been used for many decades as an easily available model system for studying mechanisms of neurotransmitter synthesis and release, specifically of cholinergic neuronal or neuroendocrine stimulus-secretion-synthesis coupling. Indeed, a recent PubMed search for "chromaffin cells" (August 2005) revealed some 3900 relevant hits.

In contrast to bona fide chromaffin cells, PC12 pheochromocytoma cells are predominantly dopaminergic, that is, they express tyrosine hydroxylase (TH) and dopamine-β-hydroxylase (DBH), and synthesize and secrete small amounts of NE [Greene and Tischler, 1976]. In the hands of most (but not all) investigators, PC12 cells do not spontaneously express PNMT or epinephrine, thus suggesting that these cells originated from the NE phenotype of chromaffin cells. In the past 25 years PC12 cells have become an enormously popular model, mainly for studying fundamental mechanisms of neuronal differentiation and mechanisms of neurotransmitter synthesis and release. Currently (August 2005) there are more than 8500 PubMed citations describing a plethora of physiologic and molecular aspects of these cells, including the inducibility of the expression of functional PNMT [Byrd et al., 1986; Kim et al., 1993; Ebert et al., 1997; Lelkes et al., 1998, Unsworth and Lelkes, 1998b].

PC12 cells are in a unique position in that they maintain some of the features of the bipotentiality of fetal/embryonic SA cells, with the added advantage that PC12 cells are readily available and rather easy to culture. Thus, depending on the appropriate environmental cues, PC12 cells can differentiate along either the sympathetic neuronal or the neuroendocrine, chromaffinergic pathway. For example, in the presence of neurotrophic agents, such as nerve growth factor (NGF) or brain-derived neurotrophic factor (BDNF), PC12 cells can differentiate into sympathetic neurons, whereas in the presence of glucocorticoids (and in organotypic culture) these cells acquire a neuroendocrine, chromaffin cell-like phenotype [Fujita et al., 1989; Mahata et al., 2002; Tischler, 2002; Vaudry et al., 2002]. This bipotentiality led to early attempts using PC12 cells as replacement cells or tissue for treating neurodegenerative disorders related to the destruction of catecholaminergic neurons, for example, in Parkinson disease [Jaeger et al., 1990; Aebischer et al., 1994; Lindner and Emerich, 1998]. The main caveat in using PC12 cells is the fact that these cells are spontaneously transformed.

About a decade ago we described that PC12 cells in coculture with adrenal medullary endothelial cells (but not endothelial cells from unrelated tissues) acquired structural and functional features reminiscent of acini of adrenal medullary chromaffin cells [Mizrachi et al., 1989, 1990]. This observation led to the extensive study of 3-D cultures of PC12 cells as a model system for tissue engineering neuroendocrine glands, specifically the adrenal medulla. [Lelkes et al., 1998; Lelkes and Unsworth, 1998, 2002]. In the wake of these studies we have been employing rotating wall vessel bioreactors (RWV, See below) as a unique cell culture environment for engineering functional tissue constructs [Unsworth and Lelkes, 1998a, 1998b, 2000; Lelkes and Unsworth, 2002]. The principles of RWV culture and the advantages of this system for generating differentiated tissue-like constructs have been described in detail in several recent reviews [Klaus, 2001; Hammond and Hammond, 2001; Lelkes and Unsworth, 2002].

In the following part of this chapter, we shall discuss the culture and 3-D assembly of PC12 cells as a source for engineering neuroendocrine tissue-like adrenal medullary organoids. Although primary chromaffin cells, especially from bovine adrenals, can be isolated in reasonable numbers and sufficient purity, and although isolated chromaffin cells can nowadays be fractionated into fairly homogeneous populations of epinephrine (E)- and norepinephrine (NE)-producing cells, no attempts have been reported, as yet, to engineer a functional adrenal medulla with primary isolates of chromaffin cells.

1.2. Preparation of Media and Reagents

1.2.1. Medium, Serum, and Plastics

In terms of the materials to be used for cell culture (Table 14.1), PC12 cells are quite tolerant with two exceptions:

i. All media formulations we have tried so far must contain at least 5% horse serum. Our current supplier is HyClone, but we have in the past equally successfully used horse serum from other vendors.

ii. We have found that PC12 cells do not grow well in tissue culture-treated flasks produced by Corning/Costar, but for optimal growth they require flasks produced by Nunc (blue cap).

Table 14.1. Reagents required for PC12 culture.

Reagent	Supplier	Cat. #	Quantity	Estimate cost
DMEM high glucose	Fisher	10-013-CV	6 × 0.5 L	$71.55
Fetal bovine serum	Hyclone	SH30071.03	0.5 L	$234.00
Horse serum	Hyclone	SH30074.03	0.5 L	$39.00
PBS without Ca^{2+} and Mg^{2+} (PBSA)	Fisher	20031CV	6 × 0.5 L	$66.00
Trypsin-EDTA	Fisher	MT25053CI	6 × 150 ml	$27.00
Flasks (T175)	Fisher	13-680-65	1 case	$70.00
L-Glutamine	Fisher	MT25005CI	6 × 150 ml	$27.00
Antibiotic/antimycotic	Fisher	MT30004CI	6 × 150 ml	$107.00
NGF	Sigma	N2393	10 μg	$180.00

1.2.2. Source of PC12 Cells

The original strain of PC12 cells is available from American Tissue Type Culture Collection (ATTC; CRL-1721). As there are conflicting results on phenotypic expression of PC12 cells, including the presence and inducibility of PNMT [Lelkes, 1991], stock cultures should be obtained from a properly validated source, such as ATCC, and an authenticated seed stock should be frozen to provide working stocks for future use.

Among the major differences between the various strains of PC12 cells are the culture conditions, specifically the choice of culture medium. Originally, Tischler and Greene grew their PC12 cells in RPMI 1640 medium with a defined set of additives (See Table 14.2). Another group of investigators, mainly using PC12 obtained from the late Gordon Guroff's laboratory at NIH, use DMEM (high glucose) supplemented with 7.5% fetal bovine serum and 7.5% horse serum. Yet other groups have changed either the medium formulation (e.g., reducing glucose levels), the amount of serum, or both, or developed novel formulations [Lelkes et al., 1997]. Important, however, seems to be the inclusion of horse serum in virtually every formulation. To date, no comparative study has been performed that would evaluate the genotypic and phenotypic differences between PC12 cells maintained in the various culture media.

1.3. Maintenance of PC12 Cells

Undifferentiated PC12 cells in either medium are round, phase-bright cells approximately 12–15 μm in diameter (Fig. 14.1). These cells, although

Table 14.2. Different culture media for PC12 cells.

Medium	ATCC (Tischler/Greene)	Guroff/Lelkes
Basal medium	RPMI 1640	DMEM
L-Glutamine	2 mM	2 mM
Glucose	4.5 g/L	4.5 g/L
Sodium pyruvate	1.0 mM	—
Sodium bicarbonate	18 mM (1.5 g/L)	26 mM (3.7 g/L)
HEPES	10 mM	—
Serum	Heat-inactivated horse serum, 10%; fetal bovine serum, 5%	Horse serum 7.5%; fetal bovine serum 7.5%

anchorage-dependent, are only loosely attached and do not spread even when cultured on tissue culture-treated surfaces. When maintained in suspension the cells readily form large, loose aggregates comprising up to hundreds of cells. In the presence of dexamethasone (0.1–10 μM), the cells remain similarly rounded and loosely attached and grow clonally in clusters. Of utmost importance, any cultures of PC12 cells should never become overcrowded and should be subcultured at 40–50% confluence. PC12 cells secrete neurotrophic factors (e.g., NGF and bFGF) and also other, as yet not fully characterized, differentiating factors, which when left to accumulate in the culture medium will a) induce PC12 cell differentiation toward the neuronal phenotype and b)give rise to a morphologically and functionally distinct phenotype of small, well-spread, anchorage-dependent cells. If these cells appear in our cultures, we abstain from any further use of the cells.

Addition of neurotrophic factors, specifically of nerve growth factor (NGF, 10–100 ng/ml), to undifferentiated PC12 cells causes a dramatic shift in their morphology: The cells rapidly flatten (<24 h) and begin (48–72 h) to extend neurites and to form (after ~120 h) neuronal networks (Fig. 14.1). Exposure of PC12 cells to NGF for <7 days results in reversible neuritogenesis, whereas after 10 days the cells have irreversibly differentiated into sympathetic neurons and thus have become dependent on NGF as a survival factor. In order to fully induce neuronal differentiation the initial seeding density of the cells must be rather low (<1000 cells/cm^2).

The key to expanding homogeneously undifferentiated PC12 cell populations is *never* to let the culture flasks become overconfluent and have differentiating growth factors accumulate in spent culture media. For this the cell cultures must be a) fed frequently (once every 48 h) and b) split weekly at a ratio of 1:8 (once the cells have reached some 50% confluence). In our laboratory, the routine protocol

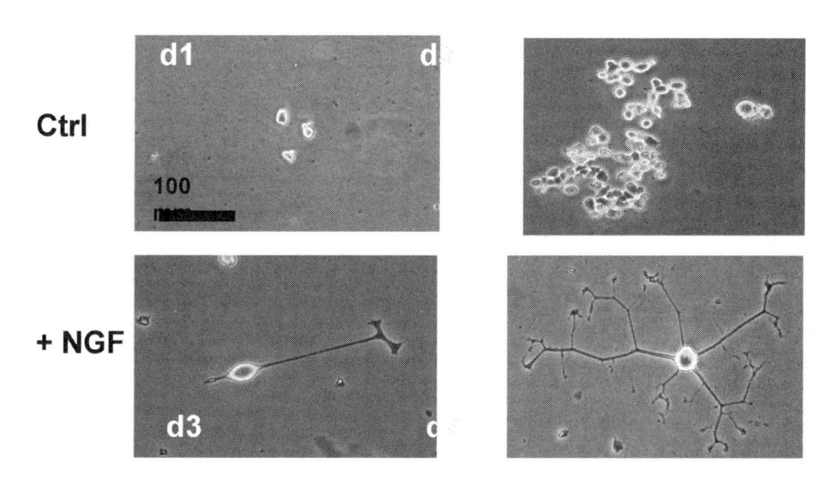

Figure 14.1. PC12 cells in 2-D culture. Top left: undifferentiated individual PC12 cells, 24 h after seeding; top right: cluster of undifferentiated PC12 cells after 7 days in culture; bottom left: single PC12 cells exposed to 25 ng/ml NGF, day 3; bottom right: neuronal network emanating from one single PC12 cell treated for 7 days with 25 ng/ml NGF. Original magnification 100×.

for passaging PC12 cells growing in T175 cell culture flasks is as described in Protocol 14.1 (all procedures are carried out under aseptic conditions).

Protocol 14.1. Culturing and Passaging Undifferentiated PC12 Cells

Reagents and Materials

Sterile

- ❑ PC12 cells, 1 T175 (175 cm^2) flask, 50% confluent
- ❑ Culture flasks, T175, 8
- ❑ Growth medium (See Table 14.1)

Protocol

(a) For each T175 flask you are about to passage, prepare 8 new T175 flasks, containing 20 ml growth medium. The medium must be prewarmed and equilibrated with the air-CO_2 mix in your incubator. Cap the flasks and place them in the back of the hood in an upright position.

(b) Remove the culture to be passaged from the incubator and observe under the inverted microscope ($<10\times$ phase-contrast objective), to ascertain the status of your culture, especially the cell density, which should not exceed $>50\%$ confluence.

(c) Place the flask containing the cells upright in the hood, carefully aspirate as much of the growth medium as possible, and close the cap.

(d) Mechanically dislodge the cells by firmly tapping the side of the culture flask 5–10 times with the flat palm of your hand (***caveat***: too little force will fail to dislodge all the cells, too much force will crack the flask).

(e) Verify under the inverted microscope that all the cells have been dislodged, if not repeat Step (d).

(f) Open the flask, quickly wash the growth surface with 2×10 ml growth medium, and cap the flask.

(g) Place the flask upright and let it stand for 2 min so that all the cells can accumulate in the growth medium at the bottom of the flask.

(h) Gently swirl the suspension, open the cap, and, keeping the flask in an upright position, use a sterile disposable 1-ml pipette to remove 1 ml cell suspension for subsequent counting of the cells in a hemocytometer or an electronic cell counter. Place the cell suspension into a sterile Eppendorf microcentrifuge tube. For determining cell numbers follow the conventional procedures.

(i) Using a 10-ml disposable pipette, carefully mix the remaining cell suspension (by gently pipetting the solution $3\times$ up and down while avoiding air bubbles) to ensure even distribution of the cells.

(j) Aspirate 2×8-ml cell suspension. Between the first and the second aspiration repeat Step (i).

(k) Uncap 4 of the flasks prepared in step (a) and quickly dispense 2 ml cell suspension into each of them.

(l) Cap each flask and place it horizontally to allow even distribution of the cells.

(m) After adding cells to all of the prepared flasks (steps i–l), return all the flasks into the CO_2 incubator. Make sure to loosen the caps (unless you work with gas-permeable caps).

(n) For feeding, fully replace the medium every 48 h.

In following these simple steps, we have been able to maintain and passage undifferentiated PC12 cells for more than 40 passages (>300 population doublings). The expected cell yield from cultures maintained in this way is approximately $1.0–1.5 \times 10^7$ cells/T175 flask. In the past, we froze a large number of early-passage cells (seed stock). Every time we have to thaw one of our seed stocks, we refreeze 30–40 vials with 1×10^6 cells/vial of early-passage cells (working stock). Using the steps outlined above has allowed us to maintain (and supply others with) undifferentiated PC12 cells for the past 20 years.

1.4. Generation of 3-D PC12 Organoids

The ultimate goal of tissue engineering is to (re)create macroscopic, 3-D tissue-like constructs, aka "organoids." In the case of PC12 cells, this can be achieved by growing the cells on biological or synthetic scaffolds, in conventional static aggregation culture or in rotating wall vessel (RWV) bioreactors (also sold as Rotatory Cell Culture Systems (RCCS) by Synthecon Inc., Houston, TX) (Fig. 14.2) in the presence or absence of microcarriers (e.g., Cytodex 3, Sigma). In the following we describe several of the standard techniques used in our laboratory for generating 3-D PC12 organoids.

1.4.1. Aggregate Cultures

In our laboratory we use aggregate cultures mainly as static controls for the organoids dynamically generated in RWV bioreactors.

Protocol 14.2. Aggregate Cultures of PC12 Cells

Reagents and Materials

Sterile

❑ Growth medium (See Table 14.1)

STLV **HARV**

Figure 14.2. Rotating wall vessel bioreactors (RWV). Left: Slow Turning Lateral Vessel (STLV); right: High Aspect Ratio Vessel (HARV). Photos courtesy of Synthecon Inc.

- ❑ Centrifuge tubes, 50 ml
- ❑ Petri dishes, non-tissue culture treated, 6.0 cm
- ❑ Petri dishes, glass or plastic, 25 cm
- ❑ Gas-permeable tissue culture bags

Nonsterile

- ❑ Gyratory shaker

Protocol

(a) Resuspend PC12 cells at an initial density of 1×10^6 cells/ml growth medium in up to 50 ml in a 50-ml centrifuge tube. Cap the centrifuge tube.

(b) Prepare 5 sterile, non-tissue culture-treated 6.0-cm Petri dishes.

(c) Carefully mix the cell suspension by inverting the capped tube 3 times

(d) Uncap and rapidly remove 10 ml cell suspension and place in a 6-cm non-TC Petri dish, replace the lid of the Petri dish, and recap the centrifuge tube.

(e) Repeat Steps (c) and (d). It is important to perform the mixing and dispensing of the cell suspension stepwise, to assure even distribution of the cells.

(f) Place 2 of the dishes inside the largest available (plastic or glass) Petri dishes (e.g., 25-cm diameter), replace the lids, and place the assembly into the incubator. The "dish in the dish" policy simplifies transport, prevents spillage and, most importantly, reduces the risk for contamination of both the cultures and of the incubators.

(g) Observe the cultures every 24 h to follow the formation of large multicellular aggregates. The formation of aggregates can be accelerated and enhanced (in terms of kinetics of aggregate formation, aggregate size, and number of cells/aggregate) if the cultures are gently shaken on a gyro-rotatory shaker placed inside the incubator for the first 24 h.

(h) As an alternative to the conventional aggregate cultures, PC12 cells at the same density as used for the dynamic cultures are placed in gas-permeable tissue culture bags (See Sources of Materials). Depending on the parameters to be analyzed, we use bags with a volume of either 7 ml or 35 ml with, respectively, one or two Luer-lock compatible inlet ports. These nonadhesive bags are placed horizontally into the incubators and serve as convenient, yet inexpensive, vessels for the aggregate cultures described above.

(i) Feed the aggregates (static controls) at the same frequency as the dynamic cultures.

1.4.2. Dynamic Formation of PC12 Organoids in RWV Bioreactors

RWV bioreactors create a suitable environment for generating macroscopic aggregates, which, because of the unique properties of this system, can differentiate into functional, tissuelike constructs, or, as we call them, "organoids" [Lelkes and Unsworth, 2002]. The principles of RWV bioreactors are described elsewhere in detail [Begley and Kleis, 2000; Hammond and Hammond, 2001] In this part of our chapter we will describe methods to generate PC12 organoids with and without cell culture beads, using, respectively, HARV and STLV type RWVs. Both systems are commercially available from Synthecon, whose instructions describe in great detail the procedures for preparing and using the RWV systems. We routinely use disposable HARVs (10 and 50 ml) and reusable STLVs (55 ml). We generate beadless aggregate cultures in HARVs only for short-term experiments (<48 h) in which we study the effects of the RWV environment on the initial stages of cell assembly and differentiation. For long-term experiments (7–30 days) the use of beadless PC12 cell cultures in HARVs (or in STLVs) is inappropriate because it will result in large organoids with necrotic cores. By contrast, when using Cytodex 3 and/or Cultisphere microcarriers as anchorage surfaces for long-term studies, we can generate and maintain large macroscopic (>5 mm) aggregates without necrotic cores (See Fig. 14.3, Color Plate 9A) for up to 30 days.

Protocol 14.3. Generation of PC12 Organoids in HARVs Without Beads

Reagents and Materials

Sterile

- ❑ HARVs, 10 or 50 ml
- ❑ Growth medium (See Table 14.1)
- ❑ Luer-lock syringes, tuberculin, 10 ml and 60 ml
- ❑ Centrifuge tubes, 50 ml

Protocol

(a) To generate PC12 organoids in the absence of beads, follow the procedures described in Protocol 14.2. Aggregation-based organoids have been generated for short-term experiments, of between 10 min and 24 h, in which we

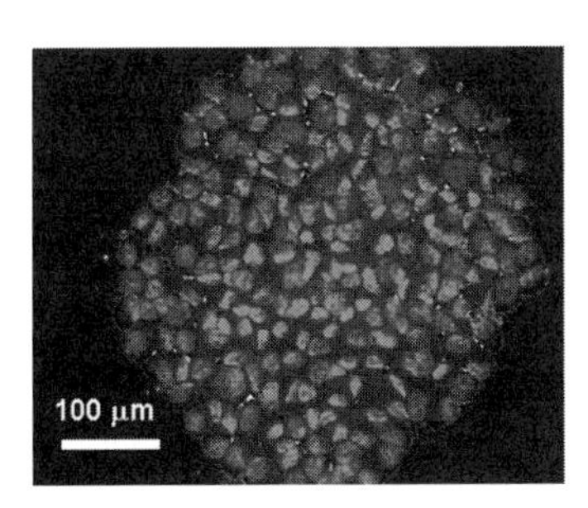

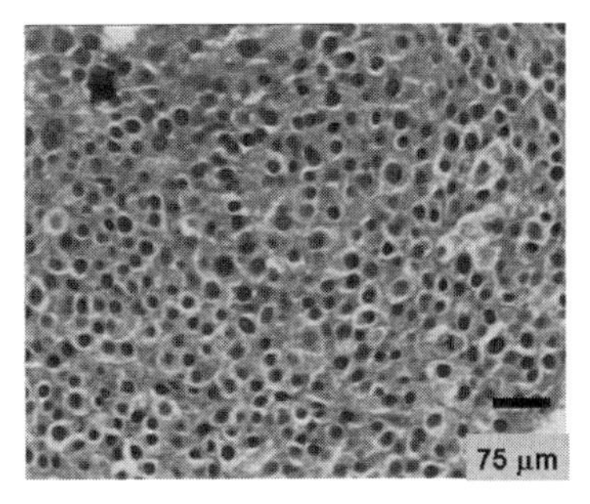

Figure 14.3. 3-D assemblies of PC12. Left: Aggregate of PC12 cells generated and maintained for 14 days under static conditions in LiveVue™ tissue culture bags. Cells/nuclei were stained with Hoechst 22358/Bisbenzimide. Right: 3-D organoid generated in the presence of Cytodex 3 beads and maintained for 20 days in STLV. Note the dense tissue like organization of the cells and the absence of a necrotic core. (See Color Plate 9A.)

have evaluated gene expression, intracellular signaling, and protein-protein interactions.

(b) Depending on the subsequent analysis and its demands for material (RNA, DNA, proteins, etc.), use either the 10-ml or the 50-ml HARVs.

(c) Use 1×10^6 cells/ml for routine experiments. We have used inoculation concentrations as low as 2×10^5 cells/ml and as high as 5×10^6 cells/ml; any concentration $<2 \times 10^5$ cells/ml will yield large number of cells that remain single and undergo apoptosis, whereas very high seeding concentrations will yield very large aggregates (5–10 mm in diameter) in 2–3 h, accompanied by rapid depletion of the nutrients in the medium and the formation of necrotic cores.

(d) Prepare an appropriate volume of cell suspension, that is, the net volume + 20% excess, i.e., 12 and 60 ml, respectively (See also Step (k)).

(e) Fill all RWVs, HARVs aseptically inside the hood, making sure that all air bubbles are removed from the vessels. Add the cell suspension via the lower syringe/Luer lock port, with the upper Luer lock having an empty syringe barrel attached. This syringe will serve as the receptacle for the excess medium and cell suspension, as the vessel is being filled through the lower port. Of utmost importance, greatest care must be exercised to remove every last air bubble from the assembly.

(f) After filling, sterilize the front (top plastic face) of the vessels with the 70% ethanol and cap the Luer-lock inlet ports of the HARVs with 1-ml tuberculin syringes, to prevent leakage and contamination.

(g) Screw the HARVs firmly onto the rotatory bases (either single or quadruple), and initiate the dynamic cell culture at a rotation speed of 10 rpm.

(h) Normally, under the above-mentioned routine conditions, feeding of the culture for up to 24 h is not necessary. The HARVs being well aerated, the levels of dissolved gases (pO_2, pCO_2) will remain identical to those in the incubator, and the glucose levels will remain normal provided the high-glucose culture medium is used. By contrast, for very high initial inoculation densities ($>2.5 \times 10^6$ cells/ml) and for experiments lasting >24 h, feed the vessels after 24 h, by replacing 75% of the cell culture medium.

(i) To feed the cultures, dismount the HARVs and aseptically place vertically in the hood, allowing the aggregates/organoids to settle to the bottom within 15 min. In the meantime prepare 2 Luer-lock syringes (either 10 ml or 60 ml), one containing medium (8 ml or 40 ml, respectively), the other empty syringe to serve as the receptacle for the spent medium. Before replacing the medium, remove a small aliquot of the spent medium, using one of the smaller syringes attached to the ports (See Step (f)) and analyze for blood gases and glucose content.

(j) To replace the medium in the HARV, remove the old syringes and attach new ones. Open the stopcocks in the front of the HARV and carefully empty the syringe containing the fresh medium (slowly, gently, so as not to disturb the aggregates/organoids sitting at the bottom), while simultaneously aspirating the spent medium into the empty syringe.

(k) To avoid creating air bubbles during the refeeding, first gently push a small volume (1 ml) of the new medium into the HARV and then open the stopcock and start manually aspirating the spent medium. Remove any air bubbles that, despite all precautions, may have found their way into the HARV, by alternatively filling and emptying in an oscillatory movement small volumes of excess medium through the two syringes. This is the main reason to prepare some 20% excess of fresh medium when filling the vessels (See Step (f)).

(l) Once all bubbles are removed, close both stopcocks and turn the HARVs upside down (the front plate facing downwards). Gently pull the pistons in both syringes (equivalent to 1–2 ml); remove both of them and replace them with small empty 1-ml tuberculin syringes. This prevents spillages that can occur when changing the syringes.

(m) To remove the organoids, open the main, central filling port and simply pour the contents into a 50-ml centrifuge tube. For further sample treatment proceed according to your specific protocol.

Protocol 14.4. Generation of PC12 Organoids with Beads in STLVs

Reagents and Materials

Sterile

- ❑ T175 flask culture of PC12 cells, semiconfluent
- ❑ Cytodex 3, collagen-coated beads
- ❑ Petri dishes, non-TC treated, 6 cm
- ❑ STLV
- ❑ Syringes, 10 ml and 60 ml
- ❑ Centrifuge tubes, 50 ml

Protocol

(a) Prepare the Cytodex 3 beads according to the manufacturer's instructions and store the suspension aseptically in the refrigerator.

(b) Attach cells to beads:

  i) To a semiconfluent cell culture (just ready for passaging, See Section 1.3) add beads aseptically at an approximate ratio of 1 bead per 10 cells. We routinely add approximately 1.75×10^6 beads to a T175 flask.

  ii) Return the flasks to the incubator and continue to feed the cultures gently for 5 days. This will allow the cells to migrate onto the beads (PC12 cells preferentially attach to collagen-coated surfaces) and to form a epitheloid monolayer on the beads.

  iii) After 5 days dislodge the cell-covered beads by firmly tapping on the side of the flask (See Protocol 14.1, Step (d)).

(c) Alternatively, to attach cells to beads:
    i) Incubate cells and beads at the ratio of 10:1 for 24 h in a non-tissue culture-treated Petri dish.
    ii) Incubate under gentle rocking on the gyratory shaker inside the incubator (See Protocol 14.2, Step (g)).

Firm attachment of the PC12 cells on the bead surface before inoculation into the STLV is critical. In contrast to other cell types, coincubation of PC12 cells with cell culture beads in the STLV will not lead to the attachment of the cells to the beads and formation of a sufficiently cell coverage to allow subsequent cell-cell contact-mediated aggregation of the beads.

(d) Resuspend the bead suspension in 5 ml medium to yield approx. 2.5×10^6 beads obtained as above (either Step (b) or (c)) covered with $\sim 2.5 \times 10^7$ cells to yield a final cell concentration of $\sim 5 \times 10^5$ cells/ml.
(e) Sterilize (autoclave) and prepare the reusable STLV as per the manufacturer's instructions. Aseptically assemble the STLV in the hood, and fill the vessel through the central filling port with 40 ml medium.
(f) Inoculate STLV with the cell-covered beads:
    i) Aspirate the bead suspension (See Step b (iii)) with a 10-ml disposable pipette.
    ii) Add the suspension through the central filling port to the vessel lumen.
    iii) Close the central filling port tightly.
(g) Open the two luer-lock compatible venting ports and firmly attach 2 empty 10-ml syringes with their plungers removed.
(h) Tilt the STLV so that the remaining air bubble rests at the top part underneath one of the venting ports. Using a 25-ml pipette gently, add up to 20 ml cell culture medium to the lower venting port, until all air has been removed inside the STLV and the medium rises up in the upper syringe (while some medium still remains in the lower syringe).
(i) At this time push the pistons into first the lower syringe and then the upper syringe and filling ports, making sure not to add any additional air bubbles into the STLV.
(j) Close the stopcocks of the venting ports, and by gently tapping the assembled STLV ascertain that there are no air bubbles inside. Frequently air is retained in some of the crevices of the STLV and dislodged by the tapping. In this case, open the stop cock and repeat Steps (i) and (j) (See also Protocol 14.3, step (k)).
(k) After all the air bubbles have been meticulously removed (which can be quite tedious and time consuming, at least in the beginning when you learn how to operate the system), mount the STLV on the aerated rotatory base and begin your experiments with a rotational speed that will prevent sedimentation of the beads. In our experience an initial rotatory speed of 14 rpm is optimal. However, be aware that as the aggregate/organoid size increases, you will have to empirically and gradually increase the rotatory speed to compensate for the enhanced sedimentation rate of the growing aggregates/organoids.

(l) Feed long-term cultures regularly, to maintain appropriate levels of pO_2, pCO_2, and nutrients, specifically glucose:
   i) Remove, while the vessels continue to rotate, 1 ml spent culture medium and determine the "vital signs" as described above (Protocol 14.3, Steps (i–k)).
   ii) If feeding is indicated, dismount the vessels from the bases, aseptically transfer the vessels to the hood, and place them upright. This allows the organoids to settle (in about 10 min) on the bottom of the STLV.
   iii) Exchange up to 75% of the spent medium (depending on the extent of glucose utilization of the growing aggregates) with new medium, essentially following the steps described above (Protocol 14.3, Steps (i–k)), using two 60-ml syringes.

(m) At the termination of the experiments:
   i) Demount the vessels, place in the incubator, and allow the aggregates/organoids to settle, as above.
   ii) Open the venting ports and the inlet ports and, using 10-ml disposable pipette, aseptically remove 20 ml of the medium (among others for one final determination of oxygen consumption and glucose utilization).

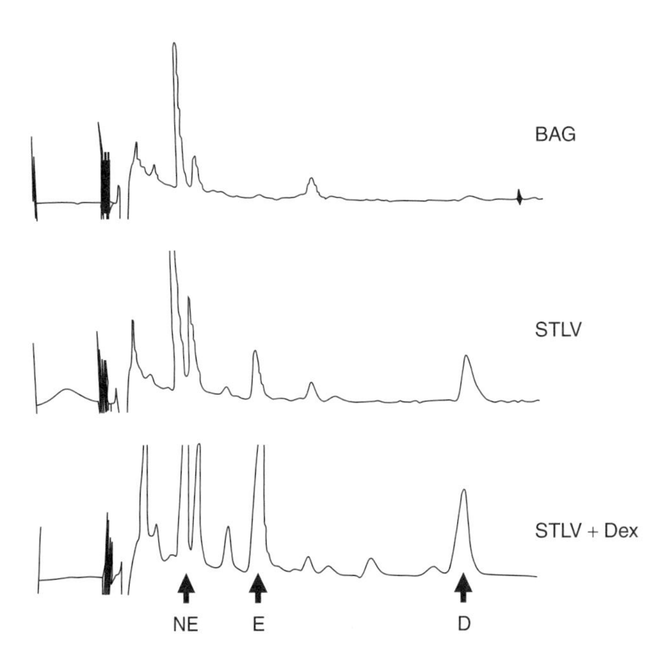

Figure 14.4. Generation of functional PC12 adrenal medullary organoids. Functionality of the paternal/maternal cells (chromaffin and PC12 cells) and of PC12 organoids grown in the absence or presence of 10 μg/ml dexamethasone was assessed by HPLC with electrochemical detection (for technical details on the HPLC methodology See Lelkes et al., 1994). In the bags (aggregates, See Fig. 14.3.), PC12 cells synthesize norepinephrine (NE) and dopamine (D) but no discernable quantities of epinephrine (E). By contrast, in the STLV, the cells synthesize epinephrine, the hallmark of adrenergic chromaffinergic differentiation; epinephrine synthesis in STLV is significantly potentiated by dexamethasone.

iii) Remove the aggregates/organoids by pouring the remaining vessel contents (~35 ml) through the central filling port into a 50-ml centrifuge tube.

Subsequent processing of the samples will follow the specific protocols for the particular analysis of protein, RNA, morphology, etc.

1.5. Conclusions

Using the techniques described above, specifically the ones detailed in Protocol 14.4, we have been able to generate functional PC12-based adrenal organoids (See Fig. 14.4). Remarkably, these organoids express significant levels of PNMT and, moreover, synthesize both norepinephrine and epinephrine, the hallmark of adrenergic differentiation, thus indicating the functional differentiation of the tissuelike constructs toward the parental, chromaffin-like phenotype. Ongoing and future studies are focused on evaluating the genotype and phenotype of the ensuing organoids and will address the role of heterotypic cocultures (e.g., addition of organ-specific mesenchymal, cortical, and endothelial cell)s as well the implantation/vascularization of the tissuelike organoids into host animals.

2. FORMATION OF SERTOLI-NT2N TISSUE CONSTRUCTS TO TREAT NEURODEGENERATIVE DISEASE

This section was contributed by Samuel Saporta and Don F. Cameron, to whom all correspondence should be addressed.

2.1. Background

Parkinson disease (PD), a neurodegenerative loss of dopaminergic neurons in the substantia nigra pars compacta, results in the loss of dopamine in the corpus striatum, among other places. This loss of dopaminergic input to the striatum results in progressively increasing tremor, bradykinesia, and rigidity, as well as alterations of cognition and affect, that, as the disease progresses, make the activities of daily living nearly impossible. Current pharmacological treatments to replace dopamine are initially effective, although palliative, and eventually lose their efficacy. Therefore, the use of dopaminergic cell replacement therapy as a long-term treatment of neurodegenerative disease has been of great interest.

Transplantation of fetal dopaminergic neurons ameliorates behavioral deficits in animal models of PD [Baker et al., 2000; Bjorklund et al., 1981; Bjorklund and Lindvall, 2000; Johnston and Becker, 1999; Palmer et al., 2001; Winkler et al., 2000]. Human fetal neurons have been transplanted into human patients with PD [Freed et al., 1990; Freeman et al., 1995; Lindvall et al., 1990], and initial controlled human clinical trials indicate the potential of this type of therapy for long-term treatment of PD, although a surgical protocol that yields consistent benefit is not yet established [Bjorklund et al., 2003; Freed et al., 2001; Olanow et al., 2003]. Alternative cell sources for cell replacement therapy have also been considered. The unique ability of embryonic stem cells to give rise to all somatic cell lineages makes them an especially promising source for cell therapy [Bjorklund et al., 2002; Freed, 2002; Kim et al., 2002]. However, it has proven difficult to reliably control their expansion and differentiation [McKay, 2002; Panchision and McKay, 2002]. Additionally, the therapeutic use of human embryonic stem cells and fetal cells is fraught with ethical issues. Some success has been reported for the expansion and differentiation of multipotent adult stem cells toward desired neuronal phenotypes in vitro [Englund et al., 2002; Espinosa-Jeffrey et al., 2002; Ourednik et al., 2001; Sanchez-Ramos et al., 2001; Uchida et al., 2000; Yang et al., 2003; Zigova et al., 2002], although, again, the extracellular instructive signals controlling their engraftment and differentiation in vivo are inadequately understood.

2.1.1. Engineered Tissue Constructs for Transplantation

The ideal cellular source for transplantation therapy in PD would be an easily expanded cell that reliably differentiates to a dopaminergic phenotype, releases dopamine, and remains dopaminergic for the life of the graft. Additionally, the ideal cellular source should elicit a minimal host-graft response in the transplant recipient. No single cell, to date, has been proven capable of demonstrating all of these desirable features. Therefore, it may be reasonable to consider combining cells for transplantation that would provide a tissue graft with these desirable properties. We have reasoned that it may be advantageous not only to combine cells with desirable properties for cell replacement therapy, but to engineer a tissue construct of these cells, ensuring aggregation and cell-to-cell contact. As will be seen below, this strategy has proven to be useful in creating a transplantable tissue construct composed of NT2N dopaminergic neurons and Sertoli cells that may, indeed, fulfill the enumerated qualities of an ideal source of dopaminergic neurons for cell replacement therapy.

2.1.2. The NT2 Neuronal Precursor Cell

The NTera-2/clone D1 (NT2) cell, a cell line derived from a human teratocarcinoma has been used extensively to study neuronal differentiation [Bani-Yaghoub et al., 2001; Guillemain et al., 2000, 2003; Iacovitti and Stull, 1997; Leypoldt et al., 2001; Misiuta et al., 2003]. This cell line is easily expanded and differentiates into immature neurons after 4–5 weeks of treatment with retinoic acid [Andrews, 1984, 1998; Pleasure and Lee, 1993]. These differentiated postmitotic neurons, designated NT2N, have phenotypic and morphological characteristics of immature neurons [Andrews, 1984; Daadi et al., 2001; Guillemain et al., 2000; Pleasure and Lee, 1993; Saporta et al., 2001]. Approximately 10% of NT2N neurons express tyrosine hydroxylase (TH), the rate-limiting enzyme in dopamine synthesis [Iacovitti and Stull, 1997; Zigova et al., 2000]. Dopaminergic NT2N neurons have been shown to engraft within the central nervous system and have proven useful in ameliorating the behavioral deficits associated with stroke [Nelson et al., 2002; Saporta et al., 1999; Watson et al., 2003; Willing et al., 2002], neurodegenerative disease [Willing et al., 1999], and spinal cord injury [Saporta et al., 2002].

2.1.3. Sertoli Cells

Sertoli cells (SC), the "nurse" cell of the testis, secrete a number of growth factors that are believed to help maturation of spermatids and protect developing sperm from immunosurveillance [Skinner, 1993]. Surprisingly, many of these growth factors, such as transforming growth factor-$\beta 1$ and -$\beta 2$, insulin-like growth factor I, glial cell line-derived neurotrophic factor, brain-derived neurotrophic factor, basic fibroblast growth factor 1 and 2, platelet derived growth factor, and neurturin, are neurotrophic. We have previously shown that coculturing rat fetal midbrain neurons with SC enhances the number of TH-positive neurons in vitro [Othberg et al., 1998a, b] and in vivo [Willing et al., 1999]. Additionally, a number of lines of evidence suggest that SC may provide immunoprotection to transplanted cells [Baker et al., 2000; Pollanen et al., 1988; Pollanen and Uksila, 1990]. They do not express major histocompatibility factor (MHC) II and express little MHC I [Pollanen and Maddocks, 1988]. Both allografts of rat and xenografts of pig SC survive in the rat brain for 2 months without immunosuppression of the recipient [Saporta et al., 1997]. Moreover, SC produce an unknown soluble factor that inhibits interleukin-2 (IL-2) production as well as IL-2-induced lymphocyte proliferation [Pollanen et al., 1990; Selawry et al., 1985], and when cotransplanted with dopaminergic neurons they enhance the survival of the transplanted neurons in recipient rats that are not systemically immunosuppressed [Willing et al., 1999].

2.2. Sertoli-NT2N-Aggregated-Cell (SNAC) Tissue Constructs

Cells cultured in the rotating wall vessel (RWV) bioreactors, originally developed by NASA to simulate conditions of microgravity, organize into tissue constructs that express tissue-specific markers [Becker et al., 1993; Lelkes et al., 1994, 1998]. This culture system has also been shown to effect the differentiation of cells grown in the RWV, as compared to those grown in conventional static cultures. For example, PC12 cells have been shown to aggregate and adopt a neuroendocrine phenotype when grown in the RWV, but not in conventional culture [Lelkes et al., 1998] (See Section 1.3). Similarly, neural stem cells [Wang and Good, 2001] and progenitor cells [Low et al., 2001] can be induced to differentiate when cultured in the RWV. We have previously reported aggregation of NT2 cells with Sertoli cells, using the High Aspect Ratio Vessel (HARV) RWV bioreactor to create a transplantable tissue construct that may provide a readily available source of dopaminergic neurons [Saporta et al., 2002].

2.2.1. NT2 Cells Rapidly Differentiate into NT2N Neurons

Immunohistochemical analysis of SNAC tissue constructs grown in the HARV RWV for 3 days reveals the presence of SC, TH-positive cells, and type III β-tubulin-positive NT2 cells (Fig. 14.5, See Color Plate 9B). These data suggest that TH-positive cells within the SNAC tissue constructs are NT2 cells that have differentiated to become NT2N dopaminergic neurons. Approximately 9% of NT2 cells differentiate to NT2N neurons after 3 days of culture in the HARV.

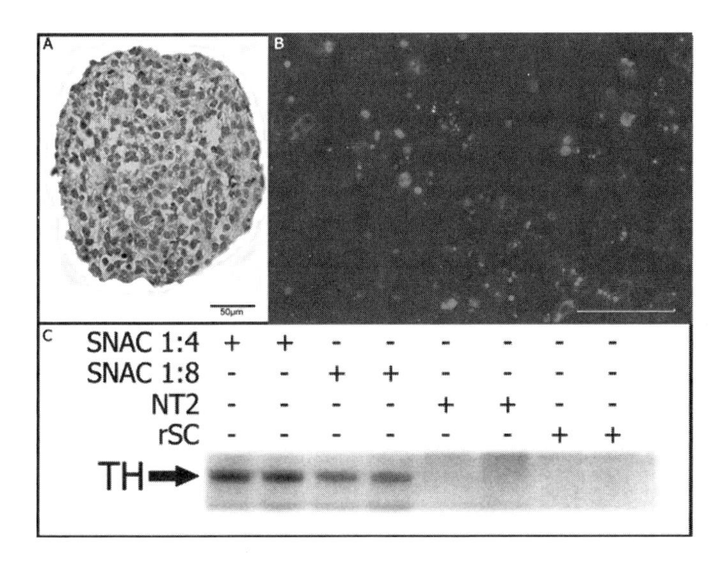

Figure 14.5. Morphology and TH content of SNAC tissue constructs. A) Photomicrograph of a SNAC tissue construct immunostained for human nuclei to show the presence of NT2 cells within the SNAC. The distribution of NT2 cells does not appear to be organized. B) Double immunostained fluorescent photomicrograph showing the presence of Sertoli cells (green) and TH-positive NT2N neurons (red) within a SNAC tissue construct. Scale bar = 100 μm C) Immunoblot of SNAC tissue constructs, NT2 cells, and Sertoli cells grown in the HARV rotating wall vessel for 3 days. The SNAC tissue construct with a starting ratio of Sertoli cells to NT2 cells of 1:4 contains the most TH. NT2 cells and Sertoli cells grown under the same conditions do not contain detectable TH. (See Color Plate 9B.)

Although TH is a good indicator of catecholaminergic neurons within the nervous system, it does not provide direct evidence that TH-positive cells contain dopamine (DA), or whether DA-containing neurons are capable of releasing DA. Therefore, we stimulated SNAC tissue constructs, NT2 cell aggregates, and SC aggregates grown in the HARV for 3 days with 60 mM KCl to stimulate release of DA into the medium, and analyzed medium from stimulated SNAC tissue constructs and cell aggregates, as well as normal medium, by high-performance liquid chromatography (HPLC) with electrochemical detection, for the presence of DA. Unstimulated SNAC tissue constructs constitutively release a small, but detectable, amount of DA into the culture medium. When stimulated with 30 mM KCl for 20 min, SNAC tissue constructs released a significant amount of DA (Table 14.3). No DA was detected in the stimulated or unstimulated NT2 aggregates or SC aggregates, or in medium.

2.2.2. Tissue Constructs Survive and Retain Their Dopaminergic Phenotype When Transplanted into the Brain

Our goal in creating the SNAC tissue construct was to develop a transplantable tissue source that would allow dopaminergic neurons to engraft within the brain, maintain the dopaminergic phenotype of the transplanted neurons, and allow survival of the graft without immunosuppression of the recipient. We have transplanted SNAC tissue constructs into nonimmunosuppressed male and female

Table 14.3. Effect of Sertoli cells on dopamine production by NT2 cells.

Sample	30 mM KCl	Dopamine concentration
SNAC	+	0.39 ± 0.07 nM
SNAC	–	0.039 nM
NT2	+	Not detectable
NT2	–	Not detectable
SC	+	Not detectable
SC	–	Not detectable
Medium	+	Not detectable
Medium	–	Not detectable

HPLC analysis of dopamine release from three independent experiments. Unstimulated SNAC released a small amount of dopamine into the medium. However, SNAC stimulated with 30 mM KCl released an order of magnitude more dopamine.

rats that have had the nigrostriatal pathway, the source of dopaminergic input to the striatum, disrupted. We have recovered viable grafts from these animals 4 weeks after transplantation and have identified dopaminergic neurons within these grafts (See Color Plate 9C). At this time point, there is little evidence of neurite extension, which may take longer than 30 days to fully develop [Saporta et al., 2002].

2.3. Conclusions

The present studies show that SC and NT2 cells aggregate into a tissue construct containing TH-positive neurons when grown in the HARV RWV under conditions of simulated microgravity in the absence of retinoic acid. These SNAC tissue constructs are a readily obtainable source of dopaminergic neurons that, when transplanted, may provide a long-lasting source of dopamine to the dopamine-deprived striatum in PD without the usual immunosuppression of the graft recipient. Longer-term transplant studies will assess the long-term survivability of these grafts and their degree of engraftment.

2.4. Preparation of Media and Reagents

2.4.1. Maintenance Medium (MM)

Dulbecco's modified minimal essential medium (DMEM)-Ham's F-12 nutrient mixture supplemented with retinoic acid, ITS+ Premix, and gentamycin sulfate. To 500 ml DMEM-F-12 add:

5.0 ml gentamycin sulfate
5.0 ml ITS+ Premix
13 μl working solution (50 mg/25 ml DMSO) of retinoic acid

Make sterile, store at 4 °C for up to 2 weeks.

2.4.2. Incubation Medium (IM)

Maintenance Medium (MM) supplemented with 10% fetal calf serum (GIBCO) and 1.0% Growth Factor Reduced Matrigel™ (MG_{GFR}). Make sterile, store at 4 °C for up to 2 weeks.

2.4.3. NT2 Medium (NT2M)

Dulbecco's modified Eagle's medium (DMEM)-Ham's F-12 nutrient mixture supplemented with 10% fetal calf serum and 5.0 ml of gentamycin sulfate.

2.4.4. Freezing Medium (FM) for NT2 Cells

DMEM supplemented with 10% fetal calf serum and 5% dimethyl sulfoxide (DMSO).

2.4.5. Trypsin Solution (for Tissue Digestion)

Dissolve 75 mg trypsin (type 1; T-8003) in 30.0 ml of MM in sterile 50.0-ml conical tube, to give 0.25%, immediately before use.

2.4.6. Trypsin-EDTA Solution (for Lifting Plated Sertoli Cells and NT2 Cells)

Trypsin, 0.25%, EDTA, 2 mM (Sigma; T-4049) is stored at −18 °C indefinitely.

2.4.7. Collagenase Solution

Dissolve 60 mg of collagenase (type IV, GIBCO) in 30.0 ml of MM in a sterile 50.0-ml conical tube, to give 0.2%, immediately before use.

2.4.8. Tris Buffer

Prepare 20 mM Tris buffer by dissolving 2.422 g of Trizma Base in 1.0 L ddH_2O and adjust pH to 7.4 with 12 N HCl. Sterilize by filtration and store at 4 °C for up to 2 months.

2.4.9. DNase Stock Solution

Dissolve 100 mg DNase in 50 ml of NaCl (0.14 M) to give 2 mg/ml. Sterilize by filtration and store at −20 °C. Dilute 1 ml with 70 ml MM as needed.

2.5. Isolation of Rat Sertoli Cells

Sertoli cells are isolated from prepubertal male rat pups (16–19 days old). Briefly, rat pups are sacrificed and testes are removed through a scrotal incision. Excised testes are placed in sterile MM and transferred to a sterile tissue culture hood. The rest of the protocol is performed observing sterile technique. The capsules (tunica albugenea) are removed from all testes, after which the parenchyma is minced into small fragments with sterile scissors and subjected to sequential enzymatic treatment at 37 °C with 0.25% trypsin and 0.2% collagenase. The resulting Sertoli cell aggregates are distributed in a volume of 50.0 ml MM into T150 cell culture flasks (Corning/Costar) and incubated at 39 °C in 5% CO_2-95% air for 48 hours. Sertoli cell-enriched monocultures are then subjected to hypotonic treatment with sterile 20 mM Tris-HCl buffer (2 min at room temperature) to expedite the removal of contaminating germ cells (15). After two washes with MM, cell culture flasks are replenished with 50.0 ml of MM and returned to the incubator at 37 °C in 5% CO_2-95% air. The resulting pretreated Sertoli cell-enriched monocultures contain greater than 95% Sertoli cells. Before coculture with NT2 cells, pretreated SC are lifted with 0.25% trypsin-EDTA (2 min at room temperature), washed (3×) with MM, counted by hemocytometric analysis, and assayed for viability by trypan blue exclusion.

Protocol 14.5. Preparation of Sertoli Cells

Reagents and Materials

Sterile

- ❑ Sodium pentobarbital (Nembutal)
- ❑ MM (See Section 2.4.1)
- ❑ IM (See Section 2.4.2)
- ❑ Trypsin, 0.25% (See Section 2.4.5)
- ❑ Collagenase, 0.2% (See Section 2.4.7)
- ❑ DNase (See Section 2.4.9)
- ❑ Tris buffer, 20 mM (See Section 2.4.8)
- ❑ Erlenmeyer flask, 125 ml
- ❑ Sterile beaker, 250 ml
- ❑ Scissors

Nonsterile

- ❑ Rats, 16–19 day old, 10
- ❑ Ethanol, 70%

Protocol

Isolation

(a) Sacrifice ten 16–19 day old rats with overdose of sodium pentobarbital (Nembutal).
(b) For each rat, wet scrotum with 70% ethanol.
(c) Remove testes and place in sterile MM in sterile tissue culture dish. Transport to sterile hood.
(d) Rest of protocol performed observing sterile technique.
(e) Remove tunica albugenea from each testis with scissors and place testicular parenchyma in clean MM.
(f) Mince parenchyma with scissors.
(g) Place minced tissue in 125-ml Erlenmeyer flask in 30 ml MM. Wash 3× with MM by resuspending and settling.
(h) Replace MM with 30 ml 0.25% trypsin solution and incubate at 37 °C in shaking water bath for 15 min.
(i) Add DNase (0.5 ml stock solution) 5.0 min before end of trypsin digestion incubation.

(j) Let tissue settle to the bottom of the flask, remove trypsin solution, and wash 3× with MM by resuspension and settling. Optional—add 1% trypsin inhibitor (Sigma) in first wash.
(k) Replace MM with 30 ml 0.2% collagenase and incubate at 37 °C in shaking water bath for at least 20 min or until most aggregates do not display peritubular cells (this is determined by microscopic evaluation of sample using 10× objective lens on inverted microscope).
(l) Let tissue settle to bottom of flask, remove collagenase solution, and wash 3× with MM by resuspension and settling.
(m) Transfer Sertoli cell aggregates to 250-ml sterile beaker and resuspend in 100 ml MM.
(n) Place beaker on stir plate, gently suspend aggregates in MM with sterile magnetic stir bar, and plate at recommended volume for culture vessels.

Pretreatment
(a) Place culture vessels with aggregates in 5% CO_2 incubator at 39 °C and incubate for 48 h.
(b) Transfer vessels to sterile hood.
(c) Remove MM and replace with 20 mM Tris-HCl buffer (just enough to cover the cells) and incubate for 2.5 min at room temperature.
(d) Remove Tris-HCl buffer, wash once with IM. and refill tissue/cell culture vessel with MM.
(e) Return tissue/cell culture vessels to 5% CO_2 incubator and incubate at 33 °C.
(f) Replace IM every 48–72 h.

2.6. Preparation of NT2 Cells

NTera-2/D1 cells are obtained from ATTC and kept frozen in liquid nitrogen until needed. Before the experiment, the frozen vial is thawed (as prescribed by the supplier), reconstituted in NT2M (See Section 2.4.3) in 15-cm cell culture dishes (Corning) and incubated at 37 °C in 5% CO_2. When the cells reach 70–80% confluence (about 2–3 days), cells are subcultured with trypsin-EDTA (See Section 2.4.6) and split 1:4 into four T150 cell culture flasks. These cells are incubated for 48 h at 37 °C, subcultured with trypsin-EDTA, pooled, washed (3×) with NT2M, counted by hemocytometer, and assayed for viability by Trypan Blue exclusion.

Protocol 14.6. Propagating NT2 Cells

Reagents and Materials

Sterile

- ❑ NT2M (See Section 2.4.3)
- ❑ Trypsin-EDTA (See Section 2.4.6)
- ❑ FM: freezing medium (See Section 2.4.4)
- ❑ Vial of NT2 cells (ATTC)

- ❑ Conical centrifuge tube, 15 ml
- ❑ Petri dish, 15 cm

Nonsterile

- ❑ Ethanol, 70%
- ❑ Water bath at 37 °C

Protocol

(a) Place frozen vial of NT2 cells in 37 °C water bath and agitate until just a "sliver" of ice is left in the vial, i.e., almost completely thawed (about 30 s).

(b) Swab vial with 70% EtOH, open, and pipette entire contents of vial (about 1.0 ml) into 10 ml NT2M in a 15-ml sterile conical centrifuge tube.

(c) Wash (3×) with NT2M by centrifugation (800 rpm [135-mm radius] for 3 min, 120 g) and plate cells into 15-cm cell culture dish.

(d) Incubate cells at 37 °C in 5% CO_2 until 70–80% confluence (3–4 days).

(e) Subculture with 0.25% trypsin-EDTA and seed total yield into four T150 cell culture flasks. Return to 37 °C in 5% CO_2-95% air.

(f) Repeat Steps (d) and (e)—this will give you 16 flasks.

(g) Subculture cells from all flasks with trypsin-EDTA, wash (3×) with NT2M and pool.

(h) Prepare 1.0-ml aliquots of 5×10^6 cells/ml in freezing medium (FM) and freeze.

(i) Store in liquid nitrogen. Each vial should be labeled "Working Stock of NT2 cells."

Protocol 14.7. Preparation of NT2 Cells for Coculturing

Reagents and Materials

Sterile

- ❑ NT2M (See Section 2.4.3)
- ❑ Trypsin-EDTA (See Section 2.4.6)
- ❑ One vial of NT2 cells (Working Stock)

Protocol

(a) Thaw and plate one vial of NT2 cells in 15-cm cell culture dish and incubate in NT2M at 37 °C in 5% CO_2.

(b) Grow to near confluence (80%) and subculture (1:4) in NT2M until near confluence.

(c) Repeat subcultures (Step (b)) if more cells are required.

(d) Harvest cells with trypsin-EDTA.

(e) Wash (3×) with NT2M. Cells are now ready to use.

2.7. Preparation of Sertoli-NT2-Aggregated-Cell (SNAC) Tissue Constructs

Sertoli cells and NT2 cells are combined at a 1:4 ratio (SC:NT2) to a total of 2.0×10^7 cells per construct. The combined cells are suspended in 10 ml of IM

(See Section 2.4.2) and placed in the HARV bioreactor. The biochamber is secured on its motor mount and the entire apparatus is placed in the incubator at 37 °C in 5% CO_2. The rotational speed is set a 62 rpm. The coculture period is for a total of 72 h, after which the aggregate is retrieved from the biochamber and washed (3×) with NT2M by resuspension and settling. The SNAC tissue construct is now ready for transplantation or assay.

Protocol 14.8. Tissue Constructs of Sertoli and NT2 Aggregated Cells (SNAC)

Reagents and Materials

Sterile or Aseptically Prepared

- ❑ SC (See Protocol 14.5)
- ❑ NT2 cells (See Protocol 14.7)
- ❑ IM
- ❑ Syringe, 5 ml or 10 ml
- ❑ HARV bioreactor

Protocol

(a) Suspend 4×10^6 SC and 1.6×10^7 NT2 cells in 5 ml IM at room temperature.

(b) Inject the suspended cells by syringe into the HARV bioreactor.

(c) Completely fill the 10-ml biochamber with additional IM, making sure that no bubbles exist within the biochamber (See Protocol 14.4 Steps (i), (j)).

(d) Secure biochamber on motor mount and place apparatus (HARV bioreactor on motor mount) in incubator at 37 °C in 5% CO_2.

(e) Connect motor mount to rheostat outside of the incubator.

(f) Turn motor on and rotate biochamber at 26 rpm. (***Note***: rotational speed at this point in time is adjusted so that the aggregate inside the biochamber remains in one spot in the biochamber. This speed will vary, depending on the size of the aggregate. We attempt to maintain multiple small aggregates, rather than one large aggregate, and the rotational speed is adjusted to maintain the position of the majority of the aggregates.)

(g) Incubate coculture for a total of 72 h.

(h) Retrieve biochamber from motor mount and place in sterile hood.

(i) Gently remove the SNAC tissue construct from the biochamber and place in 15-ml sterile conical centrifuge tube and let sediment by gravity. Wash SNAC (3×) with IM by resuspension and settling.

(j) SNAC is now ready for transplantation or assay.

---

3. THE DORSAL ROOT GANGLION AS A MODEL SYSTEM

This section was contributed by Gianluca Gallo, to whom all correspondence should be addressed.

3.1. Introduction

Dorsal root ganglia (DRG) are ball-shaped clusters of neurons, Schwann cells, and fibroblasts found outside of the dorsal portion of the spinal cord. The neurons in the DRG project axons to the periphery and into the spinal cord, thereby forming a relay system for sensory information received from skin and muscle. DRG have proven of great use in modern neurobiology and cell biology. For example, the initial discovery and elucidation of nerve growth factor used the DRG as a standard bioassay [Hamburger, 1993]. DRG are also useful in studying the effects of inhibitory molecules associated with spinal cord injury [Fournier et al., 2002]. The DRG is a reliable and robust model system for studying sensory neurons in a variety of conditions. The DRG system also provides the advantage over neuronal cell lines that it is reflective of the biology of primary neurons. Furthermore, because DRG axons project from the periphery into the spinal cord, the DRG system is of relevance to investigations of spinal cord injury and recovery.

The purpose of this section is to describe in detail the method used routinely in the Gallo laboratory to prepare both explant and dissociated cultures of DRG cells obtained from embryonic chickens and their culture on two-dimensional substrata (Fig. 14.6). The chicken DRG can also be cultured in three-dimensional gels [Pond et al., 2002]. Embryonic chickens provide an affordable and high-throughput system for experimentation. In addition, the chicken embryo is not under animal usage regulation until embryonic day 14 in the United States and day 11 in most

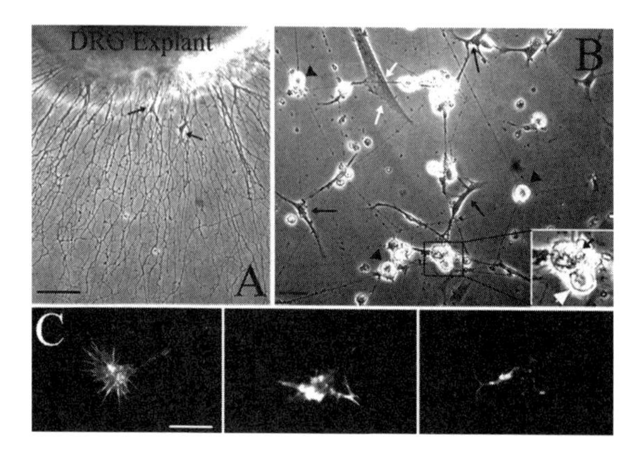

Figure 14.6. Phase-contrast microscopy examples of DRG axons and growth cones invitro. Pictures of an E10 DRG explant (A) and (B) dissociated E10 DRG cells cultured on laminin in 20 ng/ml NGF for 24 and 72 h, respectively. Black arrows in A and B denote Schwann cells. White arrows in B denote a fibroblast; note the classic fibroblast morphology relative to the multipolar Schwann cells. Black arrowheads in B denote DRG neuronal cell bodies. Inset in B shows two neuronal cell bodies. The cell body labeled with the white arrowhead is phase bright and healthy. The adjacent cell body, labeled by the black arrow, appears rough with phase imaging and represents a dying neuron. C) F-actin in growth cones of E10 DRG axons growing for 24 h from an explant (rhodamine-phalloidin stained). Note that growth cones can exhibit varied morphologies ranging from elaborate (left) to minor (right), exhibiting lamellipodia and/or filopodia. Bars = 100, 20, and 10 μm in A, B, and C, respectively.

of Europe, allowing experiments with this model system to commence and continue without the need to have approved animal use protocols.

Depending on the experiment, it is worthwhile considering whether explants or dissociated cells would be of greatest use. When explanted, chicken embryo DRG are spheres approximately 300–1000 μm in diameter depending on the embryonic age and position along the vertebral axis. Axons extend radially from explanted DRG, forming a "halo." This provides numerous growth cones, the tips of axons of which are not in contact with other cells. For this reason, explants have been of great use in studying the mechanism of growth cone collapse [Luo et al., 1993]. However, in explants, the neuronal cell bodies and initial segments of axons are not readily accessible. On the other hand, dissociated DRG cells allow investigation of cellular mechanisms acting throughout the neurons. Dissociated cells at low density also allow for the axons of specific neurons to be followed from cell body to their termini, the growth cones.

3.2. Preparation of Materials: Embryos, Reagents, and Dissection Tools

3.2.1. Eggs and Embryos

Fertilized chicken eggs can be obtained from Charles River Laboratories (www.criver.com). The maintenance of chicken eggs in the laboratory requires minimal equipment: (1) a refrigerator set at 10 °C and (2) a humid incubator set at 39 °C for the eggs. Chicken eggs can remain viable in a state of developmental stasis for up to 1 week when stored at 10 °C. Our laboratory receives 40 eggs a week, and on each day of the week seven eggs are moved from the refrigerator to the incubator. The eggs are disposed of after 14 days in the incubator because at this point they are considered animals and continued maintenance requires approved animal care protocols. A chicken egg incubator is necessary for two reasons, the maintenance of physiological temperature and the mechanized rotation of the chicken eggs (rotation of the eggs is required for normal development to occur). Various models of incubators are available depending on the needs of the laboratory. Our laboratory uses a model 1205 incubator (G.Q.F. Manufacturing Co.) that can accommodate up to 180 eggs. Additional models are available for laboratories working with smaller numbers of eggs.

3.2.2. F12H Medium

(i) Prepare 1 L of F12H medium as per manufacturer's directions, using a HEPES buffer.

(ii) To 1 L of F12H add 10 ml of 1 M HEPES and adjust pH to 7.4.

(iii) Filter sterilize.

(iv) Before use, add 5 mL of 200 mM L-glutamine to 500 ml of F12H medium and supplement with additives (See Section 3.2.3) and 5 ml of 100× Pen/Strep/Fungizone (PSF).

(v) Medium containing additives (See Section 3.2.3), PSF, and glutamine can be stored at 4 °C for up to 1 month.

3.2.3. F12HS10 Medium

F12HS10 consists of F12H medium, without the additives solution, but supplemented with 10% fetal bovine serum and filter sterilized. Storage is at 4 °C.

3.2.4. Medium Additives

Prepare the medium additives cocktail by adding to 20 ml of water:

(i) 5.1 g of phosphocreatine (vortex after adding to get into solution)

(ii) 400 mg of apo-transferrin

(iii) 20 μl of 20 mg/ml sodium selenate in water

(iv) 20 mg of insulin

(v) 20 μl of 4 μM progesterone in absolute ethanol

(vi) Filter sterilize.

The cocktail is added at 5 μl/ml of F12H medium.

(vii) Add 40 μl of 10 mg/ml sodium pyruvate in PBSA, pH 7.4, filter sterilized, per ml of F12H medium.

The additive solutions are stable for up to 3 months when stored at 4 °C.

3.2.5. Nerve Growth Factor (NGF)

Prepare a stock solution at 100 μg/ml in PBSA containing 0.1% BSA. Filter sterilize and store 100-μl aliquots at −20 °C. Working stocks of NGF are prepared by diluting to 10 μg/ml by addition of 900 μl of PBSA+0.1% BSA. Store at 4 °C for up to 3 months. The final concentration of NGF that provides optimal growth in F12H medium is 20 ng/ml. The minimum concentration suggested is 0.1 ng/ml.

3.2.6. Calcium-, Magnesium-Free Phosphate-Buffered Saline (PBSA)

PBSA (CMF-PBS, Invitrogen) is prepared according to manufacturer's directions (pH 7.4) and filter sterilized.

3.2.7. Trypsin Solution

Trypsin, 0.25%, 0.1% EDTA in Hanks' BSS PBSA is divided into 5-ml aliquots and stored at −20 °C.

3.2.8. Dissection Tools and Microscope Setup

The dissection will require two #55 forceps and a pinning forceps. The dissection requires a dissection microscope with 0.7–1.0× magnification and is vastly facilitated by the use of fiber optic illumination. For best imaging, we suggest placing the fiber optic light source above the embryo at approximately 45° to the dissection surface.

3.2.9. Laminin and Polylysine

Laminin (LN) and polylysine (PL) are commonly used culture substrata for primary neurons. PL provides a strong adhesive substratum, whereas LN is a component of the extracellular matrix and provides a more biologically relevant substratum. Prepare and store laminin stocks as suggested by the supplier. We prepare a 1 mg/ml solution of PL and store it in 100-μl aliquots at −20 °C. PL solution is prepared and diluted to the final concentration with borate buffer (add 190 mg of borax and 124 mg of boric acid to 40 ml of water). Sterilize the PL stock and borate buffer by filtration.

3.2.10. Coverslips and Substratum Coating

Explants and dissociated cells are routinely cultured on 18 × 18-mm German glass coverslips coated with either laminin or PL. Coverslips are first flamed for approximately 1 second on each side with a Bunsen burner; this removes residues from the manufacturing procedure and renders the glass hydrophilic. Prolonged exposure to the flame will crack the coverslips. Insufficient flaming will result in a hydrophobic surface, and solutions will not spread well. Each coverslip is then placed in a sterile bacteriological grade culture dish and coated overnight at 39 °C with a minimum of 100 μl of either 25 μg/ml laminin or 100 μg/ml PL. Coverslips can also be double-coated with both substrata by first coating with PL for 4 h, followed by overnight coating with laminin. When using PL it is important to suction off the solution and wash three times with PBSA to remove soluble PL that can be toxic to cultured cells. We never let the PL or laminin dry on the coverslip, and the solutions are removed just before adding culturing medium (500 μl of medium/18 × 18-mm coverslip).

3.3. Dissection Procedure

Protocol 14.9. Dissection and Culture of Dorsal Root Ganglion from Chick Embryo

Reagents and Materials

Sterile or aseptically prepared

- ❑ Embryonated eggs, 10 days
- ❑ F12HS10 (See Section 3.2.3)
- ❑ F12H medium containing 20 ng/ml NGF (See Section 3.2.5)
- ❑ CMF-PBS (PBSA)
- ❑ Trypsin (See Section 3.2.7)
- ❑ Laminin (See Section 3.2.9)
- ❑ Sterile dishes (plastic or glass; one 15 × 100 mm and one 15 × 60 mm)

- ❑ Bacteriological grade culture dish
- ❑ Conical centrifuge tube, 15 ml
- ❑ Dissection instruments (See Section 3.2.8); sterilize with 70% ethanol

Nonsterile

- ❑ Ethanol, 70%
- ❑ Pipettor with 100-μl tips
- ❑ Pipettor, 1 ml
- ❑ Microscope and light source placed in a laminar flow hood
- ❑ Coffee can with a small plastic bag inserted to collect waste during the dissection
- ❑ Water bath at 37 °C

Preparation for the dissection

UV illuminate the whole dissection setup for 15 min.

A. Removing the embryo from the egg:

(a) Add F12HS10 to the small dish.

(b) Obtain an egg that has been in the incubator for 10 days (embryonic day 10; E10).

(c) Spray the surface of the egg with 70% ethanol.

(d) With the pointed end of the pinning forceps make quick stabbing motions to crack the egg shell, proceed circumferentially and remove the "cap."

(e) You will now see a white extraembryonic membrane within the egg. Rip this membrane with your forceps and move it aside. This will expose the yolk, albumin, and embryo.

(f) The embryo will be beneath additional clear, but vascularized, extraembryonic membranes. Using the forceps, rip these membranes and move them to the side.

(g) At this point you should be able to see the embryo clearly. If the embryo is not evident; insert the forceps about 3–5 cm into the contents of the egg and gently sweep the forceps from one side of the egg to the other; this should "stir" the contents and allow you to see the embryo. Throughout this procedure make sure not to break the yolk sac, as this will obscure the remainder of the egg contents.

(h) Once you have identified the embryo, use the tips of the forceps to grab hold of the neck, just below the head, and gently lift the embryo from the remainder of the egg contents. While lifting the embryo make sure not to squeeze the tips of the forceps, as this will result in severing of the head and the body will fall back into the egg.

(i) After lifting the embryo by the head you may have to break remaining membranes that connect the embryo to the egg.

(j) Once the embryo has been removed from the egg, place it in the large dissection dish and decapitate it by squeezing the neck, just below the head, with the tips of your forceps.

B. Dissecting the embryo

(k) Wash the embryo with 1–2 ml F12HS10 to remove attached albumin and/or yolk.

(l) Place the embryo on its back and, with the dissection forceps, cut the skin over the viscera.

(m) Spread the hindlimbs apart and rotate the embryo so that the neck faces toward you.

(n) Using the tips of the dissection forceps, break the skin above the sternum and rib cage along the midline.

(o) Using the forceps, break the sternum and pull the two sides of the rib cage apart, exposing the heart, lungs, and viscera.

(p) Removing the internal organs without creating a lot of tissue debris is usually one of the biggest challenges for a novice. We suggest the following approach based on the concept that the organs are interconnected by connective tissue and can thus be thought of as a unit.

  i) Starting at the top of the rib cage, use the tips of the dissection forceps to sever connective tissue between the heart and lungs with the surrounding tissues.

  ii) Next, insert the tips of the forceps (held approximately 3–4 mm apart), between the heart/lungs and the rib cage/vertebral column. As you do this, make slight upward motions. This should allow you to lift the organs within the rib cage away from the ribs.

  iii) Continue with this procedure while moving caudally, toward the tail of the embryo.

(q) With practice you will be able to remove the majority of the organs in one continuous motion. Mastering the removal of the internal organs will greatly simplify the rest of the dissection by providing a "clean" working environment free from tissue debris. Blood from the embryo can obscure the dissection, but this is easily overcome by washing the embryo with F12HS10 by pipette.

(r) In most cases, even after removal of all the organs there will still be one obstacle to overcome before the DRG will be exposed. In the chicken embryo the late embryonic kidneys (metanephros) are in close physical apposition to the lumbosacral DRGs, and are often not removed along with all the other organs (as described above). The metanephros appears as a white, "fluffy" organ that runs from the caudal 2–3 ribs all the way down to the end of the vertebral column. The DRG will be just beneath the metanephros. In addition, in E10 embryos the metanephros interdigitates between the peripheral nerves of the DRGs. Thus it is imperative to carefully remove the metanephros without damaging the DRG. This is accomplished by lifting the metanephros away from the ribs and vertebral column, using the tips of the dissection forceps in an antero-caudal manner. You may have to remove pieces of the metanephros that remain between the DRG nerves and in the caudal-most region of the embryo.

(s) After the removal of the metanephros, the DRG and peripheral nerves should be partially visible. However, the vertebral column is covered by a connective tissue layer and it is advisable to remove this layer by picking at it with the forceps.

(t) In addition, it is important to remove the sympathetic chain ganglia before attempting the removal of individual DRG. The sympathetic chain is a strip of white nervous tissue that contains NGF-responsive neurons, and could thus contaminate the culture with non-DRG neurons. To remove the sympathetic chain, we suggest closing the tips of the forceps and gently running them over the sides of the vertebral column while moving from the caudal-most rib toward the tail of the embryo.

(u) After these obstacles have been removed, the DRG will appear as white balls of tissue lined up in a row next to the vertebral column. If the dissection was performed expertly the peripheral nerves will remain connected to the respective DRG, further facilitating identification of the DRG.

(v) The embryo is now ready for the removal of individual DRG. The DRG send nerves to the periphery and into the spinal cord. Therefore, to remove DRG from the embryo both of these nerves must be severed. Hold the tips of the forceps approximately 0.5–1 mm apart and insert them between the DRG and the vertebral column, followed by closing the tips. This should result in the severing of the DRG nerve root that projects into the spinal cord. If the peripheral nerve has been retained you should be able to the lift the DRG away from the vertebral column and observe a "ball on a chain" (i.e., DRG attached to a nerve).

(w) To free the DRG from the nerve, simply cut the nerve by crushing it with the tips of the forceps where it meets the DRG. If the nerve had been severed during earlier phases of the dissection, the DRG is ready to be removed from the embryo.

(x) Removing the isolated DRG from the embryo can be challenging at first. Resist the temptation to impale or grab the DRG with the forceps, as these procedures will damage the tissue. Instead, lift the DRG out from the embryo by "scooping" it up between the tips of the forceps held approximately 0.5 mm apart (less than the average diameter of the DRG). If the DRG tend to escape the scooping, hold them in place with the closed tips of the other forceps. With this procedure, the DRG will be caught in a drop of medium between the slightly splayed tips of the forceps. Place the DRG in the dish containing F12HS10.

(y) In our experience, after 6–10 dissections individuals develop sufficient expertise to efficiently remove a total of 12 DRG (6 from each side of the vertebral column). The lumbosacral DRG are the easiest to obtain, and are thus the focus of this description. However, with expertise, up to 40 DRG can be harvested from a single embryo along the entire axis of the vertebral column up to the cervical region. Briefly, to harvest the remaining DRG the ribs and overlying tissue/membranes must be removed along the entire axis, exposing the DRG. Removal of these DRG can be performed as described above.

(z) Clean the DRG. Before culturing, the DRG must be cleaned from attached tissue debris. During this procedure do not impale the DRG or try to grab hold of it with the forceps. It is important to clean the DRG as some of the associated tissues (e.g., the metanephros) can have contact-mediated inhibitory effects on axon extension (G. Gallo and L. Silver, unpublished observations). The "clean" DRG will have the appearance of a white sphere.

C. Culturing the DRG as an explant

(a) After the DRG are cleaned they are now ready for culturing. Remove the laminin solution (See Section 3.2.10) from the coverslip and add 500 μl F12H medium containing 20 ng/ml NGF.
(b) Pick up individual DRG, using a pipette set at 10–15 μl (100-μl tips work well).
(c) Slowly expel the DRG from the pipette tip in the center of the coverslip. If the DRG does not fall in the center, you can move the DRG around by repeated cycles of expelling medium from the pipette you used to bring the DRG into the dish.
(d) Transfer the dish to the incubator and culture overnight.
(e) After a 24-h culture period explanted DRG should exhibit a dense halo of axons extending to a mean length of 700–1000 μm (Fig. 14.6). Fibroblasts and Schwann cells will also be found migrating outward from the explant.

D. Dissociating the DRG

After the DRG have been cleaned (See Step (p) in B above) they can be dissociated as an alternative to primary explantation (See C above).

(a) Place all the DRG in a cluster and pick them up by aspiration through a Pasteur pipette, minimizing the amount of F12HS10 medium transferred with the DRG.
(b) Deposit the DRG into 5 ml CMF-PBS, in a 15-ml conical tube, and incubate at 37 °C in a water bath for 10 min.
(c) Six minutes into the CMF-PBS treatment, put a 5-ml aliquot of frozen trypsin solution into a 37 °C water bath.
(d) At the end of the 10 min, centrifuge the DRG very briefly (15 s at 1000 g).
(e) Remove the CMF-PBS, taking care not to aspirate the DRG.
(f) The trypsin solution should be fully thawed by this point, and it is ready to add to the DRG. Incubate the DRG in trypsin for 12 min in the 37 °C water bath.
(g) Briefly centrifuge the DRG, as described above (Step (d)) and remove all the trypsin solution, taking care not to aspirate the DRG.
(h) Immediately add 2 ml of F12HS10 (the proteins in the serum inactivate the trypsin).
(i) Using a pipettor set at 1 ml, mechanically triturate the DRG.

Place the tip of the pipette half way into the 2 ml of medium and repeatedly aspirate and expel 50% of the medium 15 times at a rate of approximately 1 s per aspiration/expel cycle.

Do not generate excess bubbles during this procedure.

Determine, by visual inspection, whether the DRG have fully dissociated and are no longer visible.

If remnants of the DRG are visible, continue with additional round of trituration until the DRG are no longer visible.

(j) Bring the volume up to 5 ml with additional F12HS10.
(k) Centrifuge the cells into a pellet (5 min at 1000 g). If dissociating more than four DRG, a pellet should be visible after centrifugation.
(l) Remove the F12HS10 medium, resuspend the dissociated cells in F12H medium containing 20 ng/ml NGF, and plate the cells on laminin-coated coverslips (we suggest a 500-μl volume per 18 × 18-mm coverslip). The cell density will have to be determined empirically to fit the user's experimental needs. For standard bioassay procedures we suggest plating 2–3 dissociated DRG per 18 × 18-mm coverslip.

After a 24-h culturing period axons will be 700–1000 μm in length and the neurons will be intermixed with Schwann cells and fibroblasts (Fig. 14.6). On a laminin substratum axon extension will commence between 3–6 h after plating.

ACKNOWLEDGMENTS

The research conducted in the laboratories of Drs. Lelkes and Unsworth was supported by grants from National Aeronautics and Space Administration (NASA) (P.I.L.) and NIH (B.R.U.). Drs. Lelkes and Unsworth would like to thank all their former and current collaborators/students over the past 15 years in Milwaukee, Madison, and Philadelphia; without their dedicated work and enthusiasm none of the results would have been possible. The work described in the section contributed by Saporta and Cameron was supported by grant NAG 9–1365 from NASA to S.S. Drs. Saporta and Cameron greatly appreciated the excellent technical assistance of L. Colina, J. Hushen, J. Mallery, J. Newcomb, and S. Ladd. They also wish to thank their collaborators, A.E. Willing, P. Bickford, and R. Shamekh, for help with the work described here, as well as many valuable discussions. Research performed in Dr. Gallo's laboratory was supported by NIH grant 1R01NS043251-01 and SCRF/PVA grant 2199-01.

SOURCES OF MATERIALS

Item	*Catalog #, type or size*	*Supplier*
Antibiotic/antimycotic		GIBCO, Invitrogen, 1600 Faraday Ave, Carlsbad CA
Antibiotic/antimycotic (Lelkes and Unsworth)	MT30004CI	Fisher
Bacteriological grade culture dish	08-757-100A	Fisher Scientific
Borate buffer	B-9876 and B-0252	Sigma, 3050 Spruce St, St. Louis, MO 63103

Item	*Catalog #, type or size*	*Supplier*
CO_2 incubator	Model MCO-17AIC	SANYO Scientific, distributed by Southeastern Scientific, PO Box 585147, Orlando, FL 32858
Cell culture dish	15 cm	Corning, Fisher Scientific, 3970 John's Creek Ct, Suawanee, GA 30024
Cell culture flasks	T150	Corning/Costar, Fisher Scientific
Centrifuge	CRU5000	Damon IEC Division, 300 Second Av, Needham Heights, MA 02194
Centrifuge tubes	50 ml	Falcon, B-D Biosciences
CMF-PBS	14200-075	Invitrogen
Collagenase powder	Type IV	GIBCO, Invitrogen
Conical centrifuge tube	15 ml	Nunc, Fisher Scientific
Cryogenic vial	2.0 ml	Nalgene, Fisher Scientific
Culture bags	VueLife™	American Fluoroseal Co
Dimethyl sulfoxide (DMSO)		ATCC, PO Box 1549, Manassas, VA 20108
Dissection microscope with 0.7–1.0× magnification	MX1583	Daigger, Vernon Hills
DNase II	Type V from bovine spleen	Sigma
Dulbecco's modified Eagle's medium (DMEM)		Cellgro, Fisher Scientific
DMEM high glucose (Lelkes and Unsworth)	10-013-CV	Fisher
Dulbecco's modified minimal Eagle's medium- Ham's F-12 nutrient mixture	DMEM-F12	Cellgro, Fisher Scientific
F12H medium	cat # 21700-075	Invitrogen
Fetal bovine serum		Gibco, Invitrogen
Fetal bovine serum	SH30071.03	HyClone
Fetal bovine serum	cat # MTT-350-11cv	Fisher Scientific, Pittsburgh, PA
Fiber optic illumination	12-455-20	Fischer Scientific
Flasks (T175)	13-680-65	Fisher
Forceps #55	72707-01	EMS, Hatfield, PA
Gas-permeable tissue culture bags		American Fluoroseal
Gentamycin sulfate		Sigma
German glass coverslips	ww-63-3093	Carolina Biological Supplies, Burlington NC
L-Glutamine	MT25005CI	Fisher
L-Glutamine, 200 mM	Cat # G-7513	Fisher Scientific
Growth Factor Reduced Matrigel™	MG_{GFR}	BD Biosciences, Fisher Scientific
Gyratory shaker		Stovall Belly Dancer, Fisher Scientific
HARV bioreactor		Synthecon Inc., 8054 El Rio, Houston TX 77054

Item	Catalog #, type or size	Supplier
Hen's eggs		Charles River Laboratories (www.criver.com)
HEPES buffer	Cat # H-3375	Sigma
Horse serum	SH30074.03	HyClone
Hydrochloric acid		Fisher Scientific
Incubator	Model 1205	G.Q.F. Manufacturing Co., Savannah GA
Insulin	I-5500	Sigma
ITS+ Premix culture supplement		BD Biosciences, Fisher Scientific
Laminin	23017-015	Invitrogen
NGF	256-GF-100	R&D Systems, Minneapolis, MN
NGF (Lelkes and Unsworth)	N2393	Sigma
PBS without Ca^{2+} and Mg^{2+} (PBSA)	20031CV	Fisher
Pen/Strep Fungizone mix	Cat # BW17-745E	Fisher Scientific,
Pentobarbital sodium (Sodium Nembutal)		Abbott Laboratories, North Chicago, IL 60064
Petri dishes	15 cm	Corning
Petri dishes	25 cm, glass or plastic	Corning, Falcon (B-D Biosciences), Pyrex
Petri dishes,	Non-TC grade, 6 cm	Nunc, Falcon (B-D Biosciences)
Phosphate-buffered saline,s 1× (PBS)		GIBCO, Invitrogen
Phosphocreatine	P-7936	Sigma
Pinning forceps	10-270	Fisher Scientific
Polylysine	P-9011	Sigma
Progesterone in absolute ethanol	P-8783	Sigma
Retinoic acid, sodium		Sigma
Rotatory Cell Culture Systems	RCCS	Synthecon, Inc.
Sodium pyruvate	P-2256	
Sodium selenate in water	S-8295	Sigma
apo-Transferrin	T-2252	Sigma
Trizma Base (Sigma)		Sigma
Trypan Blue		Gibco, Invitrogen
Trypsin		Sigma
Trypsin inhibitor		Sigma
Trypsin, 0.25%, 0.1% EDTA in PBSA	MT-25-053-cl	Fisher Scientific
Trypsin-EDTA solution, 0.25%		Sigma

REFERENCES

Aebischer, P., Goddard, M., Signore, A.P., Timpson, R.L. (1994) Functional recovery in hemi-parkinsonian primates transplanted with polymer-encapsulated PC12 cells. *Exp Neurol.* 126: 151–158.

Anderson, D.J. (1989) Cellular “neoteny”: a possible developmental basis for chromaffin cell plasticity. *Trends Genet.* 5: 174–178.
Anderson, D.J. (1993) Molecular control of cell fate in the neural crest: the sympathoadrenal lineage. *Annu. Rev. Neurosci.* 16: 129–158.
Anderson, D.J., and Axel. R. (1986) A bipotential neuroendocrine precursor whose choice of cell fate is determined by NGF and glucocorticoids. *Cell* 47: 1079–1090.
Anderson, D.J., and Michelson, A. (1989) Role of glucocorticoids in the chromaffin-neuron developmental decision. *Int. J. Dev. Neurosci.* 7: 475–487.
Andrews, P.W. (1984) Retinoic acid induces neuronal differentiation of a cloned human embryonal carcinoma cell line in vitro. *Dev. Biol.* 103: 285–293.
Andrews, P.W. (1998) Teratocarcinomas and human embryology: pluripotent human EC cell lines. Review article. *APMIS* 106: 158–167.
Baker, K.A., Sadi, D., Hong, M. Mendez, I. (2000) Simultaneous intrastriatal and intranigral dopaminergic grafts in the parkinsonian rat model: role of the intranigral graft. *J. Comp. Neurol.* 426: 106–116.
Bani-Yaghoub, M., Felker, J.M., Ozog, M.A., Bechberger, J.F., Naus, C.C. (2001) Array analysis of the genes regulated during neuronal differentiation of human embryonal cells. *Biochem. Cell Biol.* 79: 387–398.
Becker, J.L., Prewett, T.L., Spaulding, G.F., Goodwin, T.J. (1993) Three-dimensional growth and differentiation of ovarian tumor cell line in high aspect rotating-wall vessel: morphologic and embryologic considerations. *J. Cell. Biochem.* 51: 283–289.
Begley, C.M., and Kleis, S.J. (2000) The fluid dynamic and shear environment in the NASA/JSC rotating-wall perfused-vessel bioreactor. *Biotechnol. Bioeng.* 70: 32–40.
Bjorklund, A., Stenevi, U., Dunnett, S.B., Iversen, S.D. (1981) Functional reactivation of the deafferented neostriatum by nigral transplants. *Nature* 289: 497–499.
Bjorklund, A., and Lindvall, O. (2000) Cell replacement therapies for central nervous system disorders. *Nat. Neurosci.* 3: 537–544.
Bjorklund, A., Dunnett, S.B., Brundin, P., Stoessl, A.J., Freed, C.R., Breeze, R.E., Levivier, M., Peschanski, M., Studer, L., Barker, R. (2003) Neural transplantation for the treatment of Parkinson’s disease. *Lancet Neurol.* 2: 437–445.
Bjorklund, L.M., Sanchez-Pernaute, R., Chung, S., Andersson, T., Chen, I.Y., McNaught, K.S., Brownell, A.L., Jenkins, B.G., Wahlestedt, C., Kim, K.S., Isacson. O. (2002) Embryonic stem cells develop into functional dopaminergic neurons after transplantation in a Parkinson rat model. *Proc. Natl. Acad. Sci. USA* 99: 2344–2349.
Byrd, J.C., Hadjiconstantinou, M., Cavalla, D. (1986) Epinephrine synthesis in the PC12 pheochromocytoma cell line. *Eur. J. Pharmacol.* 127: 139–142.
Daadi, M.M., Saporta, S., Willing, A.E., Zigova, T., McGrogan, M.P., Sanberg, P.R. (2001) In vitro induction and in vivo expression of bcl-2 in the hNT neurons. *Brain Res. Bull.* 56: 147–152.
Doupe, A.J., Landis, S.C., Patterson, P.H. (1985) Environmental influences in the development of neural crest derivatives: glucocorticoids, growth factors, and chromaffin cell plasticity. *J. Neurosci.* 5: 2119–2142.
Ebert, S.N., Lindley, S.E., Bengoechea, T.G., Bain, D., Wong, D.L. (1997) Adrenergic differentiation potential in PC12 cells: influence of sodium butyrate and dexamethasone. *Brain Res. Mol. Brain Res.* 47: 24–30.
Englund, U., Fricker-Gates, R.A., Lundberg, C., Bjorklund, A., Wictorin. K. (2002) Transplantation of human neural progenitor cells into the neonatal rat brain: extensive migration and differentiation with long-distance axonal projections. *Exp. Neurol.* 173: 1–21.
Espinosa-Jeffrey, A., Becker-Catania, S.G., Zhao, P.M., Cole, R., Edmond, J., de Vellis, J. (2002) Selective specification of CNS stem cells into oligodendroglial or neuronal cell lineage: cell culture and transplant studies. *J. Neurosci. Res.* 69: 810–825.
Finotto, S., Krieglstein, K., Schober, A., Deimling, F., Lindner, K., Bruhl, B., Beier, K., Metz, J., Garcia-Arraras, J.E., Roig-Lopez, J.L., Monaghan, P., Schmid, W., Cole, T.J., Kellendonk, C., Tronche, F., Schutz, G., Unsicker, K. (1999) Analysis of mice carrying targeted mutations of the glucocorticoid receptor gene argues against an essential role of glucocorticoid signalling for generating adrenal chromaffin cells. *Development* 126: 2935–2944.

Fournier, A.E., Gould, G.C., Liu, B.P., Strittmatter, S.M. (2002) Truncated soluble Nogo receptor binds Nogo-66 and blocks inhibition of axon growth by myelin. *J. Neurosci.* 22(20): 8876–8883.

Freed, C.R., Breeze, R.E., Rosenberg, N.L., Schneck, S.A., Wells, T.H., Barrett, J.N., Grafton, S.T., Huang, S.C., Eidelberg, D., Rottenberg, D.A. (1990) Transplantation of human fetal dopamine cells for Parkinson's disease. Results at 1 year. *Arch. Neurol.* 47: 505–512.

Freed, C.R., Greene, P.E., Breeze, R.E., Tsai, W.Y., DuMouchel, W., Kao, R., Dillon, S., Winfield, H., Culver, S., Trojanowski, J.Q., Eidelberg, D., Fahn, S. (2001) Transplantation of embryonic dopamine neurons for severe Parkinson's disease. *N. Engl. J. Med.* 344: 710–719.

Freed, C.R. (2002) Will embryonic stem cells be a useful source of dopamine neurons for transplant into patients with Parkinson's disease? *Proc. Natl. Acad. Sci. USA* 99: 1755–1757.

Freeman, T.B., Olanow, C.W., Hauser, R.A., Nauert, G.M., Smith, D.A., Borlongan, C.V., Sanberg, P.R., Holt, D.A., Kordower, J.H., Vingerhoets, F.J., et al. (1995) Bilateral fetal nigral transplantation into the postcommissural putamen in Parkinson's disease. *Ann. Neurol.* 38: 379–388.

Fujita, K., Lazarovici, P., Guroff, G. (1989) Regulation of the differentiation of PC12 pheochromocytoma cells. *Environ. Health Perspect.* 80: 127–142.

Greene, L.A., and Tischler, A.S. (1976) Establishment of a noradrenergic clonal line of rat adrenal pheochromocytoma cells which respond to nerve growth factor. *Proc. Natl. Acad. Sci. USA* 73: 2424–2428.

Guillemain, I., Alonso, G., Patey, G., Privat, A., Chaudieu, I. (2000) Human NT2 neurons express a large variety of neurotransmission phenotypes in vitro. *J. Comp. Neurol.* 422: 380–395.

Guillemain, I., Fontes, G., Privat, A., Chaudieu, I. (2003) Early programmed cell death in human NT2 cell cultures during differentiation induced by all-trans-retinoic acid. *J. Neurosci. Res.* 71: 38–45.

Hamburger, V. (1993) The history of the discovery of the nerve growth factor. *J. Neurobiol.* 24(7): 893–897.

Hammond, T.G., and Hammond, J.M. (2001) Optimized suspension culture: the rotating-wall vessel. *Am. J. Physiol. Renal Physiol.* 281: F12–F25.

Huber, K., Combs, S., Ernsberger, U., Kalcheim, C., Unsicker, K. (2002) Generation of neuroendocrine chromaffin cells from sympathoadrenal progenitors: beyond the glucocorticoid hypothesis. *Ann. NY Acad. Sci.* 971: 554–559.

Iacovitti, L., and Stull, N.D. (1997) Expression of tyrosine hydroxylase in newly differentiated neurons from a human cell line (hNT). *Neuroreport* 8: 1471–1474.

Jaeger, C.B., Greene, L.A., Tresco, P.A., Winn, S.R., Aebischer, P. (1990) Polymer encapsulated dopaminergic cell lines as "alternative neural grafts." *Prog. Brain Res.* 82: 41–46.

Johnston, R.E., and Becker, J.B. (1999) Behavioral changes associated with grafts of embryonic ventral mesencephalon tissue into the striatum and/or substantia nigra in a rat model of Parkinson's Disease. *Behav. Brain Res.* 104: 179–187.

Kim, J.H., Auerbach, J.M., Rodriguez-Gomez, J.A., Velasco, I., Gavin, D., Lumelsky, N., Lee, S.H., Nguyen, J., Sanchez-Pernaute, R., Bankiewicz, K., McKay, R. (2002) Dopamine neurons derived from embryonic stem cells function in an animal model of Parkinson's disease. *Nature* 418: 50–56.

Kim, K.T., Park, D.H., Joh, T.H. (1993) Parallel up-regulation of catecholamine biosynthetic enzymes by dexamethasone in PC12 cells. *J. Neurochem.*, 60: 946–951.

Klaus, D.M. (2001) Clinostats and bioreactors. *Gravit. Space Biol. Bull.* 14: 55–64.

Lelkes, P.I. (1991) Our PC12 cells are not your PC12 cells which are not his/her PC12 cells. *6th International Symposium for Chromaffin Cell Biology*, Marburg, Germany, 1991.

Lelkes, P.I., Galvan, D.L., Hayman, G.T., Goodwin, T.J., Chatman, D.Y., Cherian, S., Garcia, R.M., Unsworth, B.R. (1998) Simulated microgravity conditions enhance differentiation of cultured PC12 cells towards the neuroendocrine phenotype. *In Vitro Cell. Dev. Biol.* 34: 316–325.

Lelkes, P.I., Ramos, E., Chick, D.M., Liu, J., Unsworth, B.R. (1994) Microgravity decreases tyrosine hydroxylase expression in rat adrenals. *FASEB J.* 8: 1177–1182.

Lelkes, P.I., Ramos, E., Nikolaychik, V.V., Wankowski, D.M., Unsworth, B.R., Goodwin, T.J. (1997) GTSF-2: a new, versatile cell culture medium for diverse normal and transformed mammalian cells. *In Vitro Cell Dev. Biol. Anim.* 33: 344–351.

Lelkes, P.I., and Unsworth, B.R. (1992) The role of heterotypic interactions between endothelial cells and parenchymal cells in organspecific differentiation: a possible trigger for vasculogenesis. In Maragoudakis, M.E., Gullino, P. and Lelkes, P.I., eds., *Angiogenesis in health and diseases*. NATO, a S I Series, Series A, Life Sciences, Plenum Publishing, New York NY, pp. 27–41.
Lelkes, P.I., and Unsworth, B.R. (2002) Neuroendocrine tissue engineering in RWV bioreactors. In Atala, A., ed., *Methods in Tissue Engineering*. San Diego, Academic Press, pp. 371–382.
Leypoldt, F., Lewerenz, J., Methner, A. (2001) Identification of genes up-regulated by retinoic-acid-induced differentiation of the human neuronal precursor cell line NTERA-2 cl.D1. *J. Neurochem.* 76: 806–814.
Lindner, M.D., and Emerich, D.F. (1998) Therapeutic potential of a polymer-encapsulated L-DOPA and dopamine-producing cell line in rodent and primate models of Parkinson's disease. *Cell Transplant.* 7: 165–174.
Lindvall, O., Rehncrona, S., Brundin, P., Gustavii, B., Astedt, B., Widner, H., Lindholm, T., Bjorklund, A., Leenders, K.L., Rothwell, J.C., et al. (1990) Neural transplantation in Parkinson's disease: the Swedish experience. *Prog. Brain Res.* 82: 729–734.
Luo, Y., Raible, D., Raper, J.A. (1993) Collapsin: a protein in brain that induces the collapse and paralysis of neuronal growth cones. *Cell* 75(2): 217–227.
Low, H.P., Savarese, T.M., Schwartz, W.J. (2001) Neural precursor cells form rudimentary tissue-like structures in a rotating-wall vessel bioreactor. *In Vitro Cell. Dev. Biol.* 37: 141–147.
Mahata, S.K., Mahapatra, N.R., Mahata, M., O'Connor, D.T. (2002) Neuroendocrine cell type-specific and inducible expression of chromogranin/secretogranin genes: crucial promoter motifs. *Ann. NY Acad. Sci.* 971: 27–38.
McKay, R. (2002) Building animals from stem cells. *Ann. NY Acad. Sci.* 961: 44.
Misiuta, I.E., Anderson, L., McGrogan, M.P., Sanberg, P.R., Willing, A.E., Zigova, T. (2003) The transcription factor Nurr1 in human NT2 cells and hNT neurons. *Brain Res. Dev. Brain Res.* 145: 107–115.
Mizrachi, Y., Lelkes, P.I., Ornberg, R.L., Goping, G., Pollard, H.B. (1998) Specific adhesion between pheochromocytoma (PC12) cells and adrenal medullary endothelial cells in co-culture. *Cell Tissue Res.* 256: 365–372.
Mizrachi, Y., Naranjo, J.R., Levi, B.Z., Pollard, H.B., Lelkes, P.I. (1990) PC12 cells differentiate into chromaffin cell-like phenotype in coculture with adrenal medullary endothelial cells. *Proc. Natl. Acad. Sci. USA.* 87: 6161–6165.
Moro, M.A., Lopez, M.G., Gandia, L., Michelena, P., Garcia, A.G. (1990) Separation and culture of living adrenaline- and noradrenaline-containing cells from bovine adrenal medullae. *Anal. Biochem.* 185: 243–248.
Nelson, P.T., Kondziolka, D., Wechsler, L., Goldstein, S., Gebel, J., DeCesare, S., Elder, E.M., Zhang, P.J., Jacobs, A., McGrogan, M., Lee, V.M., Trojanowski, J.Q. (2002) Clonal human (hNT) neuron grafts for stroke therapy: neuropathology in a patient 27 months after implantation. *Am. J. Pathol.* 160: 1201–1206.
Olanow, C.W., Goetz, C.G., Kordower, J.H., Stoessl, A.J., Sossi, V., Brin, M.F., Shannon, K.M., Nauert, G.M., Perl, D.P., Godbold, J., Freeman, T.B. (2003) A double-blind controlled trial of bilateral fetal nigral transplantation in Parkinson's disease. *Ann. Neurol.* 54: 403–414.
Othberg, A.I., Willing, A.E., Cameron, D.F., Anton, A., Saporta, S., Freeman, T.B., Sanberg, P.R. (1998a) Trophic effect of porcine Sertoli cells on rat and human ventral mesencephalic cells and hNT neurons in vitro. *Cell Transplant.* 7: 157–164.
Othberg, A.I., Willing, A.E., Saporta, S., Cameron, D.F., Sanberg, P.R. (1998b) Preparation of cell suspensions for co-transplantation: methodological considerations. *Neurosci. Lett.* 247: 111–114.
Ourednik, V., Ourednik, J., Flax, J.D., Zawada, W.M., Hutt, C., Yang, C., Park, K.I., Kim, S.U., Sidman, R.L., Freed, C.R., Snyder, E.Y. (2001) Segregation of human neural stem cells in the developing primate forebrain. *Science* 293: 1820–1824.
Palmer, M.R., Granholm, A.C., van Horne, C.G., Giardina, K.E., Freund, R.K., Moorhead, J.W., Gerhardt, G.A. (2001) Intranigral transplantation of solid tissue ventral mesencephalon or striatal grafts induces behavioral recovery in 6-OHDA-lesioned rats. *Brain Res.* 890: 86–99.

Panchision, D.M., McKay, R.D. (2002) The control of neural stem cells by morphogenic signals. *Curr. Opin. Genet. Dev.* 12: 478–487.
Pleasure, S.J., and Lee, V.M. (1993) NTera 2 cells: a human cell line which displays characteristics expected of a human committed neuronal progenitor cell. *J. Neurosci. Res.* 35: 585–602.
Pollanen, P., and Maddocks, S. (1988) Macrophages, lymphocytes and MHC II antigen in the ram and the rat testis. *J. Reprod. Fertil.* 82: 437–445.
Pollanen, P., Soder, O., Uksila, J. (1988) Testicular immunosuppressive protein. *J. Reprod. Immunol.* 14: 125–138.
Pollanen, P., and Uksila, J. (1990) Activation of the immune system in the testis. *J. Reprod. Immunol.* 18: 77–87.
Pollanen, P., von Euler, M., Soder, O. (1990) Testicular immunoregulatory factors. *J. Reprod. Immunol.* 18: 51–76.
Pond, A., Roche, F.K., Letourneau, P.C. (2002) Temporal regulation of neuropilin-1 expression and sensitivity to semaphorin 3A in NGF- and NT3-responsive chick sensory neurons. *J. Neurobiol.* 51: 43–53.
Sanchez-Ramos, J.R., Song,S., Kamath,S.G., Zigova, T., Willing, A., Cardozo-Pelaez, F., Stedeford, T., Chopp, M., Sanberg, P.R. (2001) Expression of neural markers in human umbilical cord blood. *Exp. Neurol.* 171: 109–115.
Saporta, S., Cameron, D.F., Borlongan, C.V., Sanberg, P.R. (1997) Survival of rat and porcine Sertoli cell transplants in the rat striatum without cyclosporine-A immunosuppression. *Exp. Neurol.* 146: 299–304.
Saporta, S., Borlongan, C.V., Sanberg, P.R. (1999) Neural transplantation of human neuroteratocarcinoma (hNT) neurons into ischemic rats. A quantitative dose-response analysis of cell survival and behavioral recovery. *Neuroscience* 91: 519–525.
Saporta, S., Willing, A.E., Zigova, T., Daadi, M.M., Sanberg, P.R. (2001) Comparison of calcium-binding proteins expressed in cultured hNT neurons and hNT neurons transplanted into the rat striatum. *Exp. Neurol.* 167: 252–259.
Saporta, S., Cameron, D.F., Willing, A.E., Sanberg, P.R., Colina, L.O., Hushen, J., Dejarlais, T. (2002) Characterization of Sertoli-neuron-aggregated cells (SNAC) for brain repair. *Soc. Neurosci. Abst. View. Prog.* #, 727.729.
Saporta, S., Makoui, A.S., Willing, A.E., Daadi, M., Cahill, D.W., Sanberg, P.R. (2002) Functional recovery after complete contusion injury to the spinal cord and transplantation of human neuroteratocarcinoma neurons in rats. *J. Neurosurg.* 97: 63–68.
Selawry, H., Fajaco, R., Whittington, K. (1985) Intratesticular islet allografts in the spontaneously diabetic BB/W rat. *Diabetes* 34: 1019–1024.
Skinner, M.K. (1993). Secretion of growth factors and other regulating factors. In Russel, L.D., and Griswold, M.D., eds., *Secretion of Growth Factors and Other Regulating Factors*. Clearwater, FL, Cache River Press.
Tischler, A.S. (2002) Chromaffin cells as models of endocrine cells and neurons. *Ann. NY Acad. Sci.* 971: 366–70.
Tischler, A.S., and Greene, L.A. (1978) Morphologic and cytochemical properties of a clonal line of rat adrenal pheochromocytoma cells which respond to nerve growth factor. *Lab. Invest.* 39: 77–89.
Uchida, N., Buck, D.W., He, D., Reitsma, M.J., Masek, M., Phan, T.V., Tsukamoto, A.S., Gage, F.H., Weissman, I.L. (2000) Direct isolation of human central nervous system stem cells. *Proc. Natl. Acad. Sci. USA.* 97: 14720–14725.
Unsworth, B.R., and Lelkes, P.I. (1998a) Growing tissues in microgravity. *Nat. Med.* 4: 901–907.
Unsworth, B.R., and Lelkes, P.I. (1998b) The use of rotating wall bioreactors for the assembly of differentiated tissue-like organoids. In *Advances in Tissue Engineering*. Southborough, MA, IBC Press, pp. 113–132.
Unsworth, B.R., and Lelkes, P.I. (2000) Growing tissues in microgravity. In Lanza, R., Langer, R., Vacanti, J. eds, *Principles of Tissue Engineering*, 2nd Edition. San Diego, Academic Press, pp. 157–164.
Vaudry, D., Stork, P.J., Lazarovici, P., Eiden, L.E. (2002) Signaling pathways for PC12 cell differentiation: making the right connections. *Science* 296(5573): 1648–1649.

Wang, S.S., and Good, T.A. (2001). Effect of culture in a rotating wall bioreactor on the physiology of differentiated neuron-like PC12 and SH-SY5Y cells. *J. Cell. Biochem.* 83: 574–584.
Watson, D.J., Longhi, L., Lee, E.B., Fulp, C.T., Fujimoto, S., Royo, N.C., Passini, M.A., Trojanowski, J.Q., Lee, V.M., McIntosh, T.K., Wolfe, J.H. (2003) Genetically modified NT2N human neuronal cells mediate long-term gene expression as CNS grafts in vivo and improve functional cognitive outcome following experimental traumatic brain injury. *J. Neuropathol. Exp. Neurol.* 62: 368–380.
Weiss, C., Cahill, A.L., Laslop, A., Fischer-Colbrie, R., Perlman, R.L., Winkler, H. (1996) Differences in the composition of chromaffin granules in adrenaline and noradrenaline containing cells of bovine adrenal medulla. *Neurosci. Lett.* 211: 29–32.
Willing, A.E., Othberg, A.I., Saporta, S., Anton, A., Sinibaldi, S., Poulos, S.G., Cameron, D.F., Freeman, T.B., Sanberg, P.R. (1999) Sertoli cells enhance the survival of co-transplanted dopamine neurons. *Brain Res.* 822: 246–250.
Willing, A.E., Sudberry, J.J., Othberg, A.I., Saporta, S., Poulos, S.G., Cameron, D.F., Freeman, T.B., Sanberg, P.R. (1999) Sertoli cells decrease microglial response and increase engraftment of human hNT neurons in the hemiparkinsonian rat striatum. *Brain Res. Bull.* 48: 441–444.
Willing, A.E., Saporta, S., Lixian, J., Milliken, M., Poulos, S., Bowersox, S.S., Sanberg, P.R. (2002) Preliminary study of the behavioral effects of LBS-neuron implantation on seizure susceptibility following middle cerebral artery occlusion in the rats. *Neurotoxicol. Res.* 4: 111–118.
Winkler, C., Kirik, D., Bjorklund, A., Dunnett, S.B. (2000) Transplantation in the rat model of Parkinson's disease: ectopic versus homotopic graft placement. *Prog. Brain Res.* 127: 233–265.
Wong, D.L. (2003) Why is the adrenal adrenergic? *Endocr. Pathol.* 14: 25–36.
Yang, M., Donaldson, A.E., Jiang, Y., Iacovitti, L. (2003) Factors influencing the differentiation of dopaminergic traits in transplanted neural stem cells. *Cell. Mol. Neurobiol.* 23: 851–864.
Zigova, T., Barroso, L.F., Willing, A.E., Saporta, S., McGrogan, M.P., Freeman, T.B., Sanberg, P.R. (2000) Dopaminergic phenotype of hNT cells in vitro. *Brain Res. Dev. Brain Res.* 122: 87–90.
Zigova, T., Song, S., Willing, A.E., Hudson, J.E., Newman, M.B., Saporta, S., Sanchez-Ramos, J., Sanberg, P.R. (2002) Human umbilical cord blood cells express neural antigens after transplantation into the developing rat brain. *Cell Transplant.* 11: 265–274.

15

Tissue Engineering of the Liver

Gregory H. Underhill, Jennifer Felix, Jared W. Allen,
Valerie Liu Tsang, Salman R. Khetani, and Sangeeta N. Bhatia

Division of Health Sciences and Technology (Harvard-M.I.T.), Massachusetts Institute of Technology, 77 Massachusetts Avenue, E19-502D, Cambridge, Massachusetts, 02139

Corresponding author: sbhatia@mit.edu

Culture of Cells for Tissue Engineering, edited by Gordana Vunjak-Novakovic and R. Ian Freshney

1. BACKGROUND

1.1. Liver Disease and Cell-Based Therapies

Liver failure is a significant clinical problem, representing the cause of death of over 30,000 patients in the United States every year and over 2 million patients worldwide. Organ transplantation is the only therapy to date shown to alter mortality, although the utility of organ transplantation is restricted because of the scarcity of donor organs. Surgical advances such as partial liver transplants from cadaveric or living donors have been demonstrated to be effective treatments and a means to increase the supply of donor organs [Ghobrial et al., 2000; Hashikura et al., 1994; Raia et al., 1989; Schiano et al., 2001]. These approaches take advantage of the body's role in the regulation of liver mass and the significant capacity for regeneration exhibited by the mammalian liver. This regenerative process has been extensively examined with experiments in rodent models, which demonstrate that partial hepatectomy or chemical injury induces the proliferation of the existing mature cell populations within the liver including hepatocytes, bile duct epithelial cells, and others, resulting in the replacement of lost liver mass. However, liver regeneration is not a component of all disease settings (e.g., cirrhosis) and is difficult to control clinically. Furthermore, despite surgical advances such as split liver and living donor transplantation, there is an increasing divergence between the number of patients awaiting transplantation and the number of available organs [Harper et al., 2001], suggesting that it is unlikely that liver transplantation procedures alone will meet the increasing demand. Consequently, alternative approaches are needed and are actively being pursued. These approaches include several nonbiological extracorporeal support systems, such as plasma exchange, plasmapheresis, hemodialysis, or hemoperfusion over charcoal or various resins, although these systems have achieved limited success [Allen et al., 2001; Strain and Neuberger, 2002; Yarmush et al., 1992]. It has been suggested that the inadequate effectiveness of these nonbiological schemes is due to the limited functionality of these devices. The liver exhibits a complex array of over 500 functions, including detoxification, synthetic, and metabolic processes. Thus recapitulation of a substantial number of liver functions will be required to provide sufficient liver support. To provide the myriad of known as well as currently unidentified liver functions, cell-based therapies have been proposed as an alternate approach to both organ transplantation and the use of strictly nonbiological systems. Potential cell-based therapies include the transplantation of hepatocytes, perfusion of blood through an extracorporeal device containing hepatocytes, transgenic xenografts [Costa et al., 1999; Fodor et al., 1994; Schmoeckel et al., 1997], or the implantation of hepatocellular constructs (Fig. 15.1).

1.2. Cell Sources for Liver Cell-Based Therapies

Immortalized hepatocyte cell lines such as HepG2 (human hepatoblastoma) or HepLiu (SV40 immortalized) have been utilized as readily available surrogates for hepatic tissue. However, it has been documented that these cells display an abnormal assortment of differentiated functions [Cederbaum et al., 2001; Fukaya

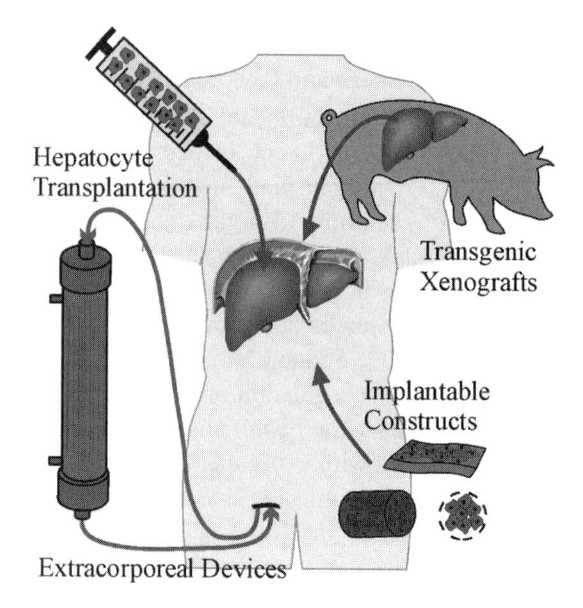

Figure 15.1. Cell-based therapies for liver disease. Extracorporeal devices perfuse patient's blood or plasma through bioreactors containing hepatocytes. Hepatocytes are transplanted directly or implanted on scaffolds. Transgenic animals are being raised in order to reduce complement-mediated damage of the endothelium. From Allen et al. [2001] (See Color Plate 10A.).

et al., 2001; Liu et al., 1999], and for clinical applications there is a risk that oncogenic factors could be transmitted to the patient. Thus the use of primary hepatocyte-based systems would eliminate these deficiencies, providing the appropriate collection of liver functions. The development of primary hepatocyte-based approaches is the focus of substantial ongoing research, yet progress has been hampered by the loss of liver-specific functions exhibited by isolated hepatocytes in vitro. Furthermore, particularly for human hepatocytes, despite the significant proliferative capacity during regenerative responses in vivo, mature hepatocyte proliferation in culture is limited. As a result, alternative primary cell sources for liver cell-based therapies are also being investigated [Allen and Bhatia, 2002], such as various stem cell populations (embryonic and adult), which retain significant proliferative ability in vitro and exhibit pluripotency (embryonic) or multipotency (adult), thereby providing a potential source of hepatocytes as well as other liver cell types. This prospective capacity to generate concurrently additional liver cell types, for example, bile duct epithelial cells from a single proliferative precursor, represents a potential means for enhancing liver-specific function of a tissue-engineered construct. However, this process may require complex differentiation cues, and many challenges remain, including the ability to control the microenvironment and organize resultant structures, strategies for delivery, etc., before stem cells can be utilized as sources of large numbers of hepatocytes or other liver cell types, particularly within multicellular systems. In addition, regardless of the cell source, the stabilization of hepatocyte functions remains a fundamental issue.

1.3. Approaches for the In Vitro Stabilization of Primary Hepatocytes

Several distinct methods have been utilized to promote hepatocyte stabilization in vitro, with a broad goal of mimicking physiologically relevant extracellular cues that are absent from standard culture models. For example, modifications in culture medium such as hormonally defined preparations with low concentrations of hormones, corticosteroids, cytokines, vitamins, or amino acids, or the addition of low levels of dimethyl sulfoxide or dexamethasone, have been shown to help promote a stabilized hepatocyte phenotype [Baribault and Marceau, 1986; Block et al., 1996; Dich et al., 1988; Isom et al., 1985; Kubota and Reid, 2000]. In addition, extracellular matrix of various compositions is also known to exhibit positive effects on hepatocyte function. These approaches include the sandwich culture of hepatocytes within collagen gel or culture on the tumor-derived basement membrane preparation Matrigel [Bissell et al., 1987; Dunn et al., 1991; Rojkind et al., 1980]. Interestingly, hepatocyte culture on Matrigel results in the formation of spheroid cellular structures [Kang et al., 2004], suggesting that cell-cell communication is a likely component of the stabilization observed in this model system. Consistent with the importance of homotypic (hepatocyte-hepatocyte) interactions, spheroidal aggregates have been shown to promote the formation of bile canaliculi, gap junctions, and tight junctions and the expression of E-cadherins and several differentiated functions [Chang, 1992; Landry et al., 1985; Saito et al., 1992].

Similar to homotypic communication, heterotypic (hepatocyte-nonparenchymal) interactions have also been shown to improve viability and differentiated function. This stabilization of hepatocyte functions has been reported for hepatocyte cocultures with both liver- and non-liver-derived cell types, and, furthermore, beneficial effects of cross-species coculture systems have also been observed [Bhatia et al., 1999]. Together, these findings suggest a highly conserved mechanism by which cocultures enhance the liver-specific function of hepatocytes. However, the precise roles of the potential regulatory factors, such as secreted signals (i.e., cytokines) or cell-associated signals (i.e., insoluble extracellular matrix or membrane-bound molecules), in the "coculture effect" have not yet been clearly defined. An understanding of the mechanisms mediating the stabilization of hepatocytes in coculture would have broad implications in liver cell biology and, specifically, would aid in the development of functional hepatic tissue constructs.

In addition to static culture systems, several bioreactor designs have also been developed as in vitro hepatocyte culture models. Perfusion systems would facilitate enhanced nutrient delivery to hepatocytes, which are highly metabolic, and additionally, in contrast to a batch process, would enable the continuous processing of blood or serum, potentially useful for extracorporeal devices. Overall, the bioreactor systems developed to date fall into the following categories: flat plate, hollow fiber, perfusion scaffolds, and packed beds, each with accompanying advantages and disadvantages [Allen et al., 2001] (Fig. 15.2). In particular, the combination of a flat plate reactor system with sandwich culture or coculture has been illustrated to improve hepatocyte stability for long-term analysis [Bader et al., 1995; Tilles et al., 2001]. Additionally, a recent three-dimensional perfusion bioreactor system has been developed based on the morphogenesis of hepatocytes into three-dimensional structures in an array of channels [Powers et al., 2002a,b]. Similar to static culture

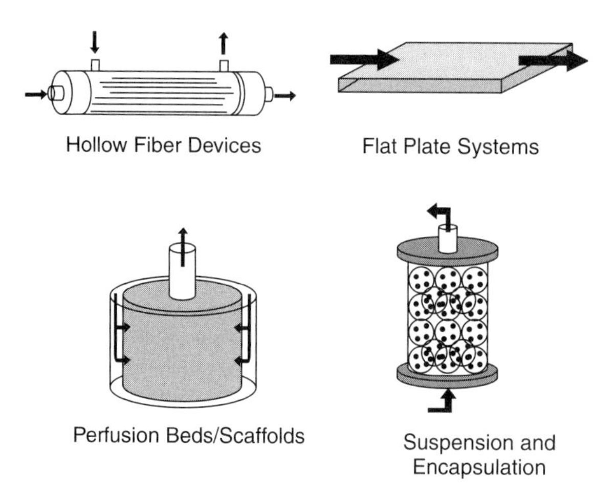

Figure 15.2. Schematics of cell-based bioreactor designs. The majority of liver cell-based bioreactor designs fall into these four general categories, each with inherent advantages and disadvantages, reviewed extensively in Allen et al. [2001].

models, preaggregation of hepatocytes into spheroids enhanced the functionality of hepatocytes in this design.

Overall, the components and characteristics of primary hepatocyte culture models have underscored the importance of microenvironmental signals, including soluble mediators, cell-extracellular matrix interactions, and cell-cell interactions, in the regulation of hepatocyte processes. Accordingly, the development of robust hepatocyte-based tissue engineering platforms will be predicated upon a fundamental knowledge and controlled reconstitution of these environmental factors.

1.4. Regulation of the Hepatocyte Microenvironment Toward the Development of Engineered Liver Tissue

The importance of extracellular signals in hepatocyte culture in vitro correlates with the significant role of the microenvironment in the liver in vivo. It is well established that the proper function of liver cells in vivo is maintained by a complex combination of extracellular cues, including tightly controlled cell-cell interactions and distribution of extracellular matrix [Gebhardt and Mecke, 1983; Michalopoulos and DeFrances, 1997; Olson et al., 1990; Reid et al., 1992]. Furthermore, alterations in the liver microenvironment are important aspects of liver pathologies [Giannelli et al., 2003] such as fibrosis and have additionally been implicated in repair processes [Kim et al., 1997; Michalopoulos and DeFrances, 1997; Rudolph et al., 1999]. Therefore, an understanding of the microenvironmental mechanisms regulating hepatocyte function represents a crucial issue for not only the improvement of in vitro culture models and extracorporeal liver devices, but also the manufacture of implantable hepatocyte constructs. Notably, in addition to tissue engineering applications, functional in vitro hepatic models would also represent ideal platforms for ADMET (adsorption, distribution, metabolism, excretion, and toxicity) testing of novel drug candidates.

Consequently, our laboratory's strategy toward the fabrication of engineered liver tissue is to develop (1) enabling platforms for the controlled recapitulation of these critical extracellular cues and (2) in vitro models to study the complexities of environmental interactions. Furthermore, current studies in our laboratory are addressing the potential utility of alternative cell sources, such as embryonic stem cells and various adult stem cell populations, for liver tissue engineering. Overall, important criteria for the development of a particular system include (1) the desired application (in vitro model, basic science, or therapeutic) and (2) the important liver properties that must be established for that certain application, such as stabilized phenotype, structural complexity (zonation), directional fluid flow, and in some cases 3-D organization. Our laboratory has developed several distinct model systems that recapitulate critical structural and/or functional aspects of the liver.

Specifically, as a means to investigate systematically the role of heterotypic contact in hepatocyte cocultures, our laboratory has developed and characterized a versatile micropatterning method, which has enabled the quantitative control of the spatial relationship of hepatocytes and fibroblasts [Bhatia, 1997]. This method is one in a repertoire of techniques that we have developed to control cell-cell interactions. These processes enable an examination of the mechanisms of interaction, including the potential importance of numerous aspects such as continuous signaling, gap junctions, and soluble factors. Furthermore, the micropatterning method described in this chapter also represents a robust platform for the in vitro functional stabilization of hepatocytes.

Moreover, to gain another level of control of environmental influences and further enhance the physiologically relevant function of hepatocytes in vitro, we have also developed additional approaches. First, we have recently described the construction of a perfusion bioreactor system that has enabled the formation of steady-state oxygen gradients, resulting in the presence of zonated hepatocyte characteristics [Allen and Bhatia, 2003), which is an important feature of the liver in vivo. For example, numerous drugs exhibit zonal toxicity due to compartmentalization of drug-metabolizing enzymes [Lindros, 1997]. As a result, the in vitro model of zonation described here represents a unique system to investigate, among many aspects, the mechanisms of zonal hepatotoxicity.

Each of these approaches (micropatterning and bioreactor systems) recapitulates two-dimensionally certain aspects of hepatocyte function. In many instances, 2-D systems are ideal because they minimize complexities in aspects such as imaging and molecular, histologic, or immunohistochemical characterization, often introduced in 3-D culture models. In contrast, for implantable systems, the development of 3-D constructs is required. Numerous reports have examined the seeding or recruitment of cells to prefabricated 3-D scaffolds [Badylak et al., 2001; Langer and Vacanti, 1993; Lee et al., 2000; Yang et al., 2001]. However, particularly for hepatocytes, which do not grow or migrate well in vitro, the development of 3-D systems would require the formation of thick constructs containing a homogeneous distribution of cells. By combining hydrogel polymerization with photolithography techniques, in which UV cross-linking is performed through a mask, we have demonstrated the generation of patterned three-dimensional hydrogels containing living cells [Liu and Bhatia, 2002]. This process is highly versatile and can be adapted to different hydrogel chemistries as well as various cell types. Consequently, this approach provides the foundation for regulating cellular orientations, including cell-cell and cell-matrix interactions, in three dimensions, important for the tissue engineering of spatially complex organs such as the liver.

In this chapter, we have assembled our expertise in several aspects of liver tissue engineering. Included are fundamental techniques such as the isolation of primary rat hepatocytes and assays of hepatocyte function, as well as novel approaches to assess and recapitulate microenvironmental influences on hepatocyte function. Although these procedures, such as the micropatterning, bioreactor, and 3-D fabrication platforms, were developed and characterized specifically for the design of hepatic systems, the basic principles outlined in this chapter are intrinsic to the fabrication of tissue-engineered constructs for numerous organ systems.

2. PREPARATION OF REAGENTS

2.1. Krebs-Ringer Buffer (KRB)

For 500 ml: add 3.57 g of NaCl, 0.21 g of KCl, 0.50 g of D-glucose, 1.05 g of $NaHCO_3$, and 2.38 g of HEPES to 500 ml of ultrapure water (UPW). Adjust pH to 7.4 with 10 N NaOH, bubble with 5% CO_2-95% N_2 for 10 min, and sterilize by filtration.

2.2. KRB with EDTA

Add 0.19 g of ethylenediaminetetracetic acid disodium salt, dihydrate (EDTA) to 500 ml of KRB (above), to give 1 mM EDTA. Stir on low heat until dissolved and sterilize by filtration.

2.3. Collagenase, 0.055% (w/v)

Make 30–90 min before surgical procedure. Add 105 mg of type IV collagenase powder to 190 ml of KRB, add 9 ml of 0.11 M $CaCl_2$, and stir to ensure that collagenase is completely dissolved (~20 min). Sterilize by filtration. To maintain reliable collagenase activity, thaw for 15 min, measure, and then immediately return to −20 °C storage.

2.4. Coating Buffer

Dissolve 1.61 g of Na_2CO_3 and 2.93 g $NaHCO_3$ in 975 ml of ddH_2O. Adjust pH to 9.6, bring volume to 1 L, and filter.

2.5. OPD Substrate Buffer

Dissolve 5.10 g of citric acid monohydrate and 13.78 g of sodium phosphate ($Na_2HPO_4{\cdot}7H_2O$) in 975 ml of ddH_2O. Adjust pH to 5.0, bring volume to 1 L, and filter.

2.6. Complete Hepatocyte Culture Medium (CHCM)

Dulbecco's modified minimal essential medium (DMEM) with Phenol Red and 21.43 mM sodium bicarbonate, supplemented with 10% fetal bovine serum (FBS), 0.5 U/ml insulin (Lilly Islets II pork regular insulin), 7.5 μg/ml hydrocortisone sodium succinate, 14.28 ng/ml glucagon, 100 U/ml penicillin, and 100 μg/ml streptomycin.

2.7. Blocking Buffer

PBSA with 1% bovine serum albumin (BSA), 50 mM glycine, and 20% normal goat serum (same species as secondary antibody in staining procedure).

2.8. Washing Buffer

PBSA with 0.05% Tween 20 (polyoxyethylene-sorbitan monolaurate) and 1% normal goat serum.

2.9. Piranha Cleaning Solution

3:1 Mixture of H_2SO_4:30% H_2O_2, prepared at time of experiment. Extreme caution should be observed because of the corrosive nature of these chemicals.

2.10. Hepatocyte Culture Medium, Serum Free (HCM/SF)

Same formulation as for CHCM (See Section 2.6), but minus FBS and plus 0.26 mM (30 μg/ml) proline.

2.11. Fibroblast Culture Medium (FCM)

DMEM with 23.81 mM sodium bicarbonate, supplemented with 10% calf serum, 100 U/ml penicillin, and 100 μg/ml streptomycin.

2.12. Hormonally Defined Hepatocyte Medium (HDHM)

DMEM-Ham's F-12, 1:1, supplemented with 5 μg/ml insulin, 100 U/ml penicillin, 100 μg/ml streptomycin, 20 mM HEPES, 10^{-8} M dexamethasone, 5 μg/ml linoleic acid, 10^{-10} M $ZnSO_4$, 10^{-7} M $CuSO_4$, and 3×10^{-10} M H_2SeO_3.

2.13. SDS Lysis Buffer

Tris-HCl, 10 mM, pH 7.4 with 0.1% sodium dodecyl sulfate (SDS).

2.14. HepG2 Culture Medium

Eagle's minimal essential medium (MEM) supplemented with 5% FBS, 100 U/ml penicillin, and 100 μg/ml streptomycin.

3. ISOLATION OF PRIMARY RAT HEPATOCYTES

The following protocols for the isolation and purification of primary rat hepatocytes are an adaptation of procedures previously described [Seglen, 1976]. Rat model systems have been widely exploited for many years to study liver development, injury, and regenerative processes [Bankston and Pino, 1980; Godlewski et al., 1997;

Michalopoulos and DeFrances, 1997; Palmes and Spiegel, 2004], and, similarly, the culture of rat hepatocytes has been extensively characterized [Mitaka, 1998].

Protocol 15.1. Surgical Procedure for Rat Hepatocyte Isolation

Reagents and Materials

Sterile

- ❑ Krebs-Ringer Buffer (KRB) (See Section 2.1)
- ❑ KRB with 1 mM EDTA (See Section 2.2)
- ❑ Collagenase, 0.055% (w/v) (See Section 2.3)
- ❑ Autoclaved surgical instruments: partial curve microdissecting forceps, rat tooth tweezers, scalpel, fine sharp-tip scissors, and Mayo blunt scissors
- ❑ Cotton-tipped applicators
- ❑ Lengths of 3-0 silk suture, 6 in., 2
- ❑ Petri dishes

Nonsterile

- ❑ Rat, Female Lewis, 5–8 weeks old, 1
- ❑ Perfusion system: Water bath to warm perfusate to 39 °C outflow temperature and 95% O_2-5% CO_2 cylinder to equilibrate perfusate through semipermeable tubing. The end point is attached to a sterile 18-gauge Angiocath™ catheter
- ❑ Veterinary-grade isoflurane anesthesia device (Model 100F)
- ❑ Betadine
- ❑ Ethanol, 70%
- ❑ Beaker, 1 L with lid
- ❑ Hair clippers
- ❑ Vacuum
- ❑ Surgical tray (perforated) with collection tub

Protocol

(a) Flush perfusion reservoir and tubing with 70% ethanol, then 500 ml UPW, and then 100 ml KRB-EDTA. Add 400 ml KRB-EDTA, clear bubbles from line, and close the perfusion circuit. Set flow rate to 20 ml/min. Allow 15 min for warming and equilibration of KRB-EDTA.

(b) Anesthetize rat with vaporizer: Adjust flow rate on the anesthesia device to 1.0 L/min 95% O_2-5% CO_2, and set the isoflurane level to 5% v/v. Put exit tubing from the device with nose-cone end in a large beaker and cover, allowing a few minutes for the concentration of isoflurane to build. Transfer rat into the beaker, cover, and monitor until fully sedated.

(c) After initial sedation, remove rat from beaker and position on paper towels with nose-cone covering its mouth. Maintain portable anesthesia, with isoflurane set to 2.5% v/v. Shave abdomen fur and vacuum loose fur.

(d) Transfer rat to surgical tray with collection tub underneath, and tape limbs to restrain. Swab abdominal region generously with ethanol, then Betadine.

(e) Perform midline incision through skin from tip of xiphoid cartilage to groin, and then dissect percutaneous tissue free of underlying musculature laterally near incision. Cut through peritoneum from the groin to xiphoid, and perform lateral cuts perpendicular to the midline through skin and abdominal fascia in order to further open the abdominal cavity.
(f) Use cotton-tipped applicators moistened with KRB-EDTA to push aside the intestines and stomach. Move the liver gently, using a rolling motion with the applicators, until the portal vein is exposed.
(g) Loosely loop a suture around the portal vein near the hilus proximal to vessel branching and another one distally between the mesenteric veins.
(h) Insert the catheter into the portal vein, and position just beyond the proximal suture. Attach the perfusate line to the catheter, and immediately cut the inferior vena cava with a scalpel. It is imperative to establish flow quickly, to prevent clotting within the liver. Tie the distal suture, and then tie the proximal suture.
(i) Cut through the left-side rib cage to access the heart, and sever the left ventricle so that the heart stops. Cut through the sternum along the midline to the clavicles, and cut along the diaphragm and ribs on both sides to create room for the liver to expand. Turn off isoflurane.
(j) Trim connective tissue between liver and surrounding organs. Prop up liver with cotton-tipped applicators so that the liver is not stretched, to permit profuse flow of collagenase. When the reservoir is almost emptied, add in the collagenase. Allow the collagenase to perfuse the liver at 18 ml/min.
(k) Turn off the pump after deep fissures appear in the liver lobes, but before disintegration of the liver capsule and before the collagenase empties (approximately 11 min). Cut the catheter. Pick up the liver with forceps and sever connective tissue to completely free liver from the abdominal cavity. Place liver in Petri dish with KRB and take to laminar flow hood.

Protocol 15.2. Purification of Primary Rat Hepatocytes

Reagents and Materials

Sterile

- ❑ Collagenase perfused rat liver (See Protocol 15.1)
- ❑ Conical centrifuge tubes, 50 ml
- ❑ KRB (See Section 2.1)
- ❑ Large bowl covered with 30 × 30-cm, 62-μm nylon mesh
- ❑ 100 × 50-mm Pyrex dish covered with 15 × 15-cm, 250-μm nylon mesh
- ❑ 10× Hanks' balanced salt solution (HBSS), without calcium or magnesium
- ❑ Percoll
- ❑ Petri dish, 10 cm

Protocol

(a) In laminar flow hood, gently agitate the liver in the Petri dish containing KRB by holding the vascular tree and shaking until most of the cells have been dispersed.

(b) Place dish with 250-μm mesh filter on ice. Prewet filter with approximately 10 ml KRB. Transfer the cellular suspension from the two Petri dishes over the 250-μm filter. Use an additional 24 ml KRB to wash cells through the filter.
(c) Place bowl with 62-μm mesh filter on ice. Prewet filter with 10 ml KRB in the horizontal direction and 10 ml vertically in order to maximize coverage. Transfer cells to 62-μm filter covering Pyrex dish. Wash filter with approximately 36 ml KRB to ensure that the majority of the cells pass through.
(d) Aliquot cell suspension equally to two 50-ml centrifuge tubes and centrifuge at 50 *g* for 3 min at 4 °C. Aspirate supernate, leaving approximately 0.5 ml to avoid aspirating hepatocytes. Bring each volume of cell suspension to 12.5 ml with KRB, and resuspend cell pellet with slow rocking.
(e) Mix 10.8 ml Percoll and 1.2 ml 10× HBSS on ice. Add 12.5 ml Percoll-HBSS mixture to each tube of cellular suspension and mix by gently inverting. Centrifuge at 50 g for 5 min at 4 °C.
(f) Aspirate supernate, including cells suspended within Percoll layer. Resuspend pellet in KRB with gentle rocking and pool cells into one 50-ml conical tube. Bring volume to approximately 40 ml with KRB. Centrifuge at 50 g for 3 min at 4 °C.
(g) Aspirate supernatant, and resuspend to a final volume of 12.5 ml KRB. Count hepatocytes with a hemocytometer. With the procedure described here, an average yield of 1.9×10^8 viable hepatocytes per rat, with 90% overall viability, is obtained.

4. ASSAYS OF HEPATOCYTE FUNCTION

Several parameters are utilized when assessing primary hepatocyte stabilization in vitro. For example, the presence in culture of the classic hepatocyte morphology including a cuboidal shape and well-defined cell borders with intact bile canaliculi is considered suggestive of a mature stable phenotype. Furthermore, there are four categories of liver function: metabolism, synthesis, bile excretion, and detoxification (phase I and phase II). Specific markers are frequently analyzed as surrogate measures of these processes in hepatocytes within various in vitro contexts. These include albumin secretion, urea synthesis, and cytochrome P-450 (phase I) activity [Bhatia et al., 1999]. In addition to being surrogate markers, each of these functions represents an important hepatocyte process, and, accordingly, each is a critical component of a potentially effective bioartificial liver device. Although not measured routinely in our laboratory, assays for bile duct excretion and phase II activity have also been performed in our laboratory as well as others to assess the degree of hepatocyte stabilization. Also, when comprehensive analysis is required, microarray analysis of the expression of a broad range of functionally important genes can be utilized. In addition to the evaluation of isolated hepatocyte populations, the measurement of hepatocyte-specific functions is important in the assessment of hepatocyte differentiation from stem cell sources such as embryonic stem cells or various adult stem cell populations [Dahlke et al., 2004; Kuai et al., 2003; Levenberg et al., 2003; Ruhnke et al., 2003; Schwartz et al., 2002]. Included in this section are protocols for determining the degree of albumin secretion, the identification of intracellular albumin, and the quantification of urea synthesis and cytochrome P-450 activity for rat hepatocytes.

4.1. Detection of Intracellular Albumin and Albumin Secretion

Protocol 15.3. Enzyme-Linked Immunosorbent Assay (ELISA) for Quantification of Albumin Secretion

Reagents and Materials

Nonsterile

- ❑ Purified rat albumin stock, 3.125 mg/ml
- ❑ Coating Buffer (See Section 2.4)
- ❑ Multiwell plates, flat bottom, 96-well
- ❑ Adhesive plate sealer
- ❑ Rabbit anti-rat albumin antibody, horseradish peroxidase (HRP) conjugated
- ❑ *o*-Phenylenediamine (OPD) tablets
- ❑ OPD Substrate Buffer (See Section 2.5)
- ❑ Tween, 0.05% in PBSA
- ❑ CHCM (See Section 2.6)
- ❑ H_2O_2
- ❑ H_2SO_4, 8 N
- ❑ Spectrophotometer

Protocol

(a) Remove supernatant from hepatocyte cultures at the time point determined by the particular experiment of interest and store at 4 °C. Calculate number of wells required for the assay based on 24 wells per plate for albumin concentration standards, and triplicates of experimental samples.

(b) For each plate in the assay, make 10 ml albumin solution (0.05 mg/ml) by adding 160 μl rat albumin stock to 10 ml coating buffer.

(c) Add 100 μl albumin solution to each well of 96-well plate, and cover with adhesive plate sealer. Incubate at 4 °C overnight.

(d) Make series of albumin standards, using serial dilution of albumin stock in same medium as experimental samples (normally CHCM). The normal standard range required is 2-fold dilutions from 100 μg/ml to 1.5625 μg/ml, as well as a 0 μg/ml control.

(e) Shake liquid out of plate, and tap plate on paper towels to thoroughly expel liquid. Wash 4× with Tween-PBSA, shaking liquid out of the plate between each wash.

(f) Load 50 μl of standards in triplicate to the appropriate wells of the 96-well plate.

(g) Load 50 μl of experimental samples also in triplicate to the 96-well plate.

(h) Prepare 1/10,000 dilution of HRP-conjugated anti-rat albumin antibody in Tween-PBSA.
(i) Add 50 μl antibody solution to each well containing either standards or experimental samples. Cover with adhesive plate sealer, and incubate at 4 °C overnight.
(j) Dissolve one OPD tablet per 25 ml OPD substrate buffer. Add 10 μl H_2O_2 per 25 ml OPD substrate buffer.
(k) Wash plate 4× with Tween-PBSA.
(l) Add 100 μl OPD solution with H_2O_2 to each well, and incubate for 5–10 min, for complete color change.
(m) Stop reaction by adding 50 μl of 8N H_2SO_4 to each well.
(n) Read absorbance at 490 nm with spectrophotometer, and, utilizing standard curve, calculate concentration of albumin within experimental samples. For many cases the OPD substrate is sufficient. However, for the detection of low concentrations of albumin (0–10 μg/ml), an alternative substrate, Ultra-TMB (3,3′,5,5′-tetramethylbenzidine), can be used.

Protocol 15.4. Immunofluorescence Assay for Detection of Intracellular Albumin

Reagents and Materials

Nonsterile

- ❑ PBSA, pH 7.4
- ❑ Paraformaldehyde, 4% in PBSA
- ❑ Triton X-100, 0.1% in PBSA
- ❑ Blocking Buffer (See Section 2.7)
- ❑ Washing Buffer (See Section 2.8)
- ❑ Rabbit anti-rat albumin antibody
- ❑ Goat anti-rabbit IgG antibody, fluorescein (FITC) conjugated
- ❑ Glass slide, 75 × 38 mm
- ❑ Vectashield
- ❑ Clear nail polish
- ❑ Fluorescence microscope

Protocol

(a) This protocol is based on hepatocyte cultures on 34-mm coverglasses in 6-well plates. Volumes given are on a per well basis, and can be easily scaled for the use with cultures of different sizes. In general, this procedure can be performed in a nonsterile manner. However, if samples will be stored at any stage before mounting, aseptic techniques should be utilized to prevent potential contamination.
(b) Aspirate culture supernatant and fix cells in 1 ml of 4% paraformaldehyde for 10 min at room temperature.

(c) Remove fixative, add 2 ml PBSA, and incubate for 5 min. Aspirate PBSA and repeat PBSA wash an additional 2× for 5 min each. Either store at 4 °C in PBSA or proceed with staining procedure.
(d) Add 1 ml 0.1% Triton X-100 for 10 min at room temperature to permeabilize cells.
(e) As previously, wash 3× with 2 ml PBSA for 5 min each.
(f) Incubate in 1 ml Blocking Buffer for 15 min at room temperature.
(g) Prepare a 1/133 dilution of rabbit anti-rat albumin antibody. For each ml, add 7.5 μl antibody to 100 μl Blocking Buffer and 892.5 μl PBSA.
(h) Aspirate Blocking Buffer, add 0.5 ml diluted anti-albumin antibody, and incubate for 2 h at 37 °C.
(i) Wash 4× with 2 ml Washing Buffer for 5 min each on orbital rocker.
(j) Prepare a 1/100 dilution of goat anti-rabbit IgG-FITC. For each ml, add 10 μl antibody to 100 μl Blocking Buffer and 890 μl PBSA.
(k) Aspirate final wash with Washing Buffer, add 0.5 ml of diluted anti-rabbit IgG antibody, and incubate in the dark for 1 h at 37 °C.
(l) Wash 3× with 2 ml Washing Buffer for 5 min each on rocker, with a final immersion in PBSA or other standard mounting buffer.
(m) Remove coverglass from buffer and mount. Hold coverglass with tweezers, touch edge to towel, and aspirate at edge to remove excess liquid. Turn coverslip upside down and add one small drop Vectashield. Slowly sandwich together coverslip and clean glass slide, and seal coverslip edges with clear nail polish.
(n) Store in the dark at 4 °C. Prepared specimens should retain fluorescent staining for several weeks when stored properly.
(o) Examine fluorescence, indicating presence of intracellular albumin, using standard fluorescence microscopy techniques with FITC filters.

Protocol 15.5. Immunohistochemistry Assay for Detection of Intracellular Albumin

Immunohistochemistry can be utilized as an alternative for immunofluorescence for the detection of intracellular albumin within hepatocytes. The following method utilizes the DAKO LSAB® horseradish peroxidase (HRP) system, and elements of this protocol are adapted from the manufacturer's procedure.

Reagents and Materials

Nonsterile

❑ PBSA, pH 7.4
❑ Rabbit anti-rat albumin antibody
❑ Bovine serum albumin (BSA)
❑ Paraformaldehyde, 4% in PBSA
❑ Triton X-100, 0.1% in PBSA

- ❑ Parafilm
- ❑ Biotin Blocking System
- ❑ Peroxidase Blocking Reagent
- ❑ Blocking Buffer (See Section 2.7)
- ❑ Washing Buffer (See Section 2.8)
- ❑ DAKO LSAB® System containing: biotinylated anti-rabbit Ig antibody, HRP-conjugated streptavidin, and chromagen substrate.

Protocol

(a) Perform Steps (a) through (e) from Protocol 15.4.
(b) Transfer coverslips to Parafilm to minimize the amount of liquid required to maintain coverage on slips.
(c) Incubate with 10 drops Avidin Blocking Agent, from Biotin Blocking System, for 20 min to suppress endogenous avidin. Wash 2×, each with 2 ml PBSA.
(d) Similarly, incubate with 10 drops Biotin Blocking Agent, from Biotin Blocking System, for 20 min to suppress endogenous biotin. Wash 2×, each with 2 ml PBSA.
(e) Incubate with 12 drops Peroxidase Blocking Reagent for 8 min to quench endogenous peroxidases. Wash 2×, each with 2 ml PBSA.
(f) Add 10 drops of Blocking Buffer and incubate for 5 min at room temperature. Use paper tissue to remove excess Blocking Buffer; do not rinse.
(g) Prepare 1/100 dilution of rabbit anti-rat albumin antibody in PBSA with 1% BSA. Add 600 μl diluted antibody solution to each coverslip. Incubate at 37 °C for 90 min, gently rocking every 30 min to ensure even coverage on substrate.
(h) Wash 3× with 2 ml Washing Buffer, for 5–10 min each on rocker.
(i) Incubate with 10 drops Biotinylated anti-rabbit Ig antibody solution for 20 min at room temperature. Wash 2× with 2 ml Washing Buffer, for 5–10 min each on rocker.
(j) Incubate with 10 drops HRP-conjugated streptavidin solution (including, according to manufacturer's instructions, 4 ml UPW, 4 drops buffer concentrate, and 1 drop concentrated HRP-streptavidin) for 20 min at room temperature. Wash 2×, each with 2 ml PBSA.
(k) Prepare substrate solution according to manufacturer's instructions: 2 ml substrate buffer, 1 drop H_2O_2, and 1 drop 3-amino-9-ethylcarbazole chromagen substrate in dimethyl formamide. Incubate for 10 min at 37 °C. Watch closely after 10 min for color change; do not overdevelop. Wash 2×, each with 2 ml PBSA.
(l) Remove coverglass from PBSA and mount. Hold coverglass with tweezers, touch edge to towel, and aspirate at edge to remove excess liquid. Turn coverslip upside-down and add one small drop Vectashield. Slowly sandwich together coverslip and clean glass slide, and seal coverslip edges with clear nail polish.
(m) Record images by photomicroscopy.

4.2. Quantitative Measure of Urea Synthesis

Protocol 15.6. Assay of Urea Synthesis in Cultured Hepatocytes

The following procedure has been adapted from Stanbio Urea Nitrogen Procedure No. 0580.

Reagents and Materials

Nonsterile

- ❑ Urea/nitrogen stock samples; 75, 50, and 25 mg/ml
- ❑ Blood Urea Nitrogen (BUN) Color Reagent
- ❑ BUN Acid Reagent
- ❑ 96-well plates, flat bottom
- ❑ Adhesive plate sealer
- ❑ Oven, 60 °C (or water bath alternatively)
- ❑ Spectrophotometer

Protocol

(a) Remove supernatant from hepatocyte cultures as indicated by the particular experiment, and store at 4 °C. Calculate number of wells required for assay based on 24 wells per plate for concentration standards, and triplicates of experimental samples.

(b) Prepare urea standards by serially diluting stock samples in same medium as experimental samples (normally CHCM). The normal range used is the same as for the albumin ELISA above (100, 50, 25, 12.5, 6.25, 3.125, 1.5625, 0 μg/ml). Alternatively, if hepatocytes are suspected to be highly functional and/or at high density prepare the following range of standards: 200, 100, 50, 25, 12.5, 6.25, 3.125, 0 μg/ml.

(c) Transfer 10 μl standards and experimental supernatants to individual wells of 96-well plate.

(d) Mix BUN Color and Acid reagents at the following ratio: 1/3 BUN Color Reagent plus 2/3 BUN Acid Reagent. Transfer 150 μl of this mixture to each well containing either standards or experimental samples. Seal plate tightly with adhesive plate sealer and incubate at 60 °C for 90 min.

(e) After color has developed, put plates on ice for 5–15 min, but no longer than 20 min.

(f) Read absorbance at 540 nm with spectrophotometer and, utilizing standard curve, calculate concentration of urea within experimental samples.

4.3. Measurement of Cytochrome P-450 Enzyme Activity

Protocol 15.7 has been adapted in our laboratory from several previous publications [Behnia et al., 2000; Burke and Mayer, 1974; Burke et al., 1985; Kelly and Sussman, 2000]. The formation of resorufin from ethoxyresorufin (EROD) reflects the cytochrome P-450 (CYP) activity of rat hepatocytes, in particular, predominantly CYP1A1 activity. The procedure for examining EROD conversion is included in this chapter as a model assay investigating hepatocyte CYP activity. Although not explicitly shown below, a similar procedure can be utilized to assess the activity of other P-450 enzymes in rat hepatocytes, for example, conversion of pentoxyresorufin (PROD) as a measure of predominantly CYP2B1 activity, benzyloxyresorufin (BROD) for CYP2B2, or methoxyresorufin (MROD) for CYP1A2.

Protocol 15.7. Assay of Cytochrome P-450 1A1 Activity in Cultured Hepatocytes

Reagents and Materials

Sterile

- ❑ Ethoxyresorufin, 1 mM stock in dimethyl sulfoxide (DMSO)
- ❑ Dicumarol, 2 mM stock in 75 mM NaOH
- ❑ 3-Methylcholanthrene (3MC), 10 mM stock in DMSO
- ❑ CHCM (See Section 2.6)
- ❑ CHCM (See Section 2.6), without phenol red

Nonsterile

- ❑ 96-Well plate
- ❑ 24-Well plate
- ❑ 0.1 M NaOH
- ❑ β-Glucuronidase, 1600 U/ml in 0.1 M sodium acetate buffer
- ❑ Resorufin, 1 mM stock in DMSO
- ❑ Spectrofluorometer

Protocol

(a) This protocol is based on hepatocyte cultures in 6-well plates. Volumes given are on a per well basis, and can be easily scaled for the use with cultures of various dimensions.

(b) At the time point indicated by the particular experiment, remove culture medium from hepatocyte cultures and add CHCM containing 2 μM 3MC. Incubate cultures in 3MC-containing medium for 48–72 h, adding fresh medium with 3MC every 24 h, to induce sufficiently the expression of CYP1A. Alternatively, a 5 μM concentration of β-napthoflavone can be used in place of 3MC to induce CYP1A expression by rat hepatocytes. For experiments requiring the induction of CYP2B expression, 1 mM phenobarbital can be used.

(c) Aspirate medium and carefully wash wells with CHCM without phenol red. Add 750 μl solution of 5 μM EROD and 10 μM dicumarol in CHCM without Phenol Red. Incubate for 20 min at 37 °C. Dicumarol is a competitive inhibitor of cytosolic oxidoreductases and prevents diaphorase-mediated metabolism of resorufin.

(d) Transfer 750 μl aliquot to 24-well plate. Add 150 μl 1600 U/ml β-glucuronidase solution and incubate at 37 °C for a minimum of 2 h (maximum overnight).

(e) Terminate β-glucuronidase reaction by adding 200 μl 0.1 M NaOH. Mix thoroughly, and transfer 200 μl from each well to the appropriate wells of a 96-well plate in triplicate.
(f) Prepare resorufin standards by diluting 1 mM resorufin stock to the following concentration range in CHCM without phenol red: 1000, 500, 250, 125, 62.5, 31.25, 15.625, and 0 nM. Add standards in triplicate to the appropriate wells of the 96-well plate containing experimental samples.
(g) Use spectrofluorometer to quantify fluorescence at 590 nm emission, with 530 nm excitation. Generate standard curve, and utilize curve to calculate resorufin concentration in the culture supernatants. The presence of resorufin in the hepatocyte culture medium, after incubation with EROD, is a reproducible indicator of CYP activity, predominantly CYP1A1, in rat hepatocytes.

5. MICROPATTERNED CELL CULTURES

Cell-cell and cell-extracellular matrix interactions are critical for proper cellular processes in numerous organ systems. Specifically, such interactions have been demonstrated to be key determinants of hepatocyte function. For example, it has been well documented that isolated hepatocytes cultured in monolayer culture display a rapid loss of phenotype [Bissell et al., 1973; Dunn et al., 1989; Leffert and Paul, 1972; Reid et al., 1980; Selden et al., 1999], preventing the effectiveness of these cells in long-term applications. Cell-cell interactions, both homotypic (hepatocyte-hepatocyte) and heterotypic (hepatocyte-nonparenchymal), have been demonstrated to exhibit positive effects on hepatocyte function. In particular, hepatocyte viability and a variety of liver functions have been shown to be stabilized for weeks in vitro on cocultivation with liver-derived cell types as well as some non-liver-derived populations [Bhatia et al., 1999]. However, despite the significant amount of data existing on potential mediators of cell communication in cocultures, the mechanisms by which coculture of hepatocytes with other cells induces and stabilizes liver-specific function and viability remain primarily undefined. Current experiments in our laboratory are examining these processes.

To study the role of cell-cell interactions in hepatocellular function, we previously developed and characterized a flexible method of micropatterning cocultures of hepatocytes and an embryonic fibroblast cell line (3T3-J2) on solid substrates (U.S. Patent No. 6,133,030) [Bhatia et al., 1997]. In this approach, photolithographic techniques were used to pattern collagen I on glass substrates, which served as an adhesive substrate for primary rat hepatocytes. After hepatocyte attachment and spreading, 3T3-J2 fibroblasts were plated, by serum-mediated adhesion, in the remaining intermixed regions. The versatility of this technique allowed the design of culture configurations that included variations in homotypic hepatocyte interactions, fibroblast cell number, as well as heterotypic interactions. Cocultures were conducted in which the heterotypic interface could be varied over three orders of magnitude, with the ratio of cell populations remaining constant (See Fig. 15.3) [Bhatia et al., 1999]. Total cell numbers were held constant by holding

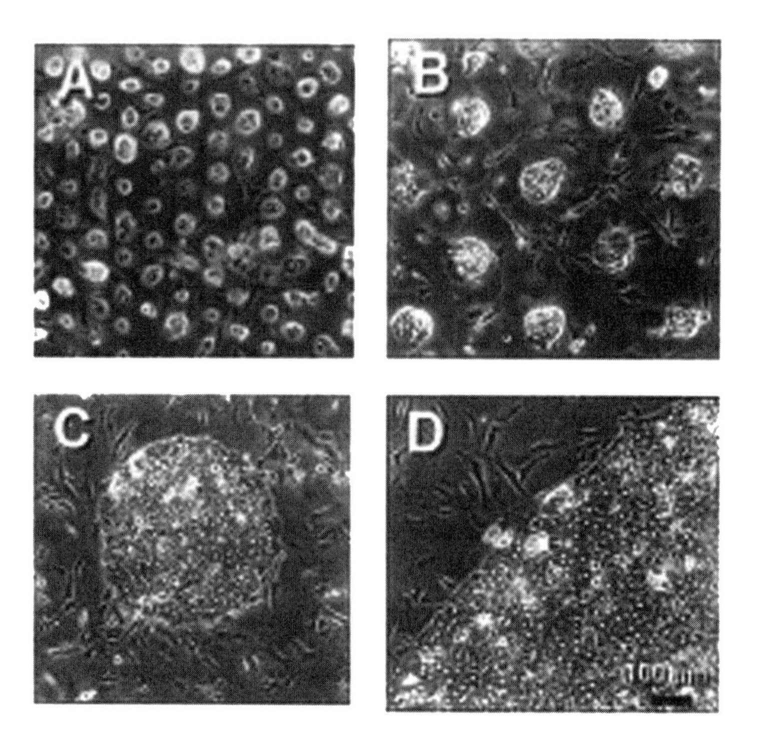

Figure 15.3. Micropatterned cocultures with constant ratio of cell populations. Phase-contrast micrographs of micropatterned cocultures indicate a broad range of heterotypic interface achieved despite similar cellular constituents. Four of five patterns used in study are shown. Diameters of hepatocyte islands were 36 μm (A), 100 μm (B), 490 μm (C), and 6800 μm (D). From Bhatia et al. [1999].

surface area of each domain (collagen I/bare glass) equal for each condition. In these experiments, liver functions, including albumin secretion and urea synthesis, were increased in cocultured configurations compared to hepatocytes alone (Fig. 15.4) (See Section 4 for the procedures for assaying albumin secretion and urea synthesis) [Bhatia et al., 1999]. The degree of upregulation varied with culture configuration, whereby cocultures with a larger initial heterotypic interface (i.e., single-cell islands) exhibited increased levels of liver-specific function.

In addition to bulk measures of hepatocyte function, in situ markers are equally important, particularly for the analysis of the role of the microenvironment on individual hepatocyte function. Immunohistochemical staining (See Protocol 15.5) indicated that hepatocytes near the heterotypic interface had a relative increase in liver-specific function (Fig. 15.5, See Color Plate 10B) [Bhatia et al., 1998b, 1999], which correlated with our data on bulk tissue function (increased heterotypic interactions equaled increased function). Notably, utilization of micropatterning techniques enabled a 12-fold reduction in fibroblast numbers with only a 50% reduction in albumin secretion, due to the ability to control the extent of heterotypic interactions [Bhatia et al., 1998a]. Reduction in the required numbers of nonparenchymal cells, which would occupy precious substrate surface area within a potential bioreactor, would essentially increase hepatocyte functionality per unit

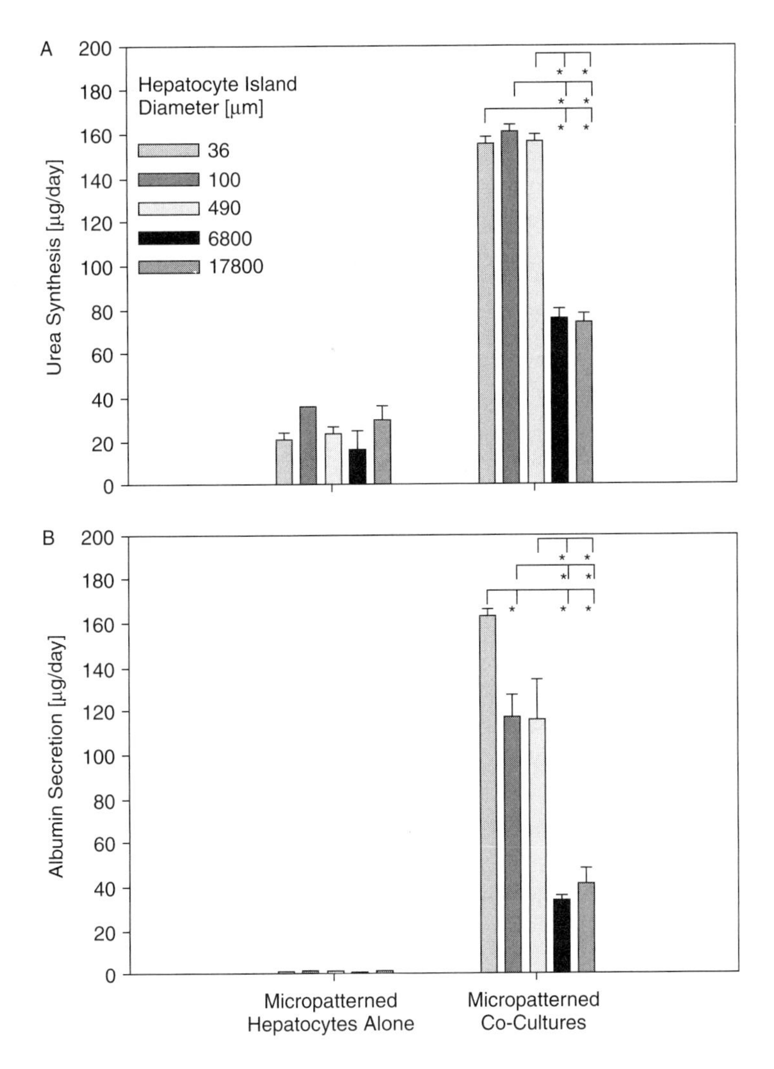

Figure 15.4. Liver-specific function of micropatterned cocultures with constant ratio of cell populations. Urea synthesis (A) and albumin secretion (B) on day 11 of culture were detected in micropatterned cocultures with varying heterotypic interactions despite similar cell numbers. Micropatterned hepatocyte-only cultures were utilized as a control. Statistical significance (*) was determined by one-way ANOVA with Tukey HSD post hoc analysis with $P < 0.05$. From Bhatia et al. [1999].

area. Thus micropatterning approaches would allow for significant improvements in the design of a projected coculture-based bioreactor.

The protocols outlined in this section describe this photopatterning procedure, which represents a multifaceted approach to investigation of complex mechanisms of cell-cell interaction, and specifically for our purposes, a means to identify the critical factors that mediate stabilization of the hepatocyte phenotype. Elucidation of these elements would have important implications in fundamental studies of cell communication as well as the development of highly functional tissue-engineered therapies for the liver.

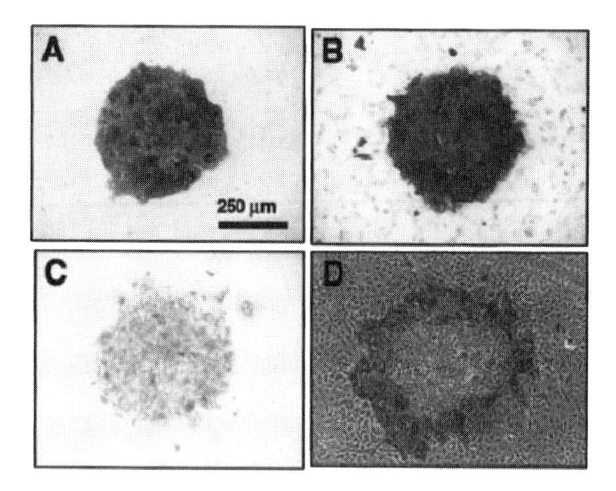

Figure 15.5. Intracellular albumin in micropatterned hepatocytes. Immunohistochemical staining of intracellular albumin in micropatterned hepatocytes in a representative (490 μm) pattern. Bright-field microscopy of hepatocytes (alone) on days 1 and 6 (A, C) and cocultures on days 1 and 6 (B, D). In coculture, albumin expression was highest near the heterotypic interface. From Bhatia et al. [1999]. (See Color Plate 10B.)

5.1. Microfabrication of Substrates

This section outlines a process for the localized modification of experimental substrates utilizing standard microfabrication techniques. In general, most integrated circuit manufacturing facilities can be used to perform these procedures.

Protocol 15.8. Photolithographic Patterning of Surface Modifications

Reagents and Materials

Nonsterile

- ❑ Circular borosilicate glass wafers, 5 cm (2 in.)
- ❑ Mask with desired pattern: Chrome masks are fabricated by a high-precision photolithographic process from Corel Draw. Alternatively, emulsion masks are commercially printed with a Linotronic-Hercules 3300 dpi high-resolution line printer.
- ❑ Piranha Cleaning Solution (See Section 2.9). Extreme caution must be observed during handling because of the corrosive nature of these chemicals. Pyrex containers must be utilized, and chemical removal should be performed via aspiration and dilution of acid mixture.
- ❑ Photoresist: Photoresist is a light-sensitive polymeric material. Any photoresist can be utilized that will adhere sufficiently to clean borosilicate, remain intact during subsequent processing, and yet be removed with relative ease after surface modification. We use one of the following positive photoresists: OCG 825–835 St, Shipley 1813, or Shipley 1818. UV exposure causes a positive photoresist to become more sensitive to developer, thereby resulting in a photoresist pattern after development that is identical to the mask pattern.

❑ Developer: Use developer appropriate for selected photoresist. We use OCG 934, MF-319, or Shipley 354.

Protocol

(a) Place borosilicate wafers in wafer carrier. Place carrier in Pyrex vat, pour Piranha Solution over wafers, and wait 10 min. Rinse wafers three times in a "dump-to-resistivity tank," a washing station that rinses wafers to an acceptable resistivity level (>10 MΩ-cm). Next, transfer wafers to a "spin-dryer." which uses air and spin cycles to dry the wafers. If a spin-dryer is not available, manual drying is done with a N_2 gas stream.

(b) Dehydrate wafers to promote adhesion of photoresist by baking for 60 min at 200 °C.

(c) Mount wafers on vacuum of a spin-coater chuck and coat with positive photoresist to a uniform layer of approximately 1 μm, by spinning at 5000 rpm for 30 s.

(d) Soft-bake for 30 min at 90 °C to drive out excess solvent and anneal any stress in the film.

(e) Expose coated substrates to 365-nm UV light in a Bottom Side Mask Aligner (Karl Suss) through patterned mask under vacuum-enhanced contact for 30–70 s at a dose of 10 mW/cm^2. Intensity and exposure time are dependent on type of photoresist and mask design.

(f) Immerse exposed photoresist in appropriate developer. Complete removal of photoresist in exposed areas is critical to subsequent surface modification. Presence of residual photoresist can be assessed by inspection under light interference or fluorescent microscopy. We develop by immersion and agitation in a bath of developer for 70 s. Surfaces should then be rinsed three times under running deionized water and cascade rinsed (if possible) for 2 min.

(g) Postbake patterned wafers for 30 min at 120 °C to drive off residual solvent and promote film adhesion.

(h) Wafers can be stored in closed containers (preferably within a nitrogen box) at room temperature for at least 1 month.

(i) Expose substrates 24 h to oxygen plasma before patterning of adhesive proteins and cell culture (See Protocol 15.8), to dry-etch (remove) a small layer of photoresist. This ensures complete removal of photoresist from exposed borosilicate; however, if substrates are well developed and pattern dimensions are larger than 10 μm, this step may be omitted. We use a parallel-plate Plasma Day Etcher at a base vacuum of 50 mtorr in an O_2 atmosphere and pressure of 100 mtorr at a power of 100 W for 2–4 min, which corresponds to an etch rate of approximately 0.1 μm/min.

The method described here was intended to be robust. In many instances, users may be able to eliminate certain elements of this process. Exposure to plasma oxygen may be unnecessary if patterns are well developed and not contaminated during

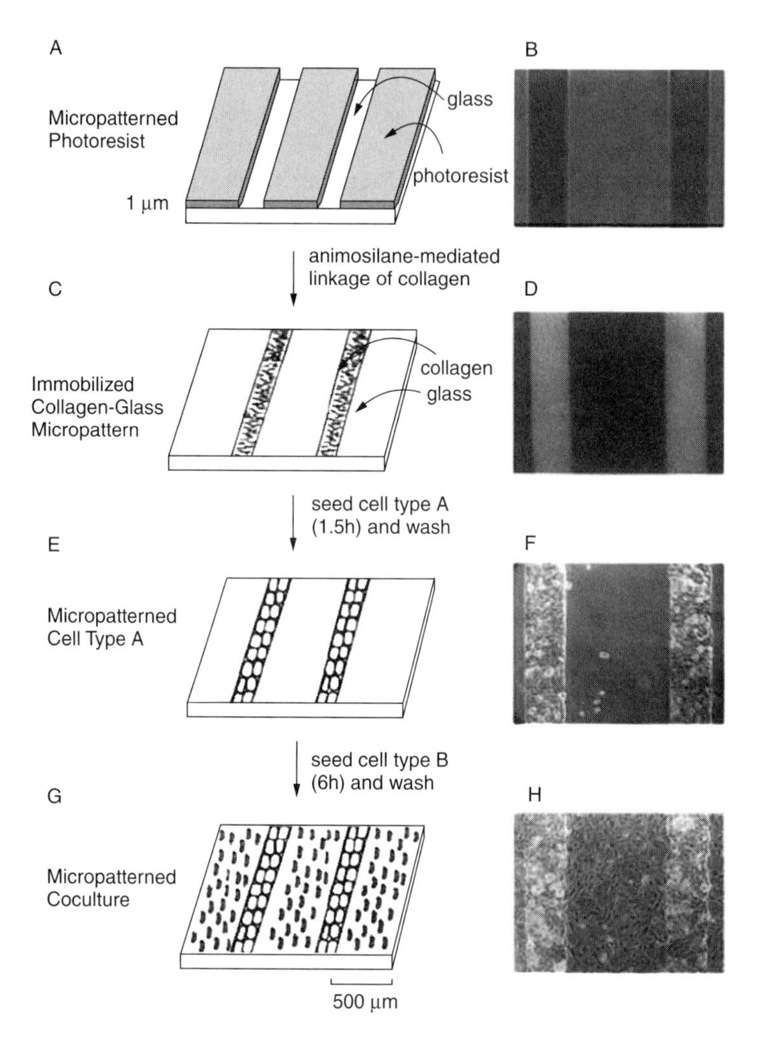

Figure 15.6. Schematic of method for generating micropatterned cocultures. Borosilicate substrates were coated with photoresist (a UV-sensitive polymer) and exposed to light through a mask, creating a photoresist pattern (A). Photoresist was visualized with epifluorescence microscopy (B) (excitation: 550 nm, emission: 575 nm). Collagen I was immobilized, followed by removal of photoresist, yielding a collagen-glass pattern (C). Indirect immunofluorescence allowed verification of collagen immobilization in appropriate locations (D). Patterned substrates were exposed to hepatocytes in serum-free media and rinsed, resulting in micropatterned hepatocytes (E). (F) Phase-contrast micrograph of 200 mm lanes of hepatocytes with 500 mm lane spacing. Addition of 3T3-J2 fibroblasts in medium supplemented with serum resulted in generation of micropatterned cocultures (G). Phase-contrast microscopy allowed morphological identification of 2 distinct cell types in "micropatterned coculture" (H). From Bhatia et al. [1999].

storage. Similarly, baking of wafers after AS modification may be unnecessary in some protocols.

Potential problems include peeling of photoresist during processing and difficulty with lift-off of photoresist. The integrity of the photoresist coating varies with solvents. In some instances, photoresist will peel away from the wafer surface prematurely. This indicates either insufficient adhesion of the photoresist to glass (often because of an insufficient dehydration bake before photoresist coating) or insufficient baking after development to harden the photoresist. Difficulty with photoresist removal resulting in extended sonication in acetone often indicates exposure of photoresist to elevated temperatures.

5.2. Cellular Micropatterning on Modified Surfaces

The localized modifications of surface chemistry described above can be utilized to pattern cells by patterning adhesive molecules. There are a number of techniques for immobilizing adhesive proteins on solid substrates [Drumheller and Hubbell, 1995]. In Protocol 15.9, we describe techniques for covalent coupling of proteins to the surface by a modified technique of Lom et al. [1993], and Britland et al. [1992], as well as direct adsorption of proteins from solution. Subsequently, these modified substrates can be utilized for the micropatterning of primary hepatocytes and an embryonic fibroblast cell line. In this approach, hepatocytes attach specifically to the collagen I patterned regions, and fibroblasts undergo nonspecific, serum-mediated attachment to the remaining unmodified areas (Fig. 15.6).

Protocol 15.9. Micropatterning of Hepatocyte/Fibroblast Cocultures

Reagents and Materials

Sterile

- ❑ Freshly isolated primary rat hepatocytes (See Section 3)
- ❑ 3T3-J2 mouse fibroblast cell line
- ❑ CHCM (See Section 2.6)
- ❑ HCM/SF (See Section 2.10)
- ❑ FCM (See Section 2.11)
- ❑ Bovine serum albumin (BSA), 0.05% (w/w) in water. Sterilize by filtration through a 0.45-μm filter and store at 4 °C.
- ❑ Deionized water (DI). Autoclave to sterilize.
- ❑ Petri dishes, 6 cm
- ❑ Mitomycin C, 10 μg/ml in fibroblast medium
- ❑ Trypsin, 0.25%, EDTA, 5 mM (0.2g/L) in CMF-HBSS

Nonsterile

- ❑ Microfabricated glass wafers (See Protocol 15.8)
- ❑ 3-[(2-Aminoethyl)amino]propyltrimethoxysilane (AS), 2% in water
- ❑ Glutaraldehyde, 2.5% (v/v) in PBSA, pH 7.4
- ❑ PBSA
- ❑ Collagen type I (rat tail), final concentration approximately 500 μg/ml
- ❑ Acetone
- ❑ 70% Ethanol
- ❑ Glass Petri dish, 10 cm
- ❑ Wafer tweezers
- ❑ Petri dish, 6 cm

Protocol

(a) Rinse microfabricated wafers (See Protocol 15.8) by immersion in distilled DI in a glass 10-cm Petri dish. Repeat. Handle substrates with wafer tweezers, and pay special attention to the orientation of the patterned surface of the wafer-transparent substrates, as they can be easily inverted without any obvious differences in appearance.

(b) To pattern a protein such as collagen by adsorption, immerse sample in 4 ml collagen solution in a 6-cm Petri dish for 1 h at 37 °C and skip to step (h).

(c) To covalently link a protein to the patterned regions, first immobilize AS by immersion of samples in AS solution for 30 s at room temperature, followed by two rinses in DI water.

(d) Dry wafers with a stream of N_2 gas to avoid drying artifacts.

(e) Bake wafers in a closed container for 10 min at 120 °C. ***Note:*** Temperatures greater than 150 °C will cause hardening of many photoresists, and removal will be difficult.

(f) Soak disks in a covered container of 2.5% glutaraldehyde in PBSA for 1 h at 25 °C, followed by two rinses in fresh PBSA. Visually inspect wafers every 15 min to evaluate the integrity of the photoresist—if the glutaraldehyde solution causes peeling of the photoresist, the process will need to be abbreviated.

(g) To immobilize collagen, immerse wafers in 4 ml collagen solution in a 6-cm Petri dish for 30 min at 37 °C.

(h) To remove photoresist and expose underlying unmodified glass, float each wafer in acetone in a glass container and sonicate the container in a bath sonicator for 1–15 min. The duration of sonication is empirically determined by observation of the first wafer in each batch—examine wafers for complete removal of photoresist (previously "pink" wafers will appear clear). Treat all wafers in an experimental batch identically to ensure comparability of immobilized protein layers on all substrates. We use 10 ml acetone in a 10-cm glass Petri dish.

(i) Rinse wafers twice by immersion in DI. As previously shown, modified areas should display differential wetting on removal of substrate from water, thereby indicating successful patterned surface modification.

(j) Wafers can be stored dry in a covered container at 4 °C for at least 2 weeks. We store wafers on a piece of filter paper (to absorb residual water and prevent sticking) in 6-cm Petri dishes. Storage of wafers in solution (i.e., PBSA or ethanol) results in transfer of patterned protein to unmodified areas, presumably via desorption from modified areas and adsorption to unmodified areas; therefore, if the immobilized protein will tolerate dehydrated storage conditions, wafers should be stored dry. If immobilized protein requires hydration to maintain its bioactivity, storage time must be empirically determined—in our case, less than 48 h.

(k) Soak premodified wafer in 70% ethanol for sterilization. Use wafer tweezers to place wafer in 6-cm dish with 5 ml ethanol solution for at least 1 h, but not more than 24 h, at room temperature in a sterile laminar flow hood.

(l) Pour autoclaved water into sterilized beaker. Flame-sterilize wafer tweezers, allow to cool, and remove wafer from 70% ethanol under sterile conditions. Immerse wafer in water and agitate gently for approximately 10 s, being sure to preserve orientation of the wafer.

(m) If using wafers modified with adhesive species, coat wafers with bovine serum albumin (BSA) to deter nonspecific cell adhesion on unmodified regions. For rat hepatocytes, BSA coating reduces nonspecific cell attachment to glass from 30% to negligible levels. Place sterilized wafers in sterile 6-cm dishes, add 4 ml BSA solution (0.1%–1.0% wt/v, in PBSA) to each dish, and place in incubator for 45 min at 37 °C.

(n) To remove residual BSA solution, use sterile tweezers to remove wafers from dishes under sterile conditions, and immerse in autoclaved water in sterile beaker, gently agitating for 10 s.

(o) Dilute stock suspension of freshly isolated rat hepatocytes in HCM/SF to a final concentration of $1-2 \times 10^6$ hepatocytes/ml.

(p) Seed hepatocytes by placing 2 ml hepatocyte solution on each wafer. Agitate solution to disperse cell suspension and place in incubator for 1–1.5 h. Wafers should be periodically agitated (i.e., every 15 min) horizontally to promote maximal cell attachment.

(q) At this point, selective cell adhesion on collagen regions should render pattern features visible. Typically, to ensure 100% confluence on adhesive areas, hepatocyte seeding is repeated two to three times. Surfaces should be rinsed twice by pipetting and then aspirating 4 ml media, reseeded with hepatocytes for 1.5 h, and rinsed again. Repeat as necessary.

(r) After seeding, aspirate HCM/SF and replace with hepatocyte culture medium containing serum (10%). Hepatocytes are then allowed to spread over the remaining modified sites. Rat hepatocytes take more than 10 h to spread, so incubate patterned hepatocytes overnight. These micropatterned cell cultures can be used in experimental studies; otherwise proceed with addition of fibroblasts.

(s) To generate hepatocyte-fibroblast coculture, fibroblasts are trypsinized, resuspended in fibroblast media, and plated in 3 ml fibroblast media per micropatterned hepatocyte culture. Typically 750,000 fibroblasts per dish are sufficient; however, in some cases it is necessary to plate growth-arrested fibroblasts in greater numbers.

(t) Growth-arrested fibroblasts are generated by incubating each 150-cm^2 fibroblast flask with 15 ml mitomycin C solution for 2 h.

(u) Incubate cocultures in fibroblast media for 24 h.

(v) Change medium to Complete Hepatocyte Media (with serum) and continue culture in this medium for duration of the experimental investigation. Preservation of pattern integrity is dependent on both cell types (competence, migration rates, etc.) and pattern dimensions—for example; our patterns are stable for several weeks for collagen-modified areas larger than a few hundred microns.

The above process is adaptable for use with many different cell types. However, the procedure will vary with the characteristics of the desired protein. In our case, collagen type I retained its bioactivity for hepatocyte attachment and spreading despite treatment with acetone, ethanol, and dehydration. Other proteins may require modifications of this protocol to retain their bioactivity. Additionally, in our experience, covalent binding of collagen is not required for hepatocyte patterning and simple adsorption to the glass substrate is adequate. Yet it should be noted that reliance on adsorption alone is likely to result in different immobilized protein conformations than covalently bound proteins, and these potential differences should be kept in mind if modifying this procedure for other adhesive molecules.

Furthermore, selection of medium for two cell types must be considered. Medium must be selected that provides adequate nutrients and buffering for each population. Also, because of the rapid mitotic rate of fibroblasts, our medium contains hydrocortisone as a growth inhibitor. Alternatively, as suggested above, under some circumstances fibroblasts can be chemically growth arrested with mitomycin C.

Potential problems include minimal cell adhesion to the substrate and, conversely, uniform cell adhesion to the substrate. Lack of cell adhesion to the substrate can indicate many problems. The most common, however, is underdevelopment of the exposed photoresist (See Protocol 15.8, Step (e)), resulting in a lack of exposed borosilicate for protein immobilization. This can be alleviated by increasing development time, or increasing exposure to oxygen plasma before surface modification. This effect is often exacerbated with small (<5 μm) pattern dimensions. Alternatively, contaminants that coat the glass and prevent protein immobilization, or commonly utilized undercoatings for promoting photoresist adhesion, could produce this effect. On the other hand, a lack of discernible pattern because of exuberant cell adhesion is frequently caused by a defect in the original photoresist coating. Photoresist can degrade and crack over time, allowing exposure of all areas of the substrate to surface modification. Also, excessive exposure to oxygen plasma will strip the photoresist from the surface of the wafer completely, allowing for homogeneous protein adsorption.

In summary, the above protocols describe a method for micropatterning hepatocyte-fibroblast cocultures based on standard microfabrication techniques. Utilization of this system enables the controlled investigation of the complex cell-cell interactions important in the stabilization of the hepatocyte phenotype.

6. BIOREACTOR SYSTEM FOR THE REGULATION OF HEPATOCYTE ZONAL HETEROGENEITY

Numerous hepatocyte functions are known to vary along the length of the liver sinusoids from the portal triad (i.e., portal vein and hepatic artery) to the central vein, a feature termed "liver zonation" (Fig. 15.7) [Kietzmann and Jungermann, 1997]. For example, urea synthesis and gluconeogenesis are localized near the portal triad (periportal region), whereas conversely, the processes of cytochrome P450 activation and detoxification, as well as glycolysis, are enriched near the

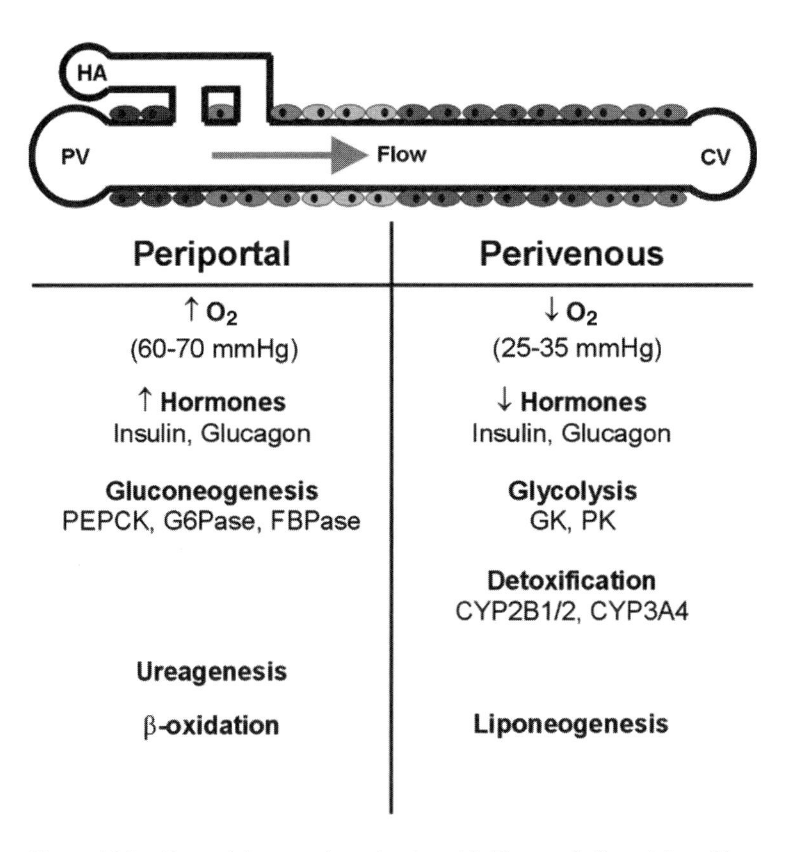

Figure 15.7. Zonated features along the sinusoid. The metabolic activity of hepatocytes along the liver sinusoid creates oxygen and hormone gradients from the periportal region to the perivenous. As a result, metabolic and detoxification functions are regionally dominant to one zone or the other, as indicated. Some specific zonated markers are listed under each process. Abbreviations: HA, hepatic artery; PV, portal vein; CV, central vein; PEPCK, phosphoenolpyruvate carboxykinase; G6Pase, glucose-6-phosphatase; FBPase, fructose-1,6-bisphosphatase; GK, glucokinase; PK, pyruvate kinase; CYP, cytochrome P-450. From Allen and Bhatia [2003], adapted from Kietzmann and Jungermann [1997].

central vein (perivenous region). Notably, the zonation of enzymes involved in carbohydrate metabolism enables a mechanism for the liver to maintain blood glucose levels relatively constant during either fasting or feeding states [Jungermann, 1992]. In addition, certain agents, such as carbon tetrachloride and acetaminophen, exhibit zonal toxicity as a result of the localized P450 induction [Lindros, 1997]. In vivo, hepatocytes experience simultaneous gradients of O_2, hormones, and extracellular matrix, which are each thought to participate to some degree in the zonation process [Jungermann and Kietzmann, 2000; Kietzmann and Jungermann, 1997]. The development of in vitro models of zonation would enable the study of these important components, thereby providing a means to predict in vivo responses such as metabolic disturbances as well as zonal hepatotoxicity.

In particular, our laboratory and others have demonstrated the critical role of O_2 concentration in lipid metabolism, urea synthesis, gluconeogenesis, and xenobiotic metabolism of isolated hepatocytes [Bhatia et al., 1996; Holzer and Maier, 1987;

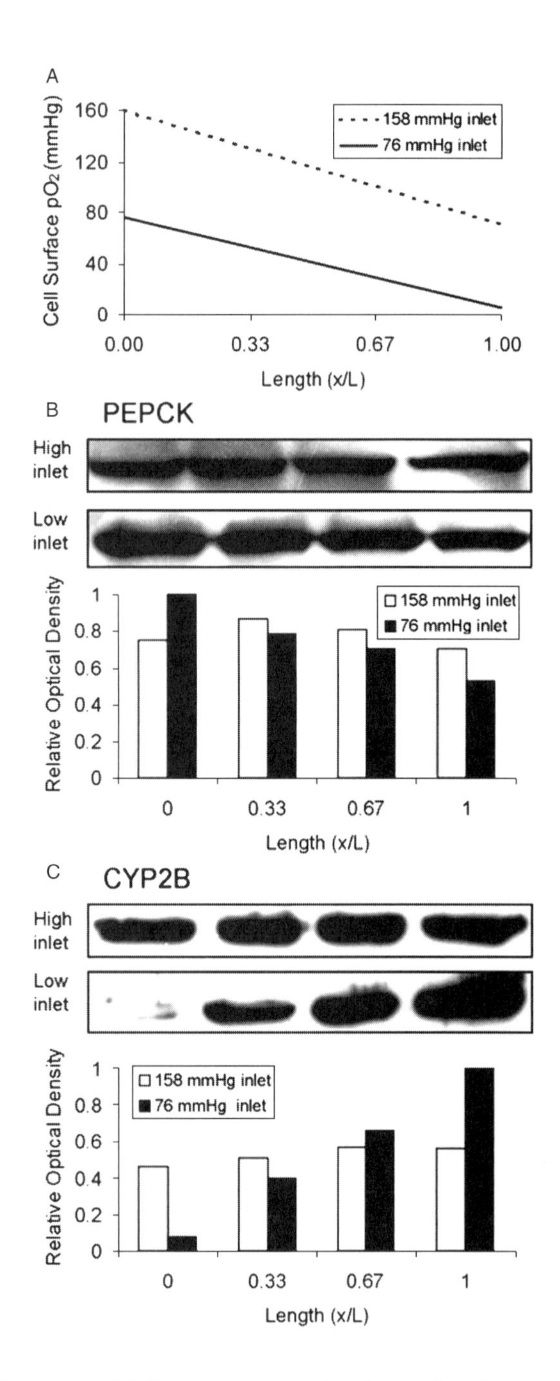

Figure 15.8. Heterogeneous induction of PEPCK and CYP2B by oxygen gradients. Bioreactors were operated with an inlet pO_2 of 76 and 158 mmHg and flow rate of 0.5 ml/min. The resulting cell surface oxygen gradients are shown schematically as calculated from the numerical model (A). Western blots of PEPCK (B) and CYP2B (C) protein levels from four regions along the chamber were analyzed to determine relative optical density. In both cases, when the bioreactor was operated with physiologic gradient (low inlet), a heterogeneous induction was observed, whereas imposing a supraphysiologic gradient (control, high inlet) resulted in a more uniform protein distribution. Blots were processed in separate experiments, enabling only qualitative comparison between conditions. Normalization of band densities to the maximal density from both experiments is meant to facilitate comparison. From Allen and Bhatia [2003].

Nauck et al., 1981; Suleiman and Stevens, 1987]. To examine the dynamic effects of O_2 gradients on hepatocyte function, we developed an in vitro perfusion system that allows exposure of cultured cells to a continuous range of O_2 tensions [Allen and Bhatia, 2003]. These controlled O_2 gradients contributed to the heterogeneous distribution of phosphoenolpyruvate carboxykinase (PEPCK), elevated upstream, and cytochrome P450 2B (CYP2B), elevated downstream (Fig. 15.8), which correlates with the distribution of these enzymes in vivo. Thus the integration of specified O_2 gradients within this bioreactor system facilitated the establishment of hepatocyte zonal variations, thereby highlighting the importance of O_2 gradients in this process. Furthermore, the incorporation of O_2 gradients represents a means for enhancing the functionality of in vitro liver models as well as bioartificial liver devices. Although the oxygen-sensing mechanism is yet to be determined, several candidates are suggested by previous studies, including various transition metals and H_2O_2 in collaboration with the transcription factor HIF-1α [Huang et al., 1996; Jungermann and Kietzmann, 2000; Kietzmann et al., 1996; Maxwell and Ratcliffe, 2002]. Ongoing experiments in our laboratory are examining the molecular mechanisms responsible for the induction of zonated gene expression and, specifically, the role of HIF-1α in the regulation of gene expression in hepatocytes in response to hypoxic conditions.

6.1. Formation of Steady-State Oxygen Gradients

The perfusion bioreactor system described below exhibits parallel-plate geometry with a uniform flow field. A schematic of this system is displayed in Fig. 15.9. Consequently, the O_2 concentration profile can be modeled as a combination of

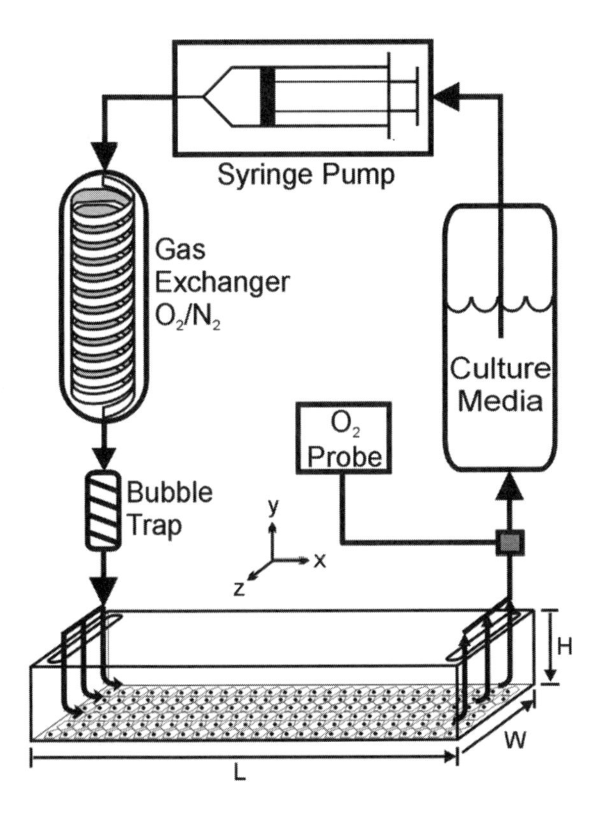

Figure 15.9. Schematic of bioreactor circuit. This parallel-plate bioreactor system containing cultured hepatocytes results in the formation of steady-state O_2 gradients, thereby generating zonated hepatocyte phenotypes.

O_2 diffusion to the cell, uptake by the cell, and convection along the length of the flow region. Both analytical and numerical models were shown to correlate closely with measured oxygen outputs. A detailed description of the mathematical modeling for this system has been reported previously [Allen and Bhatia, 2003]. Given an inlet O_2 concentration representative of physiologic periportal levels, 76 mmHg (10% O_2), a flow rate of 0.5 ml/min was determined to result in the exposure of cells in the final 50% of the chamber length to perivenous O_2 levels (<35 mmHg). Subsequently, this O_2 distribution resulted in zonated hepatocyte phenotypes within the bioreactor.

Protocol 15.10. Assembly of Bioreactor and Flow Circuit for Hepatocyte Micropatterning

Reagents and Materials

Sterile

- ❑ Freshly isolated primary rat hepatocytes (See Section 3)
- ❑ CHCM (See Section 2.6)
- ❑ HDHM (See Section 2.12)
- ❑ 70% Ethanol
- ❑ Collagen type I (rat tail), final concentration approximately 500 μg/ml

Nonsterile

- ❑ Perfusion circuit (Fig. 15.9) including: medium reservoir, gas exchanger with gas permeable Silastic tubing, O_2 probe (Clark-type electrode), syringe pump, bubble trap
- ❑ Flow deck: polycarbonate block containing inlet and outlet ports and 100-μm recess over which slide can be placed
- ❑ Stainless steel bracket, screws
- ❑ Inert silicone lubricant
- ❑ 38 mm × 75 mm glass microscope slides
- ❑ Incubator, PID controlled

Protocol

(a) Sterilize 38 mm × 75 mm microscope slide with 70% ethanol and treat with type I collagen as described in Protocol 15.9.

(b) Culture freshly isolated hepatocytes on slide to confluence with 2 seedings of 3×10^6 cells in Complete Hepatocyte Medium, followed by gentle horizontal shaking every 15 min for 1 h.

(c) After additional 2-h incubation, change medium to HDHM. Proceed with the following steps on day 1 or 2 after isolation. The hormonally defined preparation has been utilized in this procedure in order to help stabilize hepatocytes in monolayer culture, as previously described [Enat et al., 1984]. In addition, for the development of potentially clinical extracorporeal devices, it is important to optimize bioreactor parameters without the presence of nonhuman serum, which is incompatible with clinical approaches.

(d) Assemble flow chamber WITHOUT cells by adding blank microscope slide to polycarbonate block and closing with stainless steel bracket. Connect blank flow chamber to flow circuit and start flow (0.5 ml/min), in order to preequilibrate O_2 within lines and calibrate O_2 probe. Maintain all components of the flow circuit except the syringe pump within an incubator set at 37 °C. Allow an equilibration period of at least 15 min before continuing with the following steps.
(e) On a separate polycarbonate block, apply inert silicone lubricant to the outer edge of recessed section, outside the flow region.
(f) Invert slide with hepatocytes and lower onto this second flow deck. Overlay the stainless steel bracket and tighten with screws. The resultant flow field dimensions are 28 mm (width) × 55 mm (length) × 100 μm (height).
(g) Stop flow and remove flow chamber without cells. Quickly insert chamber with cultured hepatocytes into flow circuit as illustrated in Fig. 15.9, and restart flow (0.5 ml/min).

This bioreactor system represents a well-characterized platform in which O_2 gradients can be regulated by altering inlet concentration and flow rate. As expected at higher flow rates ($\geq$1.0 ml/min), given an inlet concentration of 76 mmHg, O_2 tension does not fall to perivenous levels in the length of the flow chamber. Increasing inlet O_2 can also prevent the decrease to the perivenous range. Notably, previous strategies for bioartificial liver devices have utilized preoxygenation of inlet streams to supraphysiologic levels in order to prevent inadequate O_2 delivery in parts of the bioreactor [Custer and Mullon, 1998]. However, as we have illustrated, O_2 can modulate hepatocyte function and therefore constitutes an important design criterion for an effective bioreactor system.

Another important design issue for bioreactors is the shear stress experienced by the cultured cells [Ledezma et al., 1999]. In the system described above, a flow rate of 0.5 ml/min corresponds to 1.25 dyn/cm^2, which is well below the shear stress of 5 dyn/cm^2 that was previously demonstrated to decrease rat hepatocyte function [Tilles et al., 2001]. Consequently, in our system, a flow rate of 0.5 ml/min and inlet of 76 mmHg represent the ideal parameters for preventing cellular damage due to shear stress or hypoxia, and concurrently generating adequate O_2 concentration gradients for the establishment of hepatocyte zonation.

6.2. Assessment of Zonated Features

Protocol 15.11. Examination of Zonated CYP2B Expression Resulting from Bioreactor Culture

Reagents and Materials

Sterile

- ❑ HDHM (See Section 2.12)
- ❑ Epidermal growth factor (EGF), 16 nM

- ❑ Phenobarbital (PB), 75 mM
- ❑ HDHM supplemented with 0.75 mM PB and 0.16 nM EGF [Kietzmann et al., 1999], made up just before use

Nonsterile

- ❑ Functioning perfusion bioreactor system (described above, Protocol 15.10)
- ❑ SDS lysis buffer (See Section 2.13)
- ❑ Protease inhibitor cocktail
- ❑ Microcentrifuge tubes (2 ml)
- ❑ Pestle

Protocol

(a) As described above, assemble bioreactor and flow circuit, and allow to reach steady state (approximately 15 min).
(b) Switch perfusion medium to supplemented HDHM and perfuse for 36 h.
(c) Stop flow, and remove flow chamber from flow circuit.
(d) Carefully disassemble flow chamber, and remove slide with hepatocytes.
(e) Scrape and lyse distinct sections of the chamber slide in SDS lysis buffer.
(f) Add samples to microcentrifuge tubes with protease inhibitor cocktail, homogenize with pestle, and centrifuge at 16,200 *g* for 5 min.
(g) Examine CYP2B protein expression with standard gel electrophoresis and Western blotting techniques.

Experiments examining CYP2B expression in four distinct regions of the described bioreactor have been performed (See Fig. 15.8) [Allen and Bhatia, 2003]. A similar procedure can be utilized to examine the zonated expression of other factors. For example, in place of infusion of PB and EGF, infusion of 10 nM glucagon for 8 h allows for the sufficient induction of PEPCK [Hellkamp et al., 1991], with an elevated level in the upstream section of this bioreactor system (See Fig. 15.8) [Allen and Bhatia, 2003].

Although this bioreactor system provides numerous advantages, the system is limited to short-term experimentation (3–4 days) because of the loss of differentiated functions by primary hepatocytes in monolayer culture even in the presence of the HDHM. Recent work in our laboratory has adapted hepatocyte and fibroblast coculture (described in Section 5.2), to this bioreactor system, to generate long-term stable hepatocyte cultures with defined O_2 gradients [Allen et al., 2005]. For longer-term studies, more rigorous sterilization procedures must be utilized, including the use of a polysulfone, which can be autoclaved, instead of polycarbonate for the flow deck.

In summary, the above protocols describe the assembly and utility of a perfusion bioreactor that exhibits steady-state O_2 gradients in an in vitro hepatocyte culture model. This system facilitates the investigation of the processes of liver zonation and provides a foundation for the development of bioartificial liver devices exhibiting these zonated features. In addition, future studies utilizing microscale fabrication techniques are aimed at miniaturizing this system toward the development of a parallel array of bioreactors for the high-throughput examination of hepatotoxicity.

7. PHOTOPATTERNED THREE-DIMENSIONAL HYDROGELS CONTAINING LIVING CELLS

To develop implantable tissue constructs that can be successfully incorporated into host tissue and recapitulate the desired functions, it is important to simulate sufficiently normal tissue structure, including 3-D organization. Specifically, for the design of implantable liver schemes, it would be particularly advantageous to disseminate cells three-dimensionally within a thick construct, without requiring subsequent population or recruitment of cells. Hydrogels, which are cross-linked polymer networks based on chemistry originally derived by Hubbell and colleagues [Pathak et al., 1992], can be polymerized in the presence of cells, thereby generating gels with a homogeneous cellular distribution. As a result, hydrogels represent an ideal platform for the development of these types of implantable systems. In addition, hydrogels are biocompatible and exhibit high water content, which provides them with mechanical properties similar to tissues. Furthermore, several reports indicate that cellular phenotypes are different in 2-D versus 3-D cell culture [Katz et al., 2000; Poznansky et al., 2000; Wang et al., 1998], suggesting that the development of in vitro. 3-D culture models will be required to mimic more closely in vivo environments, an application for which hydrogels are also ideally suited.

Poly(ethylene glycol) (PEG)-based hydrogels are widely used in tissue engineering applications because of their biocompatibility and hydrophilicity resulting in resistance to nonspecific protein adsorption and the ability to alter the properties of the hydrogel by modifying the PEG chain length [Peppas et al., 2000]. Specifically, encapsulation within PEG-based hydrogels has been utilized for numerous cell types, including vascular smooth muscle cells [Mann et al. 2001a,b], chondrocytes [Bryant and Anseth, 2002; Elisseeff et al., 2000], fibroblasts [Gobin and West, 2002; Hern and Hubbell, 1998], and mesenchymal stem cells [Nuttelman et al., 2004] (See also Chapter 9). In addition to cellular encapsulation, the development of tissue-engineered constructs for spatially complex organs, such as the liver, will also require the ability to form specified 3-D features and control cellular orientations.

We have recently developed a photopatterning technique to generate hydrogels containing living cells with a defined architecture [Liu and Bhatia, 2002]. Explicitly, the combined methods of hydrogel polymerization and photolithography, in which UV cross-linking is performed through an emulsion mask, enabled the localized photopolymerization of hydrogel structures containing cells. Utilizing this method, we formed single-layer hydrogels of specified dimensions containing living cells as well as composite single-layer structures containing distinct cellular domains (Fig. 15.10, See Color Plate 10C) [Liu and Bhatia, 2002]. Furthermore, we demonstrated the formation of multiple-layer hydrogel structures

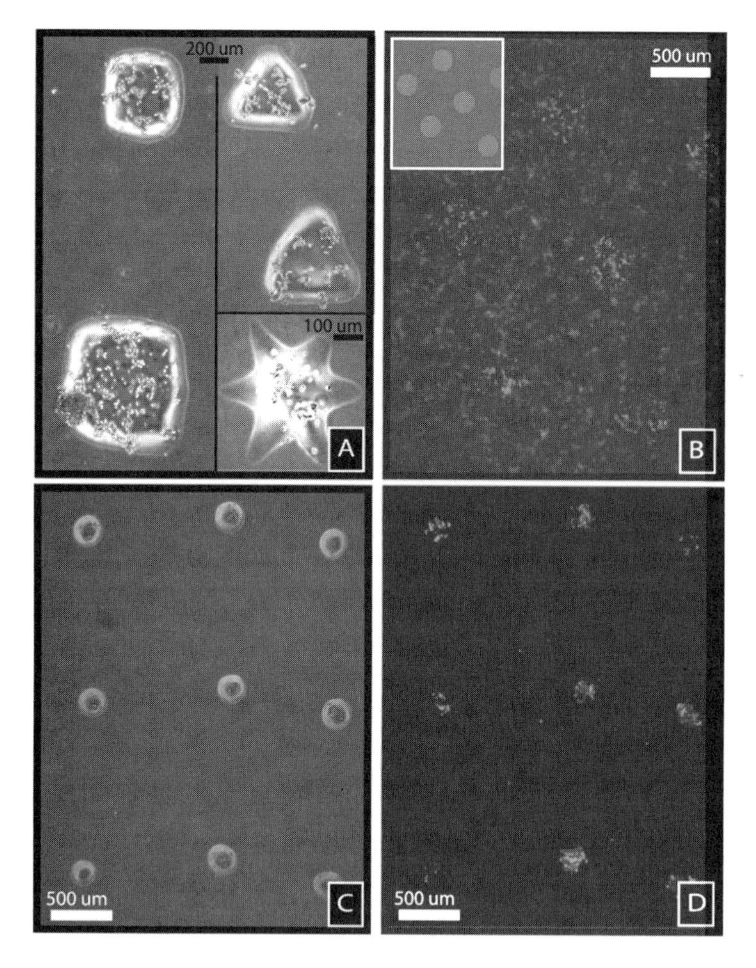

Figure 15.10. Hydrogel microstructures containing living cells. A) Cells entrapped in PEGDA hydrogels patterned in various shapes. B) Red and green labeled cells patterned within distinct domains of a single hydrogel layer. C) Phase microscopy of cellular array covalently linked to glass substrate. D) Fluorescent image of the cellular array, showing two different cell types (green, red). From Liu and Bhatia [2002]. (See Color Plate 10C.)

containing patterned cellular regions (Fig. 15.11) [Liu and Bhatia, 2002]. Protocol 15.12 describes this photopatterning procedure with HepG2 cells, a human hepatoma cell line, although this method is amenable to various cells types. Optimization parameters critical for adapting this system to other cell types are highlighted in this section.

Protocol 15.12. Photopatterning of Hydrogel with Living Cells

Reagents and Materials

Sterile

- ❑ HepG2 cells, human hepatoma cell line
- ❑ HepG2 Culture Medium (See Section 2.14)

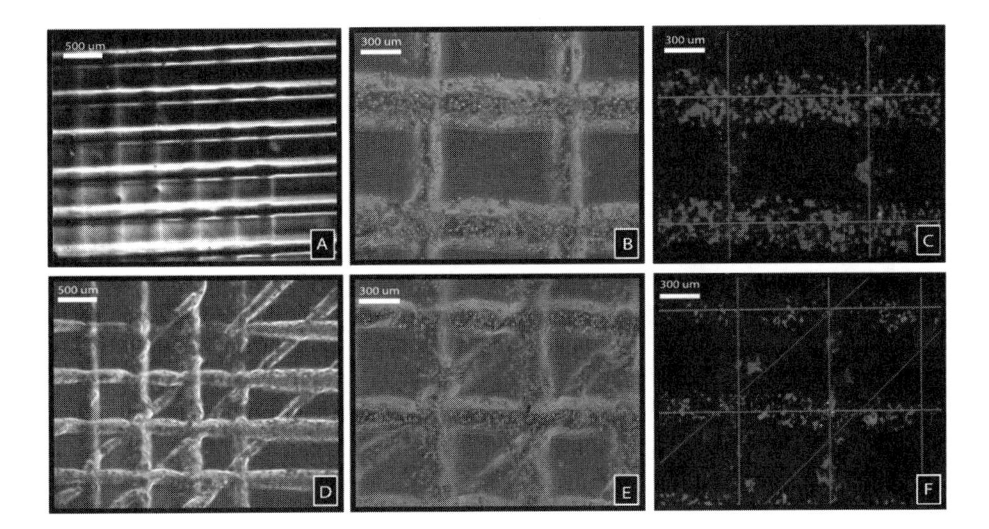

Figure 15.11. Multilayer hydrogel microstructures containing living cells. A) Two layers of patterned PEGDA lines. B) Two layers of patterned PEGDA lines containing cells. C) Fluorescent image of patterned hydrogel lines containing red and green tracked cells, demonstrating different cells in each layer. D) Three layers of patterned PEGDA lines containing cells at low magnification. E) Three layers of patterned PEGDA lines containing cells at higher magnification. F) Fluorescent image of three layers, with one layer containing red labeled cells, one layer containing green labeled cells, and all layers counterstained blue. From Liu and Bhatia [2002]. (See Color Plate 10D.)

- ❑ Microcentrifuge tubes (1.5 ml)
- ❑ Syringe filters (0.45 μm, 0.2 μm)
- ❑ Syringes (1 ml)
- ❑ Phosphate buffered saline (PBSA)

Nonsterile

- ❑ Poly(ethylene glycol) diacrylate (PEGDA), 3.4 kDa
- ❑ Photoinitiator: Irgacure 2959; 4-(2-hydroxyethoxy)phenyl-(2-hydroxy-2-propyl) ketone. Dissolved in 1-vinyl-2-pyrrolidinone to form 10% (w/v) stock solution.
- ❑ Circular borosilicate glass wafer, 2 in.
- ❑ 95% Ethanol (pH 5 with acetic acid)
- ❑ 100% Ethanol
- ❑ 70% Ethanol
- ❑ 2% (v/v) solution of 3-(trimethoxysilyl)propylmethacrylate in 95% ethanol (pH 5 with acetic acid)
- ❑ UV light source: Exfo Lite UV spot cure system with 5-mm LG Adapter, collimating single lens, and 365-nm filter
- ❑ Emulsion mask with desired pattern
- ❑ Polymerization apparatus (Fig. 15.12) including: foundation with stainless steel bracket and caliper screws, Teflon base with inlet and outlet channels, glass slide, and silicone spacers (100- to 500-μm thickness)

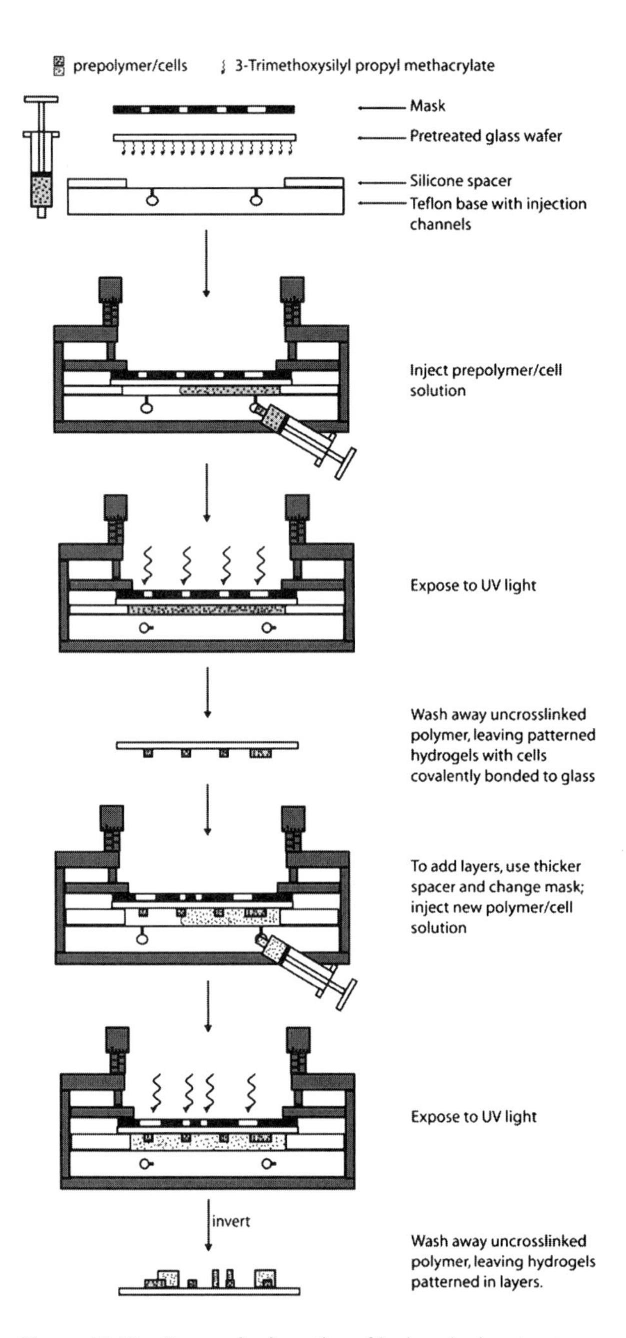

Figure 15.12. Process for formation of hydrogel microstructures containing living cells. The apparatus is assembled, including a pretreated glass wafer with reactive methacrylate groups on its surface, and a Teflon base with an inlet and outlet. Once the cells and prepolymer solution are injected, the inlet and outlet are closed, and the unit is exposed to UV light. The resulting patterned hydrogels containing cells are covalently bound to the glass wafer. At this time, a thicker spacer can be used in conjunction with a new mask to add another layer of cells. This process can be repeated several times. From Liu and Bhatia [2002].

Protocol

(a) Design emulsion mask(s) with drawing software, such as Corel Draw 11.0, and print with commercial Linotronic-Hercules 3300 dpi high-resolution line printer. For the formation of multiple layers, masks should contain reference points, such as cross hairs, for alignment purposes.

(b) Treat clean 2-in. borosilicate glass wafer with 2% 3-(trimethoxysilyl)propyl-methacrylate for 2 min. Rinse with 100% ethanol, and bake at 100 °C for 5–10 min. This process generates free methacrylate groups on the glass, which will react with the PEGDA during UV exposure and will prevent detachment of the hydrogel structures from the glass support. Alternatively, we have utilized 34-mm circular coverglasses, as opposed to 2-in. wafers, in some experiments because of their decreased thickness, for example, for situations in which confocal microscopy is required of the resultant structures.

(c) Etch marking on glass wafer outside the intended polymerization area, for alignment with cross hairs on masks.

(d) Sterilize masks, Teflon base, and silicone spacers by ultrasonic agitation (Sonicator) in 70% ethanol for 10–15 min. Let components dry in laminar flow hood.

(e) Resuspend HepG2 cells in HepG2 culture medium to 2× desired final concentration within the hydrogel. The final concentration is determined based on the particular experiment of interest.

(f) Dissolve PEGDA in PBSA to form 40% (w/v) solution.

(g) Add to the PEGDA solution the appropriate amount of photoinitiator stock (10% w/v) to achieve a photoinitiator concentration of 0.2% (w/v).

(h) Sterile filter PEGDA-photoinitiator solution, using sequentially a 0.45-μm, then a 0.2-μm syringe filter, and protect from light exposure.

(i) Assemble in laminar flow hood the polymerization apparatus illustrated in Fig. 15.12. Add sequentially the Teflon base, silicone spacer, methacrylate-treated glass wafer, mask, glass slide, and stainless steel bracket.

(j) Hold components in place by carefully tightening caliper screws on the bracket.

(k) Add cell suspension to PEGDA-photoinitiator prepolymer solution in 1:1 ratio, and mix gently.

(l) Use syringe to inject prepolymer solution with cells through the inlet port into the chamber formed by the glass wafer and silicone spacer. Alternatively, if the volume of the prepolymer solution with cells is limited, this solution can be added directly on to the Teflon base (without using inlet port), before adding the glass wafer, mask, slide, and bracket.

(m) Expose to UV light at 100 mW/cm^2 for 1.5 min. The exposed regions defined by the mask will be cross-linked to form the hydrogel. The exposure time and intensity may vary depending on the mask features.

(n) Wash away non-cross-linked polymer with PBSA, resulting in a patterned hydrogel containing cells.

To form a composite single-layer structure or cellular domains in an array format, after rinsing away the non-cross-linked polymer, a second prepolymer solution (with or without cells) can be added in the same manner and cross-linked by uniform exposure or exposure through a second mask. To form multilayer hydrogel structures, the thickness of the spacer is increased for the ensuing cross-linking steps. Alignment of the subsequent masks is achieved using the cross hairs, and the resultant structure is a complex 3-D construct containing living cells.

The above polymerization conditions have been optimized for experiments in our laboratory and can serve as a guideline, although the optimal conditions may vary depending on the cell type and desired hydrogel pattern. There are several design criteria to consider in adapting this procedure to a specific application. First, the cross-linking reaction proceeds because of the formation of free radicals from the photoinitiator, and these radicals can induce cell death. Consequently, photoinitiator concentration, as well as the intensity and duration of UV exposure, must be optimized for the particular cell type of interest. For multiple-layer structures, if possible, masks should be designed to prevent repeated UV exposure to the bottom layer. In addition, UV light is absorbed in each layer, including the glass slide on top of the apparatus and the mask. As a result, this can decrease the actual UV intensity reaching the hydrogel solutions, which is an important aspect to keep in mind when optimizing the intensity.

To increase pattern resolution, we have previously illustrated that UV exposure should be limited [Liu and Bhatia, 2002]. In addition, utilizing a collimated light source can improve pattern fidelity. Interestingly, we have found that photoinitiator concentration does not affect pattern resolution, so the lowest possible photoinitiator concentration can be used to minimize potential toxicity. Utilizing the parameters in the protocol described above, we have achieved 100-μm features with an accuracy of ± 5 μm. In general, increasing exposure intensity and decreasing the PEG chain length may enhance resolution. However, these alterations, if too extreme, can potentially decrease cell viability depending on the cell type utilized.

In summary, the protocol outlined in this section describes a method for constructing patterned 3-D hydrogel structures containing living mammalian cells. This technique can be broadly utilized for the design of implantable tissue constructs, development of defined 3-D in vitro culture models, and formation of immobilized cellular arrays as cell-based assays for pharmaceutical drug development. In particular, the formation of specified 3-D cellular structures represents an important step toward the development of tissue engineering therapies for spatially complex organs, such as the liver. Furthermore, several laboratories have demonstrated the incorporation of adhesive ligands [Hern and Hubbell, 1998; Hersel et al., 2003; Shin et al., 2003], degradable motifs [Bryant and Anseth, 2003; Lutolf et al., 2003; Metters et al., 2000; Sawhney et al., 1993; Seliktar et al., 2004], or growth factors [Gobin and West, 2003; Mann et al., 2001b; Seliktar et al., 2004; Zisch et al., 2003] into PEG-based hydrogels, providing a means to regulate cell attachment and spreading, growth, and other biological functions such as matrix deposition.

The combination of these techniques with the patterning method described here will provide a means for recapitulating within a tissue-engineered liver construct the intricate cell-cell and cell-matrix interactions critical for liver function.

8. DISCUSSION

This chapter has described several methods important for the fundamental investigation of hepatocyte biology and the development of engineered liver tissue. These included the isolation procedure for primary rat hepatocytes and numerous assays of rat hepatocyte function. Also included were protocols describing the micropatterning of hepatocyte-fibroblast cocultures, a bioreactor system for the generation of steady-state oxygen gradients and establishment of zonated features, as well as the three-dimensional fabrication of cellular constructs through photopatterning of hydrogels. Together, these procedures, and, similarly, alternative approaches developed in other laboratories, demonstrate numerous possibilities for both in vitro and clinical applications. Specifically, our laboratory's overall goals and approach for the development of engineered liver tissue are summarized in Fig. 15.13.

8.1. In Vitro Applications

The closely interconnected structure-function relationship in the liver dictates that tissue-engineered liver constructs must recapitulate, at least in part, the normal liver environment. Accordingly, much research in the field of liver tissue

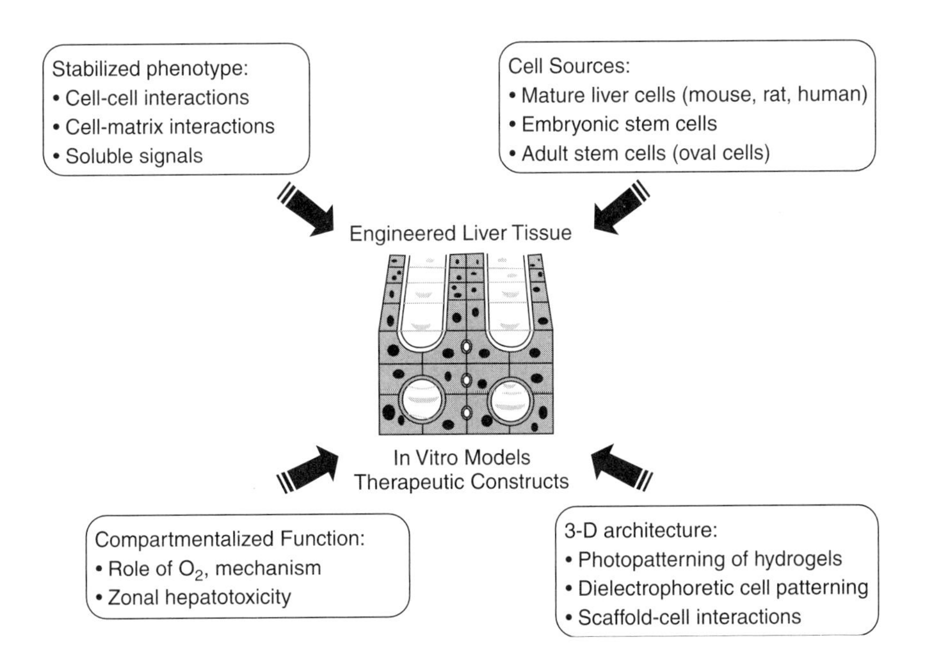

Figure 15.13. Overall approach for the design of engineered liver tissue. Each of these aspects will provide fundamental insight into liver biology and constitute important components contributing to the future development of highly functional tissue-engineered liver constructs.

engineering has initially focused on the development of in vitro culture models for liver cells, in order to characterize cellular responses in distinct environments and to systematically delineate the required extrinsic factors. For example, studies have assessed the in vitro morphogenesis of hepatocytes in pure cultures and cocultures [Berthiaume et al., 1996; Bhatia, 1997; Elcin et al., 1998; Haruyama et al., 2000; Powers and Griffith, 1998; Ranucci et al., 2000], as well as hepatocyte metabolic requirements [Balis et al., 1999; Kim et al., 2000; McClelland et al., 2003; Patzer, 2004; Rotem et al., 1994]. In addition, in our laboratory we are currently utilizing microfabrication approaches, such as the patterning procedure (See Section 5), to delineate the molecular mechanisms by which fibroblasts stabilize hepatocyte function, and the dynamics of this process. Identification of the important components involved, including soluble factors and key cell-cell and cell-matrix interactions, would enable the specified incorporation of these factors into a hepatocyte-only system, which would be more suitable for potential clinical applications.

Another important application for functional in vitro liver culture models is as a method for pharmaceutical drug development and the assessment of the risk of hepatotoxicity due to exposure to environmental toxicants. Liver toxicity is the major factor contributing to drug failures in clinical trials, and, furthermore, it has been reported that the most common cause of acute liver failure in the United States, accounting for up to one-half of cases, is drug-induced liver disease [Shakil et al., 2000; Kaplowitz and DeLeve, 2003]. Stabilized hepatocyte culture systems would represent a platform to thoroughly characterize drug metabolism, examine toxicity, and investigate drug-drug interactions. Notably, compartmentalized liver features such as zonation can regulate the degree of toxicity in response to various drugs [Kera et al., 1987; Lindros, 1997; Oinonen and Lindros, 1998; Pronko et al., 2002]. As a result, culture systems (such as the bioreactor platform described in this chapter) that not only stabilize hepatocyte function, but also reestablish certain heterogeneous elements observed in vivo, would be extremely useful in drug toxicity studies. For example, we have recently identified a fourfold decrease in the acetaminophen TD_{50} (toxic dose for 50% of cells) in the outlet region of this bioreactor [Allen et al., 2005]. Furthermore, microfabrication of functional hepatic elements would enable high-throughput analysis, in which multiple drug-drug combinations at varying doses could be analyzed in parallel.

8.2. Clinical Applications

The development of cell-based therapies for liver treatment aimed at the eventual replacement of damaged or diseased liver tissue represents a potential alternative to organ transplantation. One prospective cell-based approach is the transplantation of isolated mature hepatocytes. In experiments utilizing rodent models, transplanted hepatocytes were demonstrated to exhibit substantial proliferative capacity and the ability to replace diseased tissue, under some limited conditions [Overturf et al., 1997; Rhim et al., 1994; Sokhi et al., 2000]. However, the in vivo proliferation of transplanted hepatocytes is highly dependent on the presence of an adequate "regenerative" environment, which would be difficult to control in a clinical setting. Furthermore, survival and engraftment of transplanted hepatocytes has been reported to be quite low, only 20–30% [Gupta et al., 1999b]. Several studies have demonstrated methods to improve engraftment and enhance the selective proliferation of transplanted hepatocytes [Guha et al., 1999; Joseph et al., 2002; Laconi et al., 1998; Malhi et al., 2002; Mignon et al., 1998; Slehria et al., 2002], although the clinical utility of these approaches remains to be determined.

As an alternative to mature hepatocytes, numerous stem cell populations are currently being investigated as potential cell sources, including pluripotent embryonic stem cells and various adult stem cell types. One adult stem cell of interest is the hepatic oval cell, which mediates liver regeneration in certain cases of severe and chronic liver injury [Oh et al., 2002; Sell, 2001]. Oval cells exhibit bipotential differentiation, defined by the ability to differentiate into both hepatocytes and bile duct epithelial cells. Oval cells can be isolated and significantly expanded in culture [Petersen et al., 1998; Yang et al., 2002], and, furthermore, recent experiments in mice illustrate the oval cell-mediated therapeutic repopulation of the liver [Wang et al., 2003]. Elucidation of the mechanisms regulating the proliferation and hepatic differentiation of oval cells, as well as other stem cell populations, could lay the groundwork for the development of robust cell-based liver transplantation therapies.

Another cell-based strategy for liver treatment is the implantation of hepatocellular constructs. Similar to various tissue engineering schemes in other organ systems, for this approach in the liver, hepatocytes are cultured within or immobilized on synthetic or biological scaffolds and surgically implanted [Allen and Bhatia, 2002]. Substantial current research is focused on the development of scaffolds that promote proper hepatic function. It is clear that several hepatocyte processes, such as survival and morphogenesis, can be influenced by the scaffold characteristics [Demetriou et al., 1986a; Dixit et al., 1992; Dvir-Ginzberg et al., 2003; Elcin et al., 1998; Glicklis et al., 2000; Hasirci et al., 2001; Kim et al., 1998; Li et al., 2003; Pollok et al., 1998; Powers and Griffith, 1998; Ranucci et al., 2000; Risbud et al., 2003]. Consequently, one significant challenge for the future is the design of platforms that enable the highly specified regulation of cellular orientations in three dimensions, as well as the controlled incorporation into the host environment. A recent study utilizing a focused laser to activate locally a photolabile hydrogel structure demonstrated the patterned immobilization of an adhesive peptide, GRGDS, which subsequently exhibited a guidance effect on neurite growth [Luo and Shoichet, 2004]. Several other techniques have also been used to pattern hydrogel networks, including microfluidics [Koh et al., 2003; Tan and Desai, 2003] and photolithography [Beebe et al., 2000; Koh et al., 2002; Ward et al., 2001; Yu and Ober, 2003]. Our laboratory has developed an adaptation of standard photolithography techniques resulting in the photopatterning of hydrogel structures containing living cells [Liu and Bhatia, 2002] (See Section 7). In addition, our laboratory has recently demonstrated the dielectrophoretic positioning of cells within a hydrogel in the 10-μm range [Albrecht et al., 2002] and the combination of this technique with the photopatterning procedure described here [Albrecht et al., 2005]. Notably, scale-up of 3-D patterning of cellular constructs will likely occur via computer-aided design (CAD)-based rapid prototyping tools such as stereolithography, as part of the nascent field of "organ printing" [Jakab et al., 2004]. Taken together, the combination of these approaches allowing the spatial control of cells and bioactive factors represents an important step toward the formation of engineered liver systems more closely recapitulating normal liver structure and function.

An additional major challenge for the design of implantable liver systems is the need to overcome transport limitations within the grafted construct due to the lack of functional vasculature. Normally in vivo, hepatocytes are supplied by an extensive vasculature consisting of sinusoids with minimal extracellular matrix and a lining of fenestrated (sinusoidal) endothelial cells [Enzan et al., 1997]. Together these sinusoidal features allow for the efficient transport of nutrients to the hepatocytes, which are highly metabolic. Strategies to incorporate vasculature into engineered systems include the in vitro microfabrication of vascular units with accompanying surgical anastomosis during implantation [Griffith et al., 1997; Kaihara et al., 2000]. For example, polymer molding using microetched silicon has been shown to generate extensive channel networks with four dimensions in the order of capillaries [Borenstein et al., 2002]. An alternative approach is the incorporation of angiogenic factors within the implanted scaffolds. In particular, integration of cytokines important in angiogenesis, such as VEGF [Smith et al., 2004]. bFGF [Lee et al., 2002], and VEGF in combination with PDGF [Richardson et al., 2001], has been shown to promote the recruitment of host vasculature to implanted constructs.

An integral aspect in the development of clinically relevant cell-based liver therapies of all types is the utilization of adequate animal models in order to examine therapeutic effects as well as safety considerations. Numerous distinct large-animal models have been used to examine the efficacy of extracorporeal support systems in situations of liver failure [van de Kerkhove et al., 2004]. These systems fall into four major categories: partial hepatectomy, total hepatectomy, toxic, and ischemic models. Several extensive reviews have addressed the important criteria for developing animal models of fulminant hepatic failure [Newsome et al., 2000; Terblanche and Hickman, 1991; van de Kerkhove et al., 2004].

In addition to liver failure models, animal models have been used extensively to investigate liver regeneration. These can be broadly classified into two categories: surgically induced and chemically induced [Palmes and Spiegel, 2004]. Well-characterized models in the chemical category include exposure to toxic doses of carbon tetrachloride or acetaminophen, both situations that result in localized centrilobular necrosis [Anundi et al., 1993; Lindros et al., 1990]. Chemical-induced injury models are of particular interest for testing the utility of cell-based liver therapies, as these systems more closely mimic liver injuries commonly occurring in humans (i.e., drug toxicity). In the surgical category, the 2/3 partial hepatectomy model in rat has been widely utilized as a condition in which the injury stimulus is well-defined [Higgins and Anderson, 1931]. Interestingly, parabiotic systems have demonstrated that after partial hepatectomy, factors regulating liver cell proliferation are present in the circulation [Fisher et al., 1971; Moolten and Bucher, 1967; Roesel et al., 1989]. Consequently, although the hepatectomy model is less relevant clinically, it may serve as a well-controlled system to investigate the role of regenerative cues in the engraftment and proliferation of hepatic constructs implanted in extrahepatic sites [Jirtle and Michalopoulos, 1982]. Overall, utilization of each of these types of systems (surgical and chemical) will likely be important in testing the utility of cell-based liver therapies. In addition, knowledge of the mechanisms of liver injury and regeneration gleaned from these animal models will provide an important blueprint for the design of clinically effective engineered liver tissue.

In summary, although many challenges remain for the improvement of tissue-engineered liver systems; substantial progress has been made toward a thorough understanding of the necessary components. Technologies such as hydrogel chemistries and microfabrication represent enabling tools for investigating the critical role of the microenvironment in liver function and, subsequently, the development of structurally complex and highly functional engineered liver constructs.

ACKNOWLEDGMENTS

Funding was provided by NIH DK56966, NIH DK065152, David and Lucile Packard Foundation, NSF CAREER Award (SNB), NASA, for SRK (NSF-graduate fellowship), and for VLT (Whitaker-graduate fellowship, AAUW, ARCS).

SOURCES OF MATERIALS

Item	*Supplier*
1-Vinyl-2-pyrrolidinone	Sigma
250-μm nylon mesh (#CMN-250-D)	Small Parts
3-(Trimethoxysilyl) propyl methacrylate	Aldrich
3-0 Silk suture	Harvard Apparatus
3T3-J2 mouse fibroblast cell line	H. Green, Harvard Medical School
62-μm nylon mesh (#CMN-62-D)	Small Parts
96-Well plates, flat bottom	Nunc
Albumin, rat	ICN
Anesthesia device, veterinary grade (Model 100F)	SurgiVet
Angiocath™ catheter	Becton Dickinson
Anti-albumin antibody, rabbit anti-rat, HRP conjugated	ICN
Anti-albumin antibody, rabbit anti-rat, unconjugated	ICN
Anti-IgG antibody, goat anti-rabbit, FITC conjugated	Santa Cruz Biotechnology
Betadine	Owens and Minor, Inc.
Biotin Blocking System	DakoCytomation
Blood Urea Nitrogen (BUN) Acid Reagent	Stanbio Labs
Blood Urea Nitrogen (BUN) Color Reagent	Stanbio Labs
Borosilicate glass wafers, circular, 2 in.	Erie Scientific
Bottom Side Mask Aligner	Karl Suss
Bovine serum albumin	Sigma
Collagenase, type IV (C-5138)	Sigma
Coverglasses, circular, 34 mm	Fisher

Item	Supplier
DAKO LSAB® System-HRP	DakoCytomation
Developer appropriate for selected photoresist. We use OCG 934, MF-319, or Shipley 354	
Dulbecco's modified minimal essential medium	Invitrogen
Fetal bovine serum	Invitrogen
Glucagon	Bedford Laboratories
Goat serum	Sigma
HBSS, 10×, without calcium or magnesium	Invitrogen
HepG2 cells, human hepatoma cell line	ATCC
Hydrocortisone sodium succinate	Pharmacia
Insulin (Regular ILETIN® II)	Eli Lilly
Irgacure 2959	Ciba Specialty Chemicals
MF-319	Microchem Corp.
Minimal essential medium	Gibco
Mitomycin C	Boehringer
Nylon mesh	
O_2 probe	Microelectrodes
OCG 825–835 St	Olin-Ciba-Geigy
OCG 934	Olin-Ciba-Geigy
o-Phenylenediamine (OPD) tablets	Sigma
Paraformaldehyde	Electron Microscopy Sciences
Penicillin-streptomycin	GIBCO
Percoll	Amersham
Peroxidase Blocking Reagent	DakoCytomation
Positive photoresists, OCG 825–835 St, Shipley 1813, or Shipley 1818	
Poly(ethylene glycol) diacrylate (PEGDA), 3.4 kDa	Nektar Transforming Therapeutics
Protease inhibitor cocktail	Roche Diagnostics
Rats, female Lewis	Charles River
Shipley 1813	Microchem Corp.
Shipley 1818	Microchem Corp.
Shipley 354	Microchem Corp.
Silicone lubricant, inert	Dow Corning
Slides, glass (75 × 38 mm)	Fisher
Stanbio Urea Nitrogen Procedure No. 0580	Stanbio Labs
Syringe pump	Harvard Apparatus
Triton X-100	Fisher
Trypsin (0.25%)-EDTA (0.2 g/L)	Invitrogen
Ultra-TMB (3,3′,5,5′-tetramethylbenzidine	Research Diagnostics, Inc.
Urea/nitrogen stock samples	Stanbio Labs
UV light source: Exfo Lite	Exfo
Vectashield	Vector Laboratories
Wafer tweezers	Fluoroware

REFERENCES

Albrecht, D.R., Sah, R.L., Bhatia, S.N. (2002) Dielectrophoretic cell patterning within tissue engineering scaffolds. *IEEE Proceedings of the 24th Annual International Conference of the Engineering in Medicine and Biology Society* 2: 1708–1709.

Albrecht, D.R., Tsang, V.L., Sah, R.L., Bhatia, S.N. (2005) Photo- and electropatterning of hydrogel-encapsulated living cell arrays. *Lab Chip* 5(1): 111–118.
Allen, J.W., and Bhatia, S.N. (2002) Engineering liver therapies for the future. *Tissue. Eng.* 8(5): 725–737.
Allen, J.W., and Bhatia, S.N. (2003) Formation of Steady-state oxygen gradients in vitro: application to liver zonation. *Biotechnol. Bioeng.* 82: 253–262.
Allen, J.W., Hassanein, T., Bhatia, S.N. (2001) Advances in bioartificial liver devices. *Hepatology* 34(3): 447–455.
Allen, J.W., Khetani, S.R., Bhatia, S.N. (2005) In vitro zonation and toxicity in a hepatocyte bioreactor. *Toxicol. Sci.* 84(1): 110–119.
Anundi, I., Lahteenmaki, T., Rundgren, M., Moldeus, P., Lindros, K.O. (1993) Zonation of acetaminophen metabolism and cytochrome P450 2E1-mediated toxicity studied in isolated periportal and perivenous hepatocytes. *Biochem. Pharmacol.* 45(6): 1251–1259.
Bader, A., Knop, E., Böker, K., Frühauf, N., Schüttler, W., Oldhafer, K., Burkhard, R., Pichlmayr, R., Sewing, K.F. (1995) A novel bioreactor design for in vitro reconstruction of in vivo liver characteristics. *Artif. Organs* 19(4): 368–374.
Badylak, S.F., Park, K., Peppas, N., McCabe, G., Yoder, M. (2001). Marrow-derived cells populate scaffolds composed of xenogeneic extracellular matrix. *Exp. Hematol.* 29(11): 1310–1318.
Balis, U.J., Behnia, K., Dwarakanath, B., Bhatia, S.N., Sullivan, S.J., Yarmush, M.L., Toner, M. (1999) Oxygen consumption characteristics of porcine hepatocytes. *Metab. Eng.* 1(1): 49–62.
Bankston, P.W., and Pino, R.M. (1980) The development of the sinusoids of fetal rat liver: morphology of endothelial cells, Kupffer cells, and the transmural migration of blood cells into the sinusoids. *Am. J. Anat.* 159(1): 1–15.
Baribault, H., and Marceau, N. (1986) Dexamethasone and dimethylsulfoxide as distinct regulators of growth and differentiation of cultured suckling rat hepatocytes. *J. Cell Physiol.* 129(1): 77–84.
Beebe, D.J., Moore, J.S., Bauer, J.M., Yu, Q., Liu, R.H., Devadoss, C., Jo, B.H. (2000) Functional hydrogel structures for autonomous flow control inside microfluidic channels. *Nature* 404(6778): 588–590.
Behnia, K., Bhatia, S., Jastromb, N., Balis, U., Sullivan, S., Yarmush, M., Toner, M. (2000) Xenobiotic metabolism by cultured primary porcine hepatocytes. *Tissue Eng.* 6(5): 467–479.
Berthiaume, F., Moghe, P.V., Toner, M., Yarmush, M.L. (1996). Effect of extracellular matrix topology on cell structure, function, and physiological responsiveness: hepatocytes cultured in a sandwich configuration. *FASEB J.* 10(13): 1471–1484.
Bhatia, S., Toner, M., Foy, B., Rotem, A., O'Neil, K., Tompkins, R., Yarmush, M. (1996) Zonal liver cell heterogeneity: effects of oxygen on metabolic functions of hepatocytes. *Cell. Eng.* 1: 125–135.
Bhatia, S.N. (1997) *Controlling Cell-Cell Interactions in Hepatic Tissue Engineering Using Microfabrication* [Ph.D. Dissertation]. Cambridge, MA: Harvard/MIT 169 pp.
Bhatia, S.N., Balis, U.J., Yarmush, M.L., Toner, M. (1998a) Microfabrication of hepatocyte/fibroblast co-cultures: role of homotypic cell interactions. *Biotechnol. Prog.* 14(3): 378–387.
Bhatia, S.N., Balis, U.J., Yarmush, M.L., Toner, M. (1998b) Probing heterotypic cell interactions: hepatocyte function in microfabricated co-cultures. *J. Biomater. Sci. Polym. Ed.* 9(11): 1137–1160.
Bhatia, S.N., Balis, U.J., Yarmush, M.L., Toner, M. (1999) Effect of cell-cell interactions in preservation of cellular phenotype: cocultivation of hepatocytes and nonparenchymal cells. *FASEB J.* 13(14): 1883–1900.
Bhatia, S.N., Yarmush, M.L., Toner, M. (1997) Controlling cell interactions by micropatterning in co-cultures: hepatocytes and 3T3 fibroblasts. *J. Biomed. Mater. Res.* 34(2): 189–199.
Bissell, D.M., Arenson, D.M., Maher, J.J., Roll, F.J. (1987) Support of cultured hepatocytes by a laminin-rich gel. Evidence for a functionally significant subendothelial matrix in normal rat liver. *J. Clin. Invest.* 79(3): 801–812.
Bissell, D.M., Hammaker, L.E., Meyer, U.A. (1973) Parenchymal cells from adult rat liver in nonproliferating monolayer culture. I. Functional studies. *J. Cell Biol.* 59(3): 722–734.
Block, G.D., Locker, J., Bowen, W.C., Petersen, B.E., Katyal, S., Strom, S.C., Riley, T., Howard, T.A., Michalopoulos, G.K. (1996) Population expansion, clonal growth, and specific

differentiation patterns in primary cultures of hepatocytes induced by HGF/SF, EGF and TGF alpha in a chemically defined (HGM) medium. *J. Cell Biol.* 132(6): 1133–1149.

Borenstein, J.T., Terai, H., King, K.R., Weinberg, E.J., Kaazempur-Mofrad, M.R., Vacanti, J.P. (2002) Microfabrication technology for vascularized tissue engineering. *Biomed. Microdevices* 4(3): 167–175.

Britland, S., Perez-Arnaud, E., Clark, P., McGinn, B., Connolly, P., Moores, G. (1992) Micro-patterning proteins and synthetic peptides on solid supports: a novel application for micro-electronics fabrication technology. *Biotechnol. Prog.* 8(2): 155–160.

Bryant, S.J., and Anseth, K.S. (2002) Hydrogel properties influence ECM production by chondrocytes photoencapsulated in poly(ethylene glycol) hydrogels. *J. Biomed. Mater. Res.* 59(1): 63–72.

Bryant, S.J., and Anseth, K.S. (2003) Controlling the spatial distribution of ECM components in degradable PEG hydrogels for tissue engineering cartilage. *J. Biomed. Mater. Res.* 64A(1): 70–79.

Burke, M.D., and Mayer, R.T. (1974) Ethoxyresorufin: direct fluorimetric assay of a microsomal O-dealkylation which is preferentially inducible by 3-methylcholanthrene. *Drug Metab. Dispos.* 2(6): 583–588.

Burke, M.D., Thompson, S., Elcombe, C.R., Halpert, J., Haaparanta, T., Mayer, R.T. (1985). Ethoxy-, pentoxy- and benzyloxyphenoxazones and homologues: a series of substrates to distinguish between different induced cytochromes P-450. *Biochem. Pharmacol.* 34(18): 3337–3345.

Cederbaum, A.I., Wu, D., Mari, M., Bai, J. (2001) CYP2E1-dependent toxicity and oxidative stress in HepG2 cells. *Free Radic. Biol. Med.* 31(12): 1539–1543.

Chang, T.M.S. (1992) Artificial liver support based on artificial cells with emphasis on encapsulated hepatocytes. *Artif. Organs* 16(N1): 71–74.

Costa, C., Zhao, L., Burton, W.V., Bondioli, K.R., Williams, B.L., Hoagland, T.A., Ditullio, P.A., Ebert, K.M., Fodor, W.L. (1999) Expression of the human alpha1,2-fucosyltransferase in transgenic pigs modifies the cell surface carbohydrate phenotype and confers resistance to human serum-mediated cytolysis. *FASEB J.* 13(13): 1762–1773.

Custer, L., and Mullon, C.J. (1998) Oxygen delivery to and use by primary porcine hepatocytes in the HepatAssist 2000 system for extracorporeal treatment of patients in end-stage liver failure. *Adv. Exp. Med. Biol.* 454: 261–271.

Dahlke, M.H., Popp, F.C., Larsen, S., Schlitt, H.J., Rasko, J.E. (2004) Stem cell therapy of the liver—Fusion or fiction? *Liver Transpl.* 10(4): 471–479.

Demetriou, A.A., Whiting, J.F., Feldman, D., Levenson, S.M., Chowdhury,N.R., Moscioni,A.D., Kram, M., Chowdhury, J.R. (1986b) Replacement of liver function in rats by transplantation of microcarrier-attached hepatocytes. *Science* 233(4769): 1190–1192.

Dich, J., Vind, C., Grunnet, N. (1988) Long-term culture of hepatocytes: effect of hormones on enzyme activities and metabolic capacity. *Hepatology* 8(1): 39–45.

Dixit, V., Arthur, M., Reinhardt, R., Gitnick, G. (1992) Improved function of microencapsulated hepatocytes in a hybrid bioartificial liver support system. *Artif. Organs* 16(4): 336–341.

Drumheller, P.D., and Hubbell, J.A. (1995) Surface immobilization of adhesion ligands for investigations of cell-substrate interactions. In Bronzino, J.D., ed., *Biomedical Engineering Handbook*. Boca Raton,, CRC. pp 1583–1596.

Dunn, J.C., Tompkins, R.G., Yarmush, M.L. (1991) Long-term in vitro function of adult hepatocytes in a collagen sandwich configuration. *Biotechnol. Prog.* 7(3): 237–245.

Dunn, J.C., Yarmush, M.L., Koebe, H.G., Tompkins, R.G. (1989). Hepatocyte function and extracellular matrix geometry: long-term culture in a sandwich configuration *FASEB J.* 3(2): 174–177 [published erratum appears in *FASEB J.* 1989 May;3(7):1873].

Dvir-Ginzberg, M., Gamlieli-Bonshtein, I., Agbaria, R., Cohen, S. (2003) Liver tissue engineering within alginate scaffolds: effects of cell-seeding density on hepatocyte viability, morphology, and function. *Tissue Eng.* 9(4): 757–766.

Elcin, Y.M., Dixit, V., Gitnick, G. (1998) Hepatocyte attachment on biodegradable modified chitosan membranes: in vitro evaluation for the development of liver organoids. *Artif. Organs* 22(10): 837–846.

Elisseeff, J., McIntosh, W., Anseth, K., Riley, S., Ragan, P., Langer, R. (2000) Photoencapsulation of chondrocytes in poly(ethylene oxide)-based semi-interpenetrating networks. *J. Biomed. Mater. Res.* 51(2): 164–171.
Enat, R., Jefferson, D.M., Ruiz-Opazo, N., Gatmaitan, Z., Leinwand, L.A., Reid, L.M. (1984) Hepatocyte proliferation in vitro: its dependence on the use of serum-free hormonally defined medium and substrata of extracellular matrix. *Proc. Natl. Acad. Sci. USA* 81(5): 1411–1415.
Enzan, H., Himeno, H., Hiroi, M., Kiyoku, H., Saibara, T., Onishi, S. (1997) Development of hepatic sinusoidal structure with special reference to the Ito cells. *Microsc. Res. Tech.* 39(4): 336–349.
Fisher, B., Szuch, P., Levine, M., Fisher, E.R. (1971). A portal blood factor as the humoral agent in liver regeneration. *Science* 171(971): 575–577.
Fodor, W.L., Williams, B.L., Matis, L.A., Madri, J.A., Rollins, S.A., Knight, J.W., Velander,W., Squinto, S.P. (1994) Expression of a functional human complement inhibitor in a transgenic pig as a model for the prevention of xenogeneic hyperacute organ rejection. *Proc. Natl. Acad. Sci. USA* 91(23): 11,153–11,157.
Fukaya, K., Asahi, S., Nagamori, S., Sakaguchi, M., Gao, C., Miyazaki, M., Namba, M. (2001) Establishment of a human hepatocyte line (OUMS-29) having CYP 1A1 and 1A2 activities from fetal liver tissue by transfection of SV40 LT. *In Vitro Cell Dev. Biol. Anim.* 37(5): 266–269.
Gebhardt, R., and Mecke, D. (1983) Heterogeneous distribution of glutamine synthetase among rat liver parenchymal cells in situ and in primary culture. *EMBO J.* 2(4): 567–570.
Ghobrial, R.M., Yersiz, H., Farmer, D.G., Amersi, F., Goss, J., Chen, P., Dawson, S., Lerner, S., Nissen, N., Imagawa, D., et al. (2000) Predictors of survival after In vivo split liver transplantation: analysis of 110 consecutive patients. *Ann. Surg.* 232(3): 312–323.
Giannelli, G., Quaranta, V., Antonaci, S. (2003) Tissue remodelling in liver diseases. *Histol. Histopathol.* 18(4): 1267–1274.
Glicklis, R., Shapiro, L., Agbaria, R., Merchuk, J.C., Cohen, S. (2000) Hepatocyte behavior within three-dimensional porous alginate scaffolds. *Biotechnol. Bioeng.* 67(3): 344–353.
Gobin, A.S., and West, J.L. (2002) Cell migration through defined, synthetic ECM analogs. *FASEB J.* 16(7): 751–753.
Gobin, A.S., and West, J.L. (2003) Effects of epidermal growth factor on fibroblast migration through biomimetic hydrogels. *Biotechnol. Prog.* 19(6): 1781–1785.
Godlewski, G., Gaubert-Cristol, R., Rouy, S., Prudhomme, M. (1997) Liver development in the rat and in man during the embryonic period (Carnegie stages 11–23). *Microsc. Res. Tech.* 39(4): 314–327.
Griffith, L.G., Wu, B., Cima, M.J., Powers, M.J., Chaignaud, B., Vacanti, J.P. (1997) In vitro organogenesis of liver tissue. *Ann. NY Acad. Sci.* 831: 382–397.
Guha, C., Sharma, A., Gupta, S., Alfieri, A., Gorla, G.R., Gagandeep, S., Sokhi, R., Roy-Chowdhury, N., Tanaka, K.E., Vikram, B., et al. (1999) Amelioration of radiation-induced liver damage in partially hepatectomized rats by hepatocyte transplantation. *Cancer Res.* 59(23): 5871–5874.
Gupta, S., Rajvanshi, P., Sokhi, R., Slehria, S., Yam, A., Kerr, A., Novikoff, P.M. (1999b) Entry and integration of transplanted hepatocytes in rat liver plates occur by disruption of hepatic sinusoidal endothelium. *Hepatology* 29(2): 509–519.
Harper, A.M., Edwards, E.B., Ellison, M.D. (2001) The OPTN waiting list, 1988–2000. *Clin. Transpl.* : 73–85.
Haruyama, T., Ajioka, I., Akaike, T., Watanabe, Y. (2000) Regulation and significance of hepatocyte-derived matrix metalloproteinases in liver remodeling. *Biochem. Biophys. Res. Commun.* 272(3): 681–686.
Hashikura, Y., Makuuchi, M., Kawasaki, S., Matsunami, H., Ikegami, T., Nakazawa, Y., Kiyosawa, K., Ichida, T. (1994). Successful living-related partial liver transplantation to an adult patient. *Lancet* 343(8907): 1233–1234.
Hasirci, V., Berthiaume, F., Bondre, S.P., Gresser, J.D., Trantolo, D.J., Toner, M., Wise, D.L. (2001) Expression of liver-specific functions by rat hepatocytes seeded in treated poly(lactic-co-glycolic) acid biodegradable foams. *Tissue Eng.* 7(4): 385–394.

Hellkamp, J., Christ, B., Bastian, H., Jungermann, K. (1991) Modulation by oxygen of the glucagon-dependent activation of the phosphoenolpyruvate carboxykinase gene in rat hepatocyte cultures. *Eur. J. Biochem.* 198(3): 635–639.
Hern, D.L., and Hubbell, J.A. (1998) Incorporation of adhesion peptides into nonadhesive hydrogels useful for tissue resurfacing. *J. Biomed. Mater. Res.* 39(2): 266–276.
Hersel, U., Dahmen, C., Kessler, H. (2003) RGD modified polymers: biomaterials for stimulated cell adhesion and beyond. *Biomaterials* 24(24): 4385–4415.
Higgins, G.M., and Anderson, R.M. (1931) Experimental pathology of the liver. I: restoration of the liver of the white rat following partial surgical removal. *Arch. Pathol.* 12: 186–202.
Holzer, C., and Maier, P. (1987) Maintenance of periportal and pericentral oxygen tensions in primary rat hepatocyte cultures: influence on cellular DNA and protein content monitored by flow cytometry. *J. Cell Physiol.* 133(2): 297–304.
Huang, L.E., Arany, Z., Livingston, D.M., Bunn, H.F. (1996) Activation of hypoxia-inducible transcription factor depends primarily upon redox-sensitive stabilization of its alpha subunit. *J. Biol. Chem.* 271(50): 32,253–32,259.
Isom, H.C., Secott, T., Georgoff, I., Woodworth, C., Mummaw, J. (1985) Maintenance of differentiated rat hepatocytes in primary culture. *Proc. Natl. Acad. Sci. USA* 82(10): 3252–3256.
Jakab, K., Neagu, A., Mironov, V., Markwald, R.R., Forgacs, G. (2004) Engineering biological structures of prescribed shape using self-assembling multicellular systems. *Proc. Natl. Acad. Sci. USA* 101(9): 2864–2869.
Jirtle, R.L., and Michalopoulos, G. (1982) Effects of partial hepatectomy on transplanted hepatocytes. *Cancer Res.* 42(8): 3000–3004.
Joseph, B., Malhi, H., Bhargava, K.K., Palestro, C.J., McCuskey, R.S., Gupta, S. (2002) Kupffer cells participate in early clearance of syngeneic hepatocytes transplanted in the rat liver. *Gastroenterology* 123(5): 1677–1685.
Jungermann, K. (1992) Role of intralobular compartmentation in hepatic metabolism. *Diabetes Metab.* 18(1): 81–86.
Jungermann, K., and Kietzmann, T. (2000) Oxygen: modulator of metabolic zonation and disease of the liver. *Hepatology* 31(2): 255–260.
Kaihara, S., Borenstein, J., Koka, R., Lalan, S., Ochoa, E.R., Ravens, M., Pien, H., Cunningham, B., Vacanti, J.P. (2000) Silicon micromachining to tissue engineer branched vascular channels for liver fabrication. *Tissue Eng.* 6(2): 105–117.
Kang, Y.H., Berthiaume, F., Nath, B.D., Yarmush, M.L. (2004) Growth factors and nonparenchymal cell conditioned media induce mitogenic responses in stable long-term adult rat hepatocyte cultures. *Exp. Cell Res.* 293(2): 239–247.
Kaplowitz, N., and DeLeve, L.D. (2003) *Drug-Induced liver Disease*. New York, Marcel Dekker, xii, 773 pp.
Katz, B.Z., Zamir, E., Bershadsky, A., Kam, Z., Yamada, K.M., Geiger, B. (2000) Physical state of the extracellular matrix regulates the structure and molecular composition of cell-matrix adhesions. *Mol. Biol. Cell.* 11(3): 1047–1060.
Kelly, J.H., and Sussman, N.L. (2000) A fluorescent cell-based assay for cytochrome P-450 isozyme 1A2 induction and inhibition. *J. Biomol. Screen.* 5(4): 249–254.
Kera, Y., Sippel, H.W., Penttila, K.E., Lindros, K.O. (1987) Acinar distribution of glutathione-dependent detoxifying enzymes. Low glutathione peroxidase activity in perivenous hepatocytes. *Biochem. Pharmacol.* 36(12): 2003–2006.
Kietzmann, T., Freimann, S., Bratke, J., Jungermann, K. (1996) Regulation of the gluconeogenic phosphoenolpyruvate carboxykinase and glycolytic aldolase A gene expression by O_2 in rat hepatocyte cultures. Involvement of hydrogen peroxide as mediator in the response to O_2. *FEBS Lett.* 388(2–3): 228–232.
Kietzmann, T., Hirsch-Ernst, K.I., Kahl, G.F., Jungermann, K. (1999) Mimicry in primary rat hepatocyte cultures of the in vivo perivenous induction by phenobarbital of cytochrome P-450 2B1 mRNA: role of epidermal growth factor and perivenous oxygen tension. *Mol. Pharmacol.* 56(1): 46–53.
Kietzmann, T., and Jungermann, K. (1997) Modulation by oxygen of zonal gene expression in liver studied in primary rat hepatocyte cultures. *Cell Biol. Toxicol.* 13(4–5): 243–255.

Kim, S.S., Sundback, C.A., Kaihara, S., Benvenuto, M.S., Kim, B.S., Mooney, D.J., Vacanti, J.P. (2000) Dynamic seeding and in vitro culture of hepatocytes in a flow perfusion system. *Tissue Eng.* 6(1): 39–44.
Kim, S.S., Utsunomiya, H., Koski, J.A., Wu, B.M., Cima, M.J., Sohn, J., Mukai, K., Griffith, L.G., Vacanti, J.P. (1998) Survival and function of hepatocytes on a novel three-dimensional synthetic biodegradable polymer scaffold with an intrinsic network of channels. *Ann. Surg.* 228(1): 8–13.
Kim, T.H., Mars, W.M., Stolz, D.B., Petersen, B.E., Michalopoulos, G.K. (1997) Extracellular matrix remodeling at the early stages of liver regeneration in the rat. *Hepatology* 26(4): 896–904.
Koh, W.G., Itle, L.J., Pishko, M.V. (2003) Molding of hydrogel microstructures to create multiphenotype cell microarrays. *Anal. Chem.* 75(21): 5783–5789.
Koh, W.G., Revzin, A., Pishko, M.V. (2002) Poly(ethylene glycol) hydrogel microstructures encapsulating living cells. *Langmuir* 18(7): 2459–2462.
Kuai, X.L., Cong, X.Q., Li, X.L., Xiao, S.D. (2003) Generation of hepatocytes from cultured mouse embryonic stem cells. *Liver Transpl.* 9(10): 1094–1099.
Kubota, H., and Reid, L.M. (2000) Clonogenic hepatoblasts, common precursors for hepatocytic and biliary lineages, are lacking classical major histocompatibility complex class I antigen. *Proc. Natl. Acad. Sci. USA* 97(22): 12132–12137.
Laconi, E., Oren, R., Mukhopadhyay, D.K., Hurston, E., Laconi, S., Pani, P., Dabeva, M.D., Shafritz, D.A. (1998) Long-term, near-total liver replacement by transplantation of isolated hepatocytes in rats treated with retrorsine. *Am. J. Pathol.* 153(1): 319–329.
Landry, J., Bernier, D., Ouellet, C., Goyette, R., Marceau, N. (1985). Spheroidal aggregate culture of rat liver cells: histotypic reorganization, biomatrix deposition, and maintenance of functional activities. *J. Cell Biol.* 101(3): 914–923.
Langer, R., and Vacanti, J.P. (1993) Tissue engineering. *Science* 260(5110): 920–926.
Ledezma, G.A., Folch, A., Bhatia, S.N., Balis, U.J., Yarmush, M.L., Toner, M. (1999) Numerical model of fluid flow and oxygen transport in a radial-flow microchannel containing hepatocytes. *J. Biomech. Eng.* 121(1): 58–64.
Lee, H., Cusick, R.A., Browne, F., Ho Kim, T., Ma, P.X., Utsunomiya, H., Langer, R., Vacanti, J.P. (2002) Local delivery of basic fibroblast growth factor increases both angiogenesis and engraftment of hepatocytes in tissue-engineered polymer devices. *Transplantation* 73(10): 1589–1593.
Lee, K.Y., Peters, M.C., Anderson, K.W., Mooney, D.J. (2000) Controlled growth factor release from synthetic extracellular matrices. *Nature* 408(6815): 998–1000.
Leffert, H.L., and Paul, D. (1972) Studies on primary cultures of differentiated fetal liver cells. *J. Cell Biol.* 52(3): 559–568.
Levenberg, S., Huang, N.F., Lavik, E., Rogers, A.B., Itskovitz-Eldor, J., Langer, R. (2003) Differentiation of human embryonic stem cells on three-dimensional polymer scaffolds. *Proc. Natl. Acad. Sci. USA* 100(22): 12741–12746.
Li, J., Pan, J., Zhang, L., Yu, Y. (2003) Culture of hepatocytes on fructose-modified chitosan scaffolds. *Biomaterials* 24(13): 2317–2322.
Lin, H.M., Kauffman, H.M., McBride, M.A., Davies, D.B., Rosendale, J.D., Smith, C.M., Edwards, E.B., Daily, O.P., Kirklin, J., Shield, C.F., et al. (1998) Center-specific graft and patient survival rates: 1997 United Network for Organ Sharing (UNOS) report [See comments]. *JAMA* 280(13): 1153–1160.
Lindros, K.O. (1997) Zonation of cytochrome P450 expression, drug metabolism and toxicity in liver. *Gen. Pharmacol.* 28(2): 191–196.
Lindros, K.O., Cai, Y.A., Penttila, K.E. (1990) Role of ethanol-inducible cytochrome P-450 IIE1 in carbon tetrachloride-induced damage to centrilobular hepatocytes from ethanol-treated rats. *Hepatology* 12(5): 1092–1097.
Liu, J., Pan, J., Naik, S., Santangini, H., Trenkler, D., Thompson, N., Rifai, A., Chowdhury, J.R., Jauregui, H.O. (1999) Characterization and evaluation of detoxification functions of a nontumorigenic immortalized porcine hepatocyte cell line (HepLiu). *Cell Transplant.* 8(3): 219–232.
Liu, V.A., and Bhatia, S.N. (2002) Three-dimensional photopatterning of hydrogels containing living cells. *Biomed. Microdevices* 4(4): 257–266.

Lom, B., Healy, K.E., Hockberger, P.E. (1993) A versatile technique for patterning biomolecules onto glass coverslips. *J. Neurosci. Methods* 50(3): 385–397.
Luo, Y., and Shoichet, M.S. (2004) A photolabile hydrogel for guided three-dimensional cell growth and migration. *Nat. Mater.* 3(4): 249–253.
Lutolf, M.P., Lauer-Fields, J.L., Schmoekel, H.G., Metters, A.T., Weber, F.E., Fields, G.B., Hubbell, J.A. (2003) Synthetic matrix metalloproteinase-sensitive hydrogels for the conduction of tissue regeneration: engineering cell-invasion characteristics. *Proc. Natl. Acad. Sci. USA* 100(9): 5413–5418.
Malhi, H., Annamaneni, P., Slehria, S., Joseph, B., Bhargava, K.K., Palestro, C.J., Novikoff, P.M., Gupta, S. (2002) Cyclophosphamide disrupts hepatic sinusoidal endothelium and improves transplanted cell engraftment in rat liver. *Hepatology* 36(1): 112–121.
Mann, B.K., Gobin, A.S., Tsai, A.T., Schmedlen, R.H., West, J.L. (2001a) Smooth muscle cell growth in photopolymerized hydrogels with cell adhesive and proteolytically degradable domains: synthetic ECM analogs for tissue engineering. *Biomaterials* 22(22): 3045–3051.
Mann, B.K., Schmedlen, R.H., West, J.L. (2001b) Tethered-TGF-beta increases extracellular matrix production of vascular smooth muscle cells. *Biomaterials* 22(5): 439–444.
Maxwell, P.H., and Ratcliffe, P.J. (2002) Oxygen sensors and angiogenesis. *Semin. Cell Dev. Biol.* 13(1): 29–37.
McClelland, R.E., MacDonald, J.M., Coger, R.N. (2003) Modeling O_2 transport within engineered hepatic devices. *Biotechnol. Bioeng.* 82(1): 12–27.
Metters, A.T., Anseth, K.S., Bowman, C.N. (2000) Fundamental studies of a novel, biodegradable PEG-*b*-PLA hydrogel. *Polymer* 41: 3993–4004.
Michalopoulos, G.K., and DeFrances, M.C. (1997) Liver regeneration. *Science* 276(5309): 60–66.
Mignon, A., Guidotti, J.E., Mitchell, C., Fabre, M., Wernet, A., De La Coste, A., Soubrane, O., Gilgenkrantz, H., Kahn, A. (1998) Selective repopulation of normal mouse liver by Fas/CD95-resistant hepatocytes. *Nat. Med.* 4(10): 1185–1188.
Mitaka, T. (1998) The current status of primary hepatocyte culture. *Int. J. Exp. Pathol.* 79(6): 393–409.
Moolten, F.L., and Bucher, N.L. (1967) Regeneration of rat liver: transfer of humoral agent by cross circulation. *Science* 158(798): 272–274.
Nauck, M., Wolfle, D., Katz, N., Jungermann, K. (1981) Modulation of the glucagon-dependent induction of phosphoenolpyruvate carboxykinase and tyrosine aminotransferase by arterial and venous oxygen concentrations in hepatocyte cultures. *Eur. J. Biochem.* 119(3): 657–661.
Newsome, P.N., Plevris, J.N., Nelson, L.J., Hayes, P.C. (2000) Animal models of fulminant hepatic failure: a critical evaluation. *Liver Transpl.* 6(1): 21–31.
Nuttelman, C.R., Tripodi, M.C., Anseth, K.S. (2004) In vitro osteogenic differentiation of human mesenchymal stem cells photoencapsulated in PEG hydrogels. *J. Biomed. Mater. Res.* 68A(4): 773–782.
Oh, S.H., Hatch, H.M., Petersen, B.E. (2002). Hepatic oval "stem" cell in liver regeneration. *Semin. Cell Dev. Biol.* 13(6): 405–409.
Oinonen, T., and Lindros, K.O. (1998) Zonation of hepatic cytochrome P-450 expression and regulation. *Biochem. J.* 329 (Pt 1): 17–35.
Olson, M., Mancini, M., Venkatachalam, M., Roy, A. (1990) Hepatocyte cytodifferentiation and cell-to-cell communication. In: De Mello, W., ed., *Cell intercommunication*. Boca Raton, FL, CRC Press. pp. 71–92.
Overturf, K., al-Dhalimy, M., Ou, C.N., Finegold, M., Grompe, M. (1997) Serial transplantation reveals the stem-cell-like regenerative potential of adult mouse hepatocytes. *Am. J. Pathol.* 151(5): 1273–1280.
Palmes, D., and Spiegel, H.U. (2004) Animal models of liver regeneration. *Biomaterials* 25(9): 1601–1611.
Pathak, C.P., Sawhney, A.S., Hubbell, J.A. (1992) Rapid photopolymerization of immunoprotective gels in contact with cells and tissue. *J. Am. Chem. Soc.* 114(21): 8311–8312.
Patzer, J.F, 2nd. (2004) Oxygen consumption in a hollow fiber bioartificial liver—revisited. *Artif. Organs* 28(1): 83–98.
Peppas, N.A., Bures, P., Leobandung, W., Ichikawa, H. (2000) Hydrogels in pharmaceutical formulations. *Eur. J. Pharm. Biopharm.* 50(1): 27–46.

Petersen, B.E., Goff, J.P., Greenberger, J.S., Michalopoulos, G.K. (1998) Hepatic oval cells express the hematopoietic stem cell marker Thy-1 in the rat. *Hepatology* 27(2): 433–445.
Pollok, J.M., Kluth, D., Cusick, R.A., Lee, H., Utsunomiya, H., Ma, P.X., Langer, R., Broelsch, C.E., Vacanti, J.P. (1998) Formation of spheroidal aggregates of hepatocytes on biodegradable polymers under continuous-flow bioreactor conditions. *Eur. J. Pediatr. Surg.* 8(4): 195–199.
Powers, M.J., Domansky, K., Kaazempur-Mofrad, M.R., Kalezi, A., Capitano, A., Upadhyaya, A., Kurzawski, P., Wack, K.E., Stolz, D.B., Kamm, R. et al. (2002a) A microfabricated array bioreactor for perfused 3D liver culture. *Biotechnol. Bioeng.* 78(3): 257–269.
Powers, M.J., and Griffith, L.G. (1998) Adhesion-guided in vitro morphogenesis in pure and mixed cell cultures. *Microsc. Res. Technique* 43(N5): 379–384.
Powers, M.J., Janigian, D.M., Wack, K.E., Baker, C.S., Stolz, D.B., Griffith, L.G. (2002b) Functional behavior of primary rat liver cells in a three-dimensional perfused microarray bioreactor. *Tissue Eng.* 8(3): 499–513.
Poznansky, M.C., Evans, R.H., Foxall, R.B., Olszak, I.T., Piascik, A.H., Hartman, K.E., Brander, C., Meyer, T.H., Pykett, M.J., Chabner, K.T., et al. (2000) Efficient generation of human T cells from a tissue-engineered thymic organoid. *Nat. Biotechnol.* 18(7): 729–734.
Pronko, P.S., Jarvelainen, H.A., Lindros, K.O. (2002) Acinar distribution of rat liver arylamine *N*-acetyltransferase: effect of chronic ethanol and endotoxin exposure. *Pharmacol. Toxicol.* 90(3): 150–154.
Raia, S., Nery, J.R., Mies, S. (1989) Liver transplantation from live donors. *Lancet* 2(8661): 497.
Ranucci, C.S., Kumar, A., Batra, S.P., Moghe, P.V. (2000) Control of hepatocyte function on collagen foams: sizing matrix pores toward selective induction of 2-D and 3-D cellular morphogenesis. *Biomaterials* 21(8): 783–793.
Reid, L.M., Fiorino, A.S., Sigal, S.H., Brill, S., Holst, P.A. (1992) Extracellular matrix gradients in the space of Disse: relevance to liver biology [editorial]. *Hepatology* 15(6): 1198–1203.
Reid, L.M., Gaitmaitan, Z., Arias, I., Ponce, P., Rojkind, M. (1980) Long-term cultures of normal rat hepatocytes on liver biomatrix. *Ann. NY Acad. Sci.* 349: 70–76.
Rhim, J.A., Sandgren, E.P., Degen, J.L., Palmiter, R.D., Brinster, R.L. (1994) Replacement of diseased mouse liver by hepatic cell transplantation. *Science* 263(5150): 1149–1152.
Richardson, T.P., Peters, M.C., Ennett, A.B., Mooney, D.J. (2001) Polymeric system for dual growth factor delivery. *Nat. Biotechnol.* 19(11): 1029–1034.
Risbud, M.V., Karamuk, E., Schlosser, V., Mayer, J. (2003). Hydrogel-coated textile scaffolds as candidate in liver tissue engineering: II. Evaluation of spheroid formation and viability of hepatocytes. *J. Biomater. Sci. Polym. Ed.* 14(7): 719–731.
Roesel, J., Rigsby, D., Bailey, A., Alvarez, R., Sanchez, J.D., Campbell, V., Shrestha, K., Miller, D.M. (1989) Stimulation of protooncogene expression by partial hepatectomy is not tissue-specific. *Oncogene Res.* 5(2): 129–136.
Rojkind, M., Gatmaitan, Z., Mackensen, S., Giambrone, M.A., Ponce, P., Reid, L.M. (1980) Connective tissue biomatrix: its isolation and utilization for long-term cultures of normal rat hepatocytes. *J. Cell Biol.* 87(1): 255–263.
Rotem, A., Toner, M., Bhatia, S., Foy, B.D., Tompkins, R.G., Yarmush, M.L. (1994) Oxygen is a factor determining in vitro tissue assembly: Effects on attachment and spreading of hepatocytes. *Biotechnol. Bioengin.* 43(7): 654–660.
Rudolph, K.L., Trautwein, C., Kubicka, S., Rakemann, T., Bahr, M.J., Sedlaczek, N., Schuppan, D., Manns, M.P. (1999) Differential regulation of extracellular matrix synthesis during liver regeneration after partial hepatectomy in rats. *Hepatology* 30(5): 1159–1166.
Ruhnke, M., Ungefroren, H., Zehle, G., Bader, M., Kremer, B., Fandrich, F. (2003) Long-term culture and differentiation of rat embryonic stem cell-like cells into neuronal, glial, endothelial, and hepatic lineages. *Stem Cells* 21(4): 428–436.
Saito, S., Sakagami, K., Matsuno, T., Tanakaya, K., Takaishi, Y., Orita, K. (1992) Long-term survival and proliferation of spheroidal aggregate cultured hepatocytes transplanted into the rat spleen. *Transplant. Proc.* 24(N4): 1520–1521.
Sawhney, A.S., Pathak, C.P., Hubbell, J.A. (1993) Bioerodible hydrogels based on photopolymerized poly(ethylene glycol)-co-poly(alpha-hydroxy acid) diacrylate macromers. *Macromolecules* 26(4): 581–587.

Schiano, T.D., Kim-Schluger, L., Gondolesi, G., Miller, C.M. (2001) Adult living donor liver transplantation: the hepatologist's perspective. *Hepatology* 33(1): 3–9.
Schmoeckel, M., Bhatti, F.N., Zaidi, A., Cozzi, E., Pino-Chavez, G., Dunning, J.J., Wallwork, J., White, D.J. (1997). Xenotransplantation of pig organs transgenic for human DAF: an update. *Transplant. Proc.* 29(7): 3157–3158.
Schwartz, R.E., Reyes, M., Koodie, L., Jiang, Y., Blackstad, M., Lund, T., Lenvik, T., Johnson, S., Hu, W.S., Verfaillie, C.M. (2002) Multipotent adult progenitor cells from bone marrow differentiate into functional hepatocyte-like cells. *J. Clin. Invest.* 109(10): 1291–1302.
Seglen, P.O. (1976) Preparation of isolated rat liver cells. *Methods Cell Biol.* 13: 29–83.
Selden, C., Khalil, M., Hodgson, H.J.F. (1999) What keeps hepatocytes on the straight and narrow? Maintaining differentiated function in the liver. *Gut* 44(4): 443–446.
Seliktar, D., Zisch, A.H., Lutolf, M.P., Wrana, J.L., Hubbell, J.A. (2004) MMP-2 sensitive, VEGF-bearing bioactive hydrogels for promotion of vascular healing. *J. Biomed. Mater. Res.* 68A(4): 704–716.
Sell, S. (2001) The role of progenitor cells in repair of liver injury and in liver transplantation. *Wound Repair Regen.* 9(6): 467–482.
Shakil, A.O., Kramer, D., Mazariegos, G.V., Fung, J.J., Rakela, J. (2000) Acute liver failure: clinical features, outcome analysis, and applicability of prognostic criteria. *Liver Transpl.* 6(2): 163–169.
Shin, H., Jo, S., Mikos, A.G. (2003) Biomimetic materials for tissue engineering. *Biomaterials* 24(24): 4353–4364.
Slehria, S., Rajvanshi, P., Ito, Y., Sokhi, R.P., Bhargava, K.K., Palestro, C.J., McCuskey, R.S., Gupta, S. (2002) Hepatic sinusoidal vasodilators improve transplanted cell engraftment and ameliorate microcirculatory perturbations in the liver. *Hepatology* 35(6): 1320–1328.
Smith, M.K., Peters, M.C., Richardson, T.P., Garbern, J.C., Mooney, D.J. (2004) Locally enhanced angiogenesis promotes transplanted cell survival. *Tissue Eng.* 10(1–2): 63–71.
Sokhi, R.P., Rajvanshi, P., Gupta, S. (2000) Transplanted reporter cells help in defining onset of hepatocyte proliferation during the life of F344 rats. *Am. J. Physiol. Gastrointest. Liver Physiol.* 279(3): G631–G640.
Strain, A.J., and Neuberger, J.M. (2002) A bioartificial liver—state of the art. *Science* 295(5557): 1005–1009.
Suleiman, S.A., and Stevens, J.B. (1987) The effect of oxygen tension on rat hepatocytes in short-term culture. *In Vitro Cell Dev. Biol.* 23(5): 332–328.
Tan, W., and Desai, T.A. (2003) Microfluidic patterning of cellular biopolymer matrices for biomimetic 3D structures. *Biomed. Microdevices* 5: 235–244.
Terblanche, J., and Hickman, R. (1991) Animal models of fulminant hepatic failure. *Dig. Dis. Sci.* 36(6): 770–774.
Tilles, A.W., Baskaran, H., Roy, P., Yarmush, M.L., Toner, M. 2001. Effects of oxygenation and flow on the viability and function of rat hepatocytes cocultured in a microchannel flat-plate bioreactor. *Biotechnol. Bioeng.* V73(N5): 379–389.
van de Kerkhove, M.P., Hoekstra, R., van Gulik, T.M., Chamuleau, R.A. (2004) Large animal models of fulminant hepatic failure in artificial and bioartificial liver support research. *Biomaterials* 25(9): 1613–1625.
Wang, F., Weaver, V.M., Petersen, O.W., Larabell, C.A., Dedhar, S., Briand, P., Lupu, R., Bissell, M.J. (1998) Reciprocal interactions between beta1-integrin and epidermal growth factor receptor in three-dimensional basement membrane breast cultures: a different perspective in epithelial biology. *Proc. Natl. Acad. Sci. USA* 95(25): 14,821–14,826.
Wang, X., Foster, M., Al-Dhalimy, M., Lagasse, E., Finegold, M., Grompe, M. (2003) The origin and liver repopulating capacity of murine oval cells. *Proc. Natl. Acad. Sci. USA* 100 Suppl 1: 11,881–11,888.
Ward, J.H., Bashir, R., Peppas, N.A. (2001) Micropatterning of biomedical polymer surfaces by novel UV polymerization techniques. *J. Biomed. Mater. Res.* 56(3): 351–360.
Yang, L., Li, S., Hatch, H., Ahrens, K., Cornelius, J.G., Petersen, B.E., Peck, A.B. (2002) In vitro trans-differentiation of adult hepatic stem cells into pancreatic endocrine hormone-producing cells. *Proc. Natl. Acad. Sci. USA* 99(12): 8078–8083.

Yang, T.H., Miyoshi, H., Ohshima, N. (2001) Novel cell immobilization method utilizing centrifugal force to achieve high-density hepatocyte culture in porous scaffold. *J. Biomed. Mater. Res.* 55(3): 379–386.
Yarmush, M.L., Dunn, J.C., Tompkins, R.G. (1992) Assessment of artificial liver support technology. *Cell Transplant.* 1(5): 323–341.
Yu, T., and Ober, C.K. (2003) Methods for the topographical patterning and patterned surface modification of hydrogels based on hydroxyethyl methacrylate. *Biomacromolecules* 4(5): 1126–1131.
Zisch, A.H., Lutolf, M.P., Ehrbar, M., Raeber, G.P., Rizzi, S.C., Davies, N., Schmokel, H., Bezuidenhout, D., Djonov, V., Zilla, P., et al. (2003) Cell-demanded release of VEGF from synthetic, biointeractive cell ingrowth matrices for vascularized tissue growth. *FASEB J.* 17(15): 2260–2262.

Suppliers List

Abbott Laboratories
5440 Patrick Henry Dr., Santa Clara, CA 95054
Abbott Park, Illinois, IL 60064
Phone: 408 982 4800; 800 323 9100

Advanced Tissue Sciences
10933 North Torrey Pines Rd., La Jolla,
CA 92037-1005
Phone: 619 450 5730
Fax: 619 450 5703

Air Products
7201 Hamilton Blvd., Allentown, PA 18195
Phone: 610 481 4911; 800 654 4567
Fax: 800 880 5204
Web site: www.airproducts.com

Air Sea Atlanta
Atlanta, GA
Phone: 404 351 8600
Web site: www.airseaatlanta.com

Aire Liquide
Parc Gustave Eiffel, 8 rue Gutenberg, Bussy, Saint Georges 77607, Marme-la-Valle, Cedex 3, France
Phone: +33 1 64 76 15 00
Fax: +33 1 64 76 16 99

Albany International
Albany International Research Co.,
PO Box 9114, Mansfield, MA 02048–9114
Phone: 508 339 7300; 800 992 5017
Web site: www.airesco.com

Aldrich
1001 W. St. Paul Ave., Milwaukee, WI 53233
Phone: 414 273 3850; 800 558 9160
Fax: 414 273 4979
Web site: www.sigma.sial.com/aldrich

American Fluoroseal
431-D East Diamond Ave., Gaithersburg, MD 20877
Phone: 301 990 1407
Fax: 301 990 1472
E-mail: info@americanfluoroseal.com
Web site: www.toafc.com/

Amicon
72 Cherry Hill Dr., Beverly, MA
Phone: 508 777 3622; 800 426 4266
Fax: 508 777 6204

Appleton Electronics
205 W. Wisconsin Ave., Appleton, WI 54911
Phone: 800 877 8919
Fax: 920 734 5172
Web site: www.aedwis.com

Applied Biosciences
P.O. Box 520518, Salt Lake City, UT 84152
Phone: 800 280 7852; 801 485 4988
Fax: 801 485 4987
Web site: www.applied-biosciences.com

Applied Biosystems
850 Lincoln Centre Drive, Foster City, CA 94404
Phone: 800 327 3002; 650 638 5800
Fax: 650 638 5884
Web site: www.appliedbiosystems.com

ATCC
10801 University Boulevard, Manassas, VA
20110 2209; PO Box 1549, Manassas, VA 20108
Phone: 703 365 2700

Culture of Cells for Tissue Engineering, edited by Gordana Vunjak-Novakovic and R. Ian Freshney

Fax: 703 365 2701
Web site: www.atcc.org

Atlas Clean Air
Lomeshaye Business Village, Turner Road, Nelson, Lancashire BB9 7DR, UK
Phone: +44 (0)1282 447 666
Fax: +44 (0)1282 447 789
Web site: www.atlascleanair.com

Badger Air Brush Co.
9128 W. Belmont Ave, Franklin Park, IL 60131
Phone: 800 AIR-BRUSH; 800 247 2787; 847 678 3104
Fax: 847 671 4352
Web site: www.badger-airbrush.com

Baker Co. Inc.
Old Sandford Airport Rd., PO Drawer E, Sanford, ME 04073
Phone: 207 324 8773; 800 992 2537
Fax: 207 324 3869
Web site: www.bakerco.com

Barnstead Thermolyne Corporation
PO Box 797, Dubuque, IA 52004-0797
Phone: 319 589 0538
Fax: 319 589 0530
Web site: www.barnsteadthermolyne.com/

Baxter HealthCare
One Baxter Parkway Deerfield, IL 60015-4625
Phone: 847 948 2000; 800 422 9837; 800-4Baxter; 847 948 4770
Fax: 847 948 3642
Web site: www.baxter.com

Bayer
Diagnostics Division, 511 Benedict Ave, Tarrytown, NY 10591
Phone: 800 431 1970
Fax: 914 524 3978
Web site: www.bayermhc.com

BD Biosciences
1 Becton Drive, Franklin Lakes, NJ 07417-1886
Phone: 201 847 4222; 888 237 2762
E-mail: mail@bdl.com
Web site: www.bdbiosvciences.com

See also Fisher Scientific

Beckman-Coulter
4300 N. Harbor Blvd., Box 3100, Fullerton, CA 92834-3100
Phone: 800 742 2345; 714 871 4848
Fax: 800 643 4366; 714 773 8283
Web site: www.beckmancoulter.com/

Becton Dickinson
See BD Biosciences

Bellco Glass
340 Edrudo Rd., Vineland, NJ 08360-3493
Phone: 609 691 1075, 800 257 7043
Fax: 609 691 3247
Web site: www.bellcoglass.com

Biochrom AG
POB 46 03 09 D-12213 Berlin, Germany
Phone: +49 30 77 99 06 0
Fax: +49 30 77 10 01 2; +49 30 77 99 06 66
Web site: www.biochrom.de

Biodesign International
60 Industrial Park Road Saco, ME 04072
Phone: 207 283 6500; 888 530 0140
Fax: 207 283 4800
E-mail: info@biodesign.com
Web site: www.biodesign.com

Biofluids
See Biosource International

Biosource International
542 Flynn Road, Camarillo, CA 93012
Phone: 800 242 0607
Web site: www.biosource.com

Biovest International
8500 Evergreen Boulevard, Minneapolis, MN 55433
Phone: 763 786 0302
Fax: 763 786 0915
Web site: www.biovest.com

Cambrex
8830 Biggs Ford Road, PO Box 127, Walkersville, MD 21798
Phone: 301 898 7025; 800 638 8174
Fax: 301 845 8338
Email: biotechserv@cambrex.com
Web site: www.cambrex.com

Cambridge Biosciences
24-25 Signet Court, Newmarket Road, Cambridge, CB5 8LA, UK
Phone: +44 (0)1223 316 855
Fax: +44 (0)1223 360 732
E-mail: Tech@cbio.co.uk
Web site: www.bioscience.co.uk

Campden Instruments
Leicester, UK
Phone: 0870 2403702
E-mail: UKsales@campdeninstruments.com
Web site: www.campden-inst.com/

Lafayette, Indiana
Phone: 765 423 1505
E-mail: USsales@campdeninstruments.com

Carolina Biological Supplies
2700 York Road, Burlington, NC 27215-3398
Phone: 800 334 5551
Fax: 800 222 7112
Web site: www.carolina.com/

Cascade Biologicals
1341 Custer Drive, Portland, OR 97219
Phone: 800 778 4770
Fax: 503 292 9521
E-mail: info@cascadebio.com
Web site: www.cascadebio.com

Cellco
See Spectrum Cellco

Cellgro
See Fisher Scientific and MediaTech

Cellmark
See Orchid Cellmark

Cellmark UK
Cellmark, PO Box 265, Abingdon, Oxfordshire, OX14 1YX, UK
Phone: +44 (0)1235 528000
Web site: cellmark@orchid.co.uk

Charles River Laboratories
251 Ballardvale St., Wilmington, MA 01887
Phone: 508 658 6000, 800 LABORATORY RATS
Fax: 508 658 7132
Web site: www.criver.com

Chart Biomed
Phone: 952 882 5030; 888 683 2796
Fax: 800 232 9683
Web site: www.chartbiomed.com/

Chondrex Inc
2607 151st Place NE, Redmond, WA 98052
Phone: 425 702 6365 or 888 CHONDRE
E-mail: info@chondrex.com
Web site: www.chondrex.com

Ciba Specialty Chemicals Corporation
540 White Plains Road, P.O. Box 2005, Tarrytown, NY 10591-9005
Phone: 914 785 2000
Fax: 914 785 2211
Web site: www.cibasc.com

Cin-Made Corp
1780 Dreman Ave., Cincinnati, OH 45223
Phone: 513 681 3600
Fax: 513 541 5945
Web site: www.cin-madepackaginggroup.com

Clonetics
See Cambrex

Cole Parmer
625 E. Bunker Ct., Vernon Hills, IL 60061
Phone: 847 549 7600; 800 323 4340
Fax: 847 549 7676
Web site: www.coleparmer.com/

Corning
45 Nagog Park, Acton, MA
Phone: 978 635 2200; 800 492 1110
Fax: 978 635 2476
E-mail: ccwebmail@corning.com
Web site: www.corning.com/lifesciences

See also Fisher Scientific

Costar
See Corning

Coulter
See Beckman Coulter

Cryo-Med
51529 Birch St., New Baltimore, MI 48047
Phone: 313 725 4614
Fax: 313 725 7501

CVS GmbH
Holtsdamm 51, 20099 Hamburg, Germany
Phone: +49 40 2866 9900
Fax: +49 40 2866 9901
Web site: http://CVSciences.de

Daigger
620 Lakeview Parkway, Vernon Hills, IL 60061
Phone: 800 621 7193
Fax: 800 320 7200
E-mail: daigger@daigger.com
Web site: www.daigger.com

Dako
6392 Via Real, Carpinteria, CA 93013
Phone: 805 566 6655; 800 235 5763
Fax: 805 566 6688
Web site: www.dakousa.com/

Damon IEC
300 Second Ave., Needham Heights, MA 02194
Phone: 781 449 8060; 800 843 1113
Fax: 781 444 6743

See also Thermo Life Sciences

Davol Inc.
100 Sockanossett Crossroads, Cranston, RI
Phone: 800 556 6756
Fax: 401 946 5379
E-mail: info@davol.com
Web site: www.davol.com

Difco
See B-D Biosciences

Dow Corning
PO Box 0994, Midland, MI 48686-0994
Phone: 517 496 4000
Fax: 517 496 4586
Web site: www.dowcorning.com

ECACC
CAMR Div., Porton Down, Wilts, Salisbury SP4 0JG, UK
Phone: +44 (0)1980 612512
Fax: +44 (0)1980 611315
Web site: www.ecacc.org.uk

EMS
1560 Industry Road, Hatfield, PA 19440
Phone: 215 412 8400
Fax: 215 412 8450
E-mail: sgkcck@aol.com
Web site: www.emsdiasum.com/ems

Endotronics
See Biovest International

Falcon
See BD Biosciences

Fisher Scientific
2000 Park Ln., Pittsburgh, PA 15275
Phone: 412 490 8300; 800 766 7000
Fax: 800 926 1166
Web site: www.fisher1.com

Fison Instruments
See Fisher

Fougera
Melville, NY
Web site: www.fougera.com

GE Silicones
World Headquarters, Wilton, CT 06897
Phone: 800 255 8886
Web site: www.gesilicones.com

Gibco
See Invitrogen

Glowmark Systems
Upper Saddle River, NJ

Gow-Mac
277 Broadhead Rd. Bethlehem, PA 18017
Phone: 610 954 9000
Fax: 610 954 0599
Web site: http://www.gow-mac.com/

G.Q.F. Manufacturing Co.
P.O. Box 1552, Savannah, GA 31402-1552
Phone: 912 236 0651
Fax: 912 234 9978
E-mail: sales@gqfmfg.com
Web site: www.gqfmfg.com

Harbor Bio-Products
PO Box 464, Norwood, MA 02062
Phone: 781 344 9945
Fax: 781 341 1451

Harvard Apparatus
84 October Hill Road Holliston, MA 01746
Phone: 508 893 8999; 800 272 2775
Fax: 508 429 5732
E-mail: bioscience@harvardapparatus.com
Web site: www.harvardapparatus.com

Henry Schein
Phone: 800 711 6032; 631 843 5500 × 5117
Fax: 800 329 9109
E-mail: custserv@henryschein.com
Web site: www.henryschein.com/

Heraeus
See Vivascience AG

Heto-Holten
Gydevang 17–19, DK-3450 Allerod, Denmark
Phone: +45 48 16 62 00
Fax: +45 48 16 62 97
E-mail: info@heto-holten.com
Web site: www.heto-holten.com

Hoffmann-La Roche Ltd
Diagnostics Division, Grenzacherstrasse 124, CH-4070 Basel, Switzerland
Phone: +41 61 688 1111
Fax: +41 61 691 9391

Hotpack
See SP Industries Co

Hyclone
925 West 1800 South, Logan, UT 84321
Phone: 800 492 5663
Fax: 800 533 9450
Web site: www.hyclone.com

ICN
See MP Biomedicals

IEC (Hotpak)
10940 Dutton Rd. Philadelphia, PA 19154-3286
Phone: 215 824 1700; 800 460 7225
Fax: 215 637 0519
Web site: www.hotpack.com

Ingenieurbüro Jäckel
Hanau, Germany
Phone: +49 6181 78577
Fax: +49 6181 740368
Web site: gjaeckel@onlinehome.de

Instron
100 Royall Street, Canton, MA, 02021
Phone: 800 564 8378; 781 828 2500
Web site: www.instron.com

Integra Biosciences
Schonbühlstr. 8, CH—7000, Chur, Switzerland
Phone: +41 (0)81 286 95 30
Fax: +41 (0)81 286 95 33
Web site: info@integra-biosciences.com
E-mail: www.integra-biosciences.com

Interpore Inc.
181 Technology Drive Irvine, CA 92618-2402
Phone: 949 453 3200
Fax: 949 453 3225

Invitrogen
3175 Staley Road, Grand Island, NY 14072
Phone: 301 610 8718
Fax: 301 610 8686
Web site: www.invitrogen.com

Invitrogen
1600 Faraday Ave, Carlsbad CA
Phone: 800 955 6288
Fax: 750 603 7201
E-mail: tech_service@invitrogen.com
Web site: www.invitrogen.com

ISP Alginate
San Diego, CA
Web site: http://www.ispcorp.com/

Jackson ImmunoResearch
P.O. Box 9, 872 West Baltimore Pike, West Grove, PA 19390
Phone: 1800 FOR JAXN (367 5296); 610 869 4024
Fax: 610 869 0171
Web site: www.jacksonimmuno.com/

JEOL
11 Dearborn Road, Peabody, MA 01960
Phone: 978 535 5900
Fax: 978 536 2205
Web site: www.jeol.com

Kimble/Kontes
Vineland, NJ
Phone: 888 546 2531 Extension 1
Fax: 856 794 9762
E-mail: cs@kimkon.com
Web site: www.kimble-kontes.com/

Laboratory Impex Systems Ltd
Impex House, 15 Riverside Park, Wimborne, Dorset, BH21 1QU, UK
Phone: +44 (0)1202 840685
Fax: +44 (0)1202 840701
Web site: www.lab-impex-systems.co.uk

Lab-Line
See Barnstead Thermolyne Corporation

Leica
111 Deer Lake Rd., Deerfield, IL 60015
Phone: 847 405 0123; 800 248 0123
Fax: 847 405 0147
Web site: www.leica.com/

Leica Microsystems
Ernst Leitz Strasse, PO Box 2040, W 35530, Wetzlar-1, Germany
Phone: +49 64 41 290
Fax: +49 64 41 29 25 99
Web site: www.leica.com/

LGC Promochem
Queens Road, Teddington, Middlesex, TW11 0LY, England, UK
Phone: +44 (0)20 8943 7000; +44 (0)20 8943 8489
Fax: +44 (0)20 8943 2767; +44 (0)20 8943 8405
Web site: www.lgcpromochem.com

Life Technologies
See Invitrogen

Inamed Aesthetics
5540 Ekwill Street Santa Barbara, CA 93111
Phone: 805 683 6761
Fax: 805 967 5839
Web site: http://www.inamed.com

MediaTech
13884 Park Center Road, Herndon, VA 20171
Phone: 1800 cellgro (235 5476)
Web site: www.cellgro.com/

Medical Air Technology
Airology Centre, Mars Street, Oldham OL9 6LY, England, UK
Phone: +44 (0)161 621 6200
Fax: +44 (0)161 624 7547
Web site: www.medicalairtechnology.com/

Merck KGaA
Frankfurter Strasse 250, Postfach 4119, D-6100 Darmstadt, Germany
Phone: +49 61 51720
Fax: +49 61 5172 2000
Web site: www.merck.de

Merck Inc
PO Box 2000, RY7-220, Rahway, NJ 07065
Phone: 908 594 4600; 800 672 6372
Fax: 908 388 9778
Web site: www.merck.com/

Millipore Corp
80 Ashby Rd., Bedford, MA 01730
Phone: 781 533 6000; 800 645 5476
Fax: 617 275 5550; 800 645 5439
Web site: www.millipore.com/

Miltenyi Biotec
12740 Earhart Avenue, Auburn, CA 95602
Phone: 530 888 8871; 800 FOR MACS
Fax: 530 888 8925
E-mail: macs@miltenyibiotec.com
Web site: www.MiltenyiBiotec.com

Miltenyi Biotec GmbH
Friedrich-Ebert-Strasse 68, 51429 Bergisch Gladbach, Germany
Phone: +49 2204 83060
Fax: +49 2204 85197
E-mail: macs@miltenyibiotec.de
Web site: www.MiltenyiBiotec.com

Molecular Devices, Inc.,
MaxLine Div., 1311Orleans Dr., Sunnyvale, CA 94089
Phone: 408 747 1700; 800 635 5577
Fax: 408 747 3602
E-mail: info@moldev.com
Web site: www.moleculardevices.com

Molecular Probes
P.O. Box 22010, Eugene, OR 97402-0469
Phone: 541 465 8338
Fax: 541 344 6504
E-mail: order@probes.com
Web site: www.probes.com

MP Biomedicals
15 Morgan, Irvine, CA 92618-2005
Phone: 800 633 1352, 949 833 2500
Fax: 949 859 5989
E-mail: custserv@mpbio.com; sales@mpbio.com
Web site: www.icnbiomed.com

MVE Inc.
Biological Products Div., Two Appletree Sq., Suite 100, Bloomington, MN 55425
Phone: 612 758 4484; 888 683 2796
Fax: 612 853 9661

See also Chart Biomed

Nalge Nunc International
P.O. Box 20365, Rochester, NY 14602-0365
Phone: 716 264 3898
Fax: 716 264 3706
E-mail: intlmktg@nalgenunc.com
Web site: www.nuncbrand.com

See also Fisher Scientific

National Chemicals
1259 Seaboard Industrial Blvd., Atlanta, GA 30318
Phone: 800 237 0263
Web site: www.natchem.com/

New Brunswick
Box 4005, 44 Talmadge Road, Edison, NJ 08818-4005
Phone: 908 287 1200
Fax: 908 287 4222
Web site: www.nbsc.com/

Nikon, Inc
1300 Walt Whitman Road, Melville, NY 11747-3064
Phone: 516 547 8500
Fax: 516 547 0306
Web site: www.nikonusa.com

Nikon Europe B.V.
P.O. Box 222, 1170 AE Badhoevedorp, The Netherlands
Phone: +31 20 4496 222
Fax: +31 20 4496 298
Web site: www.nikon.co.jp/inst/

NLS Animal Health
Phone: 800 638 8620; 888 568 2825
E-mail: webadmin@nlsanimalhealth.com.
Web site: www.nlsanimalhealth.com

NuAire
2100 Fernbrook Lane, Plymouth, MN 55447
Phone: 763 553 1270; 800 328 3352
Fax: 763 553 0459
E-mail: nuaire@nuaire.com
Web site: www.nuaire.com

Nunc
See Nalge-Nunc and Fisher Scientific

Nusil Silicone Technology
2 Enterprise, Apt. 10214, Aliso Viejo, CA 92656
Phone: 949 716 6667; 949 933 7700; 805 566 4132
Fax: 949 716 6668
E-mail: ChuckM@nusil.com
Web site: www.nusil.com

Olympus America Inc.
Precision Instrument Div., Two Corporate Ctr. Dr., Melville, NY 11747-3157
Phone: 516 844 5000; 800 446 5967
Fax: 516 844 5112
E-mail: olympus@performark.com
Web site: www.olympus.com

Olympus Optical Co., (Europe) GmbH
Wendenstrasse 14–18, D-20097 Hamburg, Germany
Phone: +49 40 23 77 30
Fax: +49 40 23 77 36 47

E-mail: main@olympus.uk.com
Web site: www.olympus-europa.com

Orchid Cellmark
20271 Goldenrod Lane, Suite 120, Germantown, MD 20876
Phone: 301 428 4980; 800 872 5227
Fax: 301 428 4877
Web site: www.orchidcellmark.com

Pierce Chemical Co
3747 N. Meridan Rd., PO Box 117, Rockford, IL 61105
Phone: 815 968 0747; 800 874 3723
Fax: 815 968 7316
Web site: www.piercenet.com

Polymun Scientific
Nussdorfer Lande 11, 1190 Vienna, Austria
Phone: +43 1 36006 6202
Fax: +43 1 369 7615
E-mail: office@polymun.com
Web site: www.polymun.com

Precision Scientific
A Div. of Jouan Inc., 110-C Industrial Dr., Winchester, VA 22602
Phone: 540 869 9892; 800 621 8820
Fax: 540 869 0130
Web site: www.precisionsci.com/

PromoCell
Sickingenstrasse 63/65, D-69126 Heidelberg, Germany
Phone: +49 6221 64934 0; 0800 776 66 23
Fax: +49 6221 64934 40; 0800 100 83 06
E-mail: info@promocell.com
Web site: www.promocell.com/

Qiagen, GmbH
QIAGEN Strasse 1, 40724 Hilden, Germany
Phone: +49 (0)2103 29 12000
Fax: +49 (0)2103 29 22000

Qiagen Inc
27220 Turnberry Lane, Valencia, CA 91355
Phone: 800 426 8157; 800 DNA PREP 800 362 7737
Fax: 800 718 2056
Web site: www1.qiagen.com/

R&D Systems Inc.
614 McKinley Place N.E., Minneapolis, MN 55413
Phone: 612 379 2956; 800 343 7475
Fax: 612 656 4400
E-mail: info@RnDSystems.com
Web site: http://www.rndsystems.com/

Research Diagnostics Inc.
Research Diagnostics Inc., Pleasant Hill Road, Flanders NJ 07836
Phone: 973 584 7093; 800 631 9384
Fax: 973 584 0210
E-mail: ResearchD@aol.com

Roche Laboratories
340 Kingsland St., Nutley, NJ 07110
Phone: 201 235 5000
Fax: 201 562 2739

Ross Laboratories
Ross Consumer Relations, 625 Cleveland Avenue, Columbus, OH 43215-1724
Phone: 800 986 8510
Web site: www.ross.com/

Resolution Performance Products
1600 Smith Street, 24th Floor, P.O. Box 4500, Houston, TX 77210-4500
Phone: 877 859 2800
Web site: http://www.resins.com

Rudolph-Desco Co.
580 Sylvan Ave., PO Box 1245, Englewood Cliffs, NJ 07410
Phone: 201 568 4920
Fax: 201 568 0971

Saf-T-Pak
10807 - 182 Street, Edmonton, Alberta, Canada T5S 1J5
Phone: 780 486 0211; 800 814 7484
Fax: 780 486 0235; 888 814 7484
Web site: www.saftpak.com/

SANYO Scientific
900 N. Arlington Heights Rd., Itasca, IL 60143
Phone: 630 875 3530; 800 858 8442
Fax: 630 775 0044

See also Southeastern Scientific

Schärfe Systems
Krammerstrasse 22, D-72764, Reutlingen, Germany
Phone: +49 (0)7121 387 86 0
Fax: +49 (0)7121 387 86 99
E-mail: mail@CASY-Technology.com
Web site: www.CASY-Technology.com

Scientific Tools
Foster City, CA

Scion
Frederick, MD

Shandon Lipshaw
171 Industry Dr., Pittsburgh, PA 15275
Phone: 412 788 1133; 800 547 7429

Fax: 412 788 1138
Web site: www.shandon.com/

Shearwater Corp.
1112 Church Str., Huntsville, AL 35801
Phone: 256 533 4201
Fax: 256 533 4805
Web site: www.shearwatercorp.com

Sherwood Medical
See Sherwood, Davis & Geck

Sherwood Davis & Geck
1915 Olive St., St. Louis, MO 63103
Phone: 314 621 7788
Fax: 314 241 3127

Sigma
3050 Spruce St., St Louis, MO 63103

Silicone Specialty Fabricators
3077 Rollie Gates Drive, Paso Robles, CA 93446
Phone: 800 394 4284; 805 239 4284
Fax: 805 239 0523
Web site: www.ssfab.com/

Sony
Tokyo, Japan
Web site: www.sony.net

Southeastern Scientific
PO Box 585147, Orlando, FL 32858

Spectrum Cellco
12321 Middlebrook Road, Germantown, MD 20874
Phone: 301 916 1000
Fax: 301 916 1010
Web site: www.cellco.com/

Spectrum Medical Ind.
18617 Broadwick Street, Rancho Dominguez, CA 90220
Phone: 310 885 4600; 800 634 3300
Fax: 310 855 4666; 800 445 7330
Web site: www.spectrumlabs.com

SP Industries Co
935 Mearns Road, Warminster, PA 18974-2811
Phone: 800 HOTPACK; 215 672 7800
Fax: 215 672 7807
Web site: www.hotpack.com

SPSS Inc.,
233 S. Wacker Drive, 11th Floor, Chicago, IL 60606
Phone: 312 651 3000
Web site: www.spss.com/

Stille AB
Gårdsvägen 14, Box 709, SE-169 27 Solna, Sweden
Phone: +46 8 588 58 000
Fax: +46 8 588 58 005
E-mail: info@stille.se

Stille-Sonesta, Inc
2220 Canton Suite 209, P.O. Box 140957, Dallas, TX 75201
Phone: 800 665 1614; 214 741 2464
Fax: 214 741 2605
E-mail: stille@airmail.net
Web site: www.stille.se/

Synthecon, Inc
8054 El Rio, Houston, TX 77054
Phone: 713 741 2582
Fax: 713 741 2588
E-mail: rccs@synthecon.com
Web site: www.synthecon.com

Takara Bio Inc.
Seta 3-4-1, Otsu, Shiga, Japan
Phone: +81 77 543 7200
Fax: +81 77 543 2494
Web site: www.takara-bio.co.jp

Taylor-Wharton RDF Cryogenics
PO Box 568, Theodore, AL 36590-0568
Phone: 334 443 8680; 800 898 2657
Fax: 334 443 2250
Web site: www.taylor-wharton.com/cryohom.htm

TCS Biosciences Ltd
Botolph Claydon, Buckingham MK18 2LR, UK
Phone: +44 (0)1296 714222
Fax: +44 (0)1296 715 753
E-mail: sales@tcsgroup.co.uk
Web site: www.tcsbiosciences.co.uk

Tebu-bio APS
Forskersparken CAT, DTU Bygning 347, Denmark
Phone: +45 45 25 64 03
Fax: denmark@tebu-bio.com
Web site: www.tebu-bio.com

Thermo-Forma
P.O. Box 649, Marietta, OH 45750
Phone: 740 373 4763; 800 848 3080
Fax: 740 373 6770
Web site: www.thermoforma.com

Thermo Life Sciences
71 Bradley Road, Suite 10B, Madison, CT 06443
Phone: 203 318 8241
Fax: 203 318 8523
Web site: www.thermols.com

Thermolyne
See Barnstead Thermolyne

Upstate Biotechnology
1100 Winter Street, Suite 2300, Waltham, MA 02451
Phone: 800 233 3991, 781 890 8845

Fax: 781 890 7738
E-mail: info@upstatebiotech.com
Web site: www.upstatebiotech.com

Vector Labs
30 Ingold Road, Burlingame, CA 94010
Phone: 650 697 3600
Fax: 650 697 0339
Web site: www.vectorlabs.com

Vivascience AG
Feodor-Lynen-Strasse 21; 30625 Hannover, Germany
Phone: +49 (0)511/524 875 0
Fax: +49 (0)511/524 875 19
E-mail: info@vivascience.com
Web site: www.vivascience.de

VWR
1310 Goshen Pkwy, W.Chester, PA 19380
Phone: 610 431 1700; 800 932 5000
Fax: 610 436 1761
Web site: www.vwr.com

Wako
1600 Bellwood Road Richmond, VA 23237-1326
Phone: 804 271 7677; 800 992 9256; 804 714 1912
Fax: 804 271 7791
Web site: www.wakousa.com

Worthington
730 Vassar Avenue, Lakewood, NJ 08701
Phone: 732 942 1660 *Fax*: 732 942 9270
Web site: www.worthington-biochem.com

Zeiss
Microscope Division, One Zeiss Dr., Thronwood, NY 10594
Phone: 914 747 1800; 1800 233 2343
Fax: 914 681 7446
E-mail: micro@zeiss.com
Web site: www.zeiss.com/

Zimmer Corporation
P.O. Box 708, 1800 West Center Street, Warsaw, IN 46581-0708
Phone: 800 613 6131
Fax: 574 372 4988

Zymed Laboratories Inc.
458 Carlton Court, South San Francisco, CA 94080
Phone: 800 874 4494
Fax: 415 871 4499
E-mail: tech@zymed.com
Web site: www.zymed.com

Glossary

Abrasion: A scraping away of a portion of the surface of the cartilage to stimulate a repair response from the underlying bone.[i]

Aggrecan: A typical proteoglycan found in cartilage tissue.

Albumin secretion: Important function of hepatocytes in vivo and one of several surrogate markers for hepatocyte function in vitro. In vivo, albumin binds molecules and drugs in the blood and also contributes to the maintenance of plasma volume and blood osmotic pressure.

Alginate: A linear copolymer of β-D-mannuronic acid and α-L-guluronic acid that forms a 3-dimensional gel when exposed to divalent or trivalent cations. The gel can be used to entrap cells and can be dissolved with chelating agents, such as sodium citrate.

Alginate bead: A small ($\sim$10 μl) polymerized sphere of alginate gel, possibly containing cells, formed by dripping alginate solution into a bath containing divalent or trivalent cations, such as calcium chloride.

Alginate-recovered chondrocytes (ARC): Chondrocytes that are initially cultured in alginate gel under conditions that allow them to maintain normal phenotype and form a cell-associated matrix (CM) rich in aggrecan, and then recovered along with their CM by dissolution of the alginate gel with a chelating agent.

Alkaline phosphatase: A hydrolase enzyme, with an optimum pH of 8.6, distributed throughout the body, with concentrated forms in bone, liver, bile duct, and placenta. Elevated specific activity of the enzyme in cultures of mesenchymal stem cells treated with dexamethasone indicates osteogenic differentiation of the cells.

[i] *Stedman's Medical Dictionary, 27th edition © 2003 Lippincott Williams & Wilkins*

Culture of Cells for Tissue Engineering, edited by Gordana Vunjak-Novakovic and R. Ian Freshney

Allograft: A graft transplanted between genetically non-identical individuals of the same species.[i]

Antithrombogenic: Inhibitory to the formation of blood clots (thrombus, thrombosis); also: thrombo-resistant.

Arthroscopy: A special surgical instrument for visual examination of the interior of a joint.

Articular Cartilage: Cartilage that covers the articular surfaces of bones (i.e. at the joints).

Atherosclerosis: The disease process in which fats, cholesterol, and other substances are deposited in the lining of an artery causing the formation of a plaque which can compromise blood flow through the artery.

Autocrine: A mode of hormone or growth factor action in which a ligand binds to receptors on and affects the function of the cell type that produced it.

Autologous, autograft: Involving one individual as both donor and recipient.[ii]

BioArtificial Muscle (BAM): Tissue-engineered 3 dimensional muscle-like structures.

Biocompatibility: Compatibility with living tissue or a living system by not being toxic or injurious and not causing immunological rejection.

Bio-mimetic: Imitating a biological system.

Bioreactor: A device used for scaling up cultures, in this context for the three-dimensional culture of cells.

Burst/rupture strength: Of a blood vessel, the maximum intraluminal pressure that can be applied before rupture of the vascular wall.

Bypass, coronary artery: A surgical procedure in which blood flow is rerouted around a blockage in a coronary artery to restore blood flow to the heart.

Cardiac tissue engineering: Construction of heart tissue like structures including vessels, valves, and heart muscle.

Chondrocyte: A cartilage cell.

Chondrocytic: Of or pertaining to chondrocytes.

Chondrogenesis: The formation or development of cartilage.

Chondroprogenitor: A precursor cell that develops into a chondrocyte.

Collagen: A major protein found in the fibers of connective tissues.

Collagenase: Enzyme that cleaves peptide bonds in native collagen.

[ii] *Merriam-Webster Medical Dictionary, © 2002 Merriam-Webster, Inc.*

Confluency: When a surface is fully covered with a monolayer of cells, in which all cells are in contact with other cells all around their periphery.[iii]

Compliance: The ability of an elastic substance to yield to an applied force (the inverse of stiffness).

Cytochrome P450 enzymes: Family of enzymes that metabolize substrates such as toxins and drugs.

DAPI: (4′,6-Diamidino-2-phenylindole dihydrochloride)—Cell permeable fluorescent probe that binds to the minor groove of double-stranded DNA.

de novo: Anew.

Desmin: Protein found in intermediate filaments of muscle cells.

Dispase: Neutral protease used for cell dissociation.

Dexamethasone: A synthetic adrenocortical catabolic steroid. It is used to promote osteogenic differentiation in cultures of mesenchymal stem cells, possibly by inducing the cells to produce certain bone morphogenetic proteins.

Differentiation: The process cells undergo as they mature into normal cells. Differentiated cells have distinctive characteristics, perform specific functions and are less likely to divide.[iv] The opposite process, termed "de-differentiation" is commonly seen when chondrocytes are cultured in monolayer over several passages and they revert to a fibroblastic phenotype.

Endogenous: Originating or produced within the organism or one of its parts.

Engineered heart tissue: Artificial heart muscle constructed from heart cells, collagen, and extracellular basement membrane proteins.

ex vivo: Cultured in an artificial environment outside the living organism.

Extracellular matrix (ECM): Insoluble network of proteins and polysaccharides secreted by cells which can provide important structural support and regulatory signals to the cells. ECM consisting of laminin, collagen type IV, entactin, and several growth factors is commercially available for cell culture as an extract from mouse Engelbreth-Holm-Swarm tumors (Matrigel).

Fibrin: An insoluble protein derived from fibrinogen by the action of thrombin during blood clotting.[iv]

Fibroblastic: Resembling fibroblasts (i.e. spindle shaped).[iii]

Fibrocartilage: A kind of cartilage with a fibrous matrix (containing type I collagen fibers) that resembles fibrous connective tissue in structure.[i,iv]

[iii] *Freshney, RI. Culture of Animal Cells,* © 2005 *Wiley-Liss, Inc.*
[iv] *On-line Medical Dictionary, © 1997–98 Academic Medical Publishing & CancerWEB*

Gene transfer: Introduction of foreign DNA -"transgene"- into a cell. This can be mediated by recombinant adenovirus vectors and will in this case lead to transient expression of the transgene.

Glycosaminoglycan (GAG): Heteropolysaccharides which contain an n-acetylated hexosamine in a characteristic repeating disaccharide unit. The repeating structure of each disaccharide involves alternate 1,4- and 1,3-linkages consisting of either n-acetylglucosamine or n-acetylgalactosamine. GAG side chains (with the exception of hyaluronate) are covalently attached to a core protein at about every 12 amino acid residues to produce a proteoglycan, these proteoglycans are then non-covalently attached by link proteins to hyaluronate, forming an enormous hydrated space filling polymer found in extracellular matrix.[iv]

Graft: A material, especially a living tissue or an organ, surgically attached to or inserted into a bodily part to replace a damaged part or compensate for a defect.[i]

Growth factor: A factor (polypeptide hormone) that is involved in cell proliferation and differentiation.

Heart cells: A "physiological" mixture of cells from native hearts including cardiac myocytes and nonmyocytes (e.g. fibroblasts, endothelial cells, smooth muscle cells, pericytes, leukocytes).

Hematopoietic cell: A cell involved in the formation of blood.

Hepatocyte: Parenchymal cell of the liver responsible for numerous metabolic and synthetic functions.

Human growth hormone (hGH): Polypeptide hormone secreted by the anterior pituitary that promotes growth in humans, primarily by the release of somatomedin (insulin-like growth factor-1) from the liver.

Hydroxyproline: An amino acid ($C_5H_9NO_3$) produced during the hydrolysis of collagen.

Hyaluronan: A mucopolysaccharide made up of alternating ß1,4-linked residues of hyalobiuronic acid, forming a gelatinous material in the tissue spaces and acting as a lubricant and shock absorber.[i]

Integrin: Member of a large family of transmembrane protein receptors that are involved in cell-extracellular matrix (adhesion) and cell-cell interactions.

in vitro: Cultured outside the living body and in an artificial environment.

Isometric contraction experiment: Measurement of contractile properties at a defined preload.

Jamshidi needle: A long, tapered needle-drill combination used for obtaining bone core biopsies or bone marrow aspirates.

Liver zonation: The compartmentalization of specific functions to distinct zones along the liver sinusoid.

Mallory-Heidenhain: A histological stain that provides distinctive colors in fixed connective tissues and cells.

Matrix: A ground substance in which things are embedded or that fills a space (as for example the space within the mitochondrion). The most common usage is for a loose meshwork within which cells are embedded (e.g. extracellular matrix).[iv]

Metacarpophalangeal joint: Any of the spheroid joints between the heads of the metacarpal bones and the bases of the proximal phalanges.[i]

Mesenchymal: Derived from mesoderm or mesenchyme; mesenchymal tissue consists of undifferentiated cells loosely organized within an extracellular matrix in the embryonic mesoderm. Mesenchymal stem cells (MSCs) give rise to bone, cartilage, muscle, and other connective tissues. MSCs are undifferentiated cells within post-natal organisms that can, given the appropriate cues, differentiate along one or more of these mesenchymal lineage pathways.

Microfracture: The procedure of forming holes across the site of an articular defect, penetrating the subchondral bone marrow space to stimulate a repair reaction.[vi]

Micropatterning: Method of patterning extracellular matrix or cells on a substrate at the micron-scale.

Monolayer: A single layer of cells attached to a cell culture surface.

Multiplicity of infection (MOI): Number of infectious virus particles per cell. At an MOI of 1, all cells are infected.

Osteoblast: A bone-forming cell.

Osteocyte: A mature osteogenic cell that has become enclosed within a mineralized matrix elaborated by itself and other.

Osteochondral: Pertaining to bone and cartilage.[v]

Passage: Release of cells from monolayer, and subsequent replating. Typically, cells are passaged to expand the number of cells.

Perichondrium: The dense irregular fibrous membrane of connective tissue covering the surface of cartilage except at the endings of joints.[i]

Periosteum: The thick fibrous membrane covering the entire surface of a bone except its articular cartilage and serving as an attachment for muscles and tendons.[i]

Percoll: A proprietary suspension of colloidal silica coated with polyvinylpyrrolidone that is centrifuged to generate a density gradient capable of separating cells, viruses, or sub-cellular particles of different densities.

[v] *Dorland's Illustrated Medical Dictionary, 26th edition © 1981 W.B. Saunders*

Photolithograpy: The process used to transfer a pattern to a semiconductor wafer or other material. Photolithography typically utilizes the exposure of a light-sensitive photoresist through a mask.

Photoresist: A light-sensitive polymeric material. UV exposure causes a positive photoresist to become more sensitive to developer, thereby resulting in a photoresist pattern following development that is identical to the mask pattern. In contrast, UV exposure causes a negative photoresist to become less sensitive to developer.

Polyglycolic acid (PGA): A biodegradable polymer, with high melting point and low solubility in organic solvents. Used for biomaterial applications, including sutures and tissue engineering scaffolds.[vi]

Polylactic acid (PLA): A polymer used in biomaterial applications which is more hydrophobic and also has a slower degradation rate than polyglycolic acid.[vi]

Progenitor cells: Cells found in all tissues, and are responsible for the growth and regeneration of those tissues throughout the life of the organism.

Proliferation: The reproduction or multiplication of cells.

Proteoglycan: A high molecular weight complex of protein and polysaccharide, characteristic of structural tissues of vertebrates, such as bone and cartilage, but also present on cell surfaces. Important in determining viscoelastic properties of joints and other structures subject to mechanical deformation.[iv]

Recombinant proteins: Proteins expressed in host cells infected with cloned DNA fragments.

Resting tension: Tension at zero active force generally expressed in mN.

Scaffold: A structure made of natural or synthetic biomaterials, which provides shape and mechanical support for regeneration of a tissue from cells.

Sericin: A gelatinous protein that cements the two-fibroin filaments in a silk fiber.

Shear stress: In blood vessels, friction occurring between blood elements and the vessel wall.

Skeletal myoblasts: Precursor muscle cells capable of forming myocytes which fuse with either existing muscle fibers, or with other myocytes to form muscle fibers.

Spongialization: The complete removal of the subchondral bone plate at the lesion site, to expose the cancellous bone or spongiosa, and induce bleeding and repair response.[vii]

Stem cells: Undifferentiated cells that have the capacity to divide asymmetrically, thus generating both new stem cells and differentiated progeny.

[vi] *Ratner, BD et. al. Biomaterials Science © 1996 Academic Press*
[vii] *Hunziker, EB. " Articular cartilage repair." Osteoarthritis and Cartilage (2001) 10, 432–463.*

Strain: Expression of deformation caused by stress on an object; **cyclic strain** refers to deformation of the vascular wall caused by repeated cycles of increased blood pressure due to contraction of the heart.

Subchondral: Beneath or below the cartilage layer.

Superficial zone protein (SZP): A large glycoprotein product of the proteoglycan 4 (PRG4) gene that is secreted by the cells of the superficial zone of articular cartilage and provides lubrication at the articular surface. SZP is homologous to a molecule purified from synovial fluid, named lubricin.

Supernatant: Clear liquid overlying material deposited by centrifugation.

Suture retention strength: Maximum force that can be applied to a suture before it pulls out of the substance into which it has been placed.

Synovial fluid: A clear viscous fluid which serves as a lubricant in joints. It also helps to nourish the avascular articular cartilage.

Target validation: Identification of the role of certain proteins -"targets"- in physiologic and/or pathologic conditions. These findings might facilitate the development of new treatment strategies.

Tensile strength: The maximum amount of tensile stress (tension) that can be applied to a material before it breaks.

Tisseel: Fibrin sealant from plasma; main component is fibrinogen.

Tissue replacement therapy: Surgical replacement of diseased tissue with in vitro constructed tissue equivalents.

Transcript: A sequence of RNA produced by transcription from a DNA template.

Transduction: Process by which heritable DNA is stably transferred into the genome of dividing cells.

Tropomyosin: Contractile protein located in skeletal muscle.

Twitch tension: Contraction amplitude of a muscle generally expressed in millinewton (mN; 1 Newton $= 1$ kg $\times$ m/s^2). Used as a measure of active contractile force.

Ultimate tensile strength: The maximum resistance to fracture.

Vastus lateralis: Division of the quadriceps muscle that covers the outer anterior aspect of the femur.

VEGF: Vascular endothelial growth factor—proliferation growth factor that stimulates new blood vessel formation by triggering proliferation of vascular endothelial cells.

Vasoactive substances: Molecules which act up the cells of blood vessels to cause vascular dilation or constriction.

Yield point: The yield stress extrapolated to a shear rate of zero.

Zyderm: FDA approved, highly purified bovine dermal collagen.

Index